Analyzing Memory

Analyzing Memory

The Formation, Retention, and Measurement of Memory

Richard A. Chechile

The MIT Press
Cambridge, Massachusetts
London, England

This book was set in Stone Serif by Westchester Publishing Services. Printed and bound in the United States of America.

Library of Congress Cataloging-in-Publication Data

Names: Chechile, Richard A., author.
Title: Analyzing memory : the formation, retention, and measurement of memory / Richard A. Chechile.
Description: Cambridge, MA : MIT Press, [2018] | Includes bibliographical references and index.
Identifiers: LCCN 2018001228 | ISBN 9780262038423 (hardcover : alk. paper)
Subjects: LCSH: Memory.
Classification: LCC BF371 .C458 2018 | DDC 153.1/2—dc23 LC record available at https://lccn.loc.gov/2018001228

10 9 8 7 6 5 4 3 2 1

To Jeannette Chechile, whose help and support on all matters is boundless

Contents

Foreword

Memory is our connection with the past. It is what gives us a sense of continuity, of permanence, of time. It is crucial to language comprehension in a variety of ways. It is so much a part of our conscious experience that we cannot imagine what life would be like without it. Loss of memory is catastrophic, severely limiting one's ability to function in daily life.

There are few, if any, subjects that have received more attention from psychologists than memory. Much has been learned about it as a consequence of research, but the number of unanswered, or partially answered, questions has grown apace. Questions about memory that have evoked wonderment and stimulated research abound.

How is memory organized? How are memories clustered? Why, when one remembers an experience, does the act of remembering bring related experiences to mind? What does it mean for experiences to be related in memory? How are the contents of memory quantified? Is there one memory or more than one, perhaps several? How do the processes that underlie recognition relate to those that underlie recall? How do we deal with the hierarchical nature of recognition—with the fact that one may recognize an object as a car, as a Ford, as Joe Smith's Ford, and so on?

What triggers specific memories? What makes them sometimes pop into mind, uninvited and seemingly arbitrarily? In reflecting on one's life, what determines what comes to mind? How do we distinguish between memories of actual experiences we have had and experiences we have dreamed? Does the ability to make this distinction change with increasing age? Can the decline of memory (and cognition generally) with aging be prevented or deferred by engaging in specific activities for that purpose? Does the phenomenon of retrograde amnesia indicate that memories are somehow organized by age? How does memory search proceed? Can different strategies for searching memory be identified and taught? How does the act of recall affect the memory of what is recalled? How is it that one can know, or strongly believe, that something is in memory, even though one cannot retrieve it on demand? How can one be sure that something is not in memory? What would life be like if we could not forget, or could forget selectively? (Who among us does not have memories to which they would like to bid farewell?)

Memory researchers talk a lot about memory capacity and generally hold that capacity is a major way in which short- and long-term memory differ: the capacity of short-term memory being severely limited and that of long-term memory being much less so. But what is the capacity of long-term memory? And how close to capacity do most of us come in acquiring and retaining information?

The questions are endless, and the search for answers to them has yielded an impressive body of knowledge about this amazing phenomenon, but it has also generally produced more questions that no one had before thought to ask. With increasing knowledge has come increasing awareness of the fragmentary nature of that knowledge and of its incompleteness.

Richard Chechile has studied human memory, and memory as it manifests itself in other species, for more than 40 years. His research has touched on many of the questions just mentioned and others as well. In *Analyzing Memory: The Formation, Retention, and Measurement of Memory*, Chechile reviews much of his own research and much of the research of others on the same questions. He describes the book as an attempt "to begin the description of how human memory works, how it is organized, and how it dynamically changes." He treats memory as both a "dynamic representational system" and as a "stored representation" of what one has learned. Though he identifies the whole of memory as the focus of the book, he notes that special attention is given to three topics: "(1) how a memory representation is created, (2) how the latent cognitive processes of storage and retrieval of information can be measured and studied, and (3) how the process of encoding information varies in the degree of effectiveness."

This is an ambitious undertaking, but Chechile proves to be up to the task. He considers memory to be "the most exciting phenomenon in nature," and I am happy to say that his enthusiasm for the subject shows in his presentation of it. His review of relevant work is extensive. It treats memory at different levels of analysis and covers an impressive range of memory-related phenomena: from habituation and sensitization at the neurological level, to work with single-cell organisms, to cortical areas/systems that are involved in various aspects of memory, to the role of memory in psychological processes such as perception, categorization, and classification. With more than 900 references, the review is extremely well documented; it is a scholarly piece of work.

Chechile pays a lot of attention to methodology and analysis and does not hesitate to note what he believes to be methodological flaws in past memory research. He stresses the importance of measurement and of justification for specific measures.

He discusses ways to distinguish, and measure, storage and retrieval with a model-based measurement technique. He argues that memory researchers will be severely limited in the conclusions that they can draw from the results of experimentation in the absence of a model-based measurement process that can be applied to those results. He sees the development of metrics that can be applied to the storage of information in memory and the retrieval of information from memory, and the examination of how they change as a function of experimental conditions, as central to the issue of memory measurement. Chechile does not shy away from challenging popular views. He identifies and discusses several widely held ideas about memory that he considers to be incorrect. He examines the results of many experiments and critically evaluates the conclusions that have been drawn from them. Indeed it is not too much to say that he is downright iconoclastic in his treatment of some venerated psychological concepts. The Yerkes-Dodson law, which has been a staple in introductory psychological texts since its discovery in 1908 is, Chechile says, "a psychological myth." The double experimental dissociation experiment, a popular approach to memory research, he considers to be "hopelessly flawed." The finding of a dissociation, he contends, is "meaningless," and belief in the process-purity hypothesis, which is critical to interpretation of dissociation results, is "naive wishful thinking." He is not a fan of the use of signal-detection theory (SDT) in the measurement of memory. Noting that a considerable amount of research has been done using SDT as a means of measuring the strength of a memory, he argues that the approach is "fatally flawed." "Memory data are not well suited for a simple SDT representation on a single strength axis."

Chechile notes too what he sees as some common misconceptions and beliefs about memory: that memory records events in detail, much like a videotape; that memories remain unchanging over time; that memories may become inaccessible but are never lost; that forgetting is always a consequence of retrieval failure; that memories of traumatic events are repressed; that repressed memories can be recovered by hypnosis; that memories from a hypnosis session are accurate and therapeutic.

Memory researchers have identified several kinds of memory, among them: precategorical, episodic, semantic, short-term, long-term, working, motor, declarative, explicit, implicit, descriptive, autobiographical, conscious, and unconscious (automatic). Chechile touches on all of them and leaves no doubt that despite the great amount of attention that researchers have paid to memory, it remains enigmatic in many ways, so there is still much research to be done.

It would be unreasonable to expect a single book to address all the questions that have been, or could be, raised about memory, but if asked what I would like to have added if the book were to be expanded, I would say some discussion of the phenomenon of retrieval inhibition. When people are given a set of words to commit to memory and are later asked to recall the set, providing them with a subset of the items on the list generally does not facilitate recall of the remaining items and sometimes inhibits it. This strikes me as an interesting and important phenomenon, and I am curious to know what Chechile would say about it.

Almost forty years ago, William Estes (1980) raised the question of whether human memory was obsolete. It was an interesting question then and is arguably even more interesting today. What made, and makes, the question interesting is the breakneck speed with which information technology has been, and is, advancing. Estes compared human and computer memory and noted several important differences. He concluded that, in view of what had been learned about human memory during the two decades prior to the time of his writing, the goal of developing "mental prosthetics," in which computers take over some of the functions of the human memory system, appeared to be "ever more remote, in view of the complexities revealed regarding the organization of human memory" (p. 68). However, it is doubtful that Estes, or most other computer-savvy people, could have foreseen in 1980 the pace at which technology—especially information technology—would advance over the subsequent decades.

Human memory is not obsolete (yet at least), but there can be no doubt that more and more of the activities that depend on access to information stored in human memory can now be done with that information readily obtainable from electronic devices. As the versatility of computers continues to increase, as it is surely bound to do, the question of what, if anything, is unique about human memory and cognition will be the focus of much research, speculation, and debate.

In *Analyzing Memory*, Chechile identifies many of the directions that research will take. I venture the guess that as more and more is discovered about the nature and operation of memory, fascination with the subject will only increase. *Analyzing Memory* is not an easy read—the information in it is densely packed, and many readers will be challenged to rethink some beliefs and assumptions of long standing—but it is well worth the effort. No one with a more than casual interest in memory and in the various ways in which it has been studied can afford not to read this book.

Raymond S. Nickerson

Preface

For ten years, I served as the Book Review Editor for *The Journal of Mathematical Psychology*, and during that time period, I edited or wrote more than 600 book reviews (most were very short). I love books, and I especially treasure scholarly monographs written by a single author who has labored to provide a novel perspective about an advanced research topic. I developed a sense about the features of a successful monograph, but I have not, heretofore, written such a book. This book is my attempt to write a scientific monograph on the topic of memory. In my opinion, memory is the most exciting phenomenon in nature. I was originally trained to be a physicist, but along the way, I became consumed by questions about the processes that enabled scientists to do physics in the first place. How is it that a system of electrons, protons, and neutrons was able to discover themselves, to write papers and books about physics, and to understand so many other remarkable things? That process could not take place without the ability of biological systems to model the world. At the center of that world model is a dynamic memory system. The fascination with cognition and memory led me to stop my graduate training in physics after a master's degree and to retrain as a cognitive psychologist.

The chapters in this book are components of an overall analysis about how the memory system functions. It is my analysis. Memory research has been splintered into many separate topic areas that employ specialized tools and models. Some researchers devote the major part of their careers examining only one or two memory effects and thereby ignore the rest of the memory research. The subfields have become isolated silos that are cut off from the other research silos. This division of research into subfields has its utility, but it also has a cost. It is my belief that the time has come to synthesize the field of memory research. Many active researchers recognize the gaps in their command of the memory field, and they are interested in broadening their education, provided that there is an accessible source. This book is my attempt to synthesize memory by analyzing the many separate research areas. But clearly no one book can capture the entire story of memory. Consequently some hard choices were required in regard to the topics covered. This book is devoted primarily with (1) how a memory representation is created, (2) how the latent cognitive processes of storage and retrieval of information can be measured and studied, and (3) how the process of encoding information varies in the degree of its effectiveness. There are other issues, but these three general topics cover a vast amount of research, and the comprehensive exploration of these problems has not been addressed by other books.

The process of a synthesizing analysis is one that attempts to unpack the scientific problem from a unitary viewpoint. I have been engaged in the research and study of memory for more than forty years, so it should be expected that the islands of my understanding were used as platforms for expanding outward to other islands. Nonetheless, in many places, it was necessary for me to go outside my usual comfort zone to organize the issues, problems, and findings. Approximately two-thirds of the book is information that I did not deal with previously.

To make the information accessible for both researchers and students, highly technical jargon and language, which is understood only by an elite group of professionals, was avoided. It is my experience in editing the scientific work of others that even highly professional researchers appreciate a simple, jargon-free description. Yet terminology is unavoidable and readers can easily forget a definition or a concept that was described earlier. Hence, to aid all readers, there is a glossary that can be used to answer some questions about the meaning of key concepts. Furthermore, to make the book more accessible to a general reader, technical statistical details that are common in scientific reporting in psychological journals are avoided in the text. The relevant details and interesting related ideas are included in the footnotes. The footnotes are not annoyingly placed at the end of the book, so an interested reader can easily check the footnotes on the current page without losing his or her place. In general, it was my goal that any reader who is able to read an article of interest from a journal like *Scientific American*, which is intended for a broad audience, should also be able to read this volume. The style of writing is intentionally different from the format of scientific papers in psychological journals because professionals in fields such as medicine, computer science, neuroscience, or philosophy, who might be interested in learning more about memory, are not necessarily comfortable with the communication format of technical psychological journals. The beauty of a paper in *Scientific American* is that an article, for example about physics, can be read by a talented high school student, a physicist, or a biologist without compromising the message content. Although the information density of the writing is often quite high, the writing appears deceptively like an introductory textbook while still addressing many advanced ideas. I attempted to write this book in that fashion. The book itself has a deliberate organization. Researchers are notorious for selectively reading. Yet this book is designed to be read in the provided order of the chapters. Experts might be tempted to skip ahead to read what is written in later chapters, but I strongly encourage even experts to inhibit those impulses, to be patient, and to read the book from the beginning.

So who are the readers that could benefit from this book? I hope anyone interested in memory as a scientific topic will benefit by reading the book. Although this book is designed for the general educated reader, the book is also critically important for professional memory researchers. Given the massive specialization in the field of memory research, many of the ideas and facts discussed in the book will be unfamiliar to most researchers, especially since many chapters include new findings. Yet the book can also be used by advanced undergraduate or graduate students in a seminar course devoted to the topic of memory. I teach such a course at Tufts University, where a draft of this book was used. The articles cited and the topics discussed in the book can be a gateway toward additional class discussion. I found it helpful to ask students to prepare in advance of class a brief written response that is designed to probe their understanding of the facts and theories discussed in the chapter and to elicit from them topics and issues that they want to discuss and possibly debate in class. If a key goal of a seminar course is to advance the critical thinking of the student, then the exercise of reading a critical analysis and debating its merits is a good educational vehicle.

Finally it is hoped that the analysis provided in this book is a reasonably comprehensive introduction to the issues, problems, findings, and theories that emerged from the study of memory. Hopefully this effort toward synthesis within the field of memory will provoke additional efforts toward unification as well as inspire additional experimental studies. Debate and experimentation will ultimately sharpen our collective understanding of memory.

Acknowledgments

Many people provided help in a variety of ways toward the completion of this writing project. I very much appreciate the intellectual feedback and assistance from a number of my colleagues in the Psychology Department at Tufts University. Joseph DeBold, Ray Nickerson, and Ayanna Thomas each listened and commented on a number of the topics in the book. Robert Cook fully collaborated on the experimental research that was reported in chapter 9. Holly Taylor collaborated on the divided-viewing experiment that was reported in chapter 10. I also wish to thank many of my former students who collaborated on studies reported throughout the book. Included among these former graduate students are Jane Anderson, Daniel Barch, Keith Butler, Susan Butler, Jessica Chamberland, Steve Cohen, Susan Coley, Kim Ehrensbeck, Rebecca Fleischman, John Gerrein, William Gutowski, Stacie Krafczek, Beverly Roder, Darleen Sadoski, Lara Sloboda, Carolyn Topinka, Erin Warren, and Carolyn Youtz. I further wish to thank Alex Chechile, Cassandra Ruscz, Holly Taylor, Colin Trimmer, and Erin Warren who have read and commented on earlier draft sections of the book and also Amelia Andrews for her assistance with the graphics of certain figures. In addition, I greatly appreciate the helpful comments from Ray Nickerson as well as seven anonymous reviewers who read the entire volume. I also wish to thank the European Center of Tufts University for providing an opportunity to be a scholar-in-residency with the sole responsibility of working quietly on the book at their beautiful facility in the French Alps. Finally a special note of thanks goes to my wife Jeannette, who patiently read and thoughtfully commented on all the drafts of the whole book.

I   Building Memory Representations

I.1 Introduction to Part I

> If any one faculty of our nature may be called more wonderful than the rest, I do think it is memory. There seems something more speakingly incomprehensible in the powers, the failures, the inequalities of memory, than in any other of our intelligences. The memory is sometimes so retentive, so serviceable, so obedient; at others, so bewildered and so weak; and at others again, so tyrannic, so beyond control! We are, to be sure, a miracle every way; but our powers of recollecting and of forgetting do seem peculiarly past finding out. (Jane Austen, 1814, *Mansfield Park*, p. 141)

Upon waking each day, all that we understand about ourselves is what we can remember. Without memory, we would be lost in a state of confusion and mesmerized by a swarm of seemingly novel sensations. But with our memory system intact, we are able to understand and interpret our sensory experiences. We are able to speak and read coherently, plan a consistent set of daily activities, generate new ideas, and store an edited and interpreted copy of these experiences for future use. Memory enables us to be the individual that we think we are and to be the person that others have come to know. Given its importance, it is not surprising that memory has been discussed extensively by philosophers, neurologists, psychologists, and cognitive scientists. These scholars have contributed to the "finding out" about memory that was so inconceivable to Fanny Price in Jane Austen's *Mansfield Park*. Indeed, much has been learned, but we have also been confused about many aspects of the structure and functioning of the human memory system. Intellectual controversies abound. Although there are many clues, clear evidence has been difficult to find.

This volume is my attempt to begin the description of how human memory works, how it is organized, and how it dynamically changes. The broad architecture of the approach is developed from a number of perspectives or levels of analysis. Each framework contributes to the overall picture about memory. But before describing the organization of part I of the book, it is helpful to first clarify what is meant by levels of analysis and to discuss the role of modeling in memory research.

I.2 Levels of Scientific Inquiry

Unlike some branches of mathematics that seem to be built up from a set of initial axioms or foundational assumptions, science operates simultaneously at many different levels, and usually there is minimal linkage across levels.[1] Elementary particle physics is the lowest level of scientific description. A proton is not an elementary particle because it has the internal structure of two up quarks and one down quark (Griffiths, 1987). A neutron is also not an elementary particle because it is composed of two down quarks and one up quark (Griffiths, 1987). The quarks are elementary (irreducible) components of matter. The nucleus of an element is a complex structure of protons and neutrons, yet the fields of particle physics and nuclear physics are not required to describe properties of chemical interactions. The only pertinent fact about the nucleus of an atom that a chemist needs to know is the total number of protons because a neutral atom has the same number of electrons. Chemistry is the study of these electrons and the interactions of the electrons with the electrons from other atoms. Chemistry does not have to be concerned with the complexities of the atomic nucleus in order to develop a sophisticated and valuable science. Similarly, a single biological cell is a complex system that is made up of many molecules, yet properties of the cell can be described without knowing its detailed chemical structure. The biologist can also describe how a collection of cells functions without reducing the description to a more elementary level. Moreover, a whole collection of people can be treated in sociology and economics as a system. Each scientific field from elementary particles to a society can be treated as a system or a domain with discernible properties. Different scientific systems typically have their own language, intellectual constructs, and measurement instruments. It is pointless and foolhardy to attempt to understand a social phenomenon like *out-group discrimination* in terms of particle physics or chemistry.

Memory can be examined at several different levels. For the most part, memory will be discussed in terms of the language of cognitive psychology. Topics such as *false memory* or the *misinformation effect* make little sense at the cellular level or at the neuroanatomical level. Similarly, ascertaining the proportion of stored memories that are retrievable within a specific time limit can be treated at the cognitive level, but it would not be a suitable question at the physiological level. But some types of learning are elementary and are shared across animal species. For these types of memories, it is best to avoid a higher-level cognitive framework and to use instead simpler, mechanical models, if possible. Animal studies also enable experimental methods and measurements that are not ethically possible with humans. These elementary types of memories include *habituation* and *sensitization*. The studies at the biological level

---

1. From Gödel's incompleteness theorem, we know that even in mathematics that the complete set of all true theorems cannot be constructed from a single consistent set of axioms (Nagel & Newman, 1958).

introduce some important general principles that apply to more complex memory processes.

Many memory researchers have divided memory into different systems that are argued to be separate, autonomous entities much like the liver is different from the lungs. For example, working memory, long-term memory, procedural memory, semantic memory, and implicit memory have been hypothesized to be distinctly different systems that have different neurological foundations. The subdivision of memory has the practical value of allowing researchers to develop experimental evidence and theory within a more narrow context without the burden of accounting for how each type of memory relates to other types of memory. Narrowing the topic domain is a standard method in science. This sense of the phrase *memory systems* is quite different from what is meant in the current book by the phrase a *system-level analysis*. Although there are different types of memories, the whole of memory is the focus in this book. In fact, a case will be made later that implicit memory can sometimes arise from the loss of information that was once explicitly stored; thus, implicit memory is not a neurologically autonomous system. Consequently the system-level description of memory uses fundamental information-processing constructs rather than attempting to ground the description in a reductive fashion to biological and chemical mechanisms. Nonetheless, the systems-level theory must account for the behavioral phenomena, and it should be consistent with the known relevant facts obtained from biological studies.

1.3   Mathematical Models and the Study of Memory

It is common for scientists to develop models because they are abstractions of a process or a phenomenon of interest. Models are seemingly an inescapable feature of science. In fact, the word *model* is one of the most frequently overheard terms at a scientific conference. But models are not of one type, nor do they have the same purpose. In some cases, the model is a labeled picture or a diagram of a basic entity such as a neural cell or a dynamic process. These models are relatively straightforward to understand. There are also mathematical and computational models in psychology and biology, and these models are not as well understood or appreciated by researchers who were trained as experimentalists. For example, graduate and undergraduate training in psychology does not currently require students to be proficient in mathematical modeling. Students are routinely required to take courses about the statistical analysis of data. Thus, experimental psychologists are usually well trained to execute experiments and are able to evaluate if there are statistically significant differences among the various conditions within an experiment. However, mathematical models are quite different from the use of stock statistical procedures, which are employed by many other sciences and are free of specific psychological content. Scientific investigations in the area of mathematical psychology can be highly technical and intimidating. A different training is needed to train students to become skilled in mathematical modeling. Nonetheless, psychological processes are increasingly being understood in terms of mathematical models. The study of memory would be theoretically weak and incomplete without dealing with some of the most pertinent memory models. Moreover, the essential concepts of many mathematical models are usually simple ideas that are quite understandable. In this book, models are presented to be comprehended by the general reader, and they are only introduced when they are relevant for the larger goal of investigating memory.

I.3.1   Some Features of a Good Model

A model, like a scientific hypothesis, is an idealized representation of the scientist's account for a process or phenomenon. Scientists often see things differently when they are probing the frontier of knowledge. In fact, debate among scientists is a way for clarifying the research field. Thus, there might be competing models. A mathematical model can sharpen the debate about the underlying psychological process by providing a precise implementation of the proposed alternatives. For example, if there are two different plausible mechanisms, then it is possible to evaluate the relative merit of these proposals by implementing them in terms of competing mathematical models. Both models might be limited to specific situations, but this does not affect their utility. Thus, a model can be useful even when it is limited and has been eclipsed by a better model because it demonstrated the shortcomings of one of the possible theoretical proposals. A model might also be useful because it led scientists to conduct novel experiments in an effort to clarify the difference between rival ideas.

The topic of model comparison is complex because models can be contrasted on a number of grounds. Clearly, models should not be inaccurate, but the goodness of the fit of the model to data can be misleading. For example, a model might fit the data in a seemingly impressive fashion, but nonetheless the model might be an incorrect characterization that fails on other tests. There might be other rival models that also fit the data well but are based on principles and assumptions that are correct for most applications. The problem for researchers is to differentiate and assess similar models. This comparison should not be only based on the statistical goodness of fit of the model to data. It is possible for models to be overly complex and achieve an impressive fit in essence by modeling both some true variability along with some random statistical noise. Because the random fluctuations do not replicate, the overly complex model is likely to perform poorly when applied to other similar data sets. The degree of model complexity is a property of both the number of flexible model parameters and the mathematical structure of the model itself (Grüwald, Myung, & Pitt, 2005; Myung, Forster, & Browne, 2000). In general, models should not be more complex than required by the underlying psychological processes.[2]

In addition to avoiding model complexity, there are other desirable characteristics of a good model. It is outside the scope of this introduction to elaborate beyond a few brief comments about each of the following model attributes:

• **Validity**. Is there experimental evidence that demonstrates that the model parameters actually correspond to the stipulated component processes? For example, if there is a guessing parameter in the model, then does that parameter vary across experimental conditions in a fashion that is reasonable for a guessing process? Validity deals with questions of this type.

• **Identifiability**. If two configurations of model parameter values have the same probability for outcomes, then these two configurations are *observationally equivalent*. If there is observational equivalence, then there is ambiguity in linking data to states of the model. For an identifiable model, there are no possible cases of observational equivalence. There is a formal test for this property (Chechile, 1998).

• **Testability**. A good model should have the potential for being falsified. Testability deals with the ability to assess the model. A poor fit of the model to data is one way that a model can be rejected. But it is also possible to assess a model by experimentally testing the accuracy of one of the assumptions upon which the model is based. It is also possible to test a model by the accuracy of the predictions for an experiment.

• **Coherence**. Coherence is a special case of testability. For some models and for a particular method for estimating model parameters, it is possible to compute a probability of model coherence. If the probability of coherence is very low, then there is evidence against the model. This type of model testing will be described in more detail in part II of the book.

---

2. It is possible to detect an overly complex model by conducting a Monte Carlo computer simulation. For the simulation an idealized computer population is constructed where the model is implemented with stipulated parameter values. Next, a small amount of random measurement error is added, and a data sample is randomly selected from the population. The researcher then estimates the model parameters from the sampled data. For a good model, the accuracy for recovering the true parameter values improves with the sample size. But for an overly complex model, the parameter estimates are highly inaccurate even when there is a large amount of data sampled (Chechile & Butler, 2003).

I.3.2   Scientific Hypotheses and Models Are Approximations

I know of no case in any science where the process of discovery has ceased due to the lack of new information to be learned. Some ideas and theories might have been widely accepted for a long time before researchers realized that their previous insight was either wrong or incomplete. Consequently, even widely accepted scientific understanding should be considered an approximation. Thus, our best models are not expected to be perfect. The earlier eclipsed models could still have been valuable and a step toward the development of an improved model.

In this book, the facts, hypotheses, and models about memory are critically analyzed. At times, some models are criticized despite the fact that other researchers might still believe that the models are correct. It is unusual for all scientists to agree. The goal of the analysis in this book is to seek a deeper insight about memory and to highlight the ideas that seem to be the best current understanding. Yet this analysis itself is an approximation.

I.4   Organization of Part I

The chapters in part I deal with the building of a memory. Chapter 1 deals with lower-level nonassociative learning (i.e., habituation and sensitization). Remarkable changes take place at a cellular level with these basic memory representations. Yet principles from the study of habituation and sensitization ripple across the rest of the field of memory. But habituation and sensitization also miss other principles that become manifest with associative learning. Chapter 2 is devoted to the analysis of Pavlovian and instrumental conditioning, which are simple associative learning procedures. Readers who believe that they understand this type of learning are encouraged to read this chapter because a number of novel points bridge to other levels of analysis about memory. Chapter 3 deals with a systems analysis of the biological factors that are involved with the building of a memory representation. Again, principles from the biological analysis of learning will ground many ideas developed later at a cognitive level. The cognitive and information-processing analysis of memory is the central topic in chapter 4. In many ways, the material in chapter 4 is foundational for the ideas developed in all the rest of the book. Finally, the focus in chapter 5 is on debunking a series of common myths about memory. These myths get in the way of understanding the fundamental science of how memories are built, modified, and sometimes destroyed. Chapter 5 also deals with an overall framework for understanding the development of a memory representation.

A case is made in part I that conscious human memory involves an interaction between the memory system and sensory processing. The purpose of this interaction is to categorize and interpret noisy sensory information and to help in the formation and encoding of novel propositions about the sensory input. These processes are carried out rapidly in the context of a relentless stream of information and in a context that may require a concurrent set of motor movements. The motor responses themselves can be represented in the memory system. To enable the quick implementation of these complex cognitive tasks, it is argued that the memory system must have the property of being a modified content-addressable memory (CAM). In part I, the idea of CAM will be clarified in greater detail. Additionally, a case is made that memory has a rich semantic organization reflecting the well-learned aspects of our experience. Moreover, in part I, there is a discussion about how new information is added and how older memories are changed or deleted (i.e., how memory is a dynamic representational system).

1 Habituation and Sensitization

1.1 Overview of the Chapter

Learning and memory are interconnected. If one learns about some pattern detected from our experience, then memory is the stored representation of that learning. The simplest types of learning are changes in an innate response due to experience, and these learning processes are called *habituation* and *sensitization*. These processes are simple, but the study of them has uncovered some remarkable cellular changes that occur. So what are the biological changes that enable an organism to modify the rate of an innate response? How can we represent this type of learning and memory in terms of a model that is general across the biological world? What is the theoretical underpinning of the system model? In this chapter, we will overview the general behavioral characteristics of habituation and sensitization, examine the neurobiological changes that underlie the phenomena, and provide a system-level mathematical model for these types of memories. The representation of these simple memories further provides some basic insights about the formation and alteration of memories in general.

1.2 Introduction

In his classic paper on memory, Aristotle observed that memory is associative (Aristotle, 350 BCE). There is some merit to this observation at a macroscopic level, but simple associationism will be critiqued later in this book. Also, the phenomena of *habituation* and *sensitization* are counterexamples to Aristotle's universal claim about memory.

Habituation is a type of memory that is nonassociative. Habituation is a reduction of an innate response to the repeated presentation of a nonthreatening stimulus. A good example is the change in the innate startle response. Many animals startle when surprised by a sudden unexpected stimulus, such as a sudden loud noise. The startle reflex is a natural preparatory response that is useful for the animal's defense. Yet if the noise does not come from a threatening source, then the defensive response is a false alarm. The stimulus is not dangerous, so being startled by an irrelevant stimulus is counterproductive. Consequently, if that loud noise is repeated again from time to time, then the startle reflex is diminished. The animal becomes habituated to the stimulus. It is more efficient for the animal to reserve the startle response for unknown and potentially dangerous stimuli. This change in behavior with experience is a simple type of learning, yet the study of habituation has generated a sophisticated research literature and a Nobel Prize.[1]

Although most of the habituation studies are with a few convenient animal species, the phenomenon of habituation is quite general, and it occurs regularly for humans. For example, if a person stays overnight in a hotel near an airplane flight path, then the sound of a plane is likely to cause the person to wake up. But eventually, people living near airports habituate to the sound of planes.

1.2.1 Distinguishing Habituation from Other Phenomena

Before beginning an examination of habituation, it is important to distinguish habituation from several other phenomena and concepts. In some cases, the phenomenon is a more complex type of learning that will be examined later, whereas in other cases, the phenomenon does not involve memory.

Distinguishing Habituation from Associative Conditioning Associative memory is the linking or pairing of representations (or traces) where at least one of the traces is a neutral stimulus that is not a natural trigger for an innate reflex. For example, in Pavlovian or classical conditioning, there is a pairing between a neutral conditioned, stimulus (**CS**), such as a tone, and an unconditioned stimulus (**US**), such as an air puff to the eye (Pavlov, 1960). Normally, a tone does not elicit an eye blink. But if the tone is always followed by a strong puff of air to the eye, then one would learn to anticipant the air puff and blink to the tone. This is a learned associative response. With Pavlovian conditioning, the animal learns to produce an anticipatory conditioned response (**CR**) prior to the onset of the **US**, but with habituation, there is not the learning of an anticipatory response. Pavlovian conditioning is thus a more complex associative memory, whereas

---

1. In 2000, Eric Kandel received the Nobel Prize for his research on the biological basis of memory. Kandel studied the underlying biology of habituation and other related phenomena (Kandel, 2006).

the reduction of an innate response to a nonthreatening stimulus is regarded as a more simple nonassociative memory.

Distinguishing between Habituation and Sensory Adaptation   Habituation should not be confused with *sensory adaptation*. Sensory adaptation or *neural adaptation* is a dropping out of a sensory response when there is constant sensory input. For example, we might detect upon entering a room a distinctive odor, but we soon fail to detect the odor as we remain in the room. This phenomenon is a general principle whenever there is constant sensory input. We do not detect this effect in vision only because the eyes are hardwired to keep from being fixated on a steady input. There are three types of involuntary eye movements: drifts, tremors, and microsaccades (Martinez-Conde, Macknik, & Hubel, 2004). Drifts are a slow movement of the eyes, whereas tremors are aperiodic, wave-like eye movements, and microsaccades are small, jerky movements. A number of experimental procedures have been developed that result in a stable retinal image despite involuntary eye movements (Martinez-Conde et al., 2004). With a constant retinal stimulation, the image disappears in less than 80 ms (Coppola & Purves, 1996). Although retinal cells quickly stop responding when they have the same input stimulation, the involuntary eye movements prevent this from happening. But in general, sensory systems are designed to detect only a change in sensation. Sensory adaptation is not a learning process, and it is not a memory.

Differentiating Habituation and Sensitization   Sensitization is the opposite effect from habituation (i.e., it is an increase in an innate response after an unpleasant or aversive experience). If there is a weak response to a stimulus but the effect of that stimulus is unpleasant or aversive, then the vigor of the response to the stimulus increases. Like habituation, sensitization is a change in the rate of an innate response as a function of experience.

Distinguishing Habituation/Sensitization from Drug Habituation/Sensitization   Habituation and sensitization, as used in this chapter, refer to specific changes in behavioral reflexes and should not be confused with these same terms as it relates to drug effects. Habituation in this chapter does not refer to a habit or an addiction. Sensitization does not refer to the changes in the effect of a drug with use. Yet some other investigators hypothesized a connection between habituation and sensitization to theories of addiction and drug tolerance (Poon & Young, 2006). However, support for this speculation is weak and beyond the scope of this book about memory.

Distinguishing Habituation from Novelty Habituation   The term *habituation* is widely used
to study the attention of infants. Novel stimuli, like a blue triangle, are interesting to
a baby, but after the infant visually studies the novel stimulus for a sufficient time,
the child looks away to other stimuli. The decrease in the viewing of the blue triangle
is not the same phenomenon as with the decrease in a false-alarm response to a loud
noise. The infant does not have an innate, hardwired reflex associated for a stimulus
like a blue triangular object; rather, the object is a neutral stimulus that is new to the
infant. With the biological sense for habituation, there is a learned "turning down"
of the response to a harmless stimulus that elicited a defensive false-alarm response.
Instead, the infant is establishing a new memory. The novel stimulus is interesting
and is part of building an associative set of new memories. So the term *habituation* is
used in developmental psychology in a different sense from the biological sense of the
term. In this book, I will use the qualifying term *novelty habituation* to differentiate the
developmental-psychological sense of the word *habituation* but not use any qualifying
term for the biological sense of the term.[2] It is important to clarify the language of
memory or else confusion and useless debates will ensue. Although novelty habituation
is a powerful tool for studying infant perception, attention, cognition, and memory, it
is tapping an associative memory, and we deal with associative memory a bit later in
the book.

1.3   Habituation and Sensitization Effects

Thompson and Spencer (1966) wrote an influential paper that delineated the general
behavioral properties of habituation along with other related effects. The behavioral
effects that they discussed are listed below[3]:

1. *Habituation.* If there is a repeated presentation of a nonthreatening stimulus that
elicits an innate response, then the response strength decreases as a negative exponential function. The speed of habituation is negatively correlated with the strength of the
stimulus. Thus, weak stimuli habituate more rapidly.
2. *Frequency effect.* If the interstimulus interval is shortened (or the frequency is increased
for presenting the habituating stimulus), then the time to habituation is reduced.

2. With novelty habituation, developmental psychologists refer to the child's increase in viewing
time for a novel stimulus as "dishabituation" (Oakes, 2010). Again this term has a different meaning of the word *dishabituation* from the *habituation* literature for animals. Perhaps a more suitable
term would be *novelty dishabituation*.
3. The list by Thompson and Spencer (1966) included nine properties, but that list has been
reorganized into the present list of six behavioral properties.

3. *Spontaneous recovery*. If the stimulus that was previously habituated is withheld for some length of time, then the reintroduction of the stimulus can again elicit a response.

4. *Short-term and long-term effect*. If the cycle of habituation training followed by spontaneous recovery is repeated successively, then habituation develops in fewer trials, and more time is required for spontaneous recovery to occur.

5. *Stimulus generalization*. Following habituation training, the introduction of a similar but different nonthreatening stimulus results in an increase in responding that is proportional to the degree of the dissimilarity of the stimulus to the original training stimulus.

6. *Dishabituation*. If after habituation training to a stimulus, a novel (but harmless) stimulus is presented and followed on the next trial by the original training stimulus, then the animal will return to responding at a higher rate to the original stimulus. This phenomenon is called the dishabituation effect. Dishabituation is different from spontaneous recovery because it can occur immediately within a training session and is caused by the introduction of a new stimulus on the preceding trial, whereas spontaneous recovery is the return to responding caused simply by a lengthy temporal break in training. If there is a repeated cycling of presenting the training stimulus for a number of trials that is then followed by a single presentation of the dishabituation stimulus, which in turn is followed by another trial of the training stimulus, then there is a lessening of the dishabituation effect.

Figure 1.1A illustrates the habituation effect to the repeated, regular presentation of a mild stimulus. The dependent variable on the vertical axis is a continuous-response strength variable. Note that if habituation training is interrupted by a rest interval, then the response strength returns to a higher level. This increase in the response rate is spontaneous recovery. Figure 1.1B illustrates the dishabituation effect. Without a retention interval or a rest interval, a single presentation of a different novel stimulus reverses (to some degree) the habituation effect to the original stimulus. Note also that the horizontal axis is the number of presentation events or trials of presenting the habituating stimulus. But for any given fixed interstimulus interval, time is perfectly correlated with trials (i.e., $t = t_0 n$ where t is time, t_0 is the interstimulus interval and n denotes the number of trials).[4]

The pattern shown in figure 1.1A is a short-term habituation effect. Not shown in the figure is the effect of repeating the cycle of training followed by a rest. With more cycles

---

4. It follows that a negative exponential function of time is also equal to a power function of trials because $Ke^{-at} = KB^n$ where $B = e^{-t_0}$.

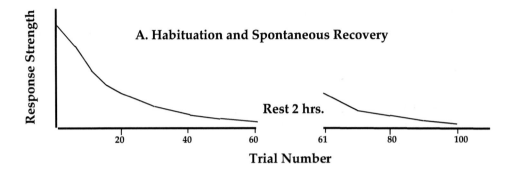

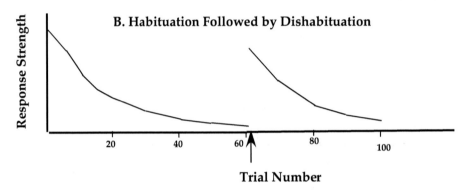

Figure 1.1

(A) Hypothetical illustration of habituation training followed by spontaneous recovery due to the rest interval. (B) An illustration of the dishabituation to the original stimulus caused by the introduction of a new stimulus after the 60th trial.

of training and rest, the response rate after the rest period is reduced, and the response habituates in fewer trials. This behavioral pattern is the long-term habituation effect.

Although stimulus generalization has been straightforward to demonstrate for Pavlovian conditioning, the effect has been more challenging to show convincingly for habituation (Grissom & Bhatnagar, 2009). Part of the problem is achieving tight stimulus control for the training stimulus. For example, consider the gill withdrawal reflex elicited by the touching of the syphon of an *Aplysia* or marine sea snail. This reflex has arguably been the most investigated habituation system studied because the animal has conveniently large neural cells for neurobiologists to study. The gill withdrawal is produced by the experimenter gently touching the syphon with a small paintbrush like the brushes used by artists. It is difficult to produce precisely the same touch stimulus on each training trial. Variation in the habituating stimulus over trials circumvents a demonstration of stimulus generalization. In contrast, in Pavlovian conditioning, a dog can be conditioned to give an anticipatory salivation response to a specific tone. The frequency of the tone can easily be controlled to be the same for each training trial. Later in extinction (the absence of the pairing of the CS and US), a tone of a different frequency can be introduced to see the effect of stimulus generalization along the frequency dimension. With tight control over the training stimulus, the effect of a change in the stimulus can be used to test for stimulus generalization. It is experimentally challenging to tightly control the habituation stimulus to see the effect of stimulus generalization.[5]

Typically, habituation is an adaptation to a mild stimulus that initially triggered an innate defensive response. In the terminology of memory research, it is a false alarm. With habituation, the animal learns to reduce the false-alarm rate. The opposite type of nonassociative memory to habituation is a process called *sensitization*. Sensitization is an increase in the innate defensive response with the repeated presentation of a more appropriate aversive stimulus (Pinsker, Hening, Carew, & Kandel, 1973). In the language of memory research, sensitization is a hit or correct detection. The initial response to a novel stimulus in sensitization training is a defensive reflex response, such the gill withdrawal of the *Aplysia*. However, this stimulus does not produce a harmless effect. For example, if the new stimulus (such as tail shock) produces a noxious reaction in the animal, then the gill withdrawal reflex is strengthened (Squire & Kandel, 1999).

5. Pavlov observed that there is a turning or an overt orientation toward a novel stimulus. He called this attentional reaction the *orienting response* (Pavlov, 1960). He considered the orienting response an innate reflex. Human babies at about three months of age demonstrate a capability of discriminating between novel and familiar visual stimuli. They exhibit this ability by means of their orienting reflex. When the babies have a choice to gaze at either a novel image or a familiar image, they spend more time gazing at the novel stimulus (Cohen, 1969; Fantz, 1964). Cohen, Gelber, and Lazar (1971) found a stimulus generalization effect for eye-gaze time. For example, if a baby saw a red circle repeatedly, then the gaze time for a green circle or a red triangle would be longer than for the familiar red circle. So far, this effect is either a demonstration of a novelty preference or a demonstration that the infant's orienting response to the familiar stimulus has habituated. But the gaze time was even longer for a green triangle because it was "more novel" because it differed in both color and shape. It should be pointed out, however, that the reduction in viewing time for a familiar object is fundamentally different from the other reflex systems studied with habituation. The difference in gaze time reported in the Cohen et al. (1971) study is likely a magnitude effect for novelty rather than a stimulus generalization effect for a habituated orienting response to red triangles. The example also demonstrates why earlier we distinguished between habituation and novelty habituation.

As with habituation, the animal changes its response rate to a stimulus as a function of experience, except with sensitization, there is a strengthening of the response, whereas with habituation, there is a weakening of the response.

Unlike habituation, there is not an accepted set of behavioral properties associated with sensitization, and this lack of agreement has resulted in some confusion in the research literature (Poon & Young, 2006). For example, some researchers do not carefully distinguish between sensitization and dishabituation. But sensitization and dishabituation are very different. For example, consider the difference between dishabituation of the startle reflex with rats and the effect of sensitization for tail shock with *Aplysia*. A rat can be trained to habituate to the loud sound of a bell that initially elicits a startle reaction. Later, if a loud bang occurs instead of a bell, then the rat tends to be startled again. Moreover, suppose if after the loud bang sound that there is the loud bell again, and then the reintroduction of the bell elicits a startle response. This case is an example of dishabituation. The bang released the habituation effect to the bell. But both the bell and the bang are not followed by anything harmful to the rat, so the repeated presentation of both stimuli will eventually result in the habituation to both stimuli (i.e., the animal will stop flinching to these loud, regularly paced sounds). Now consider the case of the *Aplysia* when the animal has habituated to the gentle touching of its syphon. If the animal's tail is now shocked, then there is a gill withdrawal. If the syphon is touched next, then there is dishabituation (i.e., there is a gill withdrawal response to the syphon touch). Tail shock releases the habituation effect to the gentle touching of the syphon. But repeating the cycle of habituation and tail shock many times does not result in the animal habituating to the shock. Instead, shock more strongly elicits the gill withdrawal. The banging sound for the rat is a generalized sound that is also harmless, whereas the tail shock is not a generalization of the syphon-touching stimulus, and it does not result in habituation.

1.3.1 Groves-Thompson Model

Groves and Thompson (1970) developed a dual-process model for the interaction of habituation and sensitization that generated considerable attention in the research literature. The essential assumption of the model is any stimulus results in two independent processes; one causes a decrement in the motor response and the other causes an excitatory response. The decremental component can be thought of as pure habituation, and the excitatory component is a pure sensitization component. The resultant behavior of a motor reflex is a mixture of these two latent components. In figure 1.2, two different theoretical outcomes are consistent with the dual-process model. The **s** and **h** dotted lines in the figure are the respective hypothetical profiles for the

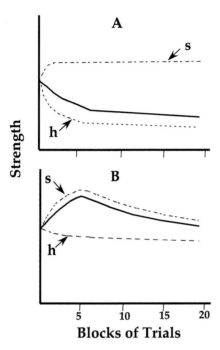

Figure 1.2
Examples of the Groves and Thompson (1970) dual-process model. (A) There is a weak sensitization latent component (the **s** dashed line) and a stronger exponential decrease for the habituation component (the **h** dashed line). The resultant strength of the motor response is the solid line. (B) There is a weaker habituation component and a peaked-shaped sensitization component.

sensitization and habituation components for two different contexts. Prescott (1998) discussed several algebraic models as to how the two components combine to produce the motor response of the reflex. The combination rule used for figure 1.2 is $K + s - h$ where K is a constant and s and h are the strength values for the respective sensitization and habituation components. The sensitization and habituation components are different in panels (A) and (B) and result in different behavioral patterns.

The dual-process model of Groves and Thompson (1970) is not a consensus model for habituation and sensitization (Boulis & Sahley, 1988). There are a number of problematic issues with the model. At a theoretical level, it is troublesome that there is not a strong rationale for how the sensitization function, habituation function, or the combination rule develops from basic principles. With two loosely defined latent functions, nearly any functional form can be predicted. Also, recent research on the habituation of the orienting reflex explicitly found results that contradict the claim of the dual-process model that dishabituation is due to an increase in sensitization (Steiner & Barry, 2014). These investigators showed that dishabituation is independent of sensitization. Finally, the particularly troublesome feature the dual-process model is the lack of clarity about what the stimulus is. If we think of a stimulus in terms of the specific pattern of sensory neurons that are activated by the event, then the experiments by Groves and Thompson (1970) on the application of electrical shock to mildly anesthetized cats with severed spinal cords activate different nerve pathways for habituation trials than for sensitization trials. The cats in this experimental preparation did not feel the effects of these different stimuli as painful due to their severed spinal cord. In other words, if $US - UR$ is the reflex that develops habituation, then there is a different reflex $US^* - UR$ that develops the sensitization response. It is unnecessary and puzzling to regard each of these separate reflexes as being modeled by two latent components—one for pure sensitization and the other for pure habituation. There is no evidence that directly supports the validity of the assumption of two latent and opposite components. Also later in this chapter, an alternative mathematical model is developed that avoids the assumption of dual processes for habituation and sensitization.

1.4   Neural Model for Habituation/Sensitization

1.4.1   Basic Neuron Anatomy

The research by Eric Kandel (along with his many associates) advanced a *synaptic plasticity* model for memory, based in large part from the detailed neurobiological study of the *Aplysia* (Kandel, 2000, 2006, 2009). The broad outline of this concept was initially suggested in 1894 by Santiago Ramòny Cajal, who hypothesized that learning might be encoded by an altered pattern of the electrical properties of neurons (Squire & Kandel, 1999).

To set the context for this research, let us first define some basic anatomical points about neurons. A simplified and idealized drawing of a multipolar neuron is shown in figure 1.3; for a comprehensive discussion of neural structure and function, see the texts by Squire et al. (2013), Purves et al. (2012), and Rudy (2008). The neuron receives chemical input by means of its dendrite. The dendritic part of the neuron also is populated by many very small protrusions that are called dendritic spines (Rudy, 2008). Dendritic spines are not shown in figure 1.3, but a sketch of a spine is shown in figure 1.4. A dendritic spine usually receives chemical input that originates from only one other neuron (as shown in figure 1.4), but there are exceptions to this general arrangement. An important feature of the physical arrangement of neurons is the gap or synapse between neurons. Neurons do not physically touch one another, but the

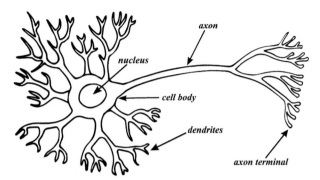

Figure 1.3
An illustration of an idealized neuron with its key components.

signal to activate the next neuron is caused by a chemical neurotransmitter that crosses the synaptic gap or cleft and penetrates the next neuron's dendrite. But to enter the dendritic spine, there needs to be a receptor gate on the membrane wall. Neurotransmitters that do not pass through the receptor gate are lost information, so there is a built-in threshold of influence caused by the arrangement of neurons. A single firing of a neuron does not mean that the next neuron will also fire. The number of receptors on the dendrite influences the likelihood of the firing of the next neuron. The structures portrayed in green in figure 1.4 are *ionotropic receptors* (Rudy, 2008). These receptors can be embedded in the spine membrane or can be floating in the watery, salty fluid within the spine. A neural cell also has a cell body that contains the cell nucleus and an axon.

When ionotropic receptors on a spine membrane contact neurotransmitters, a channel is opened that allows ions, such as sodium (Na^+) or calcium (Ca^{2+}), to flow into the spine (Rudy, 2008). There are different types of receptors; one in particular enables potassium ions (K^+) in the spine to flow out to the synaptic cleft (Purves et al., 2012). The net effect of ion transport in the spine causes a change in the electrical potential of the cell; this change can be either an increase or a decrease. If the electrical potential of the cell body exceeds a critical threshold level, then there is an all-or-none action potential discharged along the axon to its terminals (Rudy, 2008). The terminals of the axon then release chemical neurotransmitters that can be the input to a dendrite of another neuron. The neurotransmitter is not released as individual molecules but are contained in packets of about 5,000 molecules (Squire & Kandel, 1999). The packets are illustrated in figure 1.4; these bundles are called synaptic vesicles (Squire & Kandel, 1999). The neuron as a whole behaves electrically like a parallel resistor-capacitor (R-C)

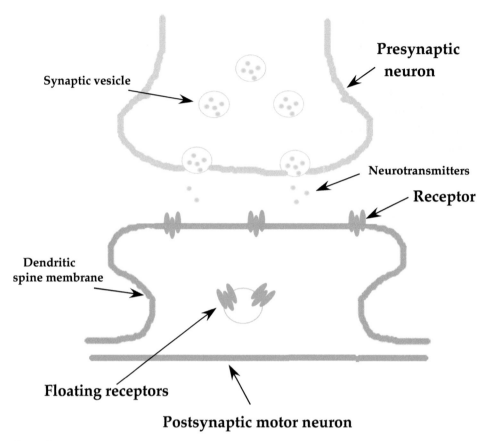

Figure 1.4

Drawing of an idealized synapse between a presynaptic neuron on an axon terminal and dendritic spine for a postsynaptic motor neuron. The green structures portray ionotropic receptors.

circuit where the capacitor stores a charge until it reaches a critical charge that results in a discharge (Katz, 1966).

A common classification of neurons is (1) sensory neurons, (2) motor neurons, and (3) interneurons (Purves et al., 2012). Sensory neurons are part of the information flow from a sensory organ to the central nervous system. Motor neurons are part of the neural linkage from the central nervous system and some muscle cell. Interneurons are within the central nervous system and deal with the communication between neurons. Primary sensory receptors, such as the rods and cone for the visual system, are not neurons (Squire et al, 2013). Sensory receptors convert energy such as light into neurotransmitters that can be detected by sensory neurons (e.g., the dendrites of a bipolar retinal cell). Similarly, muscle cells are not neural cells, but they can receive neurotransmitters from axon terminals of a motor neuron (Katz, 1966).

1.4.2   Kandel's Model for Habituation/Sensitization

Compared to the complex nervous system of humans with perhaps 100 billion neurons, the *Aplysia* has only about 20,000 neurons (Kandel, 2006). The particular species of sea snail that Kandel used as his standard test subject is the *Aplysia californica*. This species lives in the California coastal waters and is about 30 cm in length. As noted earlier, a gentle touch of its syphon triggers a reflexive contraction of its gill. There are only 40 sensory neurons for the syphon (Squire & Kandel, 1999). These 40 sensory neurons receive input from the nonneural skin cells of the syphon. These sensory neurons are labeled as **S1**, ..., **S40** in the simplified neural wiring diagram in figure 1.5. The gill muscles are controlled by only six motor neurons (Squire & Kandel, 1999). These motor neurons are labelled **M1**, ..., **M6** in figure 1.5. There is an excitatory linkage for each sensory neurons to each of the six motor neurons (Squire & Kandel, 1999). An excitatory synapse is denoted by a ◁ symbol, but an inhibitory synapse is denoted by a ◀ symbol. There are also many other interneurons that are involved with the syphon-gill reflex system. The interneurons receive dendritic input from the sensory cells and output to the various motor neurons. The output for some of the interneurons is excitatory, and other interneurons have an inhibitory output to the motor neurons. These interneurons are denoted by the **I+** and **I−** circles in figure 1.5. Squire and Kandel (1999) portray each pair of excitatory and inhibitory interneurons as connecting to all six of the motor neurons. Each interneuron is also portrayed as receiving input from one sensory neuron and outputting to each of the six motor neurons. The exact connectivity pattern of the interneurons is not fully established (Zečevi et al., 1989). In the behavioral studies for *Aplysia*, shock to the tail skin of the animal interacts with the syphon-gill withdrawal reflex. The tail skin is lateralized and represented by two neurons (Bristol, Sutton, & Carew, 2004). These neurons are called modulatory neurons and are denoted by **Mn** labeled circles in figure 1.5 (Squire & Kandel, 1999). Like sensory neurons, these neurons receive input from the non-neural tail-skin cells and output to the syphon sensory neurons. Byrne and Kandel (1996) indicate that the tail-skin neurons also link directly to gill motor neurons, but these links are ignored in figure 1.5 because the focus is on the effects of syphon skin stimulation.

The simplified neural connection pattern shown in figure 1.5 has a total of 128 neurons because there are two tail-skin modulatory neurons, forty syphon skin sensory neurons, eighty interneurons, and six gill motor neurons. The arrangement of these neurons has only 480 synaptic connections, yet this simple neural network can

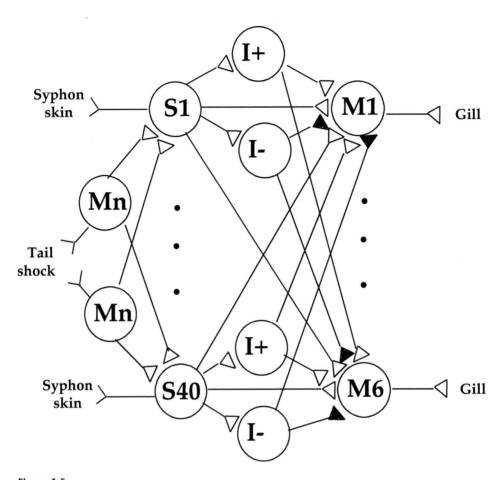

Figure 1.5
A simplified wiring diagram of the *Aplysia* syphon-gill reflex system for habituation and sensitization.

exhibit the short- and long-term behavioral changes that occur with habituation and sensitization.

The short-term habituation effect of the *Aplysia* gill withdrawal occurs in as few as fifteen trials, but with that amount of training, there is some spontaneous recovery after a 17-min rest. Squire and Kandel (1999) report that action potentials for the sensory neurons are unchanged over these trials (i.e., the electrical discharge is not altered as a function of the trial number). So the strength for the detection of the gentle touch to the syphon does not change over that time frame. Although the sensory input did not vary, the electrical potential recorded for the motor neurons did decrease over this same time period and increased again after a delay in training. Thus, the electrical potential of the *motor neurons* followed the behavioral pattern for the habituation response and spontaneous recovery of the gills. So what is happening to the motor neurons as a function of habituation training? Electron microscope studies indicate that there are changes in the distribution pattern for the receptors in the motor neurons (Bailey & Chen, 1988). See figure 1.6A for an illustration of the receptor distribution for a control condition prior to any training, whereas figure 1.6B is an illustration of the receptor distribution for short-term habituation training. Note that several of the receptors that were embedded in the dendritic spine membrane of the motor neuron have moved and are no longer attached to the spine membrane. This change in the receptor distribution within the dendritic spine occurs quickly and accounts for the decreased ionization of the motor neuron. This process is called *receptor trafficking* (Rudy, 2008).[6]

There are also short-term neural changes that take place during sensitization. If after habituation to the syphon-touching stimulus, the tail of the animal is shocked, then there is a gill withdrawal response. This stimulus causes a receptor-trafficking effect that is the opposite of the case for habituation. With habituation, the receptors on the dendritic spine move from the spine and go into the floating detached state. However, for sensitization, the floating receptors move from the state shown in figure 1.6A toward attaching to the dendritic spine membrane as illustrated in figure 1.6C (Byrne & Kandel, 1996).

6. An important molecule for receptor trafficking is called cAMP or cyclic AMP, and it is present in nonneural cells as well. E. W. Sutherland Jr. was awarded the 1971 Nobel Prize in Physiology and Medicine for his part in the discovery of cAMP as well as for his research into the chemistry triggered by cAMP. Sutherland mainly used liver cells in his research. As the Nobel press release stated, cAMP is found in bacteria and other single-cell organisms. Consequently, the complex molecular chemistry within neural cells involves substances that are common in the biological world.

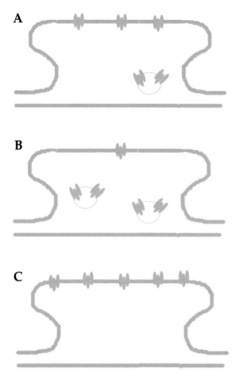

Figure 1.6
Short-term changes in motor receptor distribution in habituation (B) and sensitization (C) relative to the distribution of the receptors as shown in figure 1.4 and in panel A. The green structures portray ionotropic receptors.

In addition to the short-term change, there are also long-term neural effects of habituation. For example, with only short-term habituation training, the gill-withdrawal reflex spontaneously recovers after minutes; however, with extensive habituation training, the suppression of the gill reflex to a gentle syphon touch can last for weeks. Electron microscopic studies show that long-term habituation or sensitization training results in morphological changes in the synapses (Bailey & Chen, 1983, 1989). These investigators found that the number of sensory neuron vesicles is reduced for animals with long-term habituation training but is increased for animals with long-term sensitization training. Figure 1.7 illustrates these structural changes in the pre-and postsynapse that occur with long-term training. Relative to figure 1.7A, long-term habituation training reduces the synaptic contact with the motor neurons as illustrated in figure 1.7B (Squire & Kandel, 1999). The morphological changes with long-term

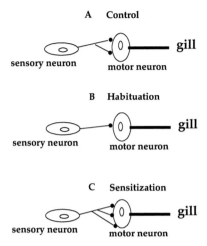

Figure 1.7

Structural changes to synaptic connection due to long-term training. (A) An illustration of a control animal, (B) the long-term effect of habituation, and (C) the long-term effect of sensitization training.

sensitization training cause the opposite effect, as illustrated in figure 1.7C. As with receptor trafficking, there is a complex molecular biology underlying the morphological changes that occur with either habituation or sensitization. These details can be found in other sources (Kandel, 2006; Rudy, 2008; Squire & Kandel, 1999), but the molecular and genetic details are not required for our purposes in this book.[7]

Note that the research on the habituation and sensitization of the *Aplysia* syphongill reflex showed that there is an alternation in the neurochemistry and structure of the synapses as a function of experience. The result of these changes is a modification of the tendency for the motor neurons to fire and thus trigger the gill muscles. The efficiency of the synapses can be either reduced or increased as a function of training. The neural plasticity that results from habituation or sensitization, even in the case of the long-term changes, can be reversed. The animal is not fixed forever with the alternated neurochemistry and structure caused by the past training. If future training is different, then the neurochemistry can be reversed and the structural synaptic connections can be altered again. The short-term receptor trafficking can move receptors either into

7. An important protein for the morphological neural changes that occur with learning is CPEB, and this protein has similarities to the prion protein that is responsible for mad cow disease (Kausik, choi, white-Grindley, Majumdar, & Kandel 2010).

or out of spine membranes depending on the training. Furthermore the "long-term" structural changes illustrated in figure 1.7 can be changed again depending on future training.

1.5 The Curious Case of the *Stentor*

While the neural plasticity model of the *Aplysia* is the standard model for the neural basis of habituation and sensitization, it is instructive to look for exceptions. The *Stentor* is a remarkable case where the synaptic neural plasticity model cannot work. The *Stentor* is a protozoan or a single-celled animal; thus, they are not a neuronal-based species. These cute little animals can be viewed through a microscope. The *Stentor* has a mouth, a bead-like cell nucleus, and many hairs (or cilia) that help the animal swim through water. Their usual form is trumpet shaped with its mouth and cilia at the wide end. Among protozoa, *Stentor* are large with a length in the range of 1 to 2 mm.

David Wood (1970, 1988) has established that the *Stentor* is capable of habituation and spontaneous recovery. The *Stentor* contracts into a spherical form when it experiences a gentle mechanical stimulus. In about 30 sec, the animal returns to its usual elongated shape. If the mechanical stimulus is repeated at a 1-min interval, then there is an exponential decrease in the contractions over trials (Jennings, 1906). The contraction into a ball shape is a false alarm because it interferes with the normal ongoing swimming and feeding behavior of the animal. Careful behavioral studies showed that the *Stentor* exhibits all of the previously discussed properties for habituation except for dishabituation (Wood, 1970, 1988). Because dishabituation is produced by the interaction of two different reflex systems, the failure to find dishabituation is not a sufficient reason for ignoring the *Stentor*.

Hinkle and Wood (1994) showed that the *Stentor* is not capable of associative learning.[8] There are consequences of having only a single cell. Nonetheless, habituation is sufficiently simple that it can be achieved by nonneural protozoa as well as by more complex neural animals. The exact biological mechanism as to how the *Stentor* implements habituation is unknown, but we can be sure that it is not the Kandel neural model because the animal lacks neurons.[9]

8. Prior to the Hinkle and Wood (1994) study, there were claims that the *Stentor* could learn an operant avoidance response. However, Hinkle and Wood showed that these claims were due to an uncontrolled factor. In a properly controlled experiment, the *Stentor* did not learn an operant response.

9. Another multicellular, nonneural organism also has been shown to exhibit habituation— *Mimosa pudica* (Gagliano, Renton, Depczynski, & Mancuso, 2014). The *Mimosa* is a plant that folds

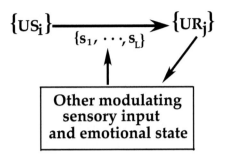

Figure 1.8
A general system representation of habituation and sensitization.

1.6   System Model for Habituation and Sensitization

All animals are born with some hardwiring for executing a response to specific stimuli. The startle reflex is a response exhibited by many species when there is an abrupt change in the stimulus environment. Sudden changes might be dangerous and indicate a predator. This hardwired response requires a genetically stored memory—not a memory based on individual experience, but rather a memory gained through the evolutionary history of the species. Species that lack this defensive response are at risk of not surviving. Hence, there is an evolutionary advantage for a species to develop the automatic defensive response as part of their genetic history. Yet memory researchers are interested in the individual memories that arise from the animal's experience. While a defensive response is advantageous, it is better still that this response is triggered only when necessary; otherwise, the response is inefficient. For example, the gentle touch on the syphon of the *Aplysia* does not result in harm to the animal. The animal thus receives feedback that the triggering event is a false alarm. The gill withdrawal reflex is not eliminated, but the intensity of the reaction to a gentle touch on the syphon is turned down. Other stimuli still can trigger a vigorous gill withdrawal response.

An illustration of a system representation of habituation is shown in figure 1.8. The unconditioned stimulus that triggers the reflex is labeled as a vector of specific sensory

---

its leaves when an experimental pot is dropped 15 cm. The leaves reopen after the disturbance. The leaf-reopening response of the plant has been shown to habituate as a function of training. Although the organism is neither neural nor an animal, it nonetheless has a sophisticated calcium ion signaling network (Yang & Poovaiah, 2003). Thus, habituation is apparently a very basic adaptation system that is advantageous for the success of many living things.

components $\{US_i\}_k$, for $i = 1, \ldots, m$. In general, a stimulus causes a number of different microevents. The k subscript denotes the fact that the stimulus occurred k times within a training session. An alternative way to express this vector is as $\{S_1, \ldots, S_m\}_k$. Both notational forms capture the fact that the Kth stimulus event is composed of a distributed set of sensory microevents.

The number of microsensory components for the stimulus event vector is considered "channel limited." By channel limited, it means that there is a restriction to the set of specific sensory fibers activated by the stimulus event. Other concurrent steady-state stimuli are not part of the target event vector. For the *Aplysia* habituation case, the target event vector would include the syphon sensory neurons but would not include other steady-state sensory neurons. The process of sensory adaptation, which was discussed earlier, works here to effectively eliminate other steady-state sensory input, so the target event is limited to the new input.

Like the stimulus vector, the unconditioned response vector has a set of microcomponents for the motor activation. Note that the dimensions of this response vector are different from the dimensions of the stimulus vector because the motor components are different from the components of the sensory vector. In the case of the *Aplysia*, there are six motor neurons for the gill withdrawal response and twenty syphon sensory neurons.

The *linkage* between the sensory vector and the unconditioned response vector is also represented by a vector. This strength vector is a L-dimensional vector where the $L \geq m$ and $L \geq n$. For example, with the neuronal arrangement shown in figure 1.5, there would be 120 components for the strength vector because each of the twenty sensory neurons for the *Aplysia* syphon skin is linked to each of the six gill motor neurons. Note that the interneurons for the *Aplysia* are not included in the strength vector because we are only concerned at the system level with the strength of the linkage between the input (sensory vector) and the output (response vector). The values for the strength vector for the Kth event are denoted as $s_k = \{s_{1(k)}, \ldots, s_{L(k)}\}$. The system model is a mathematical idealization, so the identification of either the strength components or the value of L is not required.

The components of the strength vector are a function of a number of factors. One factor that affects the strength values is the initial condition. There is a hardwired initial set of strength values that can be altered as a function of experience. Although the strength values after a long series of habituation training trials will be substantially reduced, the strength values are never reduced to zero. Learning can weaken but not extinguish altogether the hardwired linkage between the unconditioned stimulus and the unconditioned response.

In general, the initial strength values can either increase or decrease depending on the nature of other moderating sensory input. Besides the unconditioned stimulus, other sensory inputs occur during the time period following the target event. These other inputs can be either consistent or inconsistent with the target event. For example, in the case of receiving external tactile input on the syphon of the *Aplysia*, the animal might receive other hostile stimulation from other sensory receptors. Those stimuli would be consistent with the theme of a negative event occurring, so in this situation, the animal would be receiving sensory input that has the effect of increasing the current values for the strength vector. This type of experience is what occurs for a sensitization training trial. In this case, the other modulating information supports the direction of the $\{US_i\}$ to $\{UR_j\}$ linkage. However, if the other modulating sensory input is inconsistent with the hardwired reaction, then the values for components of the strength vector would decrease. The overall strength for eliciting the unconditioned response is equal to the sum of the strength values of the strength vector. This overall strength for the response is denoted as ς_k, so $\varsigma_k = \sum_{l_k=1}^{L} s_{l_k}$.

1.7   A Mathematical Representation for ς_k

The biological study of habituation and sensitization is an ongoing field of research; however, enough has been learned to develop a mathematical idealization of these processes. The system model provides a means for accounting for the learning and memory that occur with habituation and sensitization. The key variable that changes as a function of experience in the system model is the strength ς_k. In essence, the learning from the last trial event in the system model for habituation and sensitization is largely represented by the change in ς_k, from trial $k-1$ to trial k, whereas memory is represented by the value of ς_k. To see how this model functions, let us examine the fundamental habituation model.

1.7.1   Fundamental Habituation Model

Let us first focus on the dynamics of the strength variable ς as a function of habituation training. Equations (1.1) to (1.3) express how the strength variable changes as a function of habituation trials. Let the within-session trial number be denoted as k_h where $k_h = 1, 2, \ldots, n_*$ where n_* is the number of habituation trials in the full session. If training is interrupted by a long enough retention interval that results in some spontaneous recovery, then the next session is similarly characterized by the within-session trial number index k_h where $k_h = 1, 2, \ldots, n_*$. The trial index n_h denotes the overall total number of habituation training trials. For example, the total trials index $n_h = 2n_*$ after completing two training sessions of n_* trials.

$$\Delta\varsigma_{k_h} = -\alpha(\varsigma_{(k_h-1)} - \varsigma_{min(n_h)}), \tag{1.1}$$

$$\varsigma_{min(n_h)} = \varsigma_b(1 - a_0 + a_0 e^{-a_1 n_h}), \tag{1.2}$$

$$\varsigma_{k_h} = \varsigma_{(k_h-1)} + \Delta\varsigma_{k_h}. \tag{1.3}$$

All the parameters of the model are positive (i.e., α, ς_b, a_0, and a_1 are positive).

Note from equation. (1.3) that the strength after the k_h^{th} habituation trial is equal to the strength on the $(k_h - 1)$th trial plus a decrement that is denoted as $\Delta\varsigma_{k_h}$, and the value for this change is given in equation (1.1). The α parameter in equation (1.1) affects the short-term habituation rate. This parameter is on the $(0, 1)$ interval. In essence, the decrease in strength for any trial is equal to α times the difference between the strength before the k_h^{th} trial and the current lower bound $\varsigma_{min(n_h)}$. For example, if the strength from the previous trial was $\varsigma_{k_h-1} = 50$ and $\varsigma_{min(n_h)} = 20$, then a short-term learning rate parameter of $\alpha = .3$ would mean that the change in the strength for the next habituation training trial would be equal to $-.3(50 - 20)$ or a decrease of 9 strength units. Had the learning rate parameter α been equal to .5, then the decrease in strength would have been 15 strength units. Had the learning rate been at the extreme rate of $\alpha = 1$, then the change for the next trial would be 30 strength units, and this would mean that the strength would be the same as $\varsigma_{min(n_h)}$. Thus, the strength can never be less than the current limit for $\varsigma_{min(n_h)}$. Yet from equation (1.2), it follows that the value for this minimum is itself changing as a function of the total number of habituation trials. If there were no change in ς_{min} with habituation training, then the model would be the same as the "replacement learning model" of Restle and Greeno (1970) or the "linear operator model" by Bush and Mosteller (1955).[10] But with the change in ς_{min} with habituation training, the learning model differs from the classic linear operator model. It is also possible with the change in ς_{min} with n_h to distinguish between short-term (within-session) strength changes and long-term strength changes. Consequently, the mathematical model for the strength of the s_k vector captures both short-term and long-term changes that occur with habituation training.

To illustrate how the set of equations (1.1) to (1.3) operate, let us consider a hypothetical example where there are five habituation trials in a session that is followed by a break in time until eventually there is another session of five trials. For this example, let us suppose that $\varsigma_b = 20$, $\alpha = .3$, $a_0 = .7$, $a_1 = .02$, and the initial strength prior to the first session is 70, whereas the initial strength prior to second session is 63

10. This model is described in the glossary.

Table 1.1
An example of the system model for habituation

n_h	k_h	$\varsigma_{min(n_h)}$	$\Delta\varsigma_{k_h}$	ς_{k_h}
1	1	19.723	−15.083	54.917
2	2	19.451	−10.640	44.277
3	3	19.185	−7.528	36.749
4	4	18.924	−5.348	31.402
5	5	18.668	−3.820	27.581
6	1	18.417	−13.375	49.625
7	2	18.171	−9.436	40.189
8	3	17.930	−6.678	33.511
9	4	17.694	−4.745	28.766
10	5	17.462	−3.391	25.375

(i.e., $\varsigma_0 = 70$ for session 1 and $\varsigma_0 = 63$ for session 2). The value of 63 represents a 90 percent rate for spontaneous recovery. If the initial strength before session 2 had been 70, then there would have been a 100 percent spontaneous recovery. But 63 is 90 percent of 70. The values for the system model for habituation for this case are provided in table 1.1. Let us see how these numbers are computed. Based on equation (1.2), the value for $\varsigma_{min(1)} = \varsigma_b(1 - a_0 + a_0 e^{-a_1}) = 20(1 - .7 + .7e^{-.02}) = 19.723$. The decrease in strength that occurs for the first habituation trial is computed from equation (1.1), that is, $\Delta\varsigma_1 = -\alpha(\varsigma_{(0)} - \varsigma_{min(1)}) = -.3(70 - 19.723) = -15.083$. Finally, for the first trial, the resulting strength is computed from equation (1.3). Consequently, it follows that $\varsigma_1 = 70 - 15.083 = 54.917$. These same three steps are repeated for each of the habituation training trials in the first session ($1 \leq n_h \leq 5$); see table 1.1. A similar set of steps are is followed for the second session except with $\varsigma_0 = 63$ for that session.

Figure 1.9 also illustrates the short-term and long-term strength changes with habituation training. For this figure, the model parameters are again assumed to be $\varsigma_b = 20$, $\alpha = .3$, $a_0 = .7$, and $a_1 = .02$. Only the theoretical strength curves for sessions 1, 2, and 4 are shown in the figure. Now each session consists of twenty-five habituation training trials. The initial strength prior to session 1 is again set equal to 70. Following each session, there is some spontaneous recovery assumed. Spontaneous recovery would be expected if there were a lengthy retention interval between the sessions. Consequently, prior to the first trial of session 2, the strength is set equal to 63, whereas the initial strength for sessions 3 and 4 is respectively 56.7 and 51.03. Thus, 90 percent spontaneous recovery is assumed for each of the sessions except for the first session. Spontaneous recovery will be discussed in a more sophisticated way in a later subsection of this chapter.

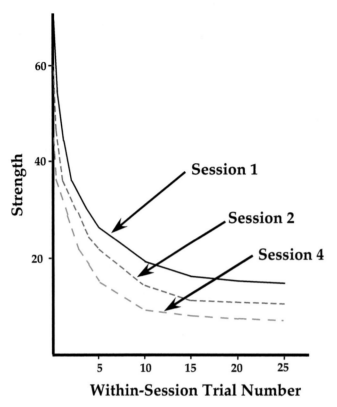

Figure 1.9
An illustration of the mathematical model for habituation for the case of four sessions of twenty-five trials each. Only the strength curves for sessions 1, 2, and 4 are shown. The initial strength prior to the four respective sessions is 70, 63, 56.7, and 51.03. The model parameters are $\varsigma_b = 20$, $\alpha = .3$, $a_0 = .7$, and $a_1 = .02$.

Finally, to assess if this model captures the exponential decay function as described by Thompson and Spencer (1966), the logarithm of the strength is correlated with the within-session trial number. If the theoretical model is approximately an exponential decay function, then the logarithm of the strength should have a high negative corre-lation with the within-session trial number.[11] These correlations over the first five trials for sessions 1, 2, and 4 (the three sessions shown in figure 1.9) are respectively −.995,

11. If the reflex strength s decreased as an exponential function of trial number n (i.e., $s = Ae^{-Bn}$), then it follows that $\log s = \log A - Cn$. Hence, the logarithm of strength should be negatively correlated with n.

$-.996$, and $-.997$. Consequently, equations (1.1) to (1.3) do approximately capture an exponential loss function. The model also captures long-term as well as short-term changes.

1.7.2 Fundamental Sensitization Model

If the modulating sensory input illustrated in figure 1.8 is consistent with the unconditioned response, then the target event is a sensitization trial. For example, the shocking of the tail of an *Aplysia* is an aversive experience to the animal, so the gill withdrawal response is an appropriate action. It is important to stress here that the unconditioned stimulus for a sensitization trial is not the same as the unconditioned stimulus for a habituation trial (e.g., tail shock is different from the gentle touching of the animal's syphon). Given that the habituation stimulus is denoted as $\{US_i\}$, let us denote the sensitization target vector as $\{US_i^*\}$, for $i = 1, \ldots, i^*$. In the case of the *Aplysia*, there is virtually no overlap between the $\{US_i\}$ and the $\{US_i^*\}$ vectors.

Equations (1.4) to (1.6) characterize the change in strength as a function of the number of sensitization trials, which are denoted as n_s.

$$\Delta\varsigma_{k_s} = \beta(\varsigma_{max(n_s)} - \varsigma_{k_s-1}), \tag{1.4}$$

$$\varsigma_{max(n_s)} = \varsigma_t(1 + b_0 - b_0 e^{-b_1 n_s}), \tag{1.5}$$

$$\varsigma_{k_s} = \varsigma_{k_s-1} + \Delta\varsigma_{k_s}. \tag{1.6}$$

All the parameters of the sensitization model are positive. The β parameter controls the sensitization learning rate, and it plays a similar role as the α parameter in habituation training. Also, the b_0, b_1, ς_t, and $\varsigma_{max(n_s)}$ parameters play an analogous role in sensitization as the respective a_0, a_1, ς_b, and $\varsigma_{min(n_h)}$ parameters for the habituation model. The sensitization model is like the habituation model but in reverse. The effect of short-term and long-term training is to increase the strength ς_{k_s}.

For both habituation and sensitization, there are four other factors that have been ignored heretofore. The first of these factors is the influence of posttraining changes in the strength due to the passage of time (i.e., the *spontaneous recovery effect*). The second factor is the effect of a *strength threshold*, and the third factor is the *similarity effect*. The fourth factor is the effect of *emotional state*.

1.7.3 Spontaneous Recovery Model

With either habituation training or sensitization training, there is typically an experimental session that consists of a regular series of trials with a fixed time interval between each trial. After the session is over, there is a more lengthy time interval before the beginning of the next session. The elapsed time between the sessions is called the retention interval, and we will see in this book that the retention interval is one of the most influential factors on memory.

The habituation strength after a retention interval of t units of time is denoted as $\varsigma_k(t)$ and given as

$$\varsigma_k(t) = \varsigma_k(0) + [\varsigma_0 - \varsigma_k(0)]\delta(1 - e^{-\eta t^{\gamma}}), \tag{1.7}$$

where the strength after habituation training is $\varsigma_k(0)$, and the pretraining strength is ς_0. All the parameters of the model are positive. The δ parameter affects the asymptotic level of spontaneous recovery, and $0 < \delta \leq 1$. If δ were 1, then for a sufficiently long retention interval, there would be a 100 percent spontaneous recovery (i.e., the strength would be equal to ς_0). If δ were .8, then the asymptotic level for spontaneous recovery would be equal to $.8\varsigma_0 + .2\varsigma_k(0)$. The η and γ parameters affect the rate of change over the retention interval. As the value of the η and γ parameters increases, the rate of approach to the asymptotic level increases. The η parameter also changes with the units of time measurement. If t were to be changed to a new time scale $t' = mt$, then the η parameter would also change to $\eta' = \frac{\eta}{m^{\gamma}}$. For example, if $\gamma = 2$, $\eta = .0576$ and time is measured in days, then to rewrite equation (1.7) in units of hours, we have $t' = 24t$ where t' is time in hours, and the corresponding value η' is equal to $\frac{.0576}{24^2} = .0001$.

To illustrate the spontaneous recovery effect predicted by equation (1.7), consider the case when $\gamma = 3$, $\eta = .00035$, $\delta = .6$, $\varsigma_0 = 120$, and $\varsigma_k(0) = 20$. The theoretical prediction is shown in figure 1.10 over a time period of 22 days. Also shown on the figure are the results from an experiment by Carew, Pinsker, and Kandel (1972).

A long retention interval following sensitization training can also cause "spontaneous recovery," except in this case, there is reduction of strength with a retention interval. The decrease in strength over the retention interval for sensitization is given by the following equation:

$$\varsigma_k(t) = \varsigma_k(0) - [\varsigma_k(0) - \varsigma_0]\delta^*(1 - e^{-\eta^* t^{\gamma^*}}). \tag{1.8}$$

The δ^*, η^*, and γ^* are positive but are potentially different from the corresponding parameters for the habituation retention function; hence, these parameters are denoted with a starred superscript.

Although equations (1.7) and (1.8) are a function of time (t), I believe that it is not time per se that causes the change in strength. It is important to remember that during training, the animal receives a well-controlled set of experiences. All the training trials are of a consistent type. Yet during the retention interval, the animal is still living, sensing stimuli, and moving about. It is possible that some of these extra-experimental experiences interfere with the thematic nature of the training. For

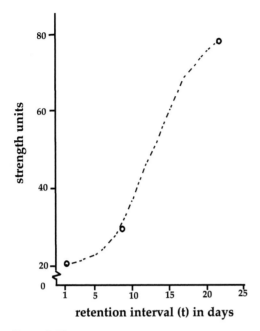

Figure 1.10
An illustration of the retention interval effect on habituation. The theoretical curve (dashed line) is based on equation (1.7) for $\gamma = 3$, $\eta = .00035$, $\delta = .6$, $\varsigma_0 = 120$, and $\varsigma_k(0) = 20$. The three data points are experimental results from the experiment by Carew et al. (1972).

example, with habituation training of the *Aplysia*, the animal might receive during the retention interval an experience that properly triggers the gill-withdrawal reflex. Such counterconditioning events were not designed by the researchers but just happened. The likely number of such events increases with the length of the retention interval. We do not know how many counterconditioning events occurred, but we do know the length of the interval, so time is used as the predictor. If the retention interval is filled with a higher density of counterconditioning events, then the value for either η or η^* would increase. Thus, the effect of time is seen here as a variable that has its predictive power by the nature of what happens in time.

1.7.4 The Strength Threshold
Up to now, we have only focused on the strength of the linkage between the unconditioned stimulus and the unconditioned response. But it is important to remember for the model of the neuron that there is a difference between increasing the ionization level of the neuron and the action potential discharge that occurs when a

critical threshold is exceeded. The system model treats the strength as the theoretical counterpart to increasing the ionization rate of neurons. It thus follows that there should also be a threshold value for strength that is sufficient to trigger a response. The threshold is denoted as ς_c, and the unconditioned response is expected to be a function of $\varsigma_k - \varsigma_c$. If $\varsigma_k - \varsigma_c = 0$, then the habituation response rate is zero, despite the fact that there is still some strength for the response (i.e., the strength is equal to ς_c). If $\varsigma_k > \varsigma_c$, then the vigor of the response depends on the magnitude of $\varsigma_k - \varsigma_c$. If $\varsigma_k < \varsigma_c$, then the motor response does not occur. But it is also possible, when there is extensive habituation training, that ς_k is substantially less than ς_c. In this case, the habituation can be decreased to a point that Humphrey (1930) called "beyond the zero point." This effect has been reported for habituation training of land snails as early as 1913 (Piéron, 1913). Piéron conducted habituation training to a point where the snails did not respond to the unconditioned stimulus and also determined how much of a delay from testing was needed in order to see some spontaneous recovery. Piéron then trained some animals many more trials past the point where they stopped responding to the unconditioned stimulus and found for those animals that they required a longer posttesting delay in order to achieve spontaneous recovery. Animals with extra habituation thus required more of a posttest retention interval in order to provide any evidence for some spontaneous recovery. Using a similar procedure, Prosser and Hunter (1936) also found a below-zero effect for the habituation training of the startle response of white rats. The below-zero effect was listed as one of the key features of habituation training in the classic paper by Thompson and Spencer (1966), and the effect is explained here in the system model by the difference between the linkage strength and the critical strength threshold.

1.7.5   The Effect of Similarity

Suppose that the strength has been reduced between the unconditioned stimulus $\{US_i\}$ vector and the unconditioned response $\{UR_j\}$ vector so that the unconditioned response does not occur, (i.e., the strength is below the threshold value). It is possible that another similar stimulus US' can still trigger the unconditioned response. This outcome can occur if the similar stimulus vector $\{US_i'\}$ elicits the unconditional response on its own even before any habituation training on the original unconditional stimulus $\{US_i\}$ vector. Because the $\{US_i'\}$ stimulus vector differs in regard to some of its components or else it would not be a different stimulus in the first place, it follows that some components of the $\{US_i'\}$ vector have not been weakened by the habituation training of the $\{US_i\}$ stimulus vector. These unaltered components of the $\{US_i'\}$ vector can have sufficient strength so as to trigger the unconditioned response. Thus, stimulus generalization of habituation is an increase in response strength relative to the habituated training stimulus.

The stimulus generalization effect for habituation occurs because the similar unconditioned stimulus has components that have not been weakened by the habituation of the original training stimulus. However, to account for the stimulus generalization effect for sensitization, the vector for the similar stimulus $\{US_i^{*\prime}\}$ is considered a subset of the sensitization training stimulus vector $\{US_i^*\}$.[12] Without additional sensory components that are also associated with eliciting the unconditioned response, there is only a subset of the components of the training $\{US_i^*\}$ vector. With fewer components activated, it thus follows that the strength for the unconditioned response is reduced. Consequently, the generalization response for sensitization is a decreased tendency for the unconditioned response.

1.7.6 Emotional State

Note that in figure 1.8, it is possible for the emotional state of the animal to influence the unconditioned response. Some stimuli directly activate an emotional response. As an example, it is not likely that hearing a new loud and abrupt noise is a pleasant experience for a rat, even for rats that have received habituation training to a different loud noise. Once the animal is left in a heightened state of agitation by the new loud sound, then the reintroduction of the previously habituated loud noise is sufficient to reactivate a startle response. The reaction of the startle response to the previously habituated stimulus is called dishabituation, and it is made possible by the carryover effect from the emotional system. In a heightened emotional state, it does not take much of a sensory input to trigger the unconditioned response.

It is important to recognize that performance in an experiment might not reflect the underlying memory. Lashley (1929) argued for the importance of distinguishing between learning and performance.[13] Although this point is obviously correct, it is still

12. Remember the stimulus that triggers sensitization is a different unconditioned stimulus from the stimulus that triggers habituation. Consequently, the training unconditioned stimulus for sensitization is denoted as US^*.

13. The Tolman and Honzik (1930b) experiment beautifully demonstrated the learning versus performance difference. In that experiment, rats were placed in a complex, multichamber maze. The control group of animals received food reward upon finding the goal box. These control animals learned the maze as demonstrated by their quick traversal of the maze from the start chamber to the goal chamber. There was also an experimental group that did not receive food early in testing (days 1–10). The experimental group animals were allowed to explore the maze. The experimental animals showed little interest in running to the goal box because there was no food in that chamber. After 10 days of free exploration, food was introduced in the goal box

easy to fall into the trap of attempting to account for all performance changes in the habituation literature as a direct consequence of a memory change. That is why in the system model framework, dishabituation of the startle response is described as an emotional carryover effect rather than as a forgetting of the previously habituated response. Dishabituation is a performance effect induced by a carryover effect from another stimulus; it is not a memory loss. One trial of a different loud noise cannot cause neuronal morphological changes from a synaptic arrangement like that of figure 1.7B back to a neural arrangement like that of figure 1.7A. It will take a more substantial counterconditioning to alter the strength values (and the memory) between the original unconditioned stimulus and the unconditioned response. Rather, the dishabituation effect (i.e., the return to responding to the original habituated training stimulus) is a performance consequence of the animal being in a different emotional state that influences the tendency to trigger the unconditioned response. In contradistinction to dishabituation, habituation and sensitization are treated here as real memory changes that occur due to consistent training.

1.7.7 Differences with Other Models

The mathematical representation of the current system model differs from several other mathematical models for habituation. For example, Wang (1994) developed a model based on two different differential equations that resulted in two separate equations for short-term and long-term effects. But the Wang model does not explicitly handle sensitization, whereas the system model has a similar structure for habituation and sensitization, and it does not assume a continuous time-base process as implied by the use of differential equations. The system model is also similar to many other general models for learning processes that have been useful in mathematical psychology.

Another mathematical representation of habituation is the Prescott and Chase (1999) model. Like the Groves and Thompson (1970) model, Prescott and Chase developed a dual-process theory. Also like the Groves and Thompson theory, which is based on data from cats with severed spinal cords, the Prescott and Chase model is based on

---

for the experimental group. Immediately, the experimental group was equal to or better than the control group and showed a rapid traversing of the maze from the start to the goal box. The Tolman and Honzik (1930b) experiment demonstrated that maze learning was being acquired by the experimental group of rats, but there was no reason for those animals to seek the goal box without there being food in that chamber. Their performance on days 1 to 10 did not reflect the real maze learning that was being acquired. Reward is important for performance, but maze learning itself only required the experience of exploring the maze (i.e., learning and performance can differ).

the severed tentacle of the *Helix aspersa* snail. Without the rest of the animal attached to the tentacle, there cannot be feedback from the other sensory detectors about the environmental event. Moreover, Prescott and Chase attempt to account for dishabituation as a mixture of excitatory and inhibitory effects. But as mentioned earlier, it is inappropriate to treat dishabituation as a memory effect.

1.8   Summarizing Nonassociative Memory

Earlier, it was noted that learning and memory are interconnected. But to be more precise, it is *acquired memories* that are the consequence of learning. Reflexes, as well as more complex behavior patterns called *species-specific behaviors*, are actually built-in memories that are coded into the genetic structure of the organism. In essence, the organism's DNA is a genetic memory that reflects the evolutionary history of the species—a memory that is available from the beginning of life for the organism. But acquired memory is the stored representation that reflects some learning about the environment, and this is the type of memory that is the focus in this book.

The most simple form of learning, which is demonstrated by all animals (even single-cell animals), is habituation of an innate response (i.e., a weakening of the response vigor). This modification of the rate of a hardwired, stimulus-response reflex occurs because of experience, and it is adaptive because the unconditioned stimulus is harmless, so the reflex response is inefficient. This type of learning is regarded as a rudimentary, nonassociative memory. Well-developed neurological studies of the *Apylsia* have shown short-term and long-term changes in the internal structure of the neurons involved as well as morphological changes to the synapses. The phenomenon of sensitization is similar to habituation, but it is also very different. It too represents a form of learning via a modification of an innate response. With sensitization, the vigor of the reflex increases because the reflex is not a harmless stimulus. There are also neural changes that take place with sensitization training. These changes include an increase in the dendritic receptors attached to the dendritic spines. There is also growth in the axon terminals. These changes make it more likely that the vigor of the response increases. Yet habituation has been demonstrated in nonneural, single-cell animals as well as in some plants. Clearly, the biological details of how these organisms modified the response rate are different from the neural model described above. Consequently, a more abstract, system-level model was advanced in the chapter. For example, at a system level, the details of the changes in the resting ionization levels of neurons and the susceptibility of the neuron to fire an all-or-none discharge are modeled by the overall strength between a sensory vector and a motor-response vector. Strength in this model is a theoretical construct that represents the short-term and the long-term neurological changes that vary with training. But strength can also apply to nonneural animals such as the *Stentor*, which shows evidence of habituation.[14] Strength thus is a more general theoretical idealization than the specific mechanisms delineated at the micro-physiological level.

Two similar, but opposite, mathematical models are provided for describing the change in the strength value as a function of habituation training and sensitization training. Other mathematical models for habituation and sensitization used instead a single dual-process approach. But it should be stressed that the sensitization response involves a different set of sensory neurons and sometimes a different set of motor neurons from the case of habituation. For this reason, dual-process models of habituation and sensitization, such as the Groves and Thompson (1970) model and the Prescott and Chase (1999) model, are seen as based on an unfounded assumption. These dual-process models treat both habituation and sensitization as having latent habituation and sensitization components. Instead, habituation is considered in the system model to involve the decrease of a *different reflex* from the reflex that is increased with sensitization training.

A key idea in the system model is the representation of the network of components associated with the reflex as a vector of L strengths values that are altered as a function of experience. The vector after the kth trial is denoted as $\{s_{1(k)}, \ldots, s_{L(k)}\}$. Importantly we also defined the overall strength for eliciting the response as $\varsigma_k = \sum_{l_k=1}^{L} s_{l_k}$. Learning is represented in the system model by a change from the current strength value due to an additional training trial. Memory, in the system model of habituation, is represented by the current strength value ς_k. Yet without additional training, the current strength value in the system does not remain stable. The retention function shown in equation (1.7) traces out the change in strength as a function of time, but it is likely that it is not time per se that causes a modification in the overall strength. Rather, it is more likely to be the result of other activities of the animal that took place during the retention interval that altered the overall strength value. But the amount of interfering activity is likely to be correlated with time, so time is the variable in equation (1.7). Dynamic changes in memory will be seen throughout this book. Neurons do not behave as if they are frozen in time. They are not built to be static.

The system model also has some properties that are shared with other more complex memories that will be discussed later. Among the shared features are (1) a distributed

---

14. Although the system model can be applied to the *Stentor*, features of the model such as the possible interaction of other input like the emotional system shown in figure 1.8 are not applicable for such a simple, one-cell animal.

representation of the stimulus and response as multicomponent vectors, (2) dynamic storage of information in a fashion that can be altered depending on subsequent events, and (3) a stimulus-driven memory elicitation. This last feature is characteristic of a content-addressable memory. With content addressability, the memory is activated by the presentation of a triggering stimulus. Consequently, for content-addressable memory, it is not necessary for the animal to search memory to ascertain what was known about the stimulus. The reaction to the unconditioned stimulus is immediate (i.e., it is a sensory-driven process).

Habituation and sensitization also have a number of differences with other types of memory that will be discussed in subsequent chapters. Habituation is a simple, nonassociative memory about a very specific unconditioned stimulus, whereas the other memories have an associative dimension. Also, more complex associative encodings are not initially hardwired. They can be destroyed, whereas the simple nonassociated memory with habituation is never completely extinguished. Barring some major neurological trauma or disease, the original hardwired reflex cannot be permanently eliminated by learned experiences. Habituation and sensitization deal with the adjustment of the degree of reactivity of reflexes. Habituation and sensitization are mechanistic processes where cognition is not needed, whereas for more complex memories, there is a substantial cognitive involvement. Finally, in the system model for habituation and sensitization, the concept of a strength is the memory, but for more complex associative memories, the concept of strength plays little or no role.

2 Associative Conditioning

2.1 Overview of the Chapter

Habituation and sensitization involve a learned modification of the strength of a stimulus-response reaction that was already in place at birth (i.e., a hardwire linkage), but all the other forms of learning deal with the acquisition of a new relationship that goes beyond the original hardwired structure of the animal. *Pavlovian* (or *classical*) conditioning is the simplest case for learning a new association between any arbitrary stimulus and another stimulus. The learning of a new stimulus-to-stimulus relationship was not part of the innate structure of the animal, so the animal becomes rewired via learning. But does this learning require anything different from a change in a strength value like what was done to account for habituation and sensitization? If there is a good association-strength model, then what effects can be predicted by the model, and which outcomes would be inconsistent with this theory?

In contrast to Pavlovian conditioning, the animal learns with *instrumental conditioning* how to solve the problem of obtaining a desired result. Hence, with instrumental conditioning, the animal learns to model the relationship between its actions and the achievement of an important goal. Instrumental conditioning is a more complex learning procedure than Pavlovian conditioning, but is there something different being acquired with instrumental condition?

The regular effects produced by Pavlovian and operant conditioning are among the most general principles in psychology. Historically, the study of conditioning was at the core of behavioral psychology, which eschewed cognitive theory. Yet evidence is advanced in this chapter against a behavioral account for conditioning (at least for mammals). Instead, a case is developed that conditioning typically has a cognitive dimension. Only under unusual conditions can there be Pavlovian conditioning of humans without cognition. To examine the necessity of cognition with conditioning, topics such as subliminal perception and learning without awareness are explored in depth. So in this chapter, there is a fresh analysis and some new insights about the memories that are established from associative conditioning procedures.

2.2   Elementary Conditioning Definitions

The essential facts about Pavlovian and operant conditions are well known; nonetheless, let us begin with some basic definitions and descriptions. Some notation is also introduced in this section to stress the idea that sensory input and subsequent responses are treated as multicomponent vectors. Traditionally in mathematics, vectors are denoted by a bold symbol. Consequently, from the previous chapter, the unconditioned stimulus was denoted as **US** and consisted of a number of sensory elements $\{S_1, \ldots, S_m\}$.

2.2.1   Classical Conditioning: Operations and Terminology

For classical or Pavlovian conditioning, the animal learns a new association between a neutral stimulus and another stimulus. The term *neutral* means that the stimulus does not have a hardwired association to the unconditioned stimulus. For example, the sound of a dog biscuit rattling in a bag is not a sound that is wired to any particular response. The neutral stimulus is defined as a conditioned stimulus. For classical conditioning, the experimenter controls the temporal arrangement between the conditioned stimulus (**CS**) and the unconditioned stimulus (**US**). For example, there is the *sound* and *sight* of a dog biscuit, which is the **CS**, and this stimulus is followed later in time by the dog biscuit (**US**) being placed in the mouth of the dog. The unconditioned stimulus (**US**) elicits a unconditioned response (**UR**) of the consumption of the biscuit. Given the consistent temporal relationship of first the **CS** and later the **US**, the animal develops a new response, the conditioned response **CR** in anticipation of the oncoming **US**. The **CR** is a vector of responses that share features in common with the **UR**, so for the example above, the **CR** would include anticipatory salivation and excitement. Thus, for Pavlovian conditioning, there is the following stimulus-response pattern:

$$CS \rightarrow CR \rightarrow US \rightarrow UR$$

Pavlov revolutionized scientific psychology with his systematic exploration of the learning illustrated by the above diagram.[1]

---

1. Pavlov investigated systematically classical conditioning from 1901 through the time of the publication of his book on this topic in 1927. In 1960, an English reprinting of this work was

A. Delay Conditioning

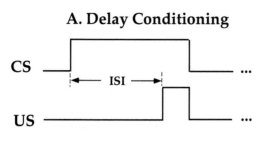

B. Trace Conditioning

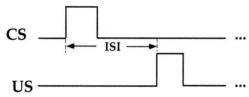

Figure 2.1
Figure shows the temporal arrangement for *delay* (A) and *trace* (B) Pavlovian conditioning. The ISI interval shown denotes the interstimulus interval.

Delayed versus Trace Conditioning   A key procedural distinction for the temporal arrangement between the conditioned stimulus and the unconditioned stimulus is between *delay* and *trace* conditioning. These arrangements are illustrated in figure 2.1. Note the interstimulus interval (ISI) can be the same for the two arrangements, but with trace conditioning, the conditioned stimulus is not on or available at the time of the unconditioned stimulus occurrence.

Higher-Order Conditioning   Pavlov also observed *higher-order* conditioning. This effect occurs by first training the animal so that a particular CS_1 elicits consistently a conditioned response **CR**. Later he paired a new stimulus CS_2 with CS_1, which was followed by the unconditioned stimulus. Higher conditioning is said to occur when the animal elicits a conditioned response to the CS_2 stimulus. Consequently, an associative chain can be learned.

---

published. Prior to his conditioning research, Pavlov was an internationally recognized authority on the physiology of digestion. In 1904, Pavlov received the Nobel Prize in Physiology and Medicine.

Extinction   While the consistent pairing of the **CS** and the **US** results in a learned antic-ipatory response **CR**, the new conditioned response can be *extinguished* when the **CS** is *not followed* by the unconditioned stimulus. The conditioned response does not extin-guish immediately, but full extinction requires a number of training trials without the **US**. In essence, extinction is a learning process where the expectancy associated with the **CS** is changed and results in the development of an active inhibition of the con-ditioned response (Robleto, Poulos, & Thompson, 2004). It is the new learning of an active inhibitory response due to a context change, that is, the **CS** no longer is followed by the **US** as it once was (Bouton, 2004).

Condition Inhibitor   Pavlov also discovered how to produce a *condition inhibitor*, which is denoted here as CS^-. Suppose whenever the **CS** (e.g., a tone) occurs, it is always followed by a **US**, but if the tone occurs along with another stimulus (e.g., a light), then the **US** never occurs. The animal will eventually learn to elicit a conditioned response to the tone but not to elicit a conditioned response to the combination of the tone and light. Pavlov argued that the light is a conditioned inhibitor. To prove this point, Pavlov conditioned the animals to learn a conditioned response to a new stimulus CS' (e.g., a ticking metronome). Later the CS^- (light) and the CS' (metronome) were paired together, and the conditioned response did not occur. This demonstration showed that the light is a general "stop" signal that inhibits any other "go" tendency associated with the other concurrent stimulus.

Stimulus Generalization   Stimulus generalization is yet another important pheno-menon discovered by Pavlov. For example, suppose a consistent conditioned response is established to a **CS** of a 1,000-hertz tone. Later the animal will also produce a con-ditioned response to other tones of frequency f that differ from the training tone (e.g., $f = 1,000 \pm \Delta$). The maximum strength of the conditioned response occurs for the training stimulus, but a weaker generalized conditioned response also occurs for other tones.[2] For sufficiently large values of Δ, the conditioned response rate is essen-tially zero. In essence, the conditioned response is not appropriate in every context. The animal learns the conditioned response when a specific neutral stimulus has a high predictive signaling value to a specific unconditioned stimulus. It is reasonable that the conditioned response is inhibited for other stimulus contexts (Bouton, 2004).

2. Pavlov measured the strength of the conditioned response by the number of drops of saliva elicited by the conditioned stimulus for his test subjects (dogs) prior to the onset of the uncon-ditioned food stimulus. Pavlov devised a method of recording the number of drops of saliva produced by the subject's salivary gland.

Sensory Preconditioning   Another particularly important finding is that of *sensory preconditioning*. This effect was discovered by Brogden (1939), and it is related to higher-order conditioning. In that experiment, dogs were first trained to experience two very different neutral stimuli that were presented together (i.e., a light and a buzzer). Later, one of the two neutral stimuli served as a conditioned stimulus, and it was followed by a shock to the leg of the animal. This training continued until the dogs demonstrated an anticipatory flexion response to the conditioned stimulus on 100 percent of the test trials. Finally, the animals were presented with the other neutral stimulus that was previously paired with the conditioned stimulus in the sensory preconditioning phase of the experiment. Compared to a control group of animals that did not receive the initial sensory preconditioning association trials, there was a significant difference between the groups, indicating that the two neutral stimuli were associated (i.e., the second neutral stimulus also elicited an immediate flexion response despite the fact that it was not directly paired with shock). Consequently, either of the two neutral stimuli could have served as a signal for the forthcoming presentation of shock.[3] A reasonable interpretation of the sensory preconditioning effect is illustrated by the following diagram:

$$S_1 \leftrightarrow S_2$$
$$\searrow$$
$$\textbf{CR}$$

In terms of this diagram, the second neutral stimulus, denoted as S_2, elicits the conditioned response via its association to the S_1 stimulus, which was paired with the **US**. From this perspective, sensory preconditioning is an alternative method for obtaining higher-order conditioning.

2.2.2   Operant Conditioning: Operations and Terminology

While Pavlov is properly linked with the invention of classical conditioning, instrumental or operant conditioning traces its roots to Thorndike (1911). For instrumental conditioning, the experimenter first induces a strong motivation so that the animals are incentivized to seek particular stimuli. Most commonly, the animal is placed on a restricted diet to induce hunger. In the experimental apparatus, the researcher controls the administration of a small amount of food contingent on the animal's behavior.

---

3. Brogden (1939) did not report statistical tests as is now standard for research reports. Nonetheless, Brogden did report all the test data per animal, and these data enable the calculation of a nonparametric randomization test (Siegel, 1956). That statistical test shows a previously paired neutral stimulus that was not paired with shock also significantly elicited a flexion response, $p < .0003$.

For example, a rat receives a food reward if the animal enters one of the chambers of a complex, multichamber maze. The critical response that leads to reward is called the instrumental response. The instrumental response is usually designed to be one with a low probability of occurrence on its own. For maze learning, the instrumental response consists of a series of correct choices at the decision points in the maze. At these points, the animals have options. The stimulus for the correct choice is called the discriminative stimulus and is denoted as S^+, and the other incorrect stimuli are denoted as S^-. The researchers measure learning by the increase in the performance (e.g., the average percent correct for the choices in traversing the maze). Another measure of learning is the speed of the animal going from the start box to the goal box. A common feature of operant conditioning is the increase in the response rate for the instrumental response; Thorndike (1911) called this phenomenon *the law of effect*. Thorndike is also responsible for the concept of a *reinforcer* as a method for increasing the response rate.

For some experiments, the reward or reinforcer is the removal of an aversive stimulus. This type of reward is called a *negative reinforcer*. For example, suppose that the floor of a maze is electrified, but there is a shock-free zone in one spot. The animals will learn the maze route to escape shock and reach the safety spot. As another example, Thorndike (1911) would place cats in a "puzzle box" from which the animals could escape confinement if they discovered the correct instrumental response. Yet another common escape-training maze is the Morse water maze (Morris, 1981). The "maze" is just a cylinder filled with colored water but with a small underwater platform located in one spot. The platform is not visible to the rat. The animal must swim to keep from drowning, but if the animal swims to the platform, then it can rest safely. Over repeated trials, the animals demonstrate learning by their speed for finding the submerged platform.

Instead of studying maze learning, investigators can also use a single box or an operant chamber to condition a low-probability response. For example, the operant chamber might have a lever that needs to be pressed in order for the animal to receive reinforcement. Learning is demonstrated by an increase in the lever-pressing rate that occurs with reinforcement. Today, the operant box is commonly equipped with a touch-sensitive screen that can be used for both the presentation of stimuli and the recording of responses. For example, a common procedure for operant discrimination learning is the *matching-to-sample task*. For this task, the pigeon might see a sample stimulus displayed on the screen at the beginning of a trial; this stimulus is the discriminative stimulus S^+. Following a variable delay period where the S^+ is not displayed, the bird is then presented with a choice between two stimuli displayed on the touch-sensitive screen where one of the stimuli is S^+, and the other stimulus is an incorrect or foil stimulus S^-. The pigeon is rewarded only if the animal pecks near the S^+ stimulus.

2.3   The Information-Theory Challenge

As a physiologist, Pavlov conceptualized classical conditioning in terms of mechanistic excitatory and inhibitory neural processes, but he did not delineate how these hypothesized mechanisms worked. In essence, behavioral psychology theorized that classical conditioning resulted in a **CS – CR – US – UR** chain where the links possessed strength values. Yet by the 1970s, a rival information-processing framework for Pavlovian conditioning emerged as a compelling theory instead of a purely mechanistic, behavioral model. Two effects raised doubts about the traditional behavioral theory of Pavlovian condition.

2.3.1   Blocking Effect

A key effect that pushed the theory of classical conditioning toward a cognitive framework was the *blocking effect* (Kamin, 1969). For this effect, the animals were first trained to elicit a consistent conditioned response to a conditioned stimulus CS_1. For the next step, the unconditioned stimulus follows the presentation of the stimulus compound of CS_1 together with CS_2. The animal still elicits a consistent conditioned response to this combination of neutral stimuli. From a purely behavioral association-strength framework, both of the neutral stimuli should have acquired a strength for the conditioned response. But from an information-theory framework, the CS_2 stimulus carries no information value because the animal already has acquired the association that the CS_1 stimulus leads to the eventual onset of the unconditioned stimulus, so the CS_2 stimulus is redundant and irrelevant.[4] Consequently, a critical test of these two conflicting interpretations of classical conditioning is to see what happens when the animal is tested with only the CS_2 stimulus. If there is still a conditioned response (although weaker because of stimulus generalization), then the purely behavioral association-strength model is valid. But if there is no conditioned response to the CS_2 stimulus, then that finding would support an information theory framework. Kamin (1969) found no conditioned response as would be expected from an information-theory perspective.

---

4. An informative stimulus reduces the uncertainty of outcomes. More details about this concept can be found in the glossary.

2.3.2 Zero-Contingency Effect

Another effect that led to a new conceptualization of classical conditioning is the *zero-contingency effect* (Rescorla, 1967). While a conditioned response develops when the conditioned stimulus occurs prior to the onset of the unconditioned stimulus, Rescorla (1967) observed that no conditioned response develops when the conditioned stimulus and unconditioned stimulus occur simultaneously. Without a temporal delay between the onset of the conditioned stimulus and the unconditioned stimulus, the conditioned stimulus cannot serve as a signaling event for the later onset of the unconditioned stimulus (i.e., it does not have predictive value).[5] One might suspect from a purely behavioral framework that there is no time for a conditioned response to occur for simultaneous conditioning. But when the conditioned stimulus is presented alone, there is still no conditioned response. The failure to find conditioning in this case is problematic for a mechanistic, behavioral framework because both stimuli were available, so an association should have developed.

2.4 Rescorla-Wagner Model

Rescorla and Wagner (1972) provided for a modification of standard association-strength theory in an effort to address the informational aspects of classical conditioning. The Rescorla-Wagner model is an attempt to preserve associationism and strength theory by accounting for the blocking effect. Their model had a similar mathematical structure to the system model for sensitization presented in the previous chapter except that the asymptotic level of strength is a constant in their model. The following equations characterize the association strength $V_n(\mathbf{A})$ for conditioned stimulus $\mathbf{A}$ on the nth pairing with the unconditioned stimulus:

$$V_n(\mathbf{A}) = V_{n-1}(\mathbf{A}) + \Delta V_{n-1}(\mathbf{A}), \tag{2.1}$$

$$\Delta V_{n-1}(\mathbf{A}) = \alpha_\mathbf{A}\beta[\lambda - V_{n-1}(total)], \tag{2.2}$$

where $\alpha_\mathbf{A}$ is a parameter for the relative salience $\mathbf{A}$, β is the parameter controlling the learning rate, λ is the asymptotic level for the strength of the association, and $V_{n-1}(total)$ is the current strength "expectation" level. Equations (2.1) and (2.2) are

---

5. Note that for an aversive **US**, there is a backward-conditioned response (i.e., if the **US** occurs before the **CS**, then there is a conditioned response). However, it is not a conditioned response of anticipatory fear. Rather, it is a relief response. Note that after an aversive unconditioned stimulus is removed, there is an unconditioned response of relief. Consequently, the finding of backward conditioning is still a case of forward conditioning. It is the forward conditioning of a relief response, that is, the conditioned stimulus is a safety signal (Moscovitch & LoLordo, 1968).

suitable when the conditioned stimulus **A** is not a compound stimulus, and in this case, the "expectation value" $V_{n-1}(total) = V_{n-1}(A)$. The learning equations for the model are more complex for compound-conditioned stimuli. To illustrate the model, let us consider the learning curve when $\alpha_A = .625$, $\beta = .4$, $\lambda = 1$, and $V_0(A) = 0$. For this example, the value for $\alpha_A \beta$ for all trials is equal to .25 because $.625 \cdot .4 = .25$. It thus follows that

$$\Delta V_0(A) = .25(1 - 0) = .25.$$

$$V_1(A) = 0 + .25 = .25.$$

$$\Delta V_1(A) = .25(1 - .25) = .1875.$$

$$V_2(A) = .25 + .1875 = .4375.$$

$$\vdots \quad \vdots$$

$$V_\infty(A) = 1.$$

Although the asymptotic value for learning strength is $\lambda = 1$, the value for $V_n(A)$ is already .999 when $n = 20$. If the animal is eventually given extinction training, then the same set of equations applies except that the value for λ is set to 0. Now if we let n denote the number of extinction training trials and $V_0(total) = 1$, then $\Delta V_0(A) = .25(0 - 1) = -.25$ and $V_1(A) = 1 - .25 = .75$.

The Rescorla-Wagner model for the case of simple classical conditioning is equivalent to the Bush-Mosteller (1955) linear-operator model, which is not really an information-theory model.[6] To see how the Rescorla-Wagner model diverges from the Bush-Mosteller model, let us consider the case for compound stimuli. One of the key concepts for compound conditioning is the term $V_{n-1}(total)$, which is the sum of the individual components of the compound. For the compound conditioned stimulus **AB**, we have $V_{n-1}(total) = V_{n-1}(A) + V_{n-1}(B)$. Also for compound conditioning, the relative salience of the components can differ (i.e., α_A and α_B do not have to be equal). Let us apply these new features to the case of a blocking-effect experiment as done by Kamin (1969). Recall for the blocking experiment there is an initial stage of training for a single conditioned stimulus that will be denoted as **A**. Later the stimulus compound of **A** along with **B** are presented together. In the Rescorla-Wagner model, if $\lambda_A = 1$, then at full conditioning for the **A** stimulus, $V_{n-1}(A) = 1$. But of course there is no strength for the **B** component prior to its first presentation, so $V_{n-1}(B) = 0$. However, the value for $V_{n-1}(total) = V_{n-1}(A) + V_{n-1}(B) = 1 + 0 = 1$. If $\alpha_A = \alpha_B = 1$, then $\Delta V_{n-1}(A) = \alpha_A \beta[\lambda_A - V_{n-1}(total)] = \alpha_A \beta[1 - 1] = 0$. The result of $\Delta V_{n-1}(A) = 0$ is not surprising because the **A** stimulus was fully conditioned in the first stage of the

6. The Bush-Mosteller model is described in the glossary.

experiment. But also note that $\Delta V_{n-1}(\mathbf{B}) = \alpha_\mathbf{B}\beta[\lambda_\mathbf{B} - V_{n-1}(total)] = \alpha_\mathbf{B}\beta[1-1] = 0$. This effect of no change in strength for the component for stimulus **B** holds for all trials; hence, the stimulus is "blocked" because the initial strength value $V_{n-1}(\mathbf{B}) = 0$.

Another finding that has an explanation from the Rescorla-Wagner framework is the *overshadowing effect*. Pavlov (1960) observed that some features of the stimulus environment have greater conditioning strength than other stimuli. For example, rats are more sensitive to auditory features of their environment than visual features. In the Rescorla-Wagner model, the overshadowing effect is explained by the relative salience for the components of a compound conditioned stimulus. Stimulus **A** overshadows stimulus **B** if $\alpha_\mathbf{A} > \alpha_\mathbf{B}$ causing $V_n(\mathbf{A}) > V_n(\mathbf{B})$.

A number of other features of classical conditioning can be explained by the Rescorla-Wagner model. Yet there are also some serious problems with the model that will be discussed in the next section.

2.4.1 Problems with the Rescorla-Wagner Model

It is commendable that the Rescorla-Wagner model is framed at a general psychological level as opposed to Pavlov's vague ideas about the irradiation of excitatory and inhibitory fields in the neural cortex. It is also a tribute to the Rescorla-Wagner model that it results in specific predictions that are easily testable. The model clearly has stimulated a resurgence of interest in classical conditioning, and it has been widely applied across a broad set of contexts beyond that of just animal learning (Siegel & Allan, 1996). Although the model offers a creditable explanation for a number of conditioning phenomena, there are nonetheless many critical effects that the model does not predict. In fact, Miller, Barnet, and Grahame (1995) identified twenty-three effects or findings that are inconsistent with the predictions of the model. It is well past the scope of this section to discuss each of these failings. Instead, one of the points raised in the Miller et al. (1995) paper will be highlighted along with two other points not mentioned in that critique. These failures of the model are important because the Rescorla-Wagner model is a theory that attempts to preserve associative-strength theory rather than adopt a cognitive model of conditioning. If the effects that are inconsistent with the Rescorla-Wagner model have an explanation from the cognitive perspective, then it will be clear that a cognitive framework for Pavlovian conditioning cannot be ignored.

One problem with the model is that it does not differentiate between memory and performance. According to the model, the "memory" of a conditioned stimulus **A** is the strength value $V_n(\mathbf{A})$. If a stimulus has been fully extinguished, then the strength $V_n(\mathbf{A}) = 0$, and the stimulus is totally unlearned or forgotten. Yet there are a number of instances where animals show some recovery from this hypothesized unlearned state. Miller et al. (1995) discuss three cases where there is recovery from extinction. One case is the effect of spontaneous recovery that is produced by a long retention interval since the last presentation of an extinction trial (Pavlov, 1960; Robbins, 1990). In the Rescorla-Wagner model, the strength $V_{n+1}(A)$ should not be greater than zero if $V_n(A) = 0$. The finding of an unaccountable spontaneous recovery is thus problematic.[7]

While spontaneous recovery is a problem for the Rescorla-Wagner model, a cognitive framework can easily account for this effect by stressing the learning versus performance distinction. From the cognitive perspective, two predictive rules can be learned and stored in memory. Rule 1 (learned during the original training) is: the US follows the CS. Rule 2 (learned during the extinction training) is: the US does not follow the CS. After a long delay since the last extinction training trial, the animal might temporarily forget the more recent rule and act as if rule 1 were applicable. This error would be a performance failure, but it does not mean that the memory of either rule 1 or rule 2 was lost from memory. Thus, temporary forgetting could account for the spontaneous recovery. Notice that this cognitive account does not employ the idea of an associative bond with a strength value. Rather, a rule is in the form of an if-then predictive relationship. How a rule can be represented neurologically is not known. Nonetheless the cognitive model argues that the CS elicits a prediction that the US follows.[8]

Another difficulty with the Rescorla-Wagner model is its failure to predict accurately for the case of simultaneous conditioning (i.e., when the onsets for the conditioned and unconditioned stimuli completely overlap). If the onset of the conditioned stimulus does not occur prior to the onset of the unconditioned stimulus, then the CS cannot serve as a signal for the oncoming US; thus, no anticipatory conditioned response develops. Yet from equation (2.2) there should be a positive increment to the strength of the conditioned response for each trial conducted with simultaneous conditioning. One might argue that the model does not apply in the case of simultaneous

---

7. Miller et al. (1995) also discuss the effects of "external disinhibition" and "reminder-induced recovery." The external disinhibition effect occurs when a neutral cue precedes the conditioned stimulus in extinction testing (Bottjer, 1982; Pavlov, 1960). The reminder-induced recovery occurs if the unconditioned stimulus is reintroduced prior to the continued extinction testing of the conditioned stimulus alone (Rescorla & Heth, 1975). Both of these effects are a challenge to the Rescorla-Wagner model.

8. There are a number of mathematical models for classical conditions from an information-processing perspective (e.g., Jamieson, Crump, & Hannah, 2012; Mackintosh, 1975; Pearce & Hall, 1980).

conditioning because the "expectation" term in the model or $V_{n-1}(\mathbf{A})$ does not exist when the conditioned stimulus does not occur prior to the onset of the unconditioned stimulus. Although Rescorla and Wagner (1972) called this term an expectation level, it is nonetheless strictly a strength value for an association between the conditioned stimulus and the conditioned response. It is not really an expectation in the same sense as the cognitive model developed by Tolman (1932, 1945). For Tolman, learning was the development of $\mathbf{S} - \mathbf{S}$ expectancies (i.e., which stimulus leads to what other stimulus). For example, the sight of our front door produces an expectation of what will be on the other side of the door. Yet in the Rescorla-Wagner model, learning is still framed in a behavioral $\mathbf{S} - \mathbf{R}$ strength theory structure rather than a cognitive framework of information processing. According to the Rescorla-Wagner model, simultaneous conditioning should result in a conditioned response, so the failure to find a conditioned response is a problem.[9]

The Rescorla-Wagner model runs into difficulty because it is still rooted in a behaviorist framework of focusing on the strength of specific $S - R$ associations. From a cognitive perspective, it is possible that there can be a *general* signal as well as a *specific* signal. A warning tone is a general signal that something important is coming later in time, whereas an all-clear sound is a general sign that nothing important is coming. It is also possible that a stimulus can signal that another specific stimulus is coming. From a cognitive framework, it is even possible that a signal, which foreshadows a specific event, is transformed to predict a different specific event (i.e., the rules of the game changed, so a different outcome is expected). Clearly, this flexibility of signal types is beyond the mechanistic Rescorla-Wagner $S - R$ association-strength system. But in fact, four well-controlled animal-learning experiments from the laboratory of Harry Fowler at the University of Pittsburgh show that such a flexible signaling framework is required to understand Pavlovian conditioning (Domber, 1971; Fowler et al., 1973; Ghiselli & Fowler, 1976; Goodman & Fowler, 1976). In each experiment, rats initially received Pavlovian training where the conditioned stimulus was a brief auditory sound.

---

9. One might argue that a memory really is established between the conditioned stimulus and the unconditioned stimulus in the case of simultaneous conditioning despite the fact that no conditioned response is observed. In the Tolman (1932, 1945) model, this memory association is predicted. However, a conditioned response does not occur because there is no anticipatory aspect involved in the case of simultaneous conditioning. In the Tolman framework, there is a careful distinction between learning and performance, but that distinction is not used in the Rescorla-Wagner model. Hence, if there is a lack of a conditioned response when the Rescorla-Wagner model predicts an associative strength, then there is a problem for the model.

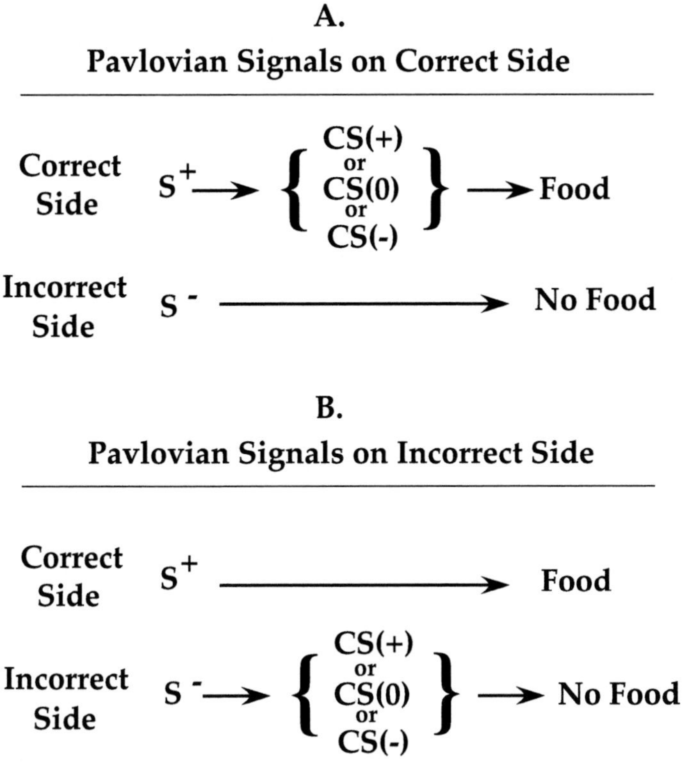

A.
Pavlovian Signals on Correct Side

Correct Side $S^+ \longrightarrow \left\{ \begin{array}{c} CS(+) \\ or \\ CS(0) \\ or \\ CS(-) \end{array} \right\} \longrightarrow$ Food

Incorrect Side $S^- \longrightarrow$ No Food

B.
Pavlovian Signals on Incorrect Side

Correct Side $S^+ \longrightarrow$ Food

Incorrect Side $S^- \longrightarrow \left\{ \begin{array}{c} CS(+) \\ or \\ CS(0) \\ or \\ CS(-) \end{array} \right\} \longrightarrow$ No Food

Figure 2.2
Designs for the research on the effect of Pavlovian fear and relief signals on later instrumental appetitive conditioning (Domber, 1971; Fowler et al., 1973; Ghiselli & Fowler, 1976; Goodman & Fowler, 1976). The design when the Pavlovian signals are on the correct arm of the T-maze is shown in panel A, whereas panel B corresponds to the design when the Pavlovian signals are on the wrong arm.

Pavlovian training resulted in (1) a $CS(+)$ linked to a shock US, (2) a $CS(-)$ linked to the absence of shock, or (3) a $CS(0)$, which is uncorrelated with either the presence or absence of a shock. The $CS(+)$ is thus a specific fear signal, whereas $CS(-)$ is a specific relief signal, and $CS(0)$ is an uninformative stimulus. Next the investigators food deprived the rats and began an instrumental T-maze discrimination-training program for food reward. The discriminative stimuli (S^+ and S^-), which signaled the respective correct and incorrect arms of the maze, were visual. But the researchers also introduced the auditory Pavlovian signals while the animals were partway through one of the arms of the T-maze. See figure 2.2 for the sequence of stimuli in the various arms of the T-maze as a function of the experimental conditions. If the Pavlovian conditioned stimuli followed the visual S^+ (i.e., it was on the correct side as shown in panel A), then the overall proportion of correct choices P_c for the various signal types yielded the following rank order:

$P_c[\mathbf{CS}(+)] > P_c[\mathbf{CS}(0)] > P_c[\mathbf{CS}(-)]$.

Thus, when a former Pavlovian "fear" signal was introduced on the correct side of the T-maze, the animals were most accurate, and when a Pavlov "relief" signal was introduced on the correct side, the animals were the least accurate. The accuracy for the uninformative Pavlovian $\mathbf{CS}(0)$ signal was between the other two conditions. One might argue that the heightened emotional reaction to a fear signal increased learning, whereas a relief signal reduced learning. However, that argument fails to account for the findings when the Pavlovian signals were introduced on the incorrect arm of the maze (see figure 2.2B). If the Pavlovian signals followed the visual S^-, then the overall discrimination accuracy for the various signal types yielded the following rank order:

$P_c[\mathbf{CS}(+)] < P_c[\mathbf{CS}(0)] < P_c[\mathbf{CS}(-)]$.

Now the former fear signal resulted in the lowest discrimination accuracy, whereas the former relief signal resulted in the highest accuracy, and the uninformative stimulus resulted in a value between the other two conditions.

These experiments from Fowler's laboratory show that the rats quickly transmuted the former Pavlovian fear $\mathbf{CS}(+)$ stimulus from a specific signal that shock is coming to a general alert signal indicating that food is coming (i.e., the rules of the game changed). Such a signaling message is in fact correct when the stimulus is placed in the alleyway just before the goal box. When the messaging is consistent, such as in this case, the discrimination learning is facilitated, and the choice accuracy is increased. However, if the general "something-is-coming" message is placed in the alleyway that leads to the chamber without food, then the message is inconsistent, and discrimination learning is impaired. Similarly, when a former Pavlovian relief signal $\mathbf{CS}(-)$ is introduced in an instrumental maze task, the specific signal that "shock is not coming" is transmuted to a general message that "nothing is coming." When this stimulus is placed on the alleyway without food, the messaging is consistent and leads to improved discrimination learning, whereas when it is placed in the alleyway leading to food, then the messaging is inconsistent and the learning is impaired. Thus, the behavior exhibited by these rats indicates that Pavlovian conditioning is more complex than a mechanistic strengthening of $S-R$ associations as in the Rescorla-Wagner model.

2.5 How Cognitive Is Pavlovian Conditioning?

2.5.1 The Temporal-Learning Hypothesis

Although it is fair to criticize "subcognitive" theories of conditioning, such as the Rescorla-Wagner model, because of the inability to handle some of the effects found for Pavlovian conditioning, it is another matter to conclude that classical conditioning is entirely explained by a cognitive framework.[10] Yet that is exactly what some researchers have argued. For example, Gallistel and Gibbon (2002) argued for a radically different basis for learning than posited by existing associationist models. They focused on the issue of the encoding of temporal information within the context of a "symbolic" framework for conditioning. A symbolic approach is one that claims that animals have a language-like understanding of the relationships among stimuli. Ward, Gallistel, and Balsam (2013) make a similar argument and further suggest that the principle of contiguity is incorrect. Contiguity is the idea that an associative bond with some strength value is formed between the mental representation of stimuli that appear together spatially or temporally. Although I share with these authors a skepticism about classic associationism, aspects of their theory are open to question. A key hypothesis for these authors is the claim that animals are learning the precise time difference between stimuli. But how robust is this hypothesis? The learning of temporal information shows that even humans do not generally have a good memory representation for time (Underwood, 1977). Underwood (1977) elaborates how memory has multiple attributes that can have different fates. Learning temporal features of an event is something that Underwood claims people do poorly. It is more likely that people learn an ordinal temporal relationship rather than learn a specific magnitude of the temporal delay between events. To learn the signaling value of a conditioned stimulus only requires the learning of a temporal order; it does not require that the animals encode the magnitude of the temporal delay between the **CS** and the **US**. Hence, if behavioral associationism is being rejected because of its inability to model temporal information, then that argument is itself worrisome because it might not be universal. Perhaps a variation of the temporal-learning hypothesis is valid for Pavlovian conditioning under some circumstances and for some animals, but does it always apply in all cases? Nonetheless, the cognitive models advanced by Gallistel and Gibbon (2002) and Ward et al. (2013) raise important issues about what was vague in the

10. Classical conditioning has been found for many species other than mammals. Mammalian brains have many similarities. Consequently, the finding that mammals are sensitive to informational aspects of classical conditioning does not mean that all species are learning a cognitive model during conditioning.

classic behavioral conception of contiguity. In chapters 3 and 4, the issue of encoding temporal information will be addressed in more detail.

2.5.2   Issue of Awareness for Pavlovian Conditioning

If it could be established that the Pavlovian conditioning of an animal necessarily implies that the animal has a *conscious* understanding of the relationship between the **CS** and **US**, then a strong case would have been made for a cognitive model of the knowledge acquired. Understandably, this position is very difficult to substantiate. Classical conditioning has been reported for many other species than just for humans or rats, so it is a daunting task to verify across all the species that the animals have an awareness of the contingency relationship between the **CS** and the **US**, especially in light of C. Lloyd Morgan's warning:

> In no case is an animal activity to be interpreted in terms of higher psychological processes if it can be fairly interpreted in terms of processes which stand lower in the scale of psychological evolution and development. (Morgan, 1894, p. 53)

Nonetheless, Lovibond and Shanks (2002), as well as Mitchell, De Houwer, and Lovibond (2009), have in fact argued that Pavlovian conditioning requires the supposition that there is a conscious awareness of the contingency relationship between the **CS** and **US**. From this viewpoint, awareness follows a "propositional learning process." In cognitive psychology, a proposition is a generalization of an association between two elements. Instead of an $A - B$ association, there is a directional, operational relationship between the components. A proposition can be symbolized as $A \xrightarrow{\text{relationship}} B$ where $\xrightarrow{\text{relationship}}$ denotes a type of relationship between the stimulus elements. For example, a proposition could mean "stimulus **A** produces or results in stimulus **B**." The proponents of the propositional conceptualization tend to focus on Pavlovian conditioning studies with humans. But classical conditioning is widely demonstrated across many species, so it is questionable to argue for the necessity of a propositional foundation for Pavlovian conditioning from mainly experiments with humans. Later, a case will be advanced for a propositional representation for another class of memory encodings based on human data. But when it comes to strictly Pavlovian conditioning, the propositional interpretation is a stronger cognitive interpretation for conditioning than is necessary in all cases. The strong cognitive interpretation of Pavlovian conditioning is controversial, and it has been criticized by a number of other researchers (Manns, Clark, & Squires, 2002; Wiens & Öhman, 2002). Rather than review their criticisms here, I will focus instead on experiments that show that under some special cases, either conscious awareness may not be necessary for classical conditioning or that unconscious experiences can facilitate later classical conditioning.

2.5.3   Pavlovian Conditioning and General Anesthesia

If it is demonstrated that unconscious animals show evidence of classical conditioning, then that finding would establish that a purely conscious model of conditioning is not universally the case. An ideal method for studying unconscious effects is to put the animal under general anesthesia and then begin conditioning training. If the animal shows evidence that this training alters performance, then conscious awareness would not be necessary for conditioning. This approach was used in a number of experiments (Edeline & Neuenschwander-El Massioui, 1988; Ghoneim, Chen, El-Zahaby, & Block, 1994; Pang, Turndorf, & Quartermain, 1996). In the Edeline and Neuenschwander-El Massioui (1988) study, rats were administered sufficient ketamine hydrochloride to induce general anesthesia, and later supplemental doses were given to maintain a deep anesthesia throughout the whole session.[11] Ketamine was selected as the general anesthetic because it was previously determined to have little effect on the processing of external stimuli that serve as a conditioned stimulus. The dosage level for ketamine was 120 mg/kg. When the animals were in an unconscious state, they received Pavlovian conditioning with a tone CS and an electric shock as the US. Another group of animals received training on two different tones where one was a CS^+ for a shock US, and the other tone was a conditioned inhibitor CS^-. Two days after the training under anesthesia, the animals were introduced to an operant bar-pressing task. Once a baseline performance level was established, the animals heard the previous Pavlovian CS^+ while in the process of doing the operant bar-pressing task. The CS^+ fear stimulus resulted in an immediate reduction of the bar-pressing rate, whereas when the CS^- stimulus was presented, it did not result in a reduction of the bar-pressing rate. Thus, the Pavlovian training administered during general anesthesia was controlling later behavior while the animals were awake.

The study by Ghoneim et al. (1994) also investigated Pavlovian conditioning for the ketamine anesthetic. In this study, two dosage levels were examined: normal (100 mg/kg) and high (200 mg/kg). The test subjects were rabbits, and the effect of conditioning under anesthesia was assessed by providing another later conditioning training three days after the animals recovered from being anesthetized. The animals that had the normal dosage of 100 mg/kg showed faster relearning compared to a control group of animals that had the same dosage level but did not have conditioning when they were unconscious. This finding with a different species and with different assessment protocols replicates the conclusions of the Edeline and Neuenschwander-El

11. The animals were monitored by an electrocardiogram, and the whiskers of the animals were also monitored to ensure there was no movement.

Massioui (1988) study. But the animals that had the high dosage of 200 mg/kg did not benefit from their conditioning trials under general anesthesia. In fact, these animals did worse than the control animals, so a critical dosage level needs to be achieved to have these animals capable of learning while unconscious. Ketamine is a drug that is known to be an antagonist of the N-methyl-D-aspartate (NMDA) receptor, which in turn has been thought to be important for memory formation (Lorrain, Baccel, Bristow, Anderson, & Varney, 2003). Nonetheless, the Ghoneim et al. (1994) study indicates that Pavlovian conditioning under the influence of ketamine positively transferred to the awake state, at least when the drug was administered in the normal dosage range.

Anesthetic drugs can have complex effects on a number of systems other than simply consciousness and pain perception. This point is illustrated by the Weinberger, Gold, and Sternberg (1984) experiment on anesthesia and classical conditioning. The study was similar in design to the Edeline and Neuenschwander-El Massioui (1988) study except that the rats were anesthetized with sodium pentobarbital (48 mg/kg). Sodium pentobarbital is used as an anesthetic agent in veterinary surgery. Classical conditioning administered during anesthesia did not transfer to the later operant behavior; however, if the animal had a mixture of sodium pentobarbital and epinephrine, then there was a significant reduction in the later bar-pressing rate, indicating the influence of the training under general anesthesia. Moreover, the dosage level of epinephrine also affected the magnitude of the suppression of bar pressing. The researchers used three dosage levels of epinephrine (.01 mg/kg, .10 mg/kg, and 1.00 mg/kg). The biggest effect of bar-pressing suppression occurred with the .01-mg/kg dosage.[12] In each case where some epinephrine was administered along with the sodium pentobarbital, there was evidence of classical conditioning occurring when the animals were anesthetized, but sodium pentobarbital alone did not result in any detectable evidence for conditioning.

Although the above studies make clear that Pavlovian conditioning can be acquired when the animal is unconscious, there is also evidence that the strength of the conditioning is weaker than when the animal is awake. From an experiment by Pang et al. (1996), a significant effect for conditioning under the influence of the anesthetic halothane was demonstrated by the later suppression of the bar-press response. However, the magnitude of the suppression was less than a control group that had Pavlovian training while the animals were in a conscious state. This positive but weaker conditioning effect was also found in the Ghoneim et al. (1994) study. Consequently,

---

12. Previously, Gold and Van Buskirk (1975) found that the dosage of .01 mg/kg was the optimum level to facilitate memory.

conscious awareness is not required to develop some level of Pavlovian conditioning, but the conditioning done while in an unconscious state is weaker than when in a conscious state.

One might argue that Pavlovian conditioning could still be based on a propositional encoding despite the evidence for Pavlovian conditioning while the animal is unconscious. That argument would have some merit if it could be shown that humans remember verbal information that occurred while they were under general anesthesia. In reviewing the research reports and case studies, Ghoneim and Block (1992) state that the literature is ambiguous, and it might be the case that errors can occur on the part of clinical anesthesiologists that result in overly light anesthesia. These experts on anesthesiology also acknowledge that there is no universally accepted physiological measure of the depth of anesthesia. Consequently, the positive reports of some patients who indicate that they remembered verbal comments during their operation might be due to the patients having some degree of wakefulness while under general anesthesia. Merikle and Daneman (1996) did a statistical meta-analysis to see if there was support for the memory of events during general anesthesia. Only two of eight studies that used a direct memory measure (such as a recognition task) showed any support for some memory retention of verbal information. Given the concern about individual differences and the possibility of overly light anesthesia, the human studies are not strong support for the hypothesis that humans can remember verbal and propositional information while under general anesthesia.[13]

Consequently, the general anesthesia studies with animals indicate that a weaker level of Pavlovian conditioning is possible while the animals are unconscious. It is doubtful that this weak conditioning is based on a propositional representation as hypothesized by advocates of the strong cognitive interpretation of classical conditioning (e.g., Lovibond & Shanks, 2002; Mitchell et al., 2009). But the general anesthesia studies do not demonstrate that the strong cognitive model is wrong for the

13. Another problem with the Merikle and Daneman (1996) meta-analysis is its use of the Rosenthal (1991) methodology for doing a meta-analysis within the context of the relative-frequency approach to statistical inference. Bayesian statisticians argue that Bayes theorem is the correct way to combine information across separate samples, and classical relative-frequency statistics is not designed to combine information across different samples. Importantly, Bayesians are very careful to only combine data sets that employ the same test protocols; otherwise, the studies are considered incommensurable. Unfortunately, the methods in the Rosenthal (1991) text combine studies regardless of their test protocols. To Bayesian statisticians, this practice is absurd and not defensible under any theory of statistical inference. See the glossary for more information about classical statistics and Bayesian statistics.

cases when the animal is conscious. Rather, these studies show that consciousness is not required for Pavlovian conditioning.

2.5.4 Critique of Research on Subliminal Conditioning

The question still remains: if people are awake, then is it necessary for them to be aware of the CS as a signaling stimulus to develop a conditioned response? Some researchers claim that a subliminal CS can effectively lead to a conditioned response (e.g., Knight, Nguyen, & Bandettini, 2003, 2006). The rationale underlying the research by Knight and his colleagues is an experimental design where there are presentations of CS^+ and CS^- stimuli at either below or above the perceptual threshold for auditory detection. The Knight research team attempted to measure absolute auditory thresholds for CS^+ and CS^- stimuli on a trial-by-trial basis rather than as fixed values for an individual. The rationale for this design decision was based on a paper by Miller (1991) that argued for this approach to increase the statistical power for detecting subliminal stimuli. The Knight team also used an adoptive-threshold method (Kaernbach, 2001) that did not have target-present and target-absent test trials to find the threshold of detection. Instead, the stimulus was either increased or decreased in sound intensity by 5 dB, depending on the response of the participant (i.e., decreased if the participant responded "yes" and increased if the participant responded "no"). Based on this methodology, Knight et al. (2003) reported an apparent significant difference between the amount of a skin conductance conditioned response to the CS^+ relative to the conditioned response to the CS^- for stimuli that were ostensible below the threshold for detection.[14] The Knight et al. (2003) experiment used a delayed conditioning arrangement (see figure 2.1). Knight et al. (2006) used both delayed and trace arrangements and only reported a significant conditioning effect for below-threshold stimuli when, the delayed procedure was used.

On the surface, the Knight et al. (2003, 2006) experiments appear to show support for subliminal classical conditioning, but these studies have design faults that mitigate against drawing that conclusion. One problem is the lack of the inclusion of "catch" trials where the stimulus is absent. Without target-absent as well as target-present test trials, we cannot rule out the hypothesis that there is a mixture of suprathreshold events as well as subthreshold events. What ostensibly is claimed to be subliminal conditioning might be simply the effect of some suprathreshold detections of the

---

14. Knight et al. (2003, 2006) used auditory CS^+ and CS^- tones of different frequency, and the US was a loud white noise stimulus of 110 dB. The conditioned response was a measure of skin conductance.

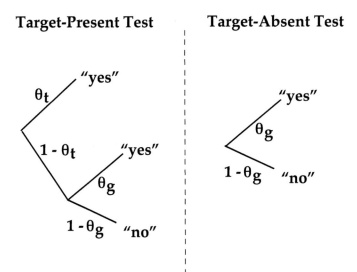

Figure 2.3
The above trees are probability representations for the target-present and target-absent tests. The θ_t parameter is the true population proportion that the stimulus is above threshold, and in such cases, the participant is assumed to respond "yes." For the $1 - \theta_t$ proportion of the target-present tests, the participant still has a conditional probability of θ_g of guessing "yes" correctly. For the target-absent catch trials, there is also a probability of θ_g for giving a "yes" response. The above trees are for a specific stimulus intensity. For subliminal perception, the stimulus intensity must be sufficiently low to result in $\theta_t = 0$.

stimulus on a subset of the test trials. To illustrate a mixture model for subthreshold and suprathreshold events, see figure 2.3. The proportion of suprathreshold events is denoted as the parameter θ_t. If on target-present tests the stimulus is suprathreshold, then it is reasonable to assume that the participant will give a "yes" response for detection. But for the $1 - \theta_t$ proportion of the test trials where the stimulus is not suprathreshold, it is possible that the participant might still respond "yes" due to guessing. The parameter θ_g is a conditional guessing probability for a "yes" response on the subset of target-present trials where the stimulus is not suprathreshold. To learn what the guessing rate is and thus also learn what the value is for θ_t, there must be catch trials where the stimulus is absent. For these trials, the response bias for saying "yes" should also be θ_g. If n_1 and n_2 are the number of target-present trials with the respective "yes" and "no" responses, and if n_3 and n_4 are the corresponding number of target-absent trials with the respective yes and no responses, then an estimate of the proportion of suprathreshold events $(\hat{\theta}_t)$ is

$$\hat{\theta}_t = \frac{n_1\,n_4 - n_2\,n_3}{(n_1 + n_2)\,n_4},$$

provided that $n_1 n_4 \geq n_2 n_3$.[15] Note that this tree model can detect the proportion of suprathreshold events for any given intensity level of the stimulus. If for some intensity level we find that θ_t is 0, then a case can be made that the target is not suprathreshold (i.e., we would have found the stimulus intensity level right below the threshold).[16] The Knight et al. (2003, 2006) studies did not have target-absent tests. Their experimental design was thus insufficient for making any claims about θ_t. Hence, it certainly might be the case that their conditioning effect was due to some suprathreshold detections of the conditioned stimulus rather than subliminal conditioning.

There is a second methodological issue that applies to not only the Knight et al. (2003, 2006) studies but also to other experiments on subliminality. Classical null-hypothesis testing within the relative-frequency framework is based on the assumption of a null effect. Yet in regard to subliminal perception, the null effect is exactly the hypothesis in question. We cannot build evidence for subliminal perception from classical null-hypothesis testing because that statistical method assumes the null effect.[17] In contrast to classical statistics, it is possible to build evidence for either the null hypothesis or an alternative hypothesis by means of Bayesian statistics and in particular by using the Bayes factor (Jeffreys, 1961; Kass & Raftery, 1995).[18] The Bayes factor is a ratio of the probabilities for the data between two rival models or values. Applied to the issue of subliminal perception, the Bayes factor could be between two values for θ_t; one model being $\theta_t = 0$ and the other model being one with a trivially small value for θ_t (e.g., a value of .02). The Bayes factor B_{01} for this example is $B_{01} = \frac{P(D|\theta_t=0)}{P(D|\theta_t=.02)}$, where $P(D|\theta_t)$ is the probability of the observed data given a value for θ_t. If the Bayes factor were ≥ 20, then Jeffreys (1961) as well as Kass and Raftery (1995) would conclude that there is "strong" support for the model that has $\theta_t = 0$ relative to the trivial effect

---

15. This value is a maximum likelihood estimate. Maximum likelihood estimation is described in more detail in the glossary.

16. For example, if the proportion of yes responses is the same on target-absent tests as for target-present tests, then $\frac{n_1}{n_1+n_2} = \frac{n_3}{n_3+n_4}$, and thus $n_1 n_4 = n_2 n_3$, which results in the estimate $\hat{\theta}_t = 0$.

17. This problem is further exacerbated by the Miller (1991) trial-by-trial threshold method because it results in larger threshold intensity estimates.

18. For more information about classical statistics, Bayesian statistics, and the Bayes factor, see the glossary.

alternative value in which $\theta_t = .02$. For these models, it follows that

$$B_{01} = \frac{\theta_g^{n_1+n_3}(1-\theta_g)^{n_2+n_4}}{(.02+.98\theta_g)^{n_1}\theta_g^{n_3}(1-\theta_g)^{n_2+n_4}(.98)^{n_2}}, \tag{2.3}$$

$$= \frac{1}{(\frac{.02}{\theta_g}+.98)^{n_1}(.98)^{n_2}}. \tag{2.4}$$

To see the sample size necessary to achieve a Bayes factor greater than 20, let us assume $n_1 = n_2 = \frac{n}{2}$ and $\theta_g = .5$. It follows from equation (2.4) that these assumptions result in $n \approx 15,000$ target-present test trials and another 15,000 observations when the target is absent.[19] Clearly, to establish that the null hypothesis of subliminality is in fact the probable case rather than just an assumption, it is necessary to obtain an enormous number of observations. Although it is hopeless to obtain this number of observations for each participant, it is possible to satisfy this requirement by combining observations over participants. There should be a nonzero stimulus intensity level such that the Bayes factor is greater than 20 when combining the observations of say 300 participants who each are given fifty target-present and fifty target-absent tests. That stimulus intensity level would rigorously satisfy the evidentiary requirements for being subthreshold.[20] Consequently, it is possible to construct an experimental design that can establish a subthreshold intensity level. It is still an open question if Pavlovian conditioning would be successful with such a rigorously established subthreshold level.

Another approach to subliminality is to present the target so quickly that the observer fails to detect the stimulus. This approach has been used by some researchers who have argued that a subliminal CS can elicit a conditioned response in humans (Öhman & Mineka, 2001; Öhman & Soares, 1993, 1994; Wiens & Öhman, 2002). The focus for most of the Öhman studies is about the subliminal perception of pictures that might or might not be phobic stimuli, but the purpose of the Öhman and Soares (1993) study was to investigate the issue of Pavlovian conditioning for a subliminal CS. The authors claim that subliminal conditioning occurs when picture stimuli are displayed for 30 msec and are immediately followed by a 100-msec masking stimulus

19. To understand this sample size in terms of classical null-hypothesis testing, one can compute the prospective power for detecting a .02 difference between two proportions when the α level is set to .05. For 15,000 observations for both target-present and target-absent tests, the power is approximately .97.

20. Some researchers might worry that model estimates are inaccurate when the data are pooled across individuals. But Chechile (2009) extensively examined models similar to the tree model in figure 2.3 and found greater accuracy for the pooling method relative to the method of fitting the model separately to each individual and then averaging estimates.

(i.e., backward masking). The masking stimuli were scrambled visual rearrangements of the targets. Masking of a briefly presented stimulus in the same spatial location is thought to make the further processing of the stimulus impossible. But the claim that the targets are undetectable can be disputed because some participants, on a subset of trials, might be able to detect the target. For the Öhman and Soares (1993) study, the CS^+ and CS^- were briefly presented pictures that were followed by a masking pattern. The CS^+ was paired with shock, but the CS^- was not followed by shock. The conditioned response was a measure of skin conductance. In extinction, the authors report a significantly elevated skin conductance elicited by the CS^+ relative to the CS^- only when the stimuli were pictures of fear objects (e.g., snakes, spiders). For nonemotional or neutral pictures (e.g., mushrooms, houses), there was not an elevated skin conductance. Thus, evidence for classical conditioning of subliminal pictures is limited to only cases when the CS^+ was a picture of a fear object.

The results from the Öhman and Soares (1993) study could be due to the fact that phobic stimuli have a very low detection threshold, very much like the high salience for certain important stimuli like your name. For such stimuli, the participants might have perceptually detected the CS^+, so the conditioned response might not be an unconscious, subliminal perception of the stimulus. Cornwell, Echiverri, and Grillion (2007) performed a clever experiment to explore this possibility. One innovative feature of this experiment is its use of a modified version of the dichoptic-masking procedure developed by Moutoussis and Zeki (2002). With dichoptic masking, the participant is shown a stimulus in one eye, such as a black cross within a red square background, while the other eye is shown a black cross within a green square background. The presentation of the images is simultaneous and brief (i.e., 50 msec). After a 150-msec blank interval, the pair of stimuli are presented again for 50 msec, but the background colors are reversed to the eyes. If the participant cannot consciously detect the cross, then he or she perceives instead a solid yellow square. Because of the use of opponent colors, there is binocular color fusion between the red and green backgrounds (i.e., red and green fuse to become yellow). If the participant can detect the cross, then it appears as a black cross within a yellow background. Moutoussis and Zeki (2002) showed that even when the stimulus disappears and is therefore not consciously detected, there was still a functional magnetic resonance imaginig (fMRI) activation pattern that is similar to when the target is not masked.[21] Thus, with dichoptic masking, it is possible to have cases where the target stimulus is "consciously invisible," but neurologically it is still represented. With dichoptic masking, it is thus possible to present stimuli that are consciously invisible, at least for some participants and for some trials. In the

21. See the glossary for a brief description of fMRI.

Cornwell et al. (2007) experiment, the researchers also ascertained a preconditioning measure of stimulus detectability under dichoptic masking conditions (i.e., they found a signal detection d' value for each participant *prior to conditioning*). The quantity d' is a measure of signal detectability that is independent of the participant's decision bias.[22] As hypothesized, there were wide individual differences, with some participants having d' values in the (.8 to 2.8) range, indicating that they were able to detect the masked target, but there were other participants with very low preconditioning d' values that were less than .3 (suggesting that these people were not consciously aware of the masked stimuli). For these low-d' individuals, there was not a difference in the later assessment of skin conductance between the CS^+ and the CS^- conditions.[23] For the high-d' individuals, there was a difference between the CS^+ and the CS^- conditions. Hence, an elevation of the skin conductance measure after conditioning training was entirely predicted by the participant's preconditioning stimulus detection score.

The Cornwell et al. (2007) experiment does not support the claim of subliminal conditioning by Öhman and Soares (1993). Remember Öhman and Soares (1993) did not find a statistical difference between the CS^+ and the CS^- for briefly presented neutral stimuli, but they did find a difference in the conditioned response for a special class of emotional (fear) stimuli. Yet it is very likely that these fear stimuli have a different rate for perceptual detection. Cornwell et al. (2007) showed that the difference between the CS^+ and CS^- is directly a function of the rate for perceptual detection (i.e., if the detection rate is low, then there is not a conditioning effect).

It should be added that the tree model in figure 2.3 can also be used for backward-masking experiments designed for studying subliminal perception. If one wants to rigorously determine when a stimulus is not detectable because the presentation time is too short, then the time can be adjusted to determine a temporal value that results in strong support for the null hypothesis via equation (2.3). Without a Bayes factor assessment, claims of subliminal perception are unsupported assertions because there could be a mixture of some suprathreshold events.[24] The examination of only low-d' individuals, as done in the Cornwell et al. (2007) experiment, is a weaker evidentiary method

22. For more information about d', see the glossary.
23. In this study the CS^+ was paired with shock, and the CS^- was paired with the absence of shock.
24. Some might wonder how it is possible to remember an event if the presentation time is so brief that it results in strong support for the null hypothesis that $\theta_t = 0$. But advocates of subliminal perception hypothesize that there is still an implicit or unconscious memory of the brief event that influences forced-choice recognition to be above chance (Schacter, 1989; Schacter, Chin, & Ochsner, 1993). Yet the methodological problem with this claim is in first building a strong case for the hypothesis that $\theta_t = 0$.

for assessing subliminal conditioning. Thus, the results from the Cornwell study are particularly negative because they demonstrate that even with a less stringent standard for evidence, there is no support for Pavlovian conditioning without target detection.

Currently, there is not a convincing demonstration that a stimulus (1) is truly subliminal and (2) still capable of developing a new conditioned response. Perhaps some experiments will be conducted that achieve these two requirements, but until such a time, it is prudent to adopt a high standard of evidence for subliminal conditioning. To firmly establish the phenomenon, the above two requirements must be established. Yet the failure to rigorously demonstrate classical conditioning for a subliminal CS still does not prove the (Lovibond & Shanks, 2002; Mitchell et al., 2009) strong cognitive position that Pavlovian conditioning involves a conscious awareness of the CS and US contingency. Currently, the only effective procedure uncovered that can successfully demonstrate unconscious Pavlovian conditioning is the administration of a general anesthetic that produces a deep unconscious state but does not interfere too strongly with the ability for forming an association between the CS and US. It would seem that barring this extraordinary procedure, the pathway for humans to learn a Pavlovian conditioned response under normal waking conditions is via a conscious process that is also likely to be understood as a propositional relationship between the CS and the US. Consequently, the strong cognitive model for Pavlovian conditioning might not be correct for all circumstances and for all species, but it is in my opinion a valid hypothesis for humans under normal waking conditions.

2.5.5   Can Postacquisition Conditioning Become Unconscious?

In general, once a task or an association has been learned to a high degree, it requires less focused attention. As William James (1892) observed, "Habit diminishes the conscious attention with which our acts are performed" (p. 6). A strong case has been made that high levels of practice result in a qualitative change in the cognitive resources and attention required to implement the task (Posner & Snyder 1975; Schneider & Shiffrin, 1977). Given the general reduced attention required for overlearned associations, is it possible to have a classical conditioned response be triggered without any conscious awareness?

Corteen and Wood (1972) did a dichotic-listening experiment that showed a conditioned response to words that the participants did not report being consciously aware. The participants were first classically conditioned to some critical words that were paired with shock; these words are the CS^+ stimuli. In a second phase of the experiment, the participants were given a dichotic-listening task with separate auditory channels presented simultaneously. The participants had to shadow (i.e., repeat aloud) the message presented to one of their ears; this ear is called the attended channel. Different words were presented to the other ear, which is called the unattended channel. Sprinkled among the words on the unattended channel were the critical CS^+ words that were previously paired with shock. Corteen and Wood reported that the critical words elicited an elevated skin conductance response on 37.7 percent of the tests, whereas there was only an elevated skin conductance response on 12.3 percent of the control word tests. They also reported that the participants did not remember the critical words when tested later (between 1 and 5 min after their occurrence). Corteen and Wood (1972) argued that the participants were not consciously aware of the critical words, but those words were still capable of eliciting a conditioned response. Although this claim might be quite accurate, it is still a very strong statement given only an informal, postexperiment memory test. Participants might have switched attention to the unattended ear momentarily and detected the critical words on some occasions but later forgot these words. Also, how should we interpret the difference in the response rate of 25.4 percent between 37.7 percent and 12.3 percent? How does this difference compare to the difference between the target and control words when these words are on the attended channel? To address this last question, von Wright, Anderson, and Stenman (1975) measured the skin conductance rate for critical words when the words were on the unattended channel versus when the words were presented on the attended channel. The authors of this experiment reported a difference of 22.0 percent between the critical words and control words for the unattended channel. Note this difference is very close to the difference of 25.4 percent for the Corteen and Wood experiment. But the corresponding difference in the percent elevated skin conductance between critical words and control words for the attended channel was 60.6 percent, which is 2.75 times the magnitude of the effect found for the unattended channel. Although there is a significant demonstration of conditioning for a CS^+ on the unattended channel, the magnitude of the effect is substantially reduced from the conditioning effect found for the attended channel.

The attentional resources required to implement the shadowing task on the attended channel reduce the cognitive capacity to process information on other channels. But is it still possible that some participants on some trials have sufficient attentional capacity to process the critical words cognitively on the secondary channel due either to attention switching or by having sufficient capacity to process both channels of information? If the conditioned responses were due to attention switching, then there ought to be problems with the participant's execution of the primary shadowing task. Dawson and Schell (1982) examined this question in careful detail. They did find more shadowing errors associated with the critical words, so they culled those suspected switched-attention events and reexamined the data. For a subgroup of participants who heard the critical words in the secondary channel in their left ear, there was a statistically significant but much reduced conditioning difference effect of 13 percent. But for the group of participants who heard the critical words on the unattended channel in their right ear, there was not a conditioning effect.

Dawson and Schell (1982) provide a convincing explanation for their pattern of conditioning effects based on the neural lateralization of auditory input. When the participants had to verbally repeat the information received via the left headphone, that information was received in the right hemisphere. The right hemisphere does receive speech input, but for most people, the right hemisphere does not play a role in speech production (Lindell, 2006). Therefore, in order for the participants to shadow the information on the left ear, it was necessary for the information to be transferred from the right hemisphere to the left hemisphere. For example, if the word that must be repeated by the participant is *lamp*, then that speech sound is ultimately received in the left hemisphere via a transfer process from the right hemisphere. This example is illustrated in the following diagram.

Left Ear		Right		Left		Right Ear
Shadow	$\Longrightarrow$	Hemisphere	$\Longrightarrow$	Hemisphere ■	$\Longleftarrow$	Critical Word
lamp		*lamp*		*lamp*		*dog*

If a critical word (like the word *dog* in the diagram) that was previously associated with shock is spoken into the right headphone, then that information has less chance of being received in the left hemisphere language-processing module because that module is busy receiving input from the right hemisphere and formulating the speech output. Dawson and Schell (1982) reported that left ear shadowing is more difficult and leads to more articulation errors and suspected switches in attentional focus. But when the shadowing-error events are removed in data processing, the remaining events resulted in the critical word (*dog*) not being processed sufficiently to evoke a conditioned response. Hence, there was not a conditioned response except when there was a switch in attention. But for the participants who had to shadow the information from the right headphone, there was a different neural-processing flow as illustrated in the diagram below.

Left Ear		Right		Left		Right Ear
Critical Word	$\Longrightarrow$	Hemisphere	$\rightarrow$ ■	Hemisphere	$\Longleftarrow$	Shadow
dog		*dog*		*lamp*		*lamp*
		$\Downarrow$				
		CR				

With this arrangement, the word that must be repeated is directly received by the left hemisphere. Whenever a critical word on the left headphone is spoken, it can be received and processed directly by the right hemisphere. Although the right hemisphere does not play a role in speech production, it does have a speech-processing module that plays an important role in the analysis of the prosodic and emotional intonation of speech (Buchanan et al., 2000; George et al., 1996; Lindell, 2006). Consequently, a conditioned response can be triggered based on only the activation in the right hemisphere, as illustrated in the neural flow diagram. Given limited attentional resources, the information on the secondary channel is not processed sufficiently for consciousness, but it is processed well enough to elicit a conditioned response from the autonomic nervous system (i.e., an increase in perspiration). Regulation of perspiration does not require conscious information processing, so a conditioned response can be demonstrated in a task where the attentional resources are needed for listening and verbally shadowing the speech on the primary channel. Research on split-brain patients (i.e., patients with surgically separated hemispheres) demonstrates that it is possible that the left and right hemispheres can have conflicting cognitive interpretations for an event, but in such cases, the single conscious interpretation that dominates is usually the one from the left hemisphere (Gazzaniga & LeDoux, 1978). The left hemisphere controls language processing, so it is reasonable that the cognitive narrative from the left hemisphere trumps the nonlinguistic interpretation of the right hemisphere. Thus, the Dawson and Schell (1982) study supports the hypothesis that a previously learned conditioned response can be triggered without apparent awareness. But this effect was produced by an unusual procedure that effectively keeps the left hemisphere too busy to process the conditioned stimulus, but the procedure still enables the right hemisphere to concurrently detect the conditioned stimulus. Although it is possible to have a conditioned response occur without conscious awareness, the magnitude of the conditioned response is larger when the participant consciously detects the conditioned stimulus (von Wright et al., 1975).

Returning to the overall question for this section, it is my assessment that Pavlovian conditioning for humans usually involves the learning of a propositional relationship between the conditioned stimulus and the unconditioned stimulus. Yet it is also possible with extraordinary procedures, such as conditioning under a general anesthetic, for a relatively weak association to be formed between the CS and the US. It is also possible, once a conditioned response has been acquired, for it to be triggered without conscious awareness, but special conditions are required to divert attention from the recognition of the conditioned stimulus. The conditioned stimulus is normally an important and salient stimulus; thus, it is likely to be consciously detected. Although the strong cognitive model of Pavlovian conditioning is considered the normal case for humans, it is premature to apply this hypothesis for all other species. There might be many species that develop a cognitive "understanding" of the relationship between the conditioned stimulus and the forthcoming unconditioned stimulus, but it is outside the scope of this book to tackle this sizable problem in comparative psychology.

2.6 Pavlovian and Instrumental Comparisons

There are a number of similar effects found for both Pavlovian and instrumental conditioning. Acquisition, extinction, spontaneous recovery, and higher-order conditioning occur for both types of learning procedures. Impressed with the similarities observed for both types of conditioning, Hull (1931) theorized that classical conditioning occurred implicitly within instrumental training and mediated the acquisition of the instrumental associations. Hull (1931) called the hypothesized conditioned response that occurred within instrumental conditioning a fractional anticipatory goal response, denoted as r_g, and theorized that this response had internal sensory feedback that he called s_g. Although it is reasonable for an anticipatory conditioned response to occur during the animal's execution of an instrumental task, it is another matter to equate the learning in these two conditioning paradigms. In fact, other theorists argued that the type of learning acquired in classical conditioning is quite different from the learning that takes place during instrumental conditioning (Miller & Konorski, 1928; Skinner, 1935).

Despite the commonality between the two conditioning procedures, there are clear procedural differences between Pavlovian and instrumental training. The strict control of the temporal sequence between the conditioned and unconditioned stimulus in Pavlovian conditioning limits the animal's behavioral choices because the experimenter needs to constrain the animal's behavior (1) to measure the conditioned response and (2) to ensure that the animal detects both stimuli. With instrumental conditioning, the animal is free to discover or implement a critical response that triggers the delivery of a desired stimulus (i.e., a reinforcing stimulus). For Pavlovian conditioning, it is not necessary to create a motivation for the animal. But with instrumental conditioning, it is necessary to motivate the animal to create a goal (e.g., depriving the animal of food so as to motivate the animal to seek a food reward for executing the instrumental response). There are also behavioral psychometric differences between the measures of classical and instrumental conditioning. Although an eyeblink response is clearly observable behavior, the conditioned response is sometimes not clearly observed with classical conditioning without special instrumentation such as the original surgical method that Pavlov created to measure the salivation of his dogs during conditioning. But with instrumental conditioning, the learned response is easily observed from the overt behavior of the animal. Beyond the procedural differences, some independent variables also have different effects for the two conditioning procedures. For example, with Pavlovian conditioning, the effect of intermittent reinforcement tends to weaken the vigor of the conditioned response, but intermittent reinforcement results in very strong performance with operant conditioning (Ferster & Skinner, 1957). For classical conditioning, a weak conditioned response is possible when the animal is under general anesthesia, but it is not possible for the animal to learn an instrumental association when unconscious. But the most important reason for a difference between the learning acquired in the two conditioning procedures is due to the fact that the animal is a passive witness to a sequence of sensory events with Pavlovian conditioning, whereas the animal is free to discover which behaviors lead to a desired outcome with instrumental conditioning. Also with Pavlovian conditioning, overt voluntary behavior is neither consequential nor possible, but with instrumental conditioning, the voluntary motor response is an essential component of the learning. With instrumental conditioning, the animal is seeking a goal, but in Pavlovian conditioning, the animal cannot alter the sequence of sensory experiences. The goal-seeking feature of instrumental conditioning results in a complex set of cognitive propositions along with a chain of learned motor actions.

A number of important differences between the two conditioning methods can be illustrated by means of an extended example of a particular maze shown in figure 2.4. The points denoted as $S_1 \cdots S_9$ in the figure correspond to the stimuli that would be encountered at key points within the maze. The start box for each trial is at S_1 and the food reward is at S_8.

After a sufficient number of trials in the maze task, the animals will come to know many things about their experiences. This learning can be represented by a linked network like the one shown in figure 2.5. The network involves propositional relationships. Propositions are denoted by a double-edged arrow along with a label for the type of relationship. For example, animals clearly will know, within the context of the apparatus and experimental room, that it is possible to obtain a food reward. Many researchers have noted that animals acquire a reward expectancy. These expectancies can be about either the type of reward (Cowles & Nissen, 1937; Tinklepaugh, 1928) or even the magnitude of the reward (Chechile & Fowler, 1973). The general and specific expectancies are cognitive concepts and are represented by propositions. For example, note the propositional link $\xrightarrow{\text{can get}}$ between the "context/apparatus" node and the "food" node. The "food" node represents the concept of the food goal.

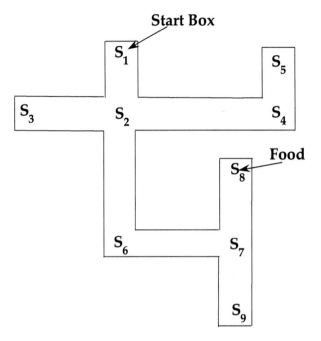

Figure 2.4
An example of an instrumental learning maze with food reward.

As another example of propositional knowledge, note that the point S_4 is east of the point S_2. This relative spatial arrangement is represented in the network as $S_2 \xrightarrow{\text{east}} S_4$ (i.e., relative to point S_2, point S_4 is in the eastward direction). The idea that animals learn spatial locations and develop a cognitive map of the maze was advanced by Tolman (1932). Although there was a vigorous debate in the research literature between the "place-learning" hypothesis and a simple S-R "response learning" hypothesis, a number of clever studies demonstrated that animals did in fact acquire a spatial or cognitive map from their free exploration of a maze (Deutsch & Clarkson, 1949; Tolman & Honzik, 1930a; Tolman, Ritchie, & Kalish, 1946, 1947). For example, Deutsch and Clarkson (1959) demonstrated for a complex maze with multiple goal boxes that rats *immediately* switched their maze route when they encountered an obstacle that blocked their path to the goal; the alternative route selected was another appropriate method for also getting the reward.

Key outcome facts are also represented in the knowledge network. For example, some maze pathways terminate at a location without food (i.e., a "nonfood termination point"), which is denoted as "NFT" in the network. A simple association in the

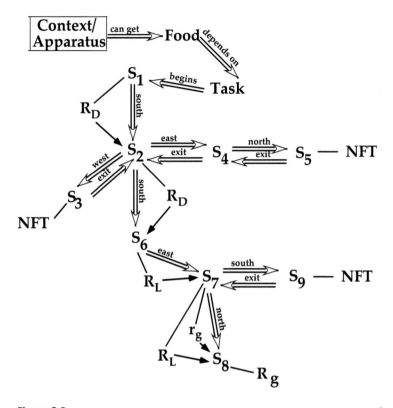

Figure 2.5
A system-level representation of the learning acquired by the instrumental training on the maze shown in figure 2.4.

network is denoted by the "—" symbol, so S_5—**NFT** means point S_5 is a maze termination location without food. This type of outcome is not represented as a stored proposition because it is perceptually apparent that point S_5 is an end-wall, and it is also apparent that the location is without food. For a proposition, there is a forward expectancy as to what will transpire from an action or in time. Because the forward expectancy property is not involved with the fact that point S_5 is a maze end-wall without food, this information is represented by an association rather than a proposition. Furthermore, note that after encountering a nonfood terminal point, the animal must turn around, exit that point, and return to a previous maze position. This idea is represented in the network by a propositional expression $S_5 \xrightarrow{\text{exit}} S_4$.

Besides propositional expressions and associations, the animal can also come to acquire with instrumental conditioning a set of motor responses. The motor response of running down the alley at point S_1 is denoted as R_D. This response leads to the expectation and experience of being at point S_2, and this learning is denoted by the expression S_1—$R_D \longrightarrow S_2$ in the network. An arrow denotes the linkage between the motoric behavior of running down the alley and the sensory experience of arriving at point S_2 (i.e., the behavior has a directional or relational property). But at point S_1, there is an association with the behavior of running down the alley, so an associative link is used between point S_1 and the running response. There is another R_D response between points S_2 and S_6, and left-turn responses (R_L) between points S_6 and S_7 and between points S_7 and S_8. In addition to the running responses, the consummatory response of eating the food is denoted in the network as R_g. Note that a classical conditioned fractional anticipatory consummatory response r_g is also learned; this Pavlovian response occurs prior to reaching point S_8.

In a maze task such as the one shown in figure 2.4, animals come to acquire the motor responses to complete the task quickly. When food is in a fixed place on each trial, it is remarkable just how fast rats can run to the goal. A well-trained rat can run at peak speeds of 90 cm/sec (Chechile & Fowler, 1973). To understand this speed, it should be noted that the mean stride length of a white rat is about 11 cm (Clarke & Parker, 1986). Consequently, the speed of 90 cm/sec is about 8 strides per second! But in other mazes, such as a T-maze, food can be in either end point of the T, so the animal runs quickly to the choice point and then selects the correct arm (usually for a highly trained animal). Once in the correct arm, the running speed is again remarkably high. For the maze in figure 2.4, animals would have made many other responses prior to mastering the maze. Earlier in training, animals might have taken a right turn at point S_2 only to discover that food was not down that pathway. Without reinforcement, these incorrect responses do not develop into well-learned habits. Consequently, only the correct responses acquire a motor memory associated with critical path points.

The extended example of the above maze task underscores just how much more complex instrumental conditioning is from Pavlovian conditioning. Part of the information learned is the cognitive map memory of the spatial relationship between maze points such as the propositional encoding $S_7 \xrightarrow{\text{south}} S_9$. This memory is the type of knowledge representation that Tolman (1932, 1945) discussed as a $S - S$ stimulus expectancy, except in this version, it is considered a cognitive proposition rather than a simple association. Later writers argued that Tolman's theory required an intervening response to be an expectancy (Guthrie, 1952; MacCorquodale & Meehl, 1953, 1954). In the MacCorquodale and Meehl formulations of expectancy theory, there is a need for (1) an elicitor stimulus such as S_7, (2) an action or response to the elicitor stimulus, and (3) an *"expectandum"* such as S_9. But contrary to MacCorquodale and Meehl (1953, 1954) theory, non-reinforced spatial information in this book is (1) not just an association and (2) does not involve intervening actions. In the current system framework, spatial knowledge is best considered a propositional relationship in general. However, for reinforced information, the animal's experiences in the maze also result in a chain that is similar to the MacCorquodale and Meehl expectation unit (e.g., note the expression S_7—$R_L \longrightarrow S_8$ in figure 2.5). This expression is separate from the propositional expression $S_7 \overset{\text{north}}{\Longrightarrow} S_8$. The propositional linkage between point S_7 and point S_8 is thought here to be a *conscious* memory, whereas the motoric chain S_7—$R_L \longrightarrow S_8$ is considered a component of the *procedural* habit acquired if the animal is sufficiently trained. Also, as noted previously, points S_7 and S_8 have a third connecting linkage (i.e., the Pavlovian-conditioned response r_g).

An implication of having different, independent representations about the maze is the possibility that there can be a different fate for the separate components. For example, a reward is critically important for maintaining responses such as the R_L and r_g responses between maze points S_7 and S_8 in figure 2.5. If a reward is withdrawn, then these responses are expected to change quickly, but there is no reason why the spatial-propositional memory should change. It is necessary for the learned responses to change because they are no longer effective for achieving the goal, but the propositional information could still be valuable. The converse dissociation can also occur. For example, it is possible that the propositional information might be difficult to remember (or become lost) if the animal were to receive many training trials that could be implemented by means of the same set of motor responses (i.e., the goal is reached by the execution of a well-rehearsed motoric procedure). The study of skilled typists provides a human counterpart to this loss of explicit-spatial information. Early in the learning of the keyboard, students develop an explicit or declarative representation for the location of the letters. But with sufficient training, modestly skilled typists come to execute about five to six keystrokes per second (Logan & Crump, 2011), and exceptionally fast typists can type about seventeen letters per second (Long, Nimmo-Smith, & Whitefield, 1983). Usually, it has been a long time since the typists first learned the keyboard layout explicitly, and in the meantime, they have been relying instead on their motoric memories for creating the letters from keystrokes. If skilled typists are asked to explicitly recall aloud a letter position, then they have difficulty answering immediately (Liu, Crump, & Logan, 2010; Snyder, Ashitaka, Shimada, Ulrich, & Logan, 2014). Instead, these typists recall the letter positions by placing their fingers on an imaginary keyboard and then reconstructing the letter position by typing the desired information on their imagined keyboard. Skilled typists do not need to rely on an explicit representation that was acquired early in their typing history.

Another important difference between Pavlovian and instrumental conditioning concerns the type of problem that the animal must solve. With Pavlovian conditioning, the learning is focused on simply the predictive regularity between the conditioned stimulus and the unconditioned stimulus, but with operant conditioning, the researcher can require many different types of problems for the animal to solve. Technological developments, such as touch-sensitive computer displays and the introduction of video displays in the operant chamber, enabled researchers to pose a rich set of problems for the animal to solve (Blough, 1977; Skinner, 1960; Wright, Cook, Rivera, Sands, & Deliux, 1988). For example, instead of a simple contingency where a displayed stimulus (the sample) must be remembered as the correct response for a later test, it is possible to display a series of stimuli and require the animal to discover a relational rule to gain a reward. Below are five hypothetical problems illustrating learning of the concept of *oddity*.

Problem 1   △   ◯   ◯
Problem 2   ⊠   ⊠   ⊡
Problem 3   △   ◯   △
Problem 4   ⊠   ⊠   ⊠
Problem 5   ◯   ◦   ◯

Three stimuli are displayed for each problem, with two stimuli being identical and one being different. The correct choice for each problem is the odd or different stimulus. There is no individual stimulus feature such as position, shape, color, or size that is always the correct choice because the stimulus that is different is the only consistent rule for reward. A stimulus that was an odd stimulus for one trial might be one of the two identical stimuli for another trial.

The oddity rule has typically been a difficult concept for animals to learn; for example, Moon and Harlow (1955) could only get rhesus monkeys to perform at a 90 percent accuracy level even after more than 250 training trials. Children with mental ages below five years typically have difficulty in mastering the oddity principle without special priming training (Ellis & Sloan, 1959; Greenfield, 1985). But there is an effective method for training the oddity concept by increasing the number of identical stimuli in the display (Gollin, Saravo, & Salten, 1967; Soraci, Alpher, Deckner, & Blanton, 1983; Zentall, Hogan, Edwards, & Hearst, 1980). When there is only a single different stimulus in a perceptual array of many identical elements, the odd stimulus is perceptually distinctive and tends to elicit immediate attention (Treisman & Gelade, 1980). In the Treisman and Gelade (1980) attention study, a single different item was detected quickly even if there were as many as thirty elements in the display, provided that the search was along a single dimension (such as color, form, or size). The different item perceptually "pops out" because it is the only item that is different. Soraci and colleagues (1983, 1987) used the idea of perceptual pop-out to induce young children to select the odd stimulus. The choice of the odd stimulus initially is not because the child understands the oddity principle but because the item is novel. Nonetheless the choice is correct and leads to reinforcement. As training progresses, the number of identical stimuli in the display is gradually reduced until the standard three-element display is used for all the rest of the training. In the Soraci studies, the children continued to select the odd stimulus, and they maintained correct performance over subsequent training days. Although the correct choices were induced initially because of the perceptual distinctiveness of the different stimulus, the children eventually learned the abstract rule of oddity.

Learning the oddity rule illustrates the complex type of learning that is possible with instrumental conditioning. Classical conditioning in contrast is a rather simple association because the animal is a passive witness to a pairing of stimuli, whereas with instrumental conditioning, the animal is free to respond in order to solve a puzzle. The puzzle can be a complex relational rule or a conceptual rule. Thus, the learning in instrumental conditioning is qualitatively different from the learning acquired in Pavlovian conditioning. Historically, however, some learning theorists argued that relational learning is simply explained by standard $S - R$ associationism along with stimulus generalization (Spence, 1937). But that account relied on a single stimulus dimension such as size for discriminating between two magnitudes of stimulus. One level was regarded as correct (e.g., the larger size), whereas the other stimulus was regarded as incorrect (e.g., the smaller size). Later in the transposition design, the specific stimulus that was previously correct was presented along with a novel stimulus that was even larger. Spence (1937) argued based on stimulus generalization of excitatory and inhibitory tendencies that the transposition phenomenon can be explained from an absolute reinforcement framework as opposed to a relational framework. But later studies have argued that the Spence $S - R$ account for transposition is problematic (Brown, Overall, & Gentry, 1959; Lazareva, 2012). Furthermore, the Spence theory cannot account for the relational learning that occurs with the oddity-learning procedure. Consequently, oddity-rule learning is important because it underscores the necessity of a cognitive model for the type of concept learning that is possible with instrumental conditioning.

2.7 Concluding Comment and Chapter Summary

The learning processes involved with Pavlovian and instrumental conditioning are quite general. Yet outside the laboratory, adult animals might not experience as many occasions of strict Pavlovian conditioning compared to the number of occasions of instrumental conditioning. More complex animals and humans are regularly engaged in a purposeful, goal-seeking activity, which is the domain of instrumental learning. The pairing of a neutral stimulus with an unconditional stimulus is less common outside of the structured environment of the psychological laboratory for adults. Yet parents can informally impose a Pavlovian learning structure on their infants. For example, the sight and sounds of mom in various contexts are likely to be regularly followed by some linking sensation. The newborn baby has a limited repertory of purposeful actions compared to adult humans, so there is a developmental shift in the relative frequency between the two types of conditioning procedures. But there is no question that both types of conditioning take place.

Historically, conditioning was the exclusive domain of behaviorists who deliberately avoided using cognitive constructs to explain learning. Yet a case is made in this chapter that it is difficult to explain conditioning without cognition. Routinely, cognition is involved in instrumental conditioning because the animal is placed in a goal-seeking setting. But Pavlovian training can develop a very weak response without conscious awareness in the special case of conditioning under general anesthesia. However, if an alert animal is given Pavlovian conditioning, then the animal learns to develop a strong response and learns to expect the onset of the unconditioned stimulus. So there is cognitive expectancy generated in normal Pavlovian conditioning.

Given the substantial cognitive component in conditioning, it is clear that we are dealing with a qualitatively different type of learning than that of the mechanistic processes underlying habituation and sensitization. With associative conditioning, new connections are made between any arbitrary neutral stimulus and other stimuli and actions. While the connections for habituation and sensitization never disappear but vary only in the rate of activation, the novel connections formed from Pavlovian and instrumental conditioning are new and can be undone. For example, the maze shown in figure 2.4 can be learned, and in doing so, the animal can rapidly run the maze from the start point to the food reward. The animal was not born with this spatial map of the maze built into its brain. Yet with enough training, the neurological connections are changed (in ways not yet discussed) so that the motor activity of running the maze is executed smoothly. Associative learning thus is not a turning up or turning down of innate connectivity, but instead the learning results in the formation of new connectivity patterns in the brain of the animal. This new connectivity can be altered at a later time if the animal is no longer in this maze context and learns other interfering tasks (i.e., associative connective patterns can change).

Conditioning results in the animal learning cognitive expectancies rather than learning a mechanical stimulus-response chain as was advanced by the early behaviorists. For example, with sufficient maze training, the animal develops a cognitive map of the maze and knows what pathways lead to what other maze locations. Although a detailed neural model is yet to be produced, at a system level, we can describe the resulting memory structure from the instrumental conditioning of the maze in figure 2.4 by means of a symbolic network like the one shown in figure 2.5. Notice that this network *cannot be summarized by an overall strength value* like what was done for the case of habituation and sensitization. Instead, there is a structured, knowledge representation with qualitatively different aspects. Moreover, this information can serve the animal on later trials. For example, if the animal were placed at point S_4 rather than S_1, then the animal would execute a modified route instead of the usual route from S_1. To represent this type of knowledge, we employed the concept of a proposition. Rather than a simple associative bond as advocated by behaviorists, the proposition does more than link two items. The proposition is like a simple declarative sentence that has a subject, a verb, and direct object, but it is more abstractly expressed as $A \xrightarrow{\text{relationship}} B$. The memory knowledge structure also includes the directed connection for a particular location, an action, and resulting location (e.g., $S_1 - R_D \longrightarrow S_2$ where locations S_1 and S_2 are the stimuli associated with a specific locations and R_D is an action to get the animal from the first point to the second point). This piece of information also has a propositional structure. Finally, it is also possible with instrumental conditioning for the animal to learn relationships among a set of stimuli such as the concept of oddity.

3 Memory: Biology to Cognition

3.1 Overview of the Chapter

Memories from conditioning are a subset of all the various memory types. Thus, in this chapter, learning and memory are examined more broadly. The chapter begins with a detailed analysis of some of the neurological and biological constraints on a cognitive model of memory. This discussion explores what brain regions are involved with the formation of a new memory. Neuroclinical evidence is discussed in some detail. From this biological background, a number of cognitive representations emerge for event memory. Semantic memories that are not tied to specific events are also discussed. Furthermore, the distinction between explicit and implicit memory is explored in some detail. In general, there is a fresh examination in this chapter about how a cognitive framework for memory and forgetting emerges from the constraints of the underlying biology.

3.2 The Biological Factors of Trace Storage

An enormous amount of research has been devoted to exploring the biological basis of memory, and a number of excellent textbooks are available that focus on this topic (e.g., Eichenbaum, 2008, 2012; Rudy, 2008; Squire & Kandel, 1999). Although surprising findings have already been detected at the neurological level, it is also clear that the story of the biological underpinnings of memory is a long way from being completed. Currently, it is known that molecules, such as acetylcholine, dopamine, glutamate, serotonin, norepinephrine, and gamma-aminobutyric acid (GABA), are neurotransmitters that carry a chemical message to postsynaptic neurons, but more than fifty other chemicals also can serve as neurotransmitters. Even more transmitters are still being discovered (Gluck, Mercado, & Myers, 2008). Neurologists have traditionally focused attention on neurons, but not all cells in the brain are neural cells. The brain also has many nonneural glial cells that can generate intercellular calcium-ion waves over considerable distances when stimulated, and these cells can release transmitters, except they are not neurotransmitters (Fiacco, Agulhon, & McCarthy, 2009). Glial cells also seem to play an important role in learning (Han et al., 2013). Furthermore, it is clear that important cognitive-level regularities about memory have not yet been understood at the neurological level. Despite these considerable limitations, it is still important to review some of the key factors at the biological level that will inform our understanding of memory trace formation.

3.2.1 Specialized Computations

Neurological studies have abundantly demonstrated that the brain solves many problems by specialized subsystems. The full explication of this point is beyond the scope of the present book, but let us consider a few examples from sensory perception. It is clearly established that vision, audition, taste, smell, and touch/proprioception are neurologically separate, parallel detection/processing systems. Yet, in the early and mid-tweniteth century, gestalt psychologists argued that within a sensory modality, the stimulus had holistic properties, so it was a mistake to examine components of the stimulus rather than examine the organizational properties of the whole stimulus. As Max Wertheimer strongly stated in a 1924 address to the Kant Society in Berlin,

There are wholes, the behavior of which is not determined by that of their individual elements, but where the part-processes are themselves determined by the intrinsic nature of the whole. (English translation from Ellis 1938, p. 2)

Yet for each modality, evidence of specialized cells was found in the neocortex that were selectively sensitive to small components (or features) of the stimulus. For example, Hubel and Wiesel (1962) discovered single cells in the visual cortex of anesthetized cats that were selectively sensitive to various types of lights of specific orientation and size. The identification of different types of feature detection cells and the organization of the visual cortex have continued to develop to a high degree (Hubel & Wiesel, 2005). Although the perception of a stimulus feels like a unitary event (a phenomenon that we will consider later), the neurological evidence is nonetheless quite clear that the brain is examining the retinal image along a parallel set of independent, elemental pathways. Brain damage can result in a syndrome called *apperceptive agnosia* where the components of the visual event are detected, but the linkage of these components is not integrated into a perceptual whole (Ferreira, Ceccaldi, Giusiano, & Poncet, 1998).

The parallel activation of the primary feature detection cells in the occipital lobe, which are triggered by the visual stimulus, does not result in an understanding of

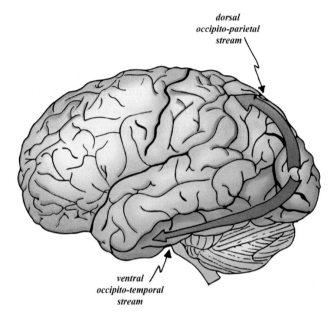

Figure 3.1
An illustration of the separate dorsal (where) stream and the ventral (what) stream for visual stimuli.

either *what* the stimulus is or *where* the stimulus is located in the environment. Postoccipital processing is required for both of these computations to be generated. These calculations are handled by different brain regions (Ungerleider & Mishkin, 1982).

The *what* pathway is the ventral stream from the occipital lobe to the temporal lobe and is shown in figure 3.1. Damage to this pathway results in a syndrome called *associative agnosia* where the patient can see the object but has difficulty with identifying or recognizing the object (Heilman, 2002). In the case of simple visual agnosia, the patient might not identify an object such as car keys, but the patient could draw a picture of the object and describe its appearance. But if the keys are jangled, then the patient would immediately gain access to their stored knowledge about this stimulus and identify it correctly as car keys because a separate auditory pathway to the conceptual knowledge is undamaged. Sometimes the agnosia is manifested for a more limited class of visual stimuli. The famous case described by Oliver Sacks (1985) of a man with *prosopagnosia* illustrates an example of a more specific visual agnosia for facial stimuli. Sacks (1995) also described a case of a painter who lost the understanding of color; this type of visual agnosia is called *achromatopsia*.

The dorsal pathway shown in figure 3.1 is responsible for depth perception and size constancy. Bilateral damage to this neural stream results in problems with depth perception and size constancy (Berryhill, Fendrich, & Olson, 2009). A patient with this type of damage would be able to identify two objects shown at different depths but would not be able to tell which one was closer. Shmuelof and Zohar (2006) have also linked the dorsal pathway to action recognition and planning. Consequently, patients with damage to this pathway would experience problems with visually guided actions.

The above few examples from neuropsychology underscore that the representation of information is massively distributed along separate channels that are processed in parallel. Eventually, the whole perception, recognition, and localization of the stimulus are integrated, but prior to that time, the processing is implemented by specialized neural regions. One neural region that is particularly important for memory is the hippocampus, which is discussed in the next section.

3.2.2 Memory and the Hippocampus

Henry Molaison (1926–2008), who was referred to as H.M. when he was still alive, is arguably the most widely studied patient suffering from the unintended consequences of a medical procedure. In 1953, Molaison had surgery that removed, from both hemispheres, most of the hippocampus along with the parahippocampal gyrus, the entorhinal cortex, and the amygdala (Hilts, 1995). While the operation did remedy his problem with epileptic seizures, the surgery also left him with a lifelong, severe anterograde amnesia (i.e., a problem in acquiring new memories) (Corkin, 2013; Hilts, 1995; Scoville & Milner, 1957). The operation left Molaison in a condition where he showed virtually no new learning as detected by a wide variety of verbally mediated tasks, such as the learning of a free-recall list, the detection of repeated patterns or faces, and the learning of a paired-associate list (Corkin, 2002). These tasks measure the acquisition and retention of novel events for a relatively long-term period. Nonetheless, Molaison was normal on procedures that tapped his encoding and short-term retention of novel items, such as repeating a short string of words, three-digit numbers, and the short retention of pure tones (Wickelgren, 1968). Yet after a delay or interruption, the information encoded on the short-term task was forgotten. Molaison was also poor on a fragment completion task for words that were novel (i.e., words that came into the language subsequent to his operation) (Postle & Corkin, 1998). If Molaison heard earlier the word COMPLIMENT and later was given a word fragment COMPL_, then he was likely to complete the fragment with the prime word *compliment* rather than an unprimed word like *completion*. *Compliment* was a word in his mental lexicon prior to his operation. But if he heard or saw the word FRISBEE and was later given the fragment FRI_, then he was not likely to complete the fragment with the primed word *frisbee* because that word was new to his vocabulary. Molaison's problem with long-term retention was not limited to just verbal material because he also exhibited anterograde amnesia for geometric patterns and nonsense figures despite being normal on perceptual-detection tasks (Kimura, 1963; Milner, Corkin, & Teuber, 1968; Teuber, Milner, & Vaughan, 1968). For normal individuals, the encoding of geometric shapes is still likely to be mediated by a subvocal propositional process.

Although most initial studies described Molaison as being unable to acquire any new information, later reports did show that he could learn some new facts and skills. Freed, Corkin, and Cohen (1987) tested Molaison's memory of vivid pictures after delays of 10 min, 24 h, 72 h, and 1 week. Control participants saw each picture once for 750 msec, whereas Molaison saw the photographs twice and had 10 sec per picture to study them. The extra time was provided to offset Molaison's profound learning dis-ability. Freed et al. (1987) examined his memory of the pictures using a two-alternative forced-choice test. With a two-alternative test, Molaison had a 50 percent chance of correctly selecting an old photograph. If a person scores above 50 percent, then that indicates some degree of memory. Chechile (2006) reanalyzed these data and showed that Molaison had an above-zero retention for 47 percent of the pictures after a delay of a week. Interestingly, the control participants had a nonzero retention rate of only 27 percent after a week. With the extra study time and the repetition of the picture set, Molaison was able to retain some knowledge about the pictures after a relatively long delay of a week. Nonetheless, Molaison's retention of the pictures was at zero after six months, whereas the control participants were still able to retain about 20 percent of their picture knowledge (Freed & Corkin, 1988). Although Molaison had trouble retain-ing new information, he still did have some retention beyond the 30-sec limit that is traditionally associated with short-term memory. Moreover, in terms of new semantic memory or general fact knowledge about important world events, other studies showed that Molaison could successfully answer some questions (Skotko et al., 2004). For exam-ple, he was aware that John F. Kennedy was a U.S. president who was assassinated. This information was retained many years after the assassination. Molaison was also able to acquire a classical conditioned eyeblink response (Woodruff-Pak, 1993). Interest-ingly, he showed no explicit memory of wearing the apparatus for the conditioning experiment despite having been trained on previous sessions. Nonetheless, the con-ditioned response did develop. Finally, Molaison was quite normal on motor-learning tasks (Corkin, 1968). Tasks such as mirror drawing, rotary pursuit, and bimanual track-ing are initially difficult, but people can improve with practice. Molaison's tracking showed improvement with practice, and he could retain his skills after a long delay (Gabrieli, Corkin, Mickel, & Growdon, 1993).

Although Molaison exhibited normal short-term memory and speaking skills, there were indications that he was not flexible in encoding novel sentences that contain ambiguities (MacKay, Burke, & Stewart, 1998). Normal individuals can understand and process sentences such as "It is the east, and Juliet is the sun" (a line from Romeo's speech in Shakespeare's *Romeo and Juliet*). Molaison had difficulty in answering who did what to whom (MacKay, James, Taylor, & Marian, 2007). MacKay and colleagues argued that Molaison's language-processing difficulty with novel sentences reflected a general problem with the binding of the components of the sentence into an integrated whole. This interpretation is consistent with the idea that the hippocampus plays an important role in the binding of the widely distributed components that are produced when encoding an event.

Although much has been learned from the extensive study of a few amnesic patients, it is important to note that brain damage is not cleanly limited to only a single neural structure. Naturally occurring lesions do not simply isolate a single neural area that is responsible for only a specific class of neural computations. For a more targeted exploration of neural functions, it is necessary to use animal studies (Good, 2002). These studies have uncovered a linkage between spatial learning and the hippocampus (Moser, Kropff, & Moser, 2008). For example, place cells have been detected (Moser et al., 2008); these cells fire when the animal moves through a particular location in space. Head-direction cells were identified that fire when the animal is facing in a particular direction even when the animal is stationary (Taube, 2007). Grid cells were also found that had a regular triangular firing pattern across the spatial field (Moser et al., 2008). Moreover, in the entorhinal cortex, which is a brain region adjacent to the hippocampus, border cells were detected intermingled among other head-direction cells and grid cells (Solstad, Boccara, Kropff, Moser, & Moser, 2008). Border cells fire when the animal is at spatial borders. In essence, researchers found that the hippocampus has a spatial-map receptive field (O'Keefe & Dostrovsky, 1971). The hippocampal spatial map is described as a representation of the locations relative to each other rather than relative to the position of the rat (Solstad et al., 2008; Wilson & McNaughton, 1993). Such a representation is called an *allocentric* spatial encoding rather than an *egocentric* spatial model.[1] Having spatial-place cells in close neural proximity to one another makes the hippocampus a good structure for building a spatial map representation.[2]

1. Note that the system-level representation for the learning of an instrumental training in a maze discussed in chapter 2 also evoked the idea of an allocentric model of the maze. See figure 2.5.
2. Although researchers agree that the hippocampus plays an important role in the development of a spatial map, it is noteworthy that Henry Molaison, who was without a hippocampus, could

There is also evidence that the hippocampus builds a temporal representation of events (Eichenbaum, 2014; Howard & Eichenbaum, 2015). It has been suggested that the time cells are in fact the very same cells as the place cells described above (Eichenbaum, 2014). But this conclusion might be limited because for the Eichenbaum experiments rats move through a spatial field, so there is a correlation between the spatial locations and the arrival times at the locations. But for a task where the rat is stationary and must wait before being allowed to react to a discriminative stimulus, different hippocampal cells were found to be selectively sensitive to the amount of time that elapsed in the delay interval (MacDonald, Lepage, Eden, & Eichenbaum, 2011). Some cells had a peak response for a short delay, whereas other cells had a peak response for various longer delays. Other investigators also identified hippocampal cells in the rat that responded selectively as a function of how much time elapsed regardless of the specific task (Pastalkova, Itskov, Amarasingham, & Buzsáki, 2008). For that study, the time cells responded in a similar fashion for the same elapsed time for completing either a maze-learning task or a running-wheel task. Working with monkeys, Naya and Suzuki (2011) recorded single hippocampal cells as well as cells in the adjacent perihinal cortex while the animals performed a temporal-order task with visual stimuli. These authors reported that the hippocampal cells provided timing signals, whereas the perihinal cells responded to the conjunction of visual-object information along with temporal information. Thus, a body of research demonstrates that the hippocampus builds a representation of the temporal component of event memory. Yet there is also some support that the frontal lobes play an important role for the memory of time and order (Kesner, Hopkins, & Fineman, 1994; Mangels, 1997; Milner, Corsi, & Leonard, 1991; Shimamura, Janowsky, & Squire, 1990). These investigators found that patients with frontal lobe damage experienced difficulties remembering the order of events. Jenkins and Ranganath (2010) argued, based on event-related fMRI brain scans of normal people, that the temporal lobe (which contains the hippocampus) and the prefrontal cortex were both active regions for a task involving temporal information.

To build a memory representation for a sequence of experiences, it is important to encode (1) item content for each event component of the sequence, (2) spatial-location information for the events, and (3) the temporal-ordinal information about the sequence. These three aspects of event memory correspond to the *what*, *where*, and

---

nonetheless demonstrate a detailed cognitive map of his housing environment that he lived in subsequent to his operation (Corkin, 2002). This observation indicates that other brain regions can also contribute to the memory of a spatial map.

when attributes of episodic encoding. There is physiological evidence that specialized neural cells within the hippocampus are involved with each of these memory attributes, although other brain regions can also play a role with event memory encoding. The sensory properties elicited by experience activate a widely distributed pattern of neocortical activity. Perhaps the hippocampus is important to memory formation because it is well located in the brain to link the overall conjunction of the separately activated neural firings. Many researchers have argued that the binding of the separate, distributed sensory representations of an episode is a key function of the hippocampus (O'Keefe & Nadel, 1978; O'Reilly & Rudy, 2001; Squire, 1992). But there is also evidence that more than one brain region can do this integrative linking. For example, O'Reilly and Rudy (2001) argue that the neocortex can represent a conjunction of event activations. In their model, the hippocampus is important for the rapid learning of an event, particularly a novel event that is not part of repeated training, whereas the neocortex plays a role in integrating the features that are part of long-term knowledge. Long-term knowledge would also include semantic memory. Semantic memory is different from episodic memory because it develops slowly from many experiences. Semantic memory is not linked to a specific event.

3.2.3 Retrograde Amnesia

A newly acquired memory can be forgotten as a function of time and postencoding events; such normal memory loss will be discussed later in this chapter. But more extraordinary events also can result in a more dramatic memory loss or amnesia. For example, a severe concussion can lead to both anterograde amnesia and retrograde amnesia. Although the memory impairments from a minor concussion might be limited, a serious injury can leave the patient in a coma, and upon waking from the coma, the patient is left with a profound memory loss. In such cases, there is both anterograde and retrograde amnesia. Anterograde amnesia is the difficulty in forming new, postinjury associations, whereas retrograde amnesia is the forgetting of the events that occurred prior to injury (McGaugh & Herz, 1972). Retrograde amnesia is an important phenomenon that has contributed to the consolidation theory of memory formation, which will be discussed in detail later in the book.

The linkage between serious head trauma and memory loss dates back to Ribot (1882). Ribot observed that it was the most recent memories that were lost with head injury; this observation has come to be called *Ribot's law*. Ribot's law is related to another memory law formulated in the nineteenth century: *Jost's law*. Jost (1897) hypothesized a number of learning and memory laws that have come to be validated by subsequent careful experiments (Youtz, 1941). Jost's second law states that when there are two memories of equal strength, then it is the most recent memory that has the greater tendency to be lost. Ribot's law and Jost's law both stress the vulnerability of recent memories relative to older memories; however, retrograde amnesia is not just the loss of an isolated fact, as would be the case for normal forgetting, but rather there is a total memory loss of all the events learned for a lengthy period prior to the accident. For example, a patient with retrograde amnesia might have a total memory loss for all the events learned for the past 2 years, but the patient might have normal memory for information older than 2 years. Given enough time after head trauma, the patient might heal and come to remember most of the events over the past 2 years except for a permanent memory loss for the events two weeks before the accident (Barbizet, 1970).

Electroconvulsive therapy (ECT) is a treatment for clinical depression. ECT is known to produce retrograde amnesia. For example, after an ECT session, patients might experience retrograde amnesia for the information acquired over the past one to three years (Squire & Cohen, 1979; Squire, Slater, & Chace, 1975). Electroconvulsive shock (ECS) produces a similar effect in animals, but with a different time scale. For example, mice only have a life expectancy of about 2 to 3 years in a protected environment. When mice experience retrograde amnesia, the forgetting is limited to the learning for the past 1 to 3 weeks (Squire & Spanis, 1984).

Another method for inducing retrograde amnesia is through the administration of a drug such as anisomycin. Anisomycin inhibits protein synthesis and is an antibiotic. Most antibiotics have difficulty crossing the blood-brain barrier, but these drugs can be injected directly into the brain of the animal. For example, Nader, Schafe, and Le Doux (2000) infused anisomycin into the amygdala of rats and discovered a subsequent retrograde amnesia for a previously learned Pavlovian fear response. Nader et al. argued that the amnesic effect was the result of a disruption of protein synthesis. Yet other researchers have pointed out that the effects of anisomycin are complex and can lead to a type of cell destruction called apoptosis (Rudy, Biedenkapp, Moineau, & Bolding, 2006). Regardless of the mechanism, it is clear that retrograde amnesia can be produced by the infusion of drugs.

3.2.4 Long-Term Potentiation (LTP) and Depression (LTD)

Given the neural plasticity delineated in chapter 1, it is reasonable to expect that memory formation is linked to the changes in the electrical properties of neurons. Conversely, electrical stimulation of neural tissues should also result in changes in neural electrical properties. In fact, even before the well-known research on habituation and sensitization in the 1970s by Kandel, Lømo found changes in synaptic efficiency that were caused by the electrical stimulation of hippocampal cells in anesthetized rabbits (Lømo, 1966). These changes in single cells have come to be called long-term potentiation (LTP) (Bliss & Lømo, 1973). Dunwiddie and Lynch (1978) also found a corresponding long-term depression (LTD) effect on synaptic responses.

There is an enormous body of research focused on the topic of LTP; see McDermott (2010) for an entertaining book about one laboratory's research in this area. The finding of an optimal frequency is among the key discoveries about LTP. Larson, Wong, and Lynch (1986) were the first to report that LTP for in vitro hippocampal brain slices occurred when there was a short burst of electrical stimulation at the theta-wave frequency, which is about 5 hertz. Larson and Lynch (1986) also linked the optimal frequency to mechanisms affecting neural receptors. Interestingly, theta waves are naturally occurring waves that are found when an animal is engaged in a learning task. But Grover, Kim, Cooke, and Holmes (2009) also found LTP in the hippocampus for stimulation near the delta-wave frequency of about 1 hertz; delta waves are typically produced during sleep.

Other research has indicated that LTP and LTD are not unique effects limited to hippocampal cells, but rather these effects are widespread in the mammalian brain (Malenka & Bear, 2004). Furthermore, LTP and LTD occur with habituation and sensitization with simple invertebrate organisms that lack a hippocampus, such as *Aplysia* (Byrne & Kandel, 1996). LTP and LTD also occur with other more neurologically complex invertebrate species like the octopus. Cephalopods (the genus that includes the octopus) do not have a hippocampus, but these animals still have considerable skill in spatial learning (Young, 1991) as well as skill on other sophisticated learning tasks (Hochner, Brown, Langella, Shomrat, & Fiorito, 2003).[3]

Long-term potentiation and long-term depression directly alter the neural-firing rate. The linkage of LTP and LTD to memory is because the recording of a memory is widely thought to correspond to changes in the underlying synaptic efficiency that alters the neural firing rate. Recall from our previous discussion of habituation and sensitization that the number of synaptic connections changes as a function of training (see figure 1.7). If there are more presynaptic terminals, then the tendency for neuron A to trigger neuron B is increased. Conversely, if there are fewer presynaptic terminals, then there is a corresponding depression of synaptic efficiency. Both the increase and decrease of the number of presynaptic terminals have been observed (Bailey & Chen,

---

3. It was 600 million years ago before mammals and cephalopods shared a common evolutionary ancestor, and that animal was microscopic with only a primitive nervous system (Vitti, 2013). This example also demonstrates that very different biological arrangements can nonetheless solve a similar computational problem, such as learning about their spatial environment. Having an intact hippocampus is not the only way for neurons to implement spatial learning.

1983, 1989). In fact, the same two neurons can either increase or decrease the number of presynaptic terminals depending on environmental events. Consequently, the idea of a permanent record of the past is a neurological illusion. The level of synaptic efficiency is not a constant.

3.2.5 Memory and Neurogenesis

But is there more to the biological underpinning of memory formation than LTP and LTD? Evidence is now mounting that learning might also be accompanied by the growth of new neural cells. Since the pioneering research by Ramón y Cajal (1913), it has been thought that the brains of mammals did not develop new neural cells once the animals matured to adulthood. Yet Altman and Das (1965) reported that adult rats grew new neural cells in the dentate gyrus of the hippocampus and in the olfactory bulb—a process that is called *neurogenesis*. Moreover, these new cells increase in number if the animal is exposed to an informationally enriched environment (Kempermann, Kuhn, & Gage, 1997; van Praag, Kempermann, & Gage, 1999). The new cells are typically interneurons, and the cells are more electrically reactive (i.e., they more easily exhibit LTP than older neural cells) (Schmidt-Hieber, Jonas, & Bischofberger, 2004; van Praag et al., 2002). A number of researchers have explored how computational models of learning and memory are altered by neurogenesis (Aimone, Wiles, & Gage, 2006; Becker, 2005; Finnegan & Becker, 2015). Becker (2005) suggested that neurogenesis can increase memory capacity. Finnegan and Becker (2015) discuss a model where neurogenesis changes the rate of interference among memory representations. For the model by Aimone et al. (2006), neurogenesis contributes to the formation of temporal clusters of event memories. In fact, Cai et al. (2016) demonstrated experimentally that hippocampal-cell ensembles do exhibit a temporal organization as expected from the Aimone et al. model. Time plays a critical role in many theories of forgetting (e.g., Ribot's law and Jost's law). We have seen in section 3.2.2 that the hippocampus contributes to the computation of the temporal dimension of events. Perhaps the neurogenesis process also is a factor in the long-term temporal organization of memory. Consequently, the linkage of neurogenesis to temporal clusters is a particularly intriguing possibility.

3.3 Emerging Cognitive Theories of Forgetting

The research described in the previous section has stimulated some important memory distinctions, and it has influenced a number of cognitive memory models—especially models of memory loss. In this section, the focus is on cognitive models of forgetting.

3.3.1   Atkinson-Shiffrin Model and Working Memory

Historical Context   The study of Henry Molaison motivated the distinction between short-term and long-term memory. This distinction was also a key property of the model developed by Atkinson and Shiffrin (1968) as well as the Baddeley and Hitch (1974) working-memory model. However, before discussing these models, let us first examine some of the other historical developments that set the stage for the Atkinson-Shiffrin model. Much earlier in the history of psychology, William James argued for two types of memory. Based on the study of conscious experience, James (1890a) divided memory into a primary (or short-term) memory and a secondary (or long-term) memory. James argued that an experienced event, or what he called a state of mind, was a primary memory.

The first point to be noticed is that *for a state of mind to survive in memory it must have endured for a certain length of time.* In other words, it must be what I call a substantive state. (James, 1890a, p. 643)

Unlike primary memory, secondary memory had to be recovered by a cognitive act.

Memory proper, or secondary memory, as it might be styled, is the knowledge of a former state of mind after it has already once dropped from consciousness; or rather *it is the knowledge of an event, or fact,* of which meantime we have not been thinking, *with the additional consciousness that we have thought or experienced it before.* (James, 1890a, p. 648)

The Jamesian distinction between primary and secondary memory was ignored during the first half of the twentieth century because it was based on an introspective analysis rather than based on behavioral evidence. The early memory experiments did not force psychologists to divide memory. Researchers in those early days used list-learning tasks such as paired-associate learning and serial-order learning. But those procedures are insensitive to the short-term retention changes that occur between the study and the test phases of a trial. Consequently, there was no experimental rationale for dividing memory. But with the invention of the Brown-Peterson task, short-term memory retention could be readily examined (Brown, 1958; Peterson & Peterson, 1959). In the Brown-Peterson task, a single, subspan memory target is briefly presented.[4] Following target presentation, the participants are kept busy so that they cannot rehearse the memory target. The target item is then tested after the rehearsal-preventing activity. It is curious why the Brown-Peterson task was not invented earlier. It seems that memory researchers clung to the methods created by Ebbinghaus and the other early

4. Memory span is the longest list length where the person is at a 50 percent level for repeating back the entire list in the right order. Typically, memory span is about five items for units such as digits and words. A subspan list simply means that the list length is less than the usual memory span, whereas a supraspan list is longer than the memory span.

experimental psychologists.[5] Those procedures were designed to study information that had to be repeated many times before a person could learn the list. In essence, the unit of analysis was the supraspan list itself. Nonetheless, the clinical reports about Henry Molaison pushed psychologists to investigate the memory retention of a single recent event, and the Brown-Peterson task was an excellent tool for this purpose.

Basic Model Structure   The Atkinson and Shiffrin (1968) model is frequently called the modal model of memory because most psychologists believe it captures their view of memory as consisting of several separate but interacting memory stores (Radvansky, 2006). These stores are illustrated in figure 3.2. New information must pass through the three stores in precisely the order shown in the figure. If information in the sensory register fails to make it to the short-term store, then it is lost from memory. Similarly, if information fails to transfer from the short-term store to the long-term store, then that memory is also lost. But if information is successfully transferred to the long-term store, then it is assumed to be permanent.

The three memory stores differ in regard to their storage capacity. At any given time, the sensory register is assumed to contain all the information that the sensory systems can currently detect. The duration for storing this information is very brief, as will be discussed below. The capacity of the short-term store (STS) is roughly limited to the memory span, so only a handful of items from the sensory register (SR) can be contained in the STS. It is possible that a person can retrieve information from the LTS and transfer it back to the STS, but the total capacity of the STS is still limited in terms of the number of items and limited in terms of the duration of those items. Without an active rehearsal of information to preserve an item in the STS and without the transferring of the information to the LTS, the information will be lost in less than 30 sec. But information in the LTS is assumed to be unlimited in time and unlimited in capacity.

Sensory Register   The sensory register corresponds to a very brief sensory/perceptual persistence. The experimental evidence for this store comes from the Sperling (1960) paper. Sperling used a partial-report testing method for studying supraspan visual arrays and found evidence for a rapidly decaying persistence. For a typical test trial, there was a visual array of nine letters that was displayed as three rows of three letters. Following a brief presentation of the array for 50 msec, the participants were directed to report

5. The serial learning task was employed by Ebbinghaus (1885). The paired-associate task was first developed by Mary Whiton Calkins, but the credit for her accomplishment was lost to many because her written account was a section of a larger paper with the original authorship listed only as Münsterberg and Campbell (1894).

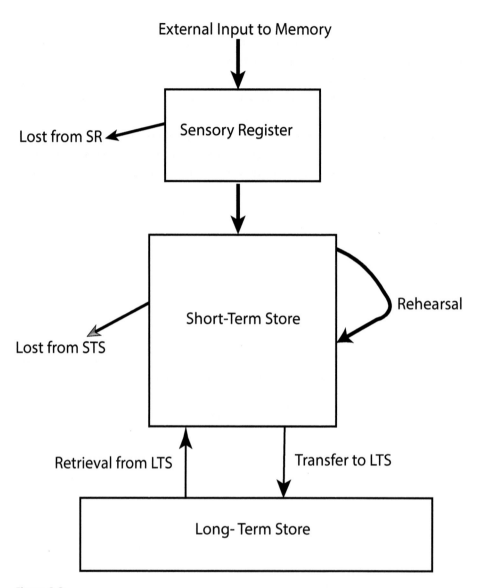

Figure 3.2
The multiple-store model of Atkinson and Shiffrin (1968).

the letters from only one of the three rows. The cue as to which row to report was based on an auditory signal. This cuing method is the partial-report procedure, and it was designed to limit the reporting requirement so that the number of items was within their memory span capacity. If the cue is delayed after the offset of the array, then the partial-report accuracy is impaired. For approximately 1 sec after the offset of the stimulus array, the accuracy rate for the partial report exceeds the accuracy for the whole report. The rapidly fading improvement is interpreted as the consequence of a sensory/perceptual memory, which has come to be called an *iconic memory*. Researchers using a backward-masking method have found that the precategorical, raw sensory afterimage is very brief (i.e., about 200 msec or less) (Liss, 1968). It is important to remember the distinction between perception, and pure sensation. Perception differs from sensation because with perception, *prior* categorical information is generated from the raw sensory stimulation.[6] According to Coltheart (1980), the iconic trace begins as a brief precategorical visual persistence, but it gains informational content based on the activation of prior knowledge. Consequently, the benefit shown on the partial-report test for cue offsets less than 200 msec is due to the fading persistence of the pure sensory information, but for offsets between 200 msec and 1,000 msec, it is due to a slower fading of the activated perceptual content.[7] After a 1-sec offset, there is no benefit for the partial-report method. Hence, for up to 1 sec after a visual stimulus is presented, there is a fading continuation of the activation despite the fact that the stimulus is no longer physically present. However, the entire iconic trace is eliminated if there is an interfering or masking stimulus presented after the target is turned off (Liss, 1968).

6. William James believed that a pure sensation was an abstraction, and he argued that "infants must go through a long education of the eye and ear before they can perceive the realities which adults perceive. *Every perception is an acquired perception*" (James, 1890b, p. 78).

7. Not all researchers accepted the idea of an iconic memory. For example, Merikle (1980), based on methodological concerns with the partial-report protocol, rejected the hypothesis of an iconic memory trace. Merikle demonstrated that the superiority of the partial-report method over the whole-report method varied with the physical arrangement of the cue. This finding led him to question the existence of an iconic memory. But a study by Di Lollo (1977) employed a novel procedure to study iconic memory, and this experiment did not use the partial-report protocol. In the Di Lollo procedure, two arrays of nonoverlapping dots were displayed consecutively with a blank screen intervening for 10 msec. The second array of dots was displayed for only 10 msec. If the iconic memory of the first array of dots fused with the second dot array, then the participant would see a location where there was a missing dot. The participants did in fact automatically detect the missing dot among the possible twenty-five locations available. This demonstration thus provides evidence of the persistence of the first array of dots.

The acoustic counterpart to the iconic trace has also been detected. This memory is referred to as the *echoic* trace. Using the partial versus whole report method, Darwin, Turvey, and Crowder (1972) found a persistence of about 4 sec for auditory stimuli. But Cowan (1984) argued that the precategorical, pure auditory time period is about 300 msec. Consequently, the duration of the memory in the sensory register is dependent on the sensory modality. In a subsequent elaboration of the sensory register, Shiffrin (1975) showed sensory input activating existing memory representations as would be expected of perception, and the resulting perceptual unit is transferred to the short-term store as in the original model.

Short-Term and Long-Term Stores   Atkinson and Shiffrin (1968) also made an important distinction between control processes and structural features of memory. The structural aspects of memory are the three memory stores with their informational and temporal properties, whereas the strategy that a person uses for encoding information is a control process. For example, a person might decide to ignore some items in the STS to elaborate the encoding of another item in the STS. The subsequent processing of the selected item might include the retrieving of related information from the LTS and combining that information with the current item to transfer back to the LTS an integrated and elaborated item. But another strategy might be to do a rote rehearsal of a set of items in the STS and not make an effort to transfer these items to the LTS. It is up to the person to control his or her LTS and STS hardware; hence, the term *control processes* is used for these strategic operations.

Finally, a key assumption made in the Atkinson-Shiffrin model is that the susceptibility to storage loss is all or none. No storage losses occur for information that is transferred to the long-term store, but if information is not transferred to the long-term store, then 100 percent of it is lost. Long-term forgetting is assumed to be strictly a failure of retrieval. This topic will be explored in greater detail later in this chapter as well as in many other chapters.

Working Memory   Another influential version of the multistore theory is the Baddeley and Hitch (1974) *working-memory* model. Working memory is largely an elaboration of the memory processes that occur within the short-term store of the Atkinson-Shiffrin model. Baddeley and Hitch argued that more was going on in short-term memory than simply holding a copy of the episodic event temporally. They felt that short-term memory or working memory must have a *central executive* function as well as separate stores depending on the stimulus. In the original model, working memory consisted of a *central executive*, a *phonological loop*, and a *visuospatial sketchpad*. Baddeley (2000) added

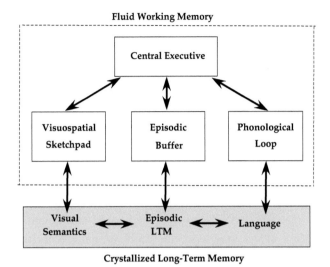

Figure 3.3
The Baddeley (2000) version of fluid working memory and "crystallized" long-term memory.

another component to working memory; this component he called the *episodic buffer*. See figure 3.3 for an illustration of the relationship between working memory, which is described as a fluid memory system, and the *crystallized long-term memory*. Working memory is thus an elaboration of the short-term memory processes in the context of the multistore framework.

The working-memory model illustrated in figure 3.3 is a general framework. Many specific mathematical models of working memory have been developed (e.g., Botvinick & Plaut, 2006; Brown, Preece, & Hulme, 2000; Burgess & Hitch, 1992, 1999; Farrell, 2006; Henson, 1998; Neath & Brown, 2006; Page & Norris, 1998). These models lead to specific testable predictions. The eight aforementioned mathematical implementations are different in a number of ways. See Lewandowsky and Farrell (2008) for a review of these models. Each model can account for some experimental phenomena, but each model also has some shortcomings accounting for other phenomena. Lewandowsky and Farrell (2011) also point out that there are 144 possibilities for models of just the phonological loop, so there are many specific mathematical implementations possible of working memory.[8] Thus, researchers in the area of working memory have not yet reached a consensus about a specific mathematical

8. The 144 models arise from all the possible combinations for the theoretical alternatives to six decisions about the phonological loop.

implementation. Yet in terms of forgetting, the working memory models generally assume that information that is not transferred to the long-term store will eventually be lost. But both the idea of a time-based decay of information as well as the idea of a storage interference process are considered creditable alternatives.

3.3.2   Forgetting for the SAM and REM Models

A number of mathematical models for memory have been developed (e.g., SAM, REM, MINERVA2, TODAM, CHARM). These models make elaborate representational and processing assumptions about memory, and it is outside the scope of this subsection about forgetting to discuss all of them. Yet one can see brief descriptions of the abovementioned models in the glossary. In this section, only two models are discussed (i.e., the SAM and REM models). These models are selected because they have been influential and because they are an elaboration of the long-term store of the Atkinson-Shiffrin model, which was discussed in the previous subsection.

The *Search of Associative Memory* or SAM model was developed initially in two key papers (Raaijmakers & Shiffrin 1980, 1981), but this work was inspired from earlier papers by Shiffrin (1970) and Raaijmakers (1979). Gillund and Shiffrin (1984) further developed the SAM model. Forgetting in the SAM model is assumed to be caused by a failure to retrieve the target information based on the cues presented at the time of test. This model hypothesizes that event knowledge is represented as a "memory image" consisting of item information, associative information, and contextual information. The memory image is assumed to be a part of the long-term store of the Atkinson-Shiffrin multiple-store model. A key feature of the SAM model is an assumption that there are stored strength values linking the memory image to the various possible cues Q_j, $j = 1, ..., m$ that can be used to probe for the memory. New information can be added to the long-term store, but the strength value of the components of the memory image does not degrade. Hence, forgetting in the SAM model is assumed to be due to strictly a retrieval failure. See the glossary for many additional details about the SAM model.[9]

Although the SAM model can account for a number of experimental findings, it nonetheless is inconsistent with the mirror effect for recognition memory. This effect is described in the glossary and in chapter 8. In an effort to deal with this problem, the *Retrieving Effectively from Memory* or the REM model was developed (Shiffrin & Steyvers, 1997). The REM model is a modification of the SAM model. Information in

9. The compound cue theory (McKoon & Ratcliff, 1992; Ratcliff & McKoon, 1988) is an extension of the SAM model. This theory is designed to account for priming effects (i.e., the facilitation of the retrieval of a word when it is preceded by the presentation of a related word).

the long-term store about a memory image is captured in this model by a vector of features. The values for the vector components are the feature-occurrence frequencies. On recognition testing, a participant must decide if a test probe is old or new. The decision is based in this model on the Bayes rule, that is, $\Phi = \frac{P(old\ item|data)}{P(new\ item|data)}$. If $\Phi > 1$, then the person tends to respond positively that the probe stimulus is an old item; otherwise, the person decides that the test probe is a new item. There is also a method within the REM model to apply the Bayes rule for a recall test. It is not necessary to delineate here all the properties for either the REM model or the SAM model; additional details about these models can be found in the glossary. These models, as well as the other previously mentioned mathematical models of memory (e.g., MINERVA2, TODAM, and CHARM), are complex (i.e., the models make many representational and processing assumptions). Given that there are 144 possible models for just the phonological loop within the working-memory framework (Lewandowsky & Farrell, 2011), it would seem that there should also be many possible variants of the complex models of long-term memory. There are many modeling decisions for an intricate model, and there is a risk that the theorist can make an error with any given modeling decision. Thus, an elaborate model is unlikely to be correct on the whole.[10] But for our purposes, the main point that is stressed here about both the SAM model and the REM model is the assumption that forgetting is considered to be only a retrieval failure. This assumption will be debated and challenged later in the book.

3.3.3 Consolidation Theory: Early Form

Consolidation theory is actually an old idea about forgetting that originated from a paper by Müller and Pilzecker (1900). They examined the retention loss for a list of nonsense syllables between two groups of learners. One group learned the set of nonsense syllables followed by the learning of an additional set of items, whereas the other group learned the nonsense syllables but rested for the same period of time. They found superior memory retention for the group that rested after the original learning. In the Müller and Pilzecker consolidation hypothesis, forgetting was based on the idea that learning was not complete at the end of training. The state of the memory representation after

---

10. There is a problem when models become so complex that it is difficult to validate and identify the various aspects of the model. It is important that models be testable and validated. Later in part II of the book, a different approach, for understanding memory is advanced. With this alternative approach there is a detailed analysis of the memory tasks in an effort to extract several important memory components such as storage and retrieval. With this approach, complex assumptions either about the representation of memory or about the retrieval of information from storage are avoided.

training was like a fresh painting. The paint is still wet, so the figure can be distorted or destroyed before the paint has hardened. Similarly, in the consolidation account, memory requires a period of time before the representation is stabilized. Learning a competing second list effectively distorts the memory of the first list.

Support for the consolidation hypothesis was also found in the Jenkins and Dallenbach (1924) classic experiment that contrasted sleep versus nonsleep. Jenkins and Dallenbach found there was an improvement in memory retention when a person slept for a period of time relative to a person who was awake for the same amount of time. The passage of time without any disruptive events slows memory changes. In consolidation theory, the storage of the memory trace is considered malleable. On one hand, it can be disrupted, but on the other hand, it can be hardened.

However, the improvement with sleep is only relative to the level of retention of a waking person after the same amount of time. For example, it is not the case that one is at an 80 percent level of mastery and is at a higher level after sleeping. Rather, after sleeping, one might be at a 70 percent level, but the corresponding control individual, who was awake but not allowed to rehearse the critical information, might be at a 60 percent level. In the Jenkins and Dallenbach study, both the sleeping and waking groups exhibited a decline with time, but the amount of the decline was less for the sleeping group. There are a number of possible explanations for this relative difference in performance. One hypothesis is that sleeping simply protects the person from being exposed to competing events that might disrupt the storage of critical information and simultaneously allows for the hardening of the memory trace. This hypothesis is the consolidation account. If this hypothesis were valid, then why is there a memory loss for the sleep condition? A second hypothesis is that there is active additional learning taking place during sleeping, and these processes improve the memory representation (Stickgold & Walker, 2013). Perhaps through dreams there is further information processing, especially in regard to solving problems that were unsolved prior to sleep, but again this hypothesis does not have a ready explanation as to why there is not an absolute sleep advantage in regard to memory retention. A third hypothesis is that after sleeping, the ongoing storage interference is less vigorous toward items from the prior temporal context (Chechile, 2006). With this hypothesis, the interference effect on the critical material is more vigorous for the waking group. The shift in temporal context, caused by sleeping, protects the critical material. The initial memory is not so much improved, but rather the storage interference is reduced with the passage of time. Also, there are fewer interference events while the person is asleep. Consequently, the sleep advantage effect per se is not compelling evidence for strictly the consolidation hypothesis because there are several creditable alternative hypotheses.

3.3.4 Classical Interference Theory

McGeoch (1932, 1942) advanced an interference theory of memory loss for experiments similar to the Müller and Plizecker study. Interference theory was based on the idea of a temporary disruption of the retrieval of information in storage, and it became the dominant model of forgetting rather than consolidation theory. The experimental design for retroactive interference studies is illustrated in figure 3.4A. For this design, all participants learn an initial A-B list of paired associates. The experimental group then learns a second A-C list. The notation of A-C expresses the fact that the stimuli used in the second list are the same as the ones used in original A-B list, but a new set of C responses is required for the second list. The control group does not learn a new list. But the control group is kept busy doing a nonlearning task that takes the same amount of time as the time needed by the experimental group to learning the A-C list. Typically, there is near-perfect performance for the control group for remembering the A-B list, but correspondingly, there is poorer memory retention for the experimental group. With interference theory, this decrease in memory retention is an indication of a retroactive interference effect where old learning is disrupted by subsequent learning. Thus, both consolidation theory and interference theory have an account for the forgetting caused by a second list. But interference theory also predicts a corresponding proactive interference effect; see figure 3.4B for the basic design for proactive interference studies.

For the proactive interference design, there is again a contrast between the performance of an experimental group and a control group. The experimental group learns two lists (i.e., lists A-B and A-C). All people in the experimental group are trained on the second list until they achieve a particular training criterion (such as one perfect trial). The control group only learns the A-C list and is trained to the same criterion level as the experimental group. Finally, after the same amount of time has elapsed in phase 3, both groups are tested on the A-C list. Consolidation theory predicts that the memory retention rate should be the same for the two groups. According to consolidation theory, the A-C traces are metaphorically written in wet paint and must be hardened by the passage of time. If the time is the same for the two groups, then the paint should harden to the same degree. Yet there is a consistent finding of a significantly lower retention rate for the experimental group. This difference is the proactive interference effect. The finding of proactive interference resulted in the widespread rejection of the consolidation hypothesis. In interference theory, forgetting is thought to be caused by a retrieval impairment rather than the lost of the association from memory storage. Interference theory accounts for proactive interference in terms of a competition process among the associations.

A.

Retroactive Interference Design

	Phase 1	Phase 2	Phase 3
Experimental Group	A - B	A - C	Test for A - B
Control Group	A - B	——	Test for A - B

B.

Proactive Interference Design

	Phase 1	Phase 2	Phase 3	Phase 4
Experimental Group	A - B	A - C	Time	Test for A - C
Control Group	——	A - C	Time	Test for A - C

Figure 3.4
(A) A diagram of the design for retroactive interference studies. (B) The corresponding design for proactive interference studies.

Despite the widespread acceptance of the interference theory account, there is nonetheless a methodological concern about the standard interference designs as illustrated in figure 3.4. Note for both types of interference, the participants in the experimental condition learn *two lists*, but the participants in the control condition only learn a *single list*. Is there still interference if we design a control condition where the participants learn two lists? For example, in the case of the proactive interference design, suppose that all participants learn an initial **A-B** list of associates, and the experimental group learns a second **A-D** list, but the control group learns a second **C-D** list. Training for the **D** response is continued until the participant reaches an established criterion level. Later, after a retention interval, we can assess the rate of recall for the **D** items when cued by the **C** stimuli versus the retention of the **D** items when cued by the **A** stimuli. If there is an effect of proactive interference from the learning of two different responses to the same **A** stimuli, then the rate of recall for the **D** items should be less when cued by the **A** items rather than when cued by the **C** items. This alternative experimental design was exactly what was done in a study by Anderson (1983). Anderson reported that the rate for recalling the **D** items after a delay of 1 week was 70 percent when cued by **A**, but it was 63 percent when cued by **C**. Using the standard proactive interference design as shown in figure 3.4B, Koppenaal (1963) found significant proactive interference after a delay of 1 week. But in the Anderson experiment, the **A** cue resulted in a *higher* recall rate. This effect is the opposite of what would be expected from the McGeoch idea of response competition between associates leading to proactive interference. Consequently, the Anderson experiment indicates that the impairment found with memory retrieval is more a function of the total amount of stimulus material learned in the experiment rather than a specific associative-interference effect. The standard proactive interference control condition only requires the learning of half the amount of material compared to the experimental condition. When both the experimental and control conditions learn the same number of associations, the support for associative interference vanishes.

So why does the standard single-list control for proactive interference result in better memory retention relative to the two-list control condition? After the retention interval of a week, a participant in the control condition from the standard single-list design (i.e., no **A-B** initial list) could have the following *error types* when given an **A** probe for the **D** response:

- Recalling another (incorrect **A**) item
- Recalling another incorrect **D** item
- Recalling an extraneous item
- Simply an omission or a blank

However, a corresponding participant in the Anderson (1983) control condition could make the following error types when given a **C** probe for the **D** response:

- Recalling an **A** item
- Recalling a **B** item
- Recalling another (incorrect) **C** item
- Recalling another incorrect **D** item
- Recalling an extraneous item
- Simply an omission or a blank

In light of the larger number of possibilities for making an error with the two-list control condition, it is not surprising that the standard single-list control condition results in better memory retention.[11]

11. Although the findings of the Anderson (1983) study call into question the response-competition interpretation of proactive interference favored by McGeoch, the demonstration of

Given that there is a difference for proactive interference between a single-list control and a two-list control, it is reasonable to see if there is a similar effect of a two-list control group for the case of retroactive interference. Suppose that all participants learn an initial **A-B** list, but the experimental group learns an **A-D** second list, whereas the control group learns a **C-D** second list. A test of specific associative interference is obtained by comparing the recall rate for the **B** items when cued by the A stimuli. This design was employed by Ehrensbeck (1979). For that experiment, the initial **A-B** learning was either to a partial-learning criterion or to an overlearning criterion.[12] The experimental group had six training trials on a second **A-D** list, whereas the control grouped had six training trials on a **C-D** list. After the training on the second list, which took about 18 min, the participants in both conditions were given a recall test for the first-list **B** responses. For the participants who studied the first list to a partial-learning criterion, the final recall rate in the experimental condition was 59.2 percent, whereas the recall rate was 60.0 percent for the control condition. This difference is not statistically significant.[13] For the participants in the overlearning training condition, the final test for the **B** items was 90 percent for the experimental group and 83.3 percent for the control group. Note that the condition means are in the opposite direction from what would be expected from a response-competition interpretation of retroactive interference, but this difference is not significant.[14] Thus, like the Anderson (1983) study, the Ehrensbeck (1979) experiment did not find interference when a two-list procedure is employed for both the control and experimental conditions. Thus, if the control condition must learn two lists, then the evidence for associative interference is substantially reduced. The learning of another list where the stimuli and responses are all different (i.e., the **C-D** list) is in some cases as disruptive to memory retention as the learning of a list where the associative cues are identical in the two lists but responses to the cues are different.[15]

Finally, it is noteworthy that classic interference theory was mainly based on the paired-associate task. In chapter 11, the learning and memory processes linked with this

---

proactive interference with the standard design is still a problem for the consolidation hypothesis. Something more than the passage of time is responsible for the greater forgetting for the experimental group relative to the standard single-list control condition.

12. For the partial-learning condition, the criterion for stopping **A-B** testing is when the participant was correct on fifteen of the eighteen paired associates. For the overlearning criterion, the training was stopped after the participant completed three perfect trials in a row.

13. A two-sample t-test was not significant, $t(38) = 0.09$.

14. A two-sample t-test was not significant, $t(38) = -1.36$.

15. McGovern (1964) found for shorter lists that the recall rate for the **B** responses for the respective *rest*, *C-D*, and *A-D* second-list conditions were 96 percent, 82 percent and 60 percent. Thus, for the McGovern experiment, there is some support for an associative interference effect because

task are examined in greater detail. The task has a methodological limitation because massive memory changes can occur between a learning event and the later testing of memory retention. In section 11.3.5, a model for the dynamic changes that occur with this task is presented. The model describes both the learning spikes that occur with the study of an event as well as the forgetting that occurs after the study of a pair of items. A memory retention function is also employed to describe the changes in memory that occur over a posttraining retention interval. The model is called the *Dynamic Encoding-Forgetting* model or the DEF model.

3.3.5   Consolidation Theory Revival

The idea that all forgetting is strictly a retrieval phenomenon eventually became a difficult hypothesis to maintain in light of amnesiacs such as Henry Molaison. His case prompted the reemergence of a storage-based theory of forgetting.[16] It was not clear how a retrieval theory could explain the fact that Molaison was virtually unable to learn associations lasting beyond some short time period.[17]

After the discovery of the disruption of memory caused by hippocampal damage, consolidation theory underwent an evolution. The standard model of a revived consolidation theory stressed different stages for the stabilization of a memory encoding.

---

of the difference between the **C-D** and the **A-D** conditions. But Bower, Thompson-Schill, and Tulving (1994) found a wide variation in the magnitude of retroactive interference depending on the properties of the words used in the two lists.

16. Some writers such as Spear (1978) have attributed the reemergence of consolidation theory to Hebb (1949). However, Hebb really did not discuss the consolidation hypothesis in his book, which was published before the appearance of reports about Henry Molaison. Hebb instead distinguished between early and late learning, but that idea is not the same as poststudy changes in the memory representation due to just the passage of time. In fact, Hebb cited McGeoch for the observation that the late stage of training was qualitatively different from early learning. Hebb did use a distributed representation for memory that he called cell assembles. That idea was stressed later in standard consolidation theory, and consolidation theory also stressed early versus later training.

17. Warrington and Weiskrantz (1970) argued that amnesic patients suffered from a retrieval impairment rather than a deficit in storage. They based this claim on the finding that amnesic patients were able to guess above chance on cued recognition tasks. Later, a case will be made that this type of finding can be better explained by a model that has implicit memory and fractional storage. In both cases, the state of the memory is degraded in storage quality relative to an explicit memory encoding. The claim by Warrington and Weiskrantz is typical of a problem that researchers have in understanding behavioral data when they did not develop separate psychometric measures of the underlying cognitive processes such as storage and retrieval. In a later chapter, I will argue that a model-based measurement method is required before claims such as those advanced by Warrington and Weiskrantz can be supported.

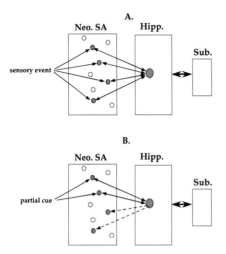

Figure 3.5

Illustration of the index theory of Teyler and DiScenna (1986). (A) Upon experiencing an event, the association areas of the neocortical sensory association areas (denoted as Neo. SA) receive activation of the features present, and these in turn activate a conjunction binding representation in the hippocampus (denoted as Hipp.). The hippocampus also has connections to subcortical areas (denoted Sub.), which includes the amygdala, medial septum, thalamus, hypothalamus, raphe nuclei, and locus coeruleus. The hippocampus activation in turn has an indexing property, so it can activate the neocortical cells that originally activated it. (B) A partial cue of the target thus can reactivate the hippocampal indexing cells which can regenerate the missing features of the target event. Thus, figure 3.5B represents the recall of the target event from an impoverished cue.

The theory is also framed in the context of the interaction between the hippocampus and the neocortex. Yet it is still important to remember that some animals with sophisticated memory systems, such as cephalopods, do not have a hippocampus, so consolidation theory has a strong mammalian bias. Nonetheless, since the focus of this book is on human memory, the limitation of consolidation theory to an interaction of the hippocampus and neocortex is not a problem. In the standard model of the revived consolidation theory, the hippocampus is hypothesized as playing a critical indexing role in the early stages of learning (Teyler & DiScenna, 1986). The Teyler-DiScenna indexing theory is consistent with the aforementioned O'Keefe and Nadel (1978) notion of the hippocampus as a neural center for generating a cognitive/spatial map, but it is also consistent with a competing model by Olton, Becker, and Handelmann (1980) that argued that the hippocampus functions as a working memory rather than as a representation of spatial information.

In the Teyler-DiScenna index theory, the hippocampus is in an ongoing interaction with the neocortical-association areas. The episodic event activates a distributed set of neocortical cells, which in turn activates automatically a set of hippocampal cells. See A in figure 3.5A for an illustration of the activation pattern. The hippocampus also receives input from subcortical areas (i.e., the amygdala, medial septum, thalamus, hypothalamus, raphe nuclei, and locus coeruleus). The hippocampus is thus ideally suited to link the distributed pattern of neural firing. Teyler and Rudy (2007) provide more details of the internal organization within the hippocampus, but those details need not be stressed here. The key notion is that the hippocampus, through long-term potentiation and long-term depression, alters the conjunction of the input activation. The hippocampus is hypothesized to model the input sensory event, and the activation of the hippocampal representation retriggers the neocortical pattern that originated it. The retriggering of the input activation pattern makes the hippocampus have an autoassociative property; this idea was introduced earlier by Marr (1971).[18,19] The degree of binding of the hippocampal representation depends on the importance of the event and on the frequency that the pattern has been activated. At early levels of training, the neocortical cells are not altered by the sensory input; only the hippocampus is altered by the experience. Through the indexing function of the hippocampal representation, the full neocortical representation of the target event can be retriggered by a cue that only contains some of the features of the event. This recall of the target from partial cues is illustrated in figure 3.5B. Of course, this property is dependent on

18. Although the hypothesized neuroanatomical structure of the hippocampus in Marr's paper is incorrect, it is nonetheless interesting to note that the paper introduced the idea of a neural network where the distributed components are bound together with changes in synaptic potentiation and can be elicited by a partial cue. Autoassociative networks have since become widely used as a neural network design. Autoassociative networks differ from heteroassociative networks previously developed by Willshaw, Buneman, and Longuet-Higgins (1969). With heteroassociative networks, some components of a distributed stimulus pattern can activate a different stimulus pattern. Autoassociative networks were later popularized through the contributions of Kohonen (1977) and Hopfield (1982). The Marr (1971) paper thus was a brilliant contribution that jumpstarted research in neural models. See the glossary for more information about autoassociative and heteroassociative neural networks.

19. The process of reconstructing the original encoding from a partial representation is called *redintegration*. Schweickert (1993) applied this idea in the context of a memory measurement model to explain how redintegration can affect the memory span in an immediate recall task. For this model, an earlier presented item in an immediate-memory string might lose some trace features, but through the redintegration process, the person can still correctly recall the whole item. Thus, redintegration results in lengthening the memory span.

the soundness of the facilitation of the projection back to the neocortex (i.e., the state of the long-term potentiation and long-term depression). If the partial cues are too impoverished, then the back projection to the neocortex fails.

In the standard model of consolidation, there is no binding of neocortical cells at early stages of learning (i.e., there is no long-term memory yet established). Evidence in support of this position is found from the Fanselow (1980) experiment. In this study, rats were transported to a test chamber, and after a delay, a shock was administered. Fanselow showed that if the shock occurred soon after being placed in the chamber (i.e., less than 6 sec), then these animals did not show single-trial learning of fear when placed back in the box on a subsequent day. But animals that were allowed to explore the chamber for 2 min before being shocked did exhibit on the next day a fear-freezing response that occurred with a single trial of training. During the 2 min of exploration on the prior day, the animals had time to establish to some degree the learning of the box itself (i.e., the context of the test chamber). Consequently, when shock was administered, there was a linkage between the learned context and the shock. Shock is a powerful stimulus that immediately demands attention and usually results in single-trial learning. In contrast to the animals that had the 2-min preexposure to the test chamber, the animals that were shocked almost immediately on placement in the box did not have time to acquire the learning of the test-chamber context. Despite the fact that shock is a powerful stimulus, these animals did not learn the fear response based on just one trial.

In the standard form of consolidation theory, the hippocampus typically cannot retain the information for an event for very long. For example Frey, Huang, and Kandel (1993) report that the duration of the LTP that results from a single train of high-frequency stimulation of hippocampal cells has a duration in the range of 1h to 3h. They also report that this activation is independent of protein synthesis. But in the case where the high-frequency stimulation is repeatedly administered to the hippocampal cells, they found that LTP lasted more than 28h and showed evidence of gene activation and protein synthesis. Apparently, hippocampal cells are especially sensitive to activation and rapidly show changes in synaptic potentiation. Buonomano (1999) found that in vitro hippocampal cells and neocortical cells differed in regard to their response to stimulation. Also, the number and structure of neocortical synaptic connections increase with learning (Greenough & Bailey, 1988). Thus, both hippocampal and neocortical cells exhibit plasticity, but their profiles are different. In the standard form of consolidation theory, the longer termed memory is thought to be due to a more slowly developing neocortical representation of the event. A key rationale as to why a neocortical representation is required for a longer termed memory over months or years comes from hippocampal lesion studies. Damage to the hippocampus results in a loss of memory of recent events, but it spares older memories (Alvarez & Squire, 1994). The older memories are assumed in standard consolidation theory to be due to a neocortical representation. In essence, the neocortex in consolidation theory plays the role of the long-term store of the Atkinson and Shiffrin (1968) model.

McClelland, McNaughton, and O'Reilly (1995) ask two excellent questions about the utility of having two complementary storage systems (i.e., one in the hippocampus and the other in the neocortex). First, if the hippocampus is necessary for the initial encoding of novel events, then why is a neocortical representation needed for a long-term representation? Second, what is the utility of the slowness for the development of the neocortical representation? From a theoretical examination of both the successes and failures of artificial neural networks such as the Rumelhart (1990) connectionist model, these investigators provide insightful answers to the above questions. They argue that the arrangement of two complementary storage systems enables an initial registration of information without the necessity of rearranging the existing long-term memory organization. Many novel events are not important to learn, so these events do not require a long-term record in memory, which in turn requires a memory reorganization. Only the important events need to be represented for a long time. They also argue that when a new event is incorporated in long-term memory, it requires an adjustment of the existing neocortical memory organization. This reorganization of memory is understandably a slower process. There is greater information-processing efficiency if only important information is recorded for later use. A bloated memory storage system that mostly contains inconsequential information would not be easy to use. Yet it is also advantageous to learn important novel events rapidly. The two complementary storage systems with different time scales for learning and forgetting can thus sharpen the information-processing ability of the organism. An animal with this type of memory structure would have an advantage of being able to have the right information available when needed but not to have a cumbersome memory that is too slow to locate the information. Improved information-processing efficiency translates to an increased chance of survival. It is not surprising then to see that humans evolved to have complementary memory systems that have different time scales for learning and forgetting.

3.3.6   Reconsolidation Theory

According to standard consolidation theory, repeated practice over days should result in a hardened memory that is not vulnerable to storage loss. But a number of researchers have questioned this hypothesis of a hardened memory representation. The first study to challenge standard consolidation theory was the paper by Misanin, Miller, and Lewis

(1968). These investigators tested the effect of electroconvulsive shock (ECS) on a conditioned fear association. If ECS was administered after the fear conditioning, then the animals showed no evidence of learning on the next day. They interpreted this effect as retrograde amnesia and hypothesized that ECS disrupted the consolidation of the learned fear association. Other animals did not receive ECS immediately, but these animals did receive ECS on the next day. These animals did exhibit a fear response even after the ECS treatment and therefore demonstrated that the fear training from the previous day had consolidated. Later, some of these animals were given further fear training so as to reinforce the initial fear conditioning. Following the additional fear training, the animals were given ECS. For those animals, the ECS resulted in the extinguishing of the learned association. The authors argued that the additional training of a consolidated response made the association pliable again and thus vulnerable to retrograde amnesia. Misanin et al. (1968) hypothesized that retrieving a learned consolidated association made it malleable, so the association had to go through a reconsolidation process to become hardened again.

The reconsolidation hypothesis was met with considerable skepticism in part because of a failure to replicate the effect (Dawson & McGaugh, 1969). Also, Miller and Springer (1973) argued that ECS influenced retrieval processes rather than disrupt a previously consolidated memory. Consequently, the reconsolidation hypothesis was not widely accepted for several decades until it reemerged as a consequence of a study by Nader, Schafe, and Le Doux (2000). Nader et al. injected anisomycin into the amygdala of rats to produce retrograde amnesia of a learned fear response. As mentioned earlier, anisomycin inhibits protein synthesis. Importantly, these investigators also showed retrograde amnesia occurred when anisomycin was injected after there was additional training of an old (and assumed consolidated) association. Because of the strong belief that memory traces require protein synthesis, the findings of Nader et al. (2000) were greeted with more acceptance than the earlier research by Misanin et al. (1968). There are a number of researchers who accepted the idea that further training of an old item makes the item malleable and thus more vulnerable to storage failure (Dudai, 2006; Nader, 2003). But other researchers have argued that the effects of drugs such as anisomycin are temporary and are reflective of a retrieval deficit rather than a storage loss (Fischer, Sananbenesi, Schrick, Speiss, & Radulovic, 2004; Lattal & Abel, 2004; Power, Berlau, McGaugh, & Steward, 2006). Without a technology for separating storage and retrieval processes, it is difficult to know what the effect of the drug is.

Both standard consolidation and reconsolidation are general physiological models of encoding and forgetting rather than a general cognitive description. The theories cannot account for retrieval losses from memory; consequently, they are limited, one-process models. Yet these theories have some good theoretical features, but they also have some problems. Both theories can deal with the broad outline of the different types of long-term potentiation (i.e., the theta and the delta waves). Also, they are consistent with the different rates by which hippocampal and neocortical cells develop long-term potentiation. Furthermore, they are consistent with the general phenomenon of retrograde amnesia. Nonetheless, there is still the troubling fact that proactive interference occurs, and that phenomenon does not fit well with either forms of consolidation theory. If proactive interference were strictly a retrieval phenomenon, then the lack of an account from consolidation theory would not be a problem, because consolidation theory is only a storage-based model. Yet Chechile and Butler (1975) and Chechile (1987) found evidence that there was a proactive interference effect on memory storage.[20] Consequently, a storage-based theory of memory is needed to account for the robust proactive interference effect on storage since consolidation theories cannot explain the effect.

3.3.7 Two-Trace Hazard Model

This model was developed to account for the effects of changes caused by time-based storage interference (Chechile, 2006). Since the original Ebbinghaus (1885) book on memory, there have been more than fifteen different mathematical functions proposed for accounting for the loss of memory as a function of the retention interval (Chechile, 2006). Moreover, there is a mathematical proof that there is a linkage between any proposed retention function and a memory hazard function (Chechile, 2006). Given an ordered dimension, such as time, and a property that can undergo a state change, the hazard function is defined as the conditional probability that the state change occurs at time t_0 given that the change did not occur already.[21] Functions that look very similar can nonetheless have qualitatively different shaped hazard functions (Chechile, 2003, 2006). It is possible that functions $u(t)$ and $v(t)$ can look similar when displayed, but the corresponding hazard functions $h_u(t)$ and $h_v(t)$ can exhibit dramatically different properties. For example, $h_u(t)$ might be monotonically decreasing with time, whereas $h_v(t)$ might be initially increasing in hazard until reaching a peak and then the hazard might decrease for all longer time values. Thus, a hazard function analysis is a powerful method for assessing very similar functions. In applying hazard theory to memory, a state change is considered the transition of a memory trace from being sufficiently stored before time t_0 to degrading at exactly time t_0 to an insufficient trace. For all the

20. These two studies employed a special psychometric procedure (like the measurement procedures described in chapter 7) to obtain separate metrics for storage and retrieval.
21. See the glossary for more information about hazard functions.

retention functions from the memory literature, the corresponding memory hazard functions were either monotonically decreasing or peak shaped (i.e., increasing to a peak and then decreasing for longer times) (Chechile, 2006). Clear experimental evidence was provided that the memory hazard function had to be peak shaped (Chechile, 2006). Only three proposals from the existing memory literature were consistent with a peak-shaped hazard function. But all three of these functions were found to be problematic for providing a suitable account across a range of experimental conditions. Consequently, the two-trace hazard model was advanced as a theory to explain the memory hazard properties and to account for the storage loss as a function of time across a wide range of conditions.

In the two-trace hazard model, all items are initially encoded with two concurrent memory representations: trace 1 and trace 2. These two traces have qualitatively different hazard properties.[22] The hazard function for trace 1 is monotonically increasing. Hence, for trace 1, the risk of a memory loss keeps increasing as a function of the retention interval. But if there were another study trial for the item, then trace 1 can be refreshed or even elaborated. If there is not a rehearsal or further study of the item, then there is an increasing risk with time of a loss. Hence, it is very likely that trace 1 will be eventually lost. Unlike trace 1, the hazard properties for trace 2 are monotonically decreasing as a function of the length of the retention interval (i.e., the risk of losing trace 2 is the largest initially and is reduced with increasing time). If an item is rehearsed or studied again, then both traces are refreshed (and perhaps even augmented), but the time clock is also reset to zero for both traces. This process is ongoing over the life of the individual, so there is always some risk that a memory is lost. The overall hazard property of the two-trace system was mathematically proven to be a peak-shaped hazard function (Chechile, 2006). If a memory survives for a long time since the last study trial, then the risk of memory loss is very small, but the risk never goes to zero. The probability of memory storage (θ_S) as a function of time (t) in the two-trace hazard model is

$$\theta_S = 1 - b(1 - exp[-a\,t^c])(1 - exp[-dt^2]), \tag{3.1}$$

where $0 < c < 1$, $0 < b < 1$, $a > 0$, and $d > 0$. The *hazard function* associated with equation (3.1) rises initially to a peak and then decreases for longer times. The peak is typically less than 15 sec, so for longer times, the risk of a loss for a surviving memory

---

22. The *fuzzy-trace theory* (Brainerd & Reyna, 1990) also advances the hypothesis that a memory encoding has two separate traces. In the fuzzy-trace theory, one of the two traces contains verbatim information, and the other trace contains gist information, whereas in the two-trace hazard model, both of the traces associated with a target event contain distinctive and incidental features.

is decreasing. This decrease in risk is consistent with Jost's second law, which stresses the increased risk of contemporary learning.

In the two-trace hazard model, time is used to characterize the memory loss, but in reality, the loss is due to the activities that occur with the passing of time (i.e., there is a storage interference process). Normally, the amount of interference is proportional to the amount of time. Because we cannot know precisely what happened during the retention interval, it is assumed that the number of interfering events n is equal to kt where k is a rate of interference. But of course, there can be experimentally induced differences in the rate of storage interference. If the interference rate and vigor are not the same between two conditions, then the values for the parameters in equation (3.1) can account for the condition differences.

In the two-trace hazard model, storage proactive interference is due to a more vigorous retroactive storage interference effect for items that share similarities to previously stored items. The more vigorous retroactive interference process could be caused when a target is encoding less efficiently. Suppose that the memory target for trial 1 of a Brown-Peterson task is to learn the pair APRICOT-PLUM, whereas on trial 2, the task is to learn the pair GRAPE-CHERRY. The use of a common category is known to result in more storage proactive interference for the retention of the trial 2 target (Chechile, 1987; Chechile & Butler, 1975). This effect is explained as the tendency to encode the second pair more in terms of their membership to the primed category of fruit. To remember the current target, it is necessary for the person to store the distinctive features of the current pair. It does not help to stress the fact that the trial 2 items are members of the fruit category; instead, the person needs to know which fruits are the current target. Thus, proactive interference for storage can be explained by the reduced encoding emphasis on storing the distinctive features of the current target. This reduced encoding elaboration is assumed to increase the risk that the storage retroactive interference during the retention interval disrupts to a greater degree the retention of the distinctive features.

Although the rationale for the two-trace hazard model is different from the rationale behind the standard form of consolidation theory, there are nonetheless similarities between the two theories. Like the Teyler-DiScenna index theory, there are two representations of the target event in the two-trace hazard model. Trace 1 is analogous to the hippocampal representation of the Teyler-DiScenna model, and trace 2 is analogous to the neocortical representation. The temporal properties for the two encodings are different for the two-trace hazard model, just as they are for the Teyler-DiScenna model.

While there are some similarities between the two-trace hazard model and consolidation theory, there are also important differences. The two-trace hazard model has an account for the finding of a storage-based proactive interference effect, whereas consolidation theory is silent about that phenomenon. This effect points out the need for stressing the encoding process in more detail. Within the multiple-store framework, there is a way to address different types of encodings. The shift in the encoding emphasis between the general versus the distinctive features of the current item is a plausible account for the storage-based proactive interference effect. This same explanation can also be evoked by the multiple-store models. But consolidation theory is a physiological description of memory loss, and it does not address the nuances of encoding. The two-trace hazard model also differs from the reconsolidation model about the consequences of additional training. Reconsolidation theory assumes that the memory is weakened by further training, whereas in the two-trace hazard model, postencoding storage loss is due to a more vigorous interference effect that operates on the most recent events. When an item is studied again, both traces are *refreshed and elaborated* rather than weakened, but the temporal clock is also reset to zero for both traces. The research on neurogenesis supports the idea that there is a temporal organization of information in the brain (Cai et al., 2016). After the encoding, the memory dynamics are again played out, so the studied item is back in a temporal zone of high hazard (i.e., the storage interference is more vigorous for recent encodings).

Finally, the two-trace hazard model also differs with the multiple-store approach in regard to the hypothesis of a permanent memory. The critical assumption in the multiple-store framework is that traces transferred to the long-term store are permanent. But in the two-trace model, there is not a single trace that is being transferred between memory stores. Rather, there are simultaneously two traces developed with different hazard properties. Yet both traces are always susceptible to disruptive loss.

3.4   Memory Types

The pattern of deficiencies and spared abilities of brain-damaged patients has contributed to the distinction of qualitatively different memories. Traditionally, memory for an event, as measured by performance variables like the percent correctly recalled or the accuracy of recognition, has come to be called *episodic memory*. Episodic memory is a type of explicit memory. Episodic memory can be described verbally; thus, it is considered a type of *declarative memory*. Episodic memory has a personal property. As William James (1890a) stated, "Memory requires more than mere dating of the fact in the past. It must be dated in *my* past" (p. 650). Yet amnesic patients like Henry Molaison lacked a long-term episodic memory for new events but still could remember general facts. This general knowledge is another type of declarative memory called *semantic memory*.

The content of semantic memory is overlearned information that is no longer linked to any specific event, and it is not autobiographical as is episodic memory. The division of declarative memory into episodic and semantic memory is illustrated in figure 3.6.

The distinction between episodic and semantic memory was originally proposed by Tulving (1972). Tulving saw these memories as being distinctively different systems. Episodic memory was autobiographical and susceptible to interference effects and memory loss, whereas semantic memory contained general information that was disconnected from autobiographical events and was not susceptible to loss. But there has been considerable evidence generated since the original proposal by Tulving that has blurred the distinction between the two types of memories (McKoon, Ratcliff, & Dell, 1986). Clearly, semantic memory information at some point in child development was an explicit event memory, but with repeated learning over many different contexts, the information became disconnected from any specific event (Chechile & Richman, 1982). Without an intact episodic memory, it is not possible to add to semantic memory. For example, the amnesic patient Henry Molaison could not expand upon his semantic memory when new concepts such as that of a frisbee were introduced into the culture (Postle & Corkin, 1998). Consequently, semantic memory is dependent on episodic memories for the development of new general information. Moreover, there can be memory loss from semantic memory if the information is not used regularly. For example, Bahrick (1984) showed in an extensive study of 733 people, over a fifty-year period, that there was forgetting of information about their Spanish language instruction acquired in high school. Although some information was preserved over the 50-year interval, there was a loss of other information. He reported an exponential

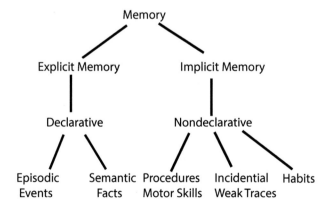

Figure 3.6
Types of memories.

decline in knowledge when Spanish was not regularly used. The loss of semantic memory content was with normal healthy individuals (i.e., a normal forgetting process). Brain-damaged patients exhibit an even more extreme loss from semantic memory. Patients with *semantic dementia* and *Alzheimer's disease* exhibit substantial degradation of their semantic memories (Rogers & Friedman, 2008). Consequently, semantic memory, like episodic memory, can undergo dynamic changes. Despite the blurring of the difference between semantic and episodic memory from Tulving's original paper, it is nonetheless useful to think of episodic memory as specific autobiographical memory and to think of semantic memory as general fact knowledge. Both semantic and episodic memories are characterized as a conscious memory of specific information that can be described by natural language (i.e., they are both explicit, declarative memories). To some the word *system* implies separate physiological structures like the respiratory system and the circulatory system. Alternatively, the indexing theory of memory consolidation might be considered different systems because there is a subcortical hippocampal component and a neocortical component. But even for these two complementary memory representations, there is a complex interaction that makes the word *system* suboptimal as a description. Rather than different memory systems, semantic memories and episodic memories are different types of memories. Both types are likely to be stored in a similar neuroanatomical fashion, but these types of memories have different informational content. Semantic memory is an interconnected set of facts that form the basis by which we come to understand the world. It affects how we encode a new event memory. But new memories can eventually become generalized facts that are added to semantic memory. Both types of memories are dynamically interacting and are changing.

Amnestic patients (like Henry Molaison) can still acquire some new learning. For example, Molaison could learn a classically conditioned eyeblink response, although when asked, he did not remember previously wearing the eyeblink conditioning apparatus (Woodruff-Pak, 1993). In his case, the conditioned response was not mediated by language, whereas to recall his prior training would require him to supply a verbal response. He could also learn a nonverbal motor task. In addition, he showed some weak, nonverbal knowledge of pictures that he previously studied when he was tested on a forced-choice recognition test (Chechile, 2006). The term *implicit* memory will be used here to capture all of these nonverbal types of learning. See figure 3.6 for an illustration of the distinction between explicit and implicit memory.[23]

23. The term *implicit memory* was introduced in psychology in the mid-1980s (e.g., Graf & Schacter, 1985); however, a similar distinction was introduced much earlier in philosophy by

Despite the widespread acceptance of the general framework shown in figure 3.6, the cognitive research on implicit memory has been challenged by a number of misconceptions and methodological difficulties. For example, there is a prevalent belief that implicit memory is an unconscious memory (Jacoby, 1991). If people recall an item, then it is clear that it is a conscious recollection. But it is neither necessary nor useful to insist that implicit memory is an unconscious memory. Establishing that an individual is unaware of some previous event because the conditions of learning are too degraded is a null hypothesis that there was no awareness of any target item under that condition. It is not possible to verify a null hypothesis with the use of classical sampling-theory statistics. Classical statistics, until quite recently, has been the near-universal method employed by experimental psychologists. For classical statistics, the null hypothesis is assumed. It is possible with classical statistics to reject an erroneous null hypothesis if the data are unlikely given that hypothesis, but it not possible to verify a true null hypothesis. However, with Bayesian methods, support for the null hypothesis can be generated. For example, the null hypothesis and some trivial non-null hypothesis can be compared with a Bayes factor analysis. If the Bayes factor ratio is suitably large, then there is support for the null hypothesis because the null hypothesis would be more probable than a small-valued alternative (nonnull) hypothesis. This type of analysis typically requires a lot of data—more data than usually obtained in experimental designs commonly used in implicit memory research. Perhaps there was a conscious awareness on some fraction of the test trials and that was enough to show a priming effect for a response-time measure. Rather than insisting on implicit memory being unconscious, it is more apt to think of implicit memory as a representation that is not good enough to produce an accurate recollection of the target event or even support a confident correct recognition of the event.

Moreover, rather than assume that all of the to-be-remembered targets are either implicit or explicit, it is more reasonable to state simply that there could be a mixture of memory types with some targets in an explicit state and other targets in an implicit state. A special case of a mixture model, where there are different proportions for the explicit and implicit states, is the case where all the items are of one type. A model with these mixture states can establish that there are implicit memories under some experimental conditions if it is estimated that all the targets are in the implicit state. This approach of thinking of memories as a mixture of types is an example of a stochastic-mixture model. This approach requires a formal psychometric model to statistically

---

Bergson (1911). Bergson stated that "the past survives under two distinct forms: first, in motor mechanisms; secondly, in independent recollections" (p. 87).

estimate the proportions in the various states. Examples of this cognitive psychometric approach toward explicit and implicit memory would include the Jacoby (1991) process-dissociation model; the Buchner, Erdfelder, and Vaterrodt-Plünnecke (1995) extended process-dissociation model; and the Chechile, Sloboda, and Chamberland (2012) implicit-explicit separation (IES) model. Each of these papers used a multinomial processing-tree (MPT) approach, and this class of models will be described in more detail in part II. Moreover, chapter 10 in part II is devoted entirely to these models of implicit and explicit memory.[24]

Instead of using a psychometric model, many researchers have argued that tasks are selectively sensitive to either implicit or explicit memory; hence, a formal model of cognitive operations can be circumvented. The approach of using a performance variable as a pure measure of a cognitive process will be referred to as the *process-purity hypothesis*. The fragment completion task has been used in this fashion (Graf, Squire, & Mandler, 1984; Schacter & Graf, 1986). For this procedure, participants are first given information to learn, and later they are asked to respond with the first word that comes to mind when cued with a three-letter stem. For example, if they studied the word pair **WINDOW-REASON**, then later they would be given the test probe **WINDOW-REA___** and be asked to complete the word fragment. Typically, amnesic patients cannot correctly answer what was the previously presented response when given a cued recall probe **WINDOW-_____**, but nonetheless these patients when given the fragment completion test tend to solve the puzzle with the word **REASON**. The rate for using the word **REASON** is elevated above a chance level if the delay between presentation and test is less than 120 min (Graf et al., 1984). In essence, the process-purity researcher is assuming that implicit memory can be measured directly by simply (1) providing a three-letter cue and (2) instructing the patient to state the first word that comes to mind.

Repetition priming on a lexical decision task has also been assumed to be a function of only implicit memory (Logan, 1990). Letter strings are displayed on each test trial, and the participant is required to quickly determine if the string is a word or not. The task is not designed to test event memory. Nonetheless, if a word was presented earlier, then the lexical decision for a repeated word is usually more rapid. This priming effect is taken as evidence of an implicit memory formed from the earlier exposure to the word.

Some memory researchers have argued, based on process-purity studies, that explicit and implicit memories are distinctly different systems (Cohen & Squire, 1980; Graf & Schacter, 1985; Parkin, 1993; Schacter, Chin, & Ochsner, 1993; Squire, 1987; Tulving,

---

24. See the glossary for more details about the IES model and the process-dissociation model.

1985a; Weiskrantz, 1987, 1989). Schacter et al. (1993) further hypothesized separate brain regions for explicit and implicit memory. They argued that the two systems are independent, so information in one system cannot move into the other system. One rationale for separate systems is the apparent difference caused by some independent variables (i.e., "experimental dissociations"). For example, Parkin (1993) claimed that implicit memory develops chronologically prior to explicit memory and lasts longer with advanced age. But this claim by Parkin is disputed by Rovee-Collier (1997), who argued that explicit and implicit memories have a common developmental timetable. Claims of experimental dissociations are another form of the process-purity approach. Both approaches have serious methodological problems. Critics of the multiple memory systems hypothesis have argued that separate systems are not required to handle experimental dissociations (Roediger, 1990). In chapter 6 (in part II), there is formal critique of the experimental dissociation claim. Moreover, in part II, a case will be advanced that no one experimental task only influences a single cognitive process; hence, a formal model, typically of a multinomial processing tree (MPT) form, is required to disentangle the contributions of all the interacting cognitive processes. This approach does not assume a null hypothesis. Another beautiful property of the MPT approach is that it can actually find evidence for a process-pure condition (not by assuming it to begin with) but by empirically estimating the proportion of the memory traces that are of a similar type. If there is an experimental condition that results in all memory traces being in a particular state, then the probability for that state will approach 1. Otherwise, the MPT model measures the probability for each possible state for the memory trace. Hence, by allowing for a stochastic mixture of states, it is possible to detect pure-state conditions. The MPT approach is a rigorous method for dealing with entangled memory processes. Chechile, Sloboda, and Chamberland (2012) and Chechile and Sloboda (2014) used the IES model to obtain separate measures for implicit and explicit memory. These researchers also demonstrated that it is more appropriate to consider explicit and implicit memories as different types of memories rather than different memory systems. In fact, they showed that for short retention intervals, nearly all the memory traces were in an explicit state. But for long retention intervals, the proportion of traces in the implicit state increased. They argued that traces that were once explicit degraded to an implicit state; *hence, the proportion in the implicit state increased.* This finding is directly contradictory to the memory systems framework that is based on the assumption that explicit and implicit memories are separate and independent systems.

The distinctions shown in figure 3.6 are useful and are a consensus. These distinctions can inform us about how memory should be measured and investigated. Yet it is important to stress that all the memory types shown in the figure are made from a common neural underpinning. It is likely that all the memory types are realized by long-term potentiation changes of neurons and by morphological changes to the synapses from the prelearned state. Yes, there is a major difference between semantic memory and episodic memory, but rather than postulating separate systems, it is better to stress that these memories differ in content and properties. They have different usage patterns, which in turn affects the dynamics of their retention. In a similar fashion, explicit and implicit memories have different features. Our ability to touch type is a motor skill that is very different from our declarative memory of either facts or events, but early in the learning history of a motor skill, declarative memories are involved. Recall earlier in our discussion of conditioning that it was difficult to avoid the concept of propositional information that is usually thought to be in the domain of declarative memory. The interconnections of ideas for the various domains of memory are appropriate and valuable. An integrative understanding of memory is less likely when researchers treat each memory type in figure 3.6 as an independent systems.[25]

3.5 Chapter Summary

The study of learning and memory has been greatly influenced by a number of biological and neurological discoveries. For example, specialized neural computations and pathways have been detected. But some discoveries about the neural processes underlying learning were accidental. For example, the bilateral removal of the hippocampus was a personal tragedy for Henry Molaison, but it provided an important insight about the necessary structures for the learning of new information. Also in the chapter, there is a discussion of the fundamental neural changes that support memory at a cellular

---

25. Some have argued that autobiographical memory is yet another memory type because of finding some rare individuals who apparently can recall what they did on any particular day (Ally, Hussey, & Donahue, 2012; Parker, Cahill, & McGaugh, 2006). But on a more careful examination of these individuals, a case can be made that they were more fixated on recalling their past than solving problems of today. In fact, Parker et al. (2006) showed that one such person who had a remarkable autobiographical memory was very poor at learning new information. She also kept a diary and spent much of her day recalling events of the past. In my opinion, a few unusual individuals do not require us to conclude that there is an independent memory system for autobiographical events. Nonetheless, the ability of some individuals to have an excellent autobiographical memory is interesting because most people lose some aspects of the event (e.g., the temporal and other incidental aspects of the experience can be easily lost). People with superior autobiographical memory seem to do well holding the separate incidental aspects of memory together with the central features of the experience.

level (i.e., long-term potentiation (LTP), long-term depression (LTD), and neurogenesis). If memory is encoded in terms of a neural representation, then it is likely to be a dynamic storage system because neurons do not have the ability to remain frozen like statues in their structure. Also in this chapter, a number of physiological and cognitive models of learning and forgetting were introduced. These theoretical frameworks included the Atkinson-Shiffrin model, working memory, the SAM model, the REM model, consolidation theory, interference theory, the Teyler-DiScenna model, reconsolidation theory, and the two-trace hazard model. Also in this chapter, the idea of postulating separate memory systems was debated. The proposals for a separate memory system included short-term memory, long-term memory, sensory memory, explicit memory, implicit memory, autobiographical memory, procedural memory, and semantic memory. The suggestion was advanced that instead of different systems, many of these proposed memory systems might better be thought of as different types of memory.

4 Information-Processing Framework

4.1 Chapter Overview

Historically, the physical world was initially conceptualized in terms of forces and energy, but the emergence of fields such as statistical thermodynamics, communication theory, computer science, and cognitive psychology has provided the idea of *information* as a powerful alternative conceptualization. In this chapter, memory is examined in terms of an information-processing framework. From this prospective, *propositions* play an important role in representing and accessing stored information. Also in this chapter, evidence is presented that only a small fraction of the available information in the environment is encoded into memory and an even smaller fraction is remembered.

4.2 Propositional Encoding

Propositions are the fundamental basis for expressing relationships, and they have the structure of a simple declarative sentence, that is, (1) a subject, (2) an action or relationship, and (3) a direct object. Propositions have a natural directionality, and they are denoted in this book as **subject** $\xrightarrow{\text{relationship}}$ **direct object**. To facilitate the discussion of propositions, let us call the subject **A**, the direct object **B**, and the relationship **R**. Previously, we employed propositions in the discussion of instrumental conditioning (see figure 2.5). But much of what we learn about the world is based upon observed relationships rather than conditioning. For example, at the time of writing this section, I am in an apartment in the French Alps, and I happen to notice that the name of the watercolor painting of a sailboat, which is hanging above my desk, has the label *Le Vercingétorix*, a boat named after a famous Gallic general who lived at the time of Julius Caesar. This event does not fit the conceptual template of conditioning, but observations like this make up much of the memory modeling of our environment that we do on a regular basis.[1] The encoding of this event is best formulated in the form of a set of propositions. Propositions are at the core of human declarative memory and are a vital part of the building of a memory engram. But there is an internal *conceptual* structure to a proposition, so it is necessary to deal with the concept of *concepts* in order to understand propositions at a general level. Thus, there is a web of interrelated ideas that needs to be carefully untangled.

4.2.1 Framework for Propositions and Concepts

Psychologists borrowed the idea of propositions from philosophers who have utilized this notation for millennia. A key property of a proposition in philosophy is its truth value. But in psychology it does not matter if the proposition is actually correct or not. Propositions are used to express the relationship between two things. Yet there are constraints on permissible propositions because usually the **A** and **B** components of a proposition are themselves conceptual units that possess a substructure. Consequently, the proposition "copper (on-the-right-of) gravity" is an absurd proposition because it violates the conceptual grounding of the **A** and **B** entities. The units *copper* and *gravity* both have a rich conceptual substructure as does the relationship *(on-the-right-of)*. These conceptual substructures impose constraints on the formation of meaningful propositions. Consequently, it is acceptable to say that "copper is (on-the-right-of) nickel" as a true fact about the periodic table of elements or as a true statement about the arrangement of objects on a lab bench. But copper cannot be (on-the-right-of) laughter. Thus, propositions represent a concept, but the units of the proposition are themselves built from other concepts. For a proposition to be true, it is necessary to avoid violating the meaning constraints imposed by the components of the proposition.

Propositions are a tool for representing information about the world, and it developed as a generalization of another philosophical concept, namely that of the association. Since the time of Aristotle, the association has been the foundation for building a mental model (or a memory) of the external word. However, observed stimuli S_1 and S_2 are more than two sensory events in either spatial or temporal proximity. There usually is a relationship between these stimulus events. Yet in associationism, there was

---

1. Social learning is known to occur in other animals as well. By social learning, it is meant the learning by example from other members of the same species. Primates are known to be capable of social learning (Reader & LaLand, 2002). But recently there is also evidence that a reptilian species, the bearded dragon *Pogona vitticeps*, is capable of observational learning (Kis, Huber, & Wilkinson, 2015). Moreover Noble, Byrne and Whiting (2014) have also found social learning with another lizard, the skink (*Eulamprus quoyii*), so social learning may be more widespread than previously thought.

only a "strength" value for the association. A strength cannot capture the richness or the qualitative aspect of the relationship between the two stimuli. But with a proposition, the two stimuli can have a functional relationship. In essence, the proposition is a generalization of the concept of an association. Because of the longstanding utilization of associations in behavioral psychology, it has not been an easy transition for psychologists to leave the safe grounding of that limited conceptual tool. For example, the Anderson and Bower (1973) book was an early cognitive model in psychology that used propositions, but later Anderson admitted that he was, at the time of the writing of the book, partially in agreement with radical behaviorists who were suspicious of the idea of any intervening theoretical construct between the stimulus and the response, other than an association strength value (Anderson, 1980). Nonetheless it became clear to Anderson and others that propositions had a substructure that was more than a value on a strength-intensity scale. If there were a substructure to a proposition, then memory recall of a proposition could be partial because the recall of a simple declarative sentence in the form of S (subject), V (verb), and O (direct object) can be partial. Anderson (1976) reported that the probability of recall of the verb increases if there is also the recall of the direct object, that is, $P(V|O) > P(V|\neg O)$, where $\neg O$ denotes the fact that the direct object was *not* recalled. Anderson also reported that $P(V|O) < 1$ and $P(V|\neg O) > 0$. If the propositions were unitary and did not possess a substructure, then the conditional probabilities would be at one of the extremes of either $P(V|O) = 1$ or $P(V|\neg O) \approx 0$. But the conditional probabilities were not at the two extremes (i.e., parts of propositions can be recalled without the remainder). Hence, the idea of a proposition as a nondecomposable entity can be rejected because if propositions were unitary, then they could only be recalled in an all-or-none fashion.

Concepts usually contain a web of propositions. Consider the concepts captured by the dictionary definition of the English word *band*. Hofstadter and Sander (2013) developed a much more complete discussion of the multiple senses of the word *band*, but a brief examination, based on my computer dictionary, is sufficient to motivate the points needed for a memory application. The dictionary includes many different concepts under the definition of *band*. Among the many concepts linked to the word is the idea of a thin loop of material, or a stripe of color, or a wedding ring, or a range of wavelengths in the electromagnetic spectrum, or a group of musicians, or a group of people, or the action of joining. Each of these senses of the word is a concept or a set of concepts in its own right. For example, for the musical sense of band, there are many types of bands (i.e., jazz, blues, rock, brass, swing, marching, drum, etc.). Within each of these types of musical bands, there are specific named groups (e.g., Beatles, U2) for just classical rock bands, and there are many other subgroups or genres for rock bands.

It would be a vast undertaking to code into memory all that is known about the *band* concept, and it would not be readily achieved by several of the scholarly theories for concepts. Before advancing a suitable approach for understanding concepts, let us first examine some of the other ideas about the topic of concepts.

Much of the enormous psychological literature about *concepts* deals with the issue of categorization. Items in a common category are considered a concept (Murphy, 2004). But before tackling the psychological findings about categorization, it is important to understand what is *not* a valid basis for determining a category (i.e., the failure of the classical philosophical basis of a category). From the philosophical tradition, a conceptual category, like that of a *cat*, is captured by a carefully crafted definition that delineates the concept along with all of its boundaries. This idea dates back to Aristotle (Apostle, 1980). But there is considerable reason to believe that this idea is rarely achieved. For example, I believe that many people are overly impressed from their training in geometry and have come to believe mistakenly that mathematics begins with axioms and continues until an elaborate edifice is constructed. In reality, that is not the way mathematics is conducted. For example, the axioms of probability theory are now accepted to be the principles articulated by Kolmogorov (1933).[2] But these axioms came 179 years after probability theory began, and a considerable body of work had already been done prior to the formulation of the Kolmogorov axioms. This example is not an isolated case. In other academic fields, scholars have their own difficulties defining key concepts. Pond (1987) described the lack of agreement in the field of metallurgy regarding the concept of a metal. No agreement has been reached among metallurgists as to the set of properties that a substance has to possess in order for it to be a metal. Usually long before a definition is established, people can effectively communicate and use the category.

Murphy (2004) traced the death of the classic definitional approach to categorization to a number of studies published in the 1970s in the psychological literature (e.g., Rips, Shoben, & Smith, 1973; Rosch, 1973; Rosch & Mervis, 1975). For example, Rips et al. (1973) found that the ease with which people confirm that an item belongs to a category depends on a typicality value. In the *bird* category, robin has a high typicality rating, whereas chickens are judged as less typical. These investigators also found that the speed of categorization is faster for the more typical members of the category. The

---

2. Independent of probability theory, mathematicians struggled with problems associated with various methods for integrating functions. The upshot of that issue was the development of mathematical measure theory. Kolmogorov recognized that probability was a mathematical measure that summed to 1 over the sample space of all possible outcomes. This idea had to wait until the development of mathematical measure theory.

classic definitional approach to categories does not have a ready account for the degree of belonging to a category. An item either fits the category definition (whatever that is) or it does not.

Three theoretical approaches for modeling the categorization process have attracted considerable attention in psychology: one is based on prototypes (Rosch, 1973), one is based on exemplars (Medin & Schaffer, 1978), and one is based on using existing world knowledge (Keil, 1989; Murphy & Medin, 1985). For the *prototype* approach, the decision about the categorizing of a novel stimulus is a function of the similarity between the feature-space representation of the new item and the feature-space representation of an idealized prototype of the category. The prototype might not be any of the stored memories from the category; rather, it is the optimal, best-case representation. For the *exemplar* model, the categorization decision for a new item is based on the aggregate value of the similarity between the new item's representation in feature space and all the other stored members of the conceptual class. Thus, with both the prototype and exemplar approaches, there is a feature-space representation of items in memory, and both theories employ the idea that the distance between items is a function of the similarity between the items. If items are close in feature space, then the items are assumed to be similar. Dissimilarity is thus a metric of the distance between items in the theoretical feature space. Consequently, memories for exemplars within a conceptual category are points in a space. Mathematically, if there are m feature dimensions and v_i, $i = 1, \ldots, m$ are the values on the various dimensions, then an item is considered to be the point $(v_1, \ldots, v_m)$. Neither the prototype model nor the exemplar model delineate what the features are. These models also have not been applied to a realistic web of concepts such as the case of the *band* concept. Also, these models have almost always been limited to the noun class of categories rather than dealing with the verb or an operator class.

In addition to the above limitations of the feature-space approach, Tversky (1977) provided compelling evidence that the judged similarity between stimuli is inconsistent with a geometric representation. Participants in one experiment were asked to select (from a choice set of three stimuli) the one stimulus that was the most similar to a given target. All stimuli in the experiment were schematic faces. Let us denote the *target stimulus* as F_a. For one group of participants, the three alternatives were face stimuli F_b, F_c, and F_p, whereas for a second group, the three alternatives were F_b, F_c, and F_q. Thus, the choice set of alternatives only differed by the third stimulus, which was either F_p or F_q. This experimental design cleverly avoids the problem of how to scale similarity judgments. Participants simply had to select one of the three alternatives that they felt was *most similar* to the target face. If the faces were points in a psychological feature space, then the ratio of the choice probabilities $r = \frac{P_{ac}}{P_{ab}}$ would be the same for two groups because the spatial dissimilarity distance among the faces F_a, F_b, and F_c should be independent of the fourth face, which was either F_p or F_q. The choice probability P_{ac} is the proportion of the participants who selected stimulus F_c as the most similar face to the target face F_a. For example, if face F_c and F_b are equally distance to the target face F_a, then $P_{ac} = P_{ab}$, so $r = 1$ regardless of the third stimulus in the choice set. But contrary to this prediction, which is based on a geometric representation of similarity, Tversky found that the value for r dramatically differed between the two groups (i.e., $r = 0.95$ for the group that had the F_p face in the choice set, whereas $r = 6.67$ for the group that had F_q in the choice set). Consequently, when participants are actually asked to assess similarity, their judgments exhibit a context dependency that is inconsistent with the type of spatial representation assumed in feature-space models. For all of the above reasons, I have been skeptical of both the prototype and exemplar models.

The world-knowledge model for categorization stresses the point that previously learned propositions are the usual basis for classifying novel events. This theoretical position has emerged slowly in the psychological literature in part because experiments frequently use stimuli that have no preexperimental meaning. This type of design is intentional because the investigator does not want preexperimental learning to vary widely and swamp the effects of learning in the laboratory. But if prior knowledge is the typical basis for making classification decisions, then the experimental design has controlled away the process of interest. The elimination of prior knowledge is a more serious design fault than the concern about the potential variability of preexperimental learning.

Murphy (2004) describes the world-knowledge model for classification by means of a hypothetical example of a person going to the zoo and seeing a strange new animal that is called a *whingelow*. The person has never seen a creature like this before, but still there is knowledge about other animals that helps infer (or at least hypothesize) properties of the animal. Suppose it has four legs, two symmetrical eyes, and one nose. Given these observations, it is reasonable to suspect that the animal is a mammal because those features are common properties of other mammals. It is also reasonable to suspect that this species will have males, females, and young. An enormous set of inferences is generated by the rich knowledge previously acquired. If another smaller, different looking animal were following a bigger whingelow, then it is reasonable to suspect that the smaller animal is a baby whingelow, and the bigger animal is its mother. The concept of the young offspring being smaller and following their mothers enables an inference about the whingelow. In some cases, these inferences about the objects and actions that we perceive are mistaken, but often they are good enough to be useful. If a previous hypothesis is incorrect, then it is replaced eventually with the new understanding.

The world-knowledge model for understanding concepts and the categorization process seems to be the most accurate, but this conclusion means that there is an apparent circularity. To understand propositions, we need to acknowledge that they often are made up of conceptual units, but conceptual categories are best understood in terms of a set of propositions learned previously. What are the irreducible elements that support this whole web of propositions? This question can be resolved by considering memory events as acquiring propositional knowledge directly. Consider a child's first experience with seeing an interesting novel stimulus—a cow. The cow has many properties that are directly observable. It is big—bigger than Mom and Dad and bigger than the family dog, Buster. Like Buster, the cow has fur. Like Buster, the cow has four legs. Like Buster, the cow has a tail and two eyes. Like Buster, the cow does not cause Mom or Dad to worry. Unlike Buster, the cow is eating grass. Unlike Buster, the cow is disinterested in humans. Unlike Buster, the cow is not moving rapidly or jumping. Mom points at the animal and calls it COW. In time, the child will come to observe many events such as this. The acquired propositions are linked with other propositions and all are part of the COW concept. These episodic events construct a pattern of knowledge that will eventually become less tied to original events, but the observed facts about the cow were, at the time of encoding, events that built upon prior learning. Each of the facts acquired involves the cow, a relationship, and a direct object. The facts encoded into memory are propositional in nature.

Concepts continue to build throughout life. New ideas are generated all the time. Douglas Hofstadter describes an amusing experience of developing a concept that he calls Danny-at-the-Grand-Canyon (Hofstadter & Sander, 2013). Hofstadter recalls an experience of seeing his small son (Danny) playing with ants and a leaf while being within a few feet of the North Rim of the Grand Canyon. Fifteen-month-old Danny was disinterested with the expansive view of the canyon, and he focused instead on a few common small objects. Many years later, Hofstadter recalls being at the grand temple of Karnak in Egypt and finding some discarded bottlecaps in the dirt. He was fascinated by these modern discarded items in the grand temple, so he stopped to pick them up, studied them, and kept them as if they were treasures. He then remembered Danny-at-the-Grand-Canyon—the concept of focusing on small incongruous items while in the presence of a much grander scene. Danny-at-the-Grand-Canyon had become a conceptual category. The reactivation of this conceptual category was a basis for linking across time to an earlier event and remembering the image of his son playing with ants near the Grand Canyon. Note that categories such as the Danny concept can be created from an encoding that includes the elaborations of a set of propositions.

4.3   Temporal/Feature-Binding Memory Framework

A number of facts about the building of a memory trace have been developed in previous sections. At this point, a few integrative points should be discussed, and these ideas are illustrated in figure 4.1. Note that the current event involves the parallel and widely distributed activation of many microfeatures. Yet we do not perceive or remember the current event as a set of separate microfeatures, but rather there are different groupings. These higher-order groupings serve the role of binding the features into a meaningful perceptual group. This integration is represented by higher-order nodes in the figure. Note that the current event is likely to have many different higher-order

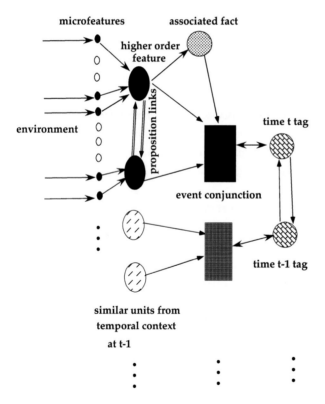

Figure 4.1
A representation of an elaborated encoding within a temporal context.

perceptual groups. In some cases, the higher-order feature unit also activates an associated fact from previous experience (i.e., a semantic memory activation). Moreover, a higher-order feature grouping is not in isolation, but it has a relationship to one or more other higher-order feature units. These relationships can be simple positional relationships such as one is to the left of the other. This relationship is nonetheless in the form of a proposition as is denoted with the directional double-edged arrow (i.e., $A \xrightarrow{\text{left of}} B$). But the relationships between higher-order groups can also be an action or some other type of relationship. The whole ensemble of activated units is further linked together into an event conjunction. Presumably, this grouping involves the contribution of the complementary hippocampal memory system. The event encoding can have other components not shown in the figure, such as the activation of the expectation of an oncoming unconditioned stimulus and the activation of an associated conditioned response.

A key feature stressed in the figure is the idea that events in memory have an important temporal organization. The memory for time is a complex issue that will be discussed later. But it is still important to understand here that the collection of activated features, propositions, conditioned responses, and associated facts from semantic memory all shares in the current NOW. In essence, there is a rough time tagging of the ensemble of units currently active in memory. Moreover, the events of the current NOW activate the prior NOWs because memory of the current experience is within a temporal context. The events are not disconnected in a random list, but rather they are integrated into a temporal context. The memories of the present thus interact with the memories of the recent past. The interactions are not shown in the figure, except for a connection between time tags.

Despite the fact that there are no direct sensory detectors for time (unless we happen to be looking at a clock), it is nonetheless clear that memories have a temporal organization, and this property has important consequences. In chapter 3, the neural foundation for the processing of the ordinal and temporal aspects for a sequence of events was discussed. The temporal flow of memory can also have jumps and breaks (Howard, Shankar, Aue, & Criss, 2015). Waking up after sleeping resets the temporal memory clock. Also, when we recall an event from the past, we engage in a temporal jump back to that point in time. In essence, memory is a method for time travel. Yet the events stored during any given day are not in a random temporal order. If the events were randomly stored, then it would be remarkably difficult to recall the sequence of events. Notice in figure 4.1 that the time tags encode the order of events. Without clock-time information being stored directly, the memory system can nonetheless encode the relative order of the events. A number of investigators have proposed computational models that delineate how the temporal component of event memory is generated (Howard & Eichenbaum, 2015; Howard et al., 2014, 2015). Thus, having the events ordered can achieve an organized temporal dimension without knowing the actual clock time.

4.4 Syntactic-Pattern Recognition

4.4.1 Is Visual-Pattern Memory a Feature-Space Model?

Feature-space models treat information in memory as points within a vector space—a theoretical framework that was criticized previously in the section dealing with concepts and propositions. A vector space is a wonderful mathematical representation for a number of problems. For example, in physics, vectors are a valuable tool for solving problems dealing with objects that are being acted upon by a number of different forces. But there are operations that are not permissible with a vector alone.[3] The concept of oddity can be expressed with propositions, but oddity is not a property that vector spaces can represent directly. One might argue that the oddity concept can be represented by processing a set of vectors. But such an encoding would consist of instructions about how to examine and compare the vectors in the set. The instructions or operations on the set of vectors is like a series of propositions rather than a point in a space. But even more important, the concept of *oddity* is an abstract concept that is independent of any given set of stimulus vectors.

In general, a vector is a point of values. Vectors can have magnitude and direction, and there can be angles between vectors. But vectors cannot be true or false. Vectors cannot represent structural relationships between subsections of a visual scene. Matrices were invented later to generalize vectors and to provide a more powerful tool for solving a new set of problems. Other mathematical objects continued to be invented.

3. Physical laws generally require more than just a vector representation. Physicists, who along with mathematicians, developed vector analysis because there was a need for operations between vectors. Concepts such as the dot and cross products of vectors along with the divergence and the curl operations were required to express physical laws. For example, Maxwell's equations for electromagnetism are four interlocking vector-operator equations. Consider, for example, Maxwell's extension of Ampere's law. This equation is $\mathbf{curl}\,\mathbf{H} = \mathbf{J} + \frac{\partial \mathbf{D}}{\partial t}$, where $\mathbf{H}$, $\mathbf{J}$, and $\mathbf{D}$ are respectively vectors for the magnetic field strength, the current density, and the electric displacement (Reitz & Milford, 1960). The equation connecting these vectors involved the **curl** operator on a vector. The key idea here is that Maxwell's equations express more than just a point in a vector space. The equations describe relationships among vectors, so these equations can be expressed by a set of propositions.

In theoretical psychology, vectors have been extensively used. In my opinion, vectors have been overly used, and they have led to some misunderstandings.[4]

The area of visual-pattern recognition has heavily employed the idea that recognition occurs when there is a bottom-up activation of the features of the old stimulus. These models can be said to be a feature-space memory (i.e., the stored memory of the pattern is a point for the combination of the features). Beside the previously discussed prototype and exemplar models for categorization, numerous other models for recognition memory also assume a vector-space representation (Ashby, 1992; Biederman, 1986, 1987, 1990; Estes, 1994; Kruschke, 1992; Nosofsky, 1986). One such model is the recognition-by-components or RBC model by Biederman (1986, 1987, 1990), which postulates that all visual forms are made up of a small number of elementary visual features that are called *geons*. These hypothesized features were designed to be the visual counterpart to the phonemes used in speech perception. Linguists believe that all languages are made from the limited set of phonemes.

4.4.2 A Divergent Approach: Syntactic-Pattern Recognition

The term *syntactic-pattern recognition* was used by a number of theoretical computer scientists who were developing compilers for processing visual patterns (e.g., Fu, 1985; Gonzalez & Thomason, 1978; Winston, 1975). The compiler is at the core of higher-order computer languages because it translates the computer program, written within a fixed set of programming rules, to a set of elementary machine-code operations. Normal compilers are aided by the ordinal nature of the programming instructions. But multidimensional visual patterns pose a challenge for developing an automatic encoding, so that the pattern is parsed into a set of operations for producing a structural description of it. These computer scientists were inspired by the work of Noam Chomsky in computational linguistics, and they developed algorithms for parsing a

---

4. As a former physicist, I have come to believe that when the scientific process is not well described by the available mathematical tools, new tools need to be employed or even invented. In the area of memory, there has been an attempt to adopt a vector-space representation, and that tool fails to code adequately the full information content of events. Consider the coding of an auditory track of information. This problem is challenging despite the fact that there is an ordered temporal dimension. Fortunately, mathematicians, physicists, and electrical engineers have developed and refined powerful methods for encoding and processing acoustic information. For example, discrete Fourier transforms can be used to render elaborate acoustic information into a set of frequency and intensity values for each unit in time (see Smith, 2007). There is a temporal ordering of a rich pattern of information. Fourier methods can also be used for visual-spatial information, but the analysis is more difficult. The mathematics involved with this type of coding is beyond what can be captured by a point in a vector space.

visual scene into language-like data structures such as trees and strings. The computer record for the scene was a language-like data structure. Pattern matches were determined by contrasting a newly parsed scene with one of the data structures previously stored in the computer's memory. The computer scientists were building automated pattern-recognition systems for applications such as detecting the identity of high-energy particles from photographic tracks that were recorded by a bubble chamber detector.

In essence, the syntactic pattern-recognition approach maps the raw visual features of a pattern into a language-like representation that represents the relationship among the parts of the pattern. The approach taken by some computer scientists thus raises the possibility that memory storage of visual patterns might be quite different from a feature-detection model where the familiarity decision about a visual pattern is hypothesized to be strictly a function of the activation of a point in feature space.

4.4.3   Is There Merit to Syntactic-Pattern Recognition?

It is clear that a syntactic system can be implemented to make recognition decisions. But do humans operate in this fashion? The idea of computing the structural relationships among the features of a pattern is not foreign to the thinking of some perceptual psychologists (Buffart, Leeuwenberg, & Restle, 1981; Palmer, 1977; van der Helm & Leeuwenberg, 1991). For example, rather than a feature-space representation, Palmer stressed the need for a hierarchical network of propositions for the perceptual comprehension of visual patterns. For Palmer, pattern perception requires the integration of the structural components of the pattern; thus, perception is more than the parallel activation of features. Moreover, Leeuwenberg and van der Helm (1991) pointed out that there were patterns that could not be produced from the set of geons proposed by Biederman. These authors also found patterns that were perceptually different but had the same geon coding. The studies by Buffart et al. (1981), Palmer (1977), and van der Helm and Leeuwenberg (1991) demonstrated that perceptual understanding required the structural configuration of the features. But those studies did not take a position about when the structural information was computed. Were the structural aspects of the pattern computed after the pattern was judged to be familiar or was the computation of structure a necessary step prior to the recognition decision?

Chechile, Anderson, Krafczek, and Coley (1996) directly tackled this issue and showed the importance of regenerating the structural and propositional information associated with a pattern before a familiarity decision could be made. This study is particularly problematic for a feature-space model because it showed that pattern recognition was more than just the bottom-up activation of features.

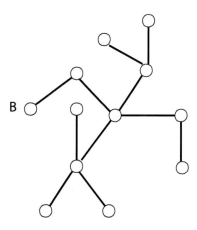

Figure 4.2
An example of the stimuli used in Chechile et al. (1996).

To evaluate the syntactic-recognition hypothesis for pattern recognition, Chechile et al. (1996) used stimuli such as the one in figure 4.2. This figure can be translated to a string format representation. The string language used eight elemental directions denoted by the numbers $1, \ldots, 8$ where $1 = \uparrow$, $2 = \nearrow$, $3 = \rightarrow, \ldots, 8 = \nwarrow$. Also, there were B, T, $+$, and v symbols for denoting respectively string beginning, terminal point, concatenation, and an OR branch in the expression. The string language also used the symbol (to denote the beginning of a subunit and the symbol) for the closure of a subunit. Consequently, the pattern shown in figure 4.2 is encoded by the following string:

$$B2 + 4 + ((3 + 5T) \, v \, (2 + (8T \, v \, 1T)) \, v \, (6 + (1T \, v \, 4T \, v \, 6T)))$$

To understand how the string captures the pattern, let us start with the begin-string symbol B and note that the line segment goes off in the northeast direction (i.e., the 2 symbol corresponds to northeast direction or $\nearrow$). Next the path switches to the southeast direction (i.e., the 4 symbol for the $\searrow$ direction) and concatenates to a branch point where the pattern goes in three separate paths. Hence, the string structure is of the form $B2 + 4 + (S_A \, v \, S_B \, v \, S_C)$, where S_A, S_B, and S_C are the three diverging path segments. Path-segment S_A is the shortest and is represented by the pattern going in the east direction, which is denoted by the 3 symbol, and then concatenates to a south-going path, which is denoted by the 5 symbol, and finally this path terminates, hence the T symbol. Consequently, the S_A path segment is coded as $(3 + 5T)$. The associative parentheses are used to contain this path as a pattern segment that can be related to the other parts of

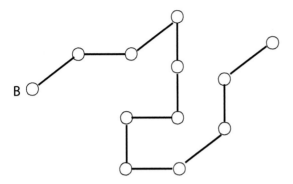

Figure 4.3
An example of a syntactically less complex pattern.

the pattern. The S_B segment is similarly coded by the string $(2 + (8T \lor 1T))$, and the S_C path segment is denoted as $(6 + (1T \lor 4T \lor 6T))$.

Chechile et al. (1996) showed that it is possible for the string description to vary in complexity for the same number of circles and links. For the pattern in figure 4.2, the string is forty symbols long, has five $\lor$ symbols for the OR branches, and has three levels of nested parentheses. To contrast this pattern, consider the following string for the pattern shown in figure 4.3:

$$B2 + 3 + 2 + 5 + 5 + 7 + 5 + 3 + 2 + 1 + 2T$$

This string has twenty-three symbols, no $\lor$ symbols for the OR branches, and no nested parentheses, but the pattern still has twelve circles and eleven connecting-line elements. Note for each metric of syntactic complexity, the pattern in figure 4.3 has reduced complexity relative to the pattern shown in figure 4.2. No claim is being made here that the above-described string language is the encoding process used by people. Rather, the string language is simply a convenient tool for creating stimuli with different levels of syntactic complexity without varying the physical features of the pattern.[5] The pattern for figure 4.2 is more likely encoded with a set of propositions. Perhaps the following set of propositions might have been produced.

The pattern has a central set of three dots in the northeast direction. From the upper dot, there is a left-shifted V shape. From the middle dot, there is a slanted upside-down W shape. The B is on the far left of

---

5. Actually, I initially did believe in a feature-space representation and was very skeptical of the syntactic approach taken by the aforementioned computer scientists. The initial experiment was designed to see if recognition memory was equivalent in recognition speed after participants learned the patterns.

this part of the pattern. From the bottom dot, there is an upside-down Y shape with the dot at the center of the Y.

In a similar fashion, the pattern in figure 4.3 might have resulted in the following set of propositions during learning.

From the B, there is an upward nearly straight northeast orientation for four dots and the line goes straight down for three dots. The rest of the pattern looks like a baby stroller.

The first pattern description is roughly captured with a seventy-one-word narrative, whereas the second pattern description is encapsulated in only thirty-two words. The string language does covary with the propositional complexity and structural complexity of the patterns. In the Chechile et al. (1996) study, all the patterns consisted of twenty circles and nineteen connecting lines, and the patterns varied over four levels of syntactic complexity. The four levels corresponded to the number of branching *v* symbols needed in the string language description of the patterns (i.e., 0, 3, 6, or 9 *v* symbols used in the string description). In each of three experiments, the participants had to learn the pattern initially by means of a paired-associate training procedure where the pattern was the stimulus member of the pair. The participants had to demonstrate at least three consecutive perfect paired-associate trials for all the items prior to undertaking the speeded-recognition test. Each person had to demonstrate that he or she knew the patterns before the timed recognition test was begun. On the recognition test, half the patterns were old patterns that were used during the paired-associate training, and the other half were novel foils. The foils, like the targets, varied over four levels of syntactic complexity. Although the participants could correctly accept the old patterns as familiar and reject the novel foils, the speed to make these decisions varied as a function of syntactic complexity. See figure 4.4 for the mean response time for targets and foils as a function of syntactic complexity.

There was a highly reliable difference in the mean response time as a function of syntactic complexity for both old targets as well as for novel foils. Importantly, Chechile et al. (1996) reported that this robust effect was replicated in two other experiments, and the effect could not be accounted for by fourteen other properties. Among the properties explored were perceptual and learning factors such as the horizontal and vertical visual angle, local and global pixel density, fractal dimension, the Fourier-power spectrum, the number of angles, target-foil similarity, degree of pattern learning, pattern symmetry, and the specific set of stimuli. Although the patterns did not statistically vary on these physical properties, the patterns did vary in syntactic complexity and thereby in the number of propositions that were likely a part of the encoding of the learned patterns.

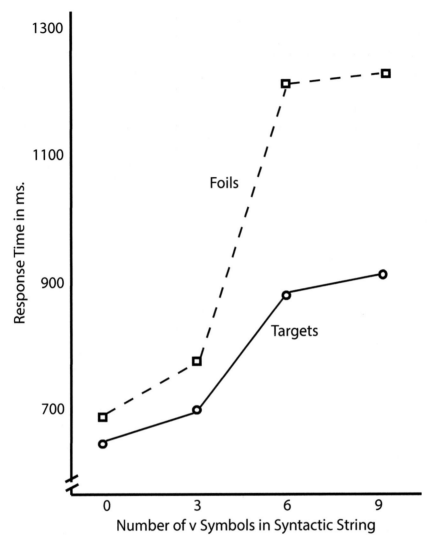

Figure 4.4
Mean recognition time as function of syntactic complexity for old items (solid line) and for foil patterns (dashed line). Data taken from experiment 1 of Chechile, Anderson, Krafcek, and Coley (1996).

The syntactic complexity effect has important implications for the format of the memory representation in general. If recognition familiarity were a function of only the bottom-up activation of the pattern features, then this complexity effect on response time should not have occurred. Clearly, the organizational structure of the pattern is important. If the structural properties of the pattern were computed subsequent to the familiarity decision, then again a syntactic complexity effect would not have occurred. The syntactic complexity effect implies that the pattern structure is not automatically activated but is computed upon the viewing of the pattern or a scene. If the structure is more elaborate, then there is a temporal cost associated with the computation of the configurative properties of the stimulus. If the analysis was done previously, then the participant would be able to reactivate the other stored properties of the stimulus. Note that the old pattern recognition decisions were faster than the corresponding foil pattern decisions. As with the indexing model illustrated in figure 3.5B and the temporal/feature-binding model illustrated in figure 4.1, the partial cue or incomplete analysis becomes sufficient to recovering the stored information about the event. Previously stored information is not forthcoming for the foils, but the participants cannot be sure that the pattern is novel until they complete the analysis. Thus, there is a complexity effect for the foil patterns. Chechile et al. (1996) further observed that the syntactic complexity effect for visual pattern recognition shares a dual-code framework with the Kosslyn (1994) model for imagery because for both cases, there is the use of propositional information as well as pictorial elements.

The parallel activation of the pattern features does not result in a recognition decision, so feature-space models for the memory representation, like the RBC model by Biederman (1986, 1987, 1990), are incorrect. The RBC model eventually evolved into the JIM model developed by Hummel and Biederman (1992). With this model, relational information is computed within a neural network context. But with the JIM model, pattern structure is automatically activated in a bottom-up fashion, so this model also does not predict a syntactic complexity effect.

The importance of syntactic/propositional relationships in pattern recognition is consistent with the finding of recognition difficulty with inverted faces (Thompson, 1980). The inverted face and the upright face have the same collection of features, but the inverted face is difficult to recognize. When the face is inverted, many of the propositional properties are incorrect. For example, the mouth is above the eyes. Recognition is less impaired by inversion for stimuli other than faces, although recognition for the upright figure is still better (Yin, 1969, 1970). Diamond and Carey (1986) showed that the weak benefit for upright to inverted dog pictures became a strong impairment for the recognition of inverted dog pictures if the person had higher levels of expertise about the topic. With increased expertise, there is in general more propositional information, so the inversion of the picture should produce a larger effect. In general, we are all experts about faces; hence, distorting the structural relationship within a face leads to problems of recognizing an inverted face that we normally detect easily when not inverted. Nonetheless, it is important to remember that Yin (1969, 1970) found that the recognition of other stimuli also showed an impairment when tested in an inverted orientation. Inverting any stimulus should disrupt some of the relationships among the components in the pattern despite the fact that all the features of the object are unchanged.[6]

4.5   How Much Is Remembered?

We are surrounded all the time by a sea of information, so it is a valuable exercise to estimate roughly the amount of information that we experience per second and to estimate how much of this information is remembered. Information is measured in bits where bits are the number of binary digits needed to represent a stimulus. For example, a single character of English text can be any one of twenty-six letters plus eight other symbols for either spacing or punctuation. The number 34 in binary is 100010, so a single text character can be expressed by a binary string of length six. Thus, there are six bits of information per character.[7]

Clearly, we are exposed to far more than a single character per second. To roughly estimate the magnitude of the information available at any arbitrary unit of time, consider the scene of a French village shown in figure 4.5. The photograph is from a JPEG file that has about five million bits of information. The retinal units activated by this scene are not the same as the units of information recorded by the digital camera. There are about 100 million primary photoreceptors in a single eye, but these are grouped into approximately 1.25 million retinal ganglion cells (Blake & Sekuler, 2006). A ganglion cell corresponds to a particular type of receptor field. But there are many different types of responses to a receptor field, so there is more than a single information bit for each ganglion cell. Consequently, an estimate of five million bits of information (the

6. The issue of the special status of facial stimuli has been questioned by Loftus, Oberg, and Dillon (2004). But the analysis by Loftus et al. dealt with the issue of the inversion effect being greater for facial rather than nonfacial stimuli. Moreover, these investigators still found a strong inversion effect for famous faces.

7. Alternatively, the information content of N items in a search set is b where $N = 2^b$. For $N = 34$, the value for b is computed as $b = \frac{\log 34}{\log 2} = 5.087$; hence, it requires more than fives bits to represent thirty-four items.

Figure 4.5
A scene from a French village.

same value as in the JPEG file) is actually a reasonable estimate for the visual information content. We can get an even more conservative estimate of the information value
for this scene by generating a verbal description. A verbal narrative greatly reduces
the total information value because it does not describe each small region. A verbal
narrative is an approximation to a perceptual encoding as opposed to a pure sensory
representation. The description below is a rough encoding of the scene.

*There are 7 parked cars on the right side of an empty cobblestoned street that is slightly uphill. These
cars are in an enlargement of the street for the purpose of parking. The street ahead narrows so that only
people and 2-wheeled vehicles can pass. The cars in the parking area are parked head-in toward a 15-foot
stonewall that is covered with green vegetation. It is late afternoon, and the near part of the wall is in
shadows. The colors of the cars respectively from nearest to farthest are light blue, blue, dark gray, dark
blue, light gray, blue, and dark blue. The 6th car down in this parking area is faced toward the street.
Each car is a 4-door sedan with a hatch back. An orange license plate is visible for 5 of the cars. On the
left side of the street there are three parallel-parked cars. The colors for these cars are respectively dark
blue, gray, and white. The first 2 cars are sedans, and the 3rd is a van. There is vegetation covering the
wall on the left side of the street. This wall is in the light, and the vegetation is a brighter green. At the
far end of the street there is a stone building on the left side of the road. This house borders the left side
of the narrow street. The building has 3 floors, and a street-side chimney. At the far end of the narrow*

passageway, on the opposite side of the intersecting street, there is a 3-story stone building. This building has white shutters showing on the 2nd and 3rd floors. On the right side of the narrow passageway, there is a wall around the yard of a 4-floor corner house that has its entrance on the intersecting cross street. The wall around the yard for this house connects in the back, at a right angle, to the wall alongside the enlarged parking area. The back wall is covered in vegetation. The 3rd floor of the corner house has two long rectangular windows and a third smaller rectangular window. On the 4th floor there are two small dormers. On the roof, there are 2 chimneys and 5 pink-red flue pipes. Above the roof of the corner house, the top of the town-center tower is showing. There are windows on the top two floors of the tower. The tower narrows to a peak. For 75% of the sky from the right, there is a solid but thin cloud covering. The solid covering ends with wispy white clouds. The sky on the left is a clear bright blue.

The above narrative contains approximately 14,000 bits of information. The narrative representation thus reduced the information content by a factor of 357. But a retinal input changes many times in a second because the eyes are moving in space, blinking, and undergoing saccadic movements. Consequently, the estimate of 14,000 bits per second is an extremely conservative estimate of the input rate, especially since all the other sensory modalities have been ignored. The question now arises as to the amount of information that is remembered later. This question has been convincingly addressed in a beautiful paper by Landauer (1986).

Based on a number of picture memory studies, Landauer was able to compute a per memory item information measure. To understand how he found this, let us consider (after some retention interval like 30 min) the number of memory items in storage, and let us denote this number as M item from a total list of n items. Landauer pointed out that the information content in memory (in bits) is b where $M = 2^b$. He next argued that the rate of correctly rejecting a novel foil is equal to $(1 - \frac{1}{2^b})^n$. Data on the correct rejection rate can thus be used to estimate the value for b. One particularly valuable database is the memory study by Standing (1973), where he examined picture memory for lists of different lengths. Standing varied the value of n across lists from 20 to 10,000. Regardless of the size of the memory list, Landauer computed a relatively stable b value of about 11 bits. The retention intervals also varied in the different experiments. The retention interval was as short as 30 min and as long as 48 h, but the corresponding estimates for b with retention interval only varied from 13 to 10 bits. Using data from the Nickerson (1968) picture-memory study, which tested memory after a 24-h delay, Landauer found a similar value for b of 12.9 bits. Landauer also reported similar b values for studies where the target items were sentences, nonsense syllables, musical segments, auditory words, and visual words. To obtain the final per second memory information rate, Landauer divided the value of b in each study by the time each stimulus was presented during learning. The resulting memory information rate was only 2 bits per second! Given the conservative value of 14,000 bits per second for the input rate of information and the memory retention rate of only 2 bits per second, we can thus conclude that far more than 99.986 percent of the information content that we perceive is forgotten. Although our information cup is only less than 0.014 percent full, we can still remember from the previous day—about the equivalence of about thirteen single-spaced pages of text. It is interesting to observe that *Ulysses* by James Joyce is a narrative account of one day of the life of a fictional character—Leopold Bloom. Joyce's book is more than 750 pages long. If Bloom were a real person, and if he tried to recall the events from the prior day, then his own account would likely be considerably shorter.

Landauer's remarkable conclusion about the number of bits of information that can be added to the memory system is based on long-term memory studies and on a particular model that links information content to memory performance for correctly rejecting lures or foils. But these two features of the estimation method are not obvious, so additional analyses are needed to estimate the information content by an alternative procedure. Seemingly, the most straightforward method to ascertain the throughput capacity of the memory system is by means of an immediate test of memory span. For studies of the memory span, the participant must repeat back immediately, in the correct order, a stimulus string. This task has been extensively studied for a wide variety of stimuli (e.g., Crannell & Parrish, 1957). Memory span is defined as the stimulus string length where participants are at a 50 percent accuracy rate. It is well known that human memory span for letters is about six. Although it can be difficult to convert some stimuli to the number of information bits, letters are easily computed.[8] Memory span studies for letters are constrained so that the same letter does not repeat in consecutive positions; thus, the information for a six-letter string is $\frac{\log 26 \cdot 25^5}{\log 2} = 27.92$ bits. But on 50 percent of the test trials, people cannot correctly reproduce an input sequence of six letters. Shorter strings of four letters result in almost perfect performance, but the accuracy rate begins to decrease for strings of five or more letters (Crannell & Parrish, 1957). The near-perfect performance with four-letter strings, but the reduced accuracy for longer

---

8. Crannell and Parrish (1957) showed that the memory span for familiar three-letter words was less than four. Warren (2015) showed that the memory span for familiar shapes was also four, but if the shapes were unfamiliar pictograms that could be drawn easily in one or two fluid line strokes, then the memory span was only one. Consequently, complex pictures that are equivalent of many line strokes contain more information than can be recalled immediately (i.e., they are beyond the memory span). The Warren study showed that the human memory span is exceeded by a single novel figure made with three or more strokes. Thus, the complex visual scene shown earlier with 14,000 bits of information is clearly beyond the human memory span capacity. The Warren study also showed that having existing information in memory about shapes increases the memory span.

strings, indicates that four-letter strings are at the capacity limit for the acquisition of new information. The corresponding estimate of the immediate memory capacity is $\frac{\log 26 \cdot 25^3}{\log 2} = 18.6$ bits. This calculation is based on the number of four-letter strings where a letter is not repeated in consecutive positions. The resulting initial memory encoding capacity is far more than two bits of information, but it is also many orders of magnitude less than the full informational amount present in the environment. Moreover, this initial amount of information is fragile and undergoes a rapid loss. The best method for studying the rapid change in stored information is by means of the Brown-Peterson experimental procedure where typically the memory target for any given test trial is three nonrepeating consonants. The memory retention function (in terms of bits of information) is displayed for several Brown-Peterson-style experiments in figure 4.6A. For each experiment, the initial information is multiplied by the recognition hit rate. There is a clear decline in the stored information over the first 30 sec after target presentation. The memory changes that result from a list-learning experiment for longer retention intervals is shown in figure 4.6B. Note that after a 24-h delay, the memory information rate is 1.87 bits, which is remarkably close to the rate reported by Landauer (1986).

In the above analyses, it is important to note that only some components of the total information environment are in the attentional spotlight (i.e., attention narrows the information from more than 14,000 bits down to approximately 19 bits of information). The selected features for further processing are not random, but they are likely to be goal objects, dangerous stimuli, or other high-priority stimuli. The greater attention given to the selected features comes at the cost of not processing the nonselective features in enough detail so that they can be encoded into memory. An example of the cost of narrowing the attentional spotlight is the *weapon focus* phenomenon where an eyewitness attends to a dangerous weapon at the cost of noticing the identity of the criminal (Cutler, Penrod, & Martens, 1987; Kramer, Buckhout, & Eugenio, 1989; Loftus, Loftus, & Messo, 1987). There is also evidence from eye movement studies that people fixate for a longer time on unusual or highly important stimuli (Antes, 1974; Loftus & Mackworth, 1978). Yet even when the observers are highly motivated to attend to the stimulus array, they can be distracted and miss important events; these failures of attention are the stock and trade of magicians and pickpockets. Macknik et al. (2008) discuss a wide range of stage tricks that can divert the attention of viewers. For example, curved hand gestures are more likely to draw attention than a straight-line hand gesture. The face and vocal message of the magician can in itself be part of the distraction. Magicians are far more expert in the art of distraction than cognitive scientists because they have tried and practiced behaviors for centuries in an effort to see what works to

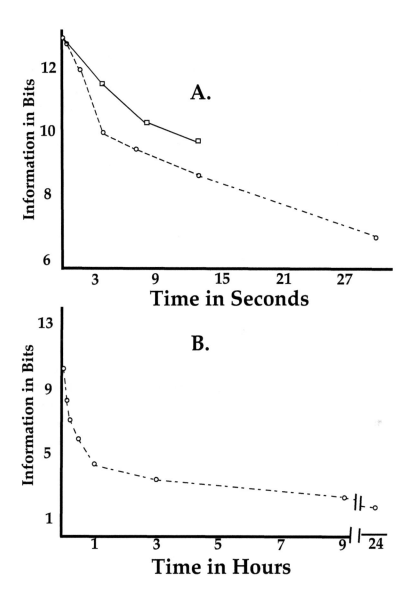

Figure 4.6
Information loss as a function of time is shown in panel A over a short-term time scale. The circles were computed based on the Sloboda (2012, Exp. 4) study, whereas the squares were based on the Chechile (1973) experiment. In panel B, the information loss is shown over a longer time scale. The information values were computed from the Sloboda (2012, Exp. 3) study.

conceal their slight-of-hand tricks. Their insights can be useful clues for possible new discoveries about attentional processes. For the purpose of this book, it is enough to recognize that most of the total information available is never processed long enough or complete enough to create a memory in the first place. If features of the stimulus array can be switched without detection, then the information was not recorded in the memory system (i.e., it was filtered out by means of attentional processes).

Some researchers continue to be impressed by the excellent performance of their participants and mistakenly argue that the memory capacity for learning visual patterns is vast. For example, Brady, Konkle, Alvarez, and Oliva (2008) had participants view 2,500 visual stimuli of objects from familiar conceptual categories for 3 sec each. The researchers reported that after a 10-min memory lag, the participants could correctly detect the items presented earlier at an 86.4 percent rate on a two-alternative forced-choice test. Brady et al. further argued that previously reported memory capacity values vastly underestimated the ability of people to learn visual forms. But there was no correction for random guessing in the Brady et al. study. Making the necessary guessing correction results in the finding that 72.86 percent of the 2,500 pictures were stored well enough for participants to correctly pick the memory target over a foil. This performance means that the information storage value is $\frac{\log(.7286 \cdot 2500)}{\log 2} = 10.83$ bits. This value is slightly larger than the values shown in figure 4.6, but remember that the visual stimuli in the Brady et al. experiment were familiar objects as opposed to nonsense syllables. Moreover, Landauer computed the memory information rate on a per second basis, so the value of 10.83 bits translates to a rate of 3.62 bits per second. This value is within the range found in other experiments, so the Brady et al. experiment really does not require a reconceptualization of the memory-encoding capacity as those authors suggested.

4.6 Memory Is Encoded Information

From an information theory perspective, *all memory is encoded information*. Also recall from chapter 1 that the following was stated:

Reflexes, as well as more complex behavior patterns called *species-specific behaviors*, are actually built-in memories that are coded into the genetic structure of the organism. In essence, the organism's DNA is a genetic memory that reflects the evolutionary history of the species—a memory that is available from the beginning of life for the organism. But acquired memory is the stored representation that reflects some learning about the environment, and this is the type of memory that is the focus in this book.

In light of information theory, it is interesting to ask how much information is contained in genetic memory (i.e., the DNA code for a person). It is well known that

DNA consists of a double helix of bonded base pairs where the pairs consist of either cytosine (C) with guanine (G) or thymine (T) with adenine (A) (Klug & Cummings, 1991). Thus, for each base-pair site along the DNA molecule, there are four possible pairings: (1) C-G, (2) G-C, (3) T-A, and (4) A-T. Given that there are about 3.2 billion base-pair sites for humans, it follows that the total number of combinations is equal to $4^{3.2 \times 10^9} = 2^{6.4 \times 10^9}$. Thus, the number of bits of information in the DNA molecule I_g for a person is about 6.4 billion bits. This stored code is the genetic memory for the species and for the person.

It is also interesting to ask how much information is possibly acquired over a lifetime of about 70 years. If we assume from the previous section that only about 2 bits per second can be remembered after a 24-h duration, then the total acquired information I_a would be computed as

$$I_a = 70 \, yrs. \times 365 \, days/yr. \times 16 \, waking \, hrs./day \times 3,600s./hr. \times 2 \, bits/s.$$

$$\approx 2.94 \times 10^9 \, bits.$$

Thus, the total amount of acquired memory in 70 years is estimated to be approximately 2.94 billion bits of information; however, this value is likely to be too high (1) because of additional memory loss with age and (2) because of information redundancy. Although the information content of genetic memory is greater than the acquired-memory information, it is nonetheless impressive to see that acquired memory is still a vast store of knowledge. Moreover, from the previous section, the total information that was briefly experienced but not remembered was more than 7,000 times greater because the memory retention cup is less than 0.014 percent full, as was noted in the previous section.

4.7 Summary of Chapter

The information-processing framework advanced in this chapter has stressed the role of propositions for the encoding of events. Propositions have a structure similar to a simple declarative sentence, that is, they have (1) a subject, (2) a relationship or action, and (3) a direct object. Thus, propositions have constituent parts that can degrade to fragments. In terms of visual-pattern recognition, a bottom-up feature-detection model based on the activation of a point in a feature space was rejected as unsuitable. Instead, a case is made for a syntactic pattern-recognition model. The analysis in this chapter also demonstrates that there is utility in (1) recording temporarily a detailed representation of the environment and (2) pruning down the total informational content to only the most salient information. Without a detailed representation of the external world, the organism is vulnerable to miss the detection of a dangerous predator or other hazards or miss the opportunities of obtaining a prized goal object. Important stimuli require rapid detection in order to implement a speedy response. But these important stimuli are a small subset of the total informational environment, and they may not be present at all. The memory system does not need to record, for the long term, all the unimportant information. If the memory system did include all the information perceptually detected, then retrieving the appropriate important information later would be slow and cumbersome. Thus, having a vigorous information-pruning process results in the storage of the most important events, which in turn increases effectiveness of memory retrieval. As noted earlier in discussing the complementary memory systems of standard consolidation theory, there is an increased chance of survival by having two complementary memory systems with different time scales for learning and forgetting. Similarly, it also increases the chance of survival to have a perceptual system capable of representing a vast amount of information from the environment along with a complementary memory system that only stores the most salient and important information for later use. The memory system itself affects perceptual processes. Because of our prior learned experiences, the information in the raw information field is quickly interpreted, and memory also influences what is selected for greater attentional focus. Attentional processes are guided by our goals and prior learning (Wolfe, Cave, & Franzel, 1989). Yet what is in the current stimulus array might be remembered for some later time. Thus, the perceptual and the memory systems are interrelated moment to moment by a bottom-up and top-down dance. The memory system is driven by sensory input, but the memory system guides attentional processes and helps in the interpretation of the sensory input. Finally, it was stressed in this chapter that memory in general is encoded information.

5 Memory: Myths and Truths

5.1 Chapter Overview

Numerous misconceptions about memory have emerged from a variety of sources. These misconceptions in the most part were initially creditable hypotheses that were shown later to be incorrect. But in some cases, these flawed notions are still believed. If scholars fall prey to these mistaken concepts, then the research literature is unnecessarily confused. But if the public-at-large believes in some of these myths, then there can be dangerous personal consequences to some unfortunate individuals. In this chapter, many of the memory myths are identified and debunked to prepare for the work ahead. Also in this chapter, there is a discussion of content-addressable memory.

5.2 A Few Misconceptions about Memory

Although the focus of part I is about the building of a memory representation, this is a good place, nonetheless, to address some major misconceptions about memory. In most cases, the misconceptions or myths are easily refuted by the facts already developed in the previous chapters.

5.2.1 Myth of Static Memory

Simon and Chabris (2011) conducted a telephone survey assessing people's beliefs about memory. Their sample included 1,490 people living in the United States, and they also asked the same questions to a group of eighty-nine professionals working in the area of memory and experimental psychology. One question dealt with the stability of memory. People were asked: if a memory is formed for an event, then does it remain constant? A total of 47.6 percent of the general public believed memories are static, whereas *none* of the memory professionals agreed with that statement. There is good reason why the professionals disagree with the idea that a memory remains stable. There has been a long history of studies in psychology that documents the changing nature of memory. Without retracing all the research on this point, it is sufficient to simply mention the work of Bartlett (1932). For example, in his famous "War of the Ghosts" experiment, Bartlett investigated the memory of his participants after their reading and studying of a Canadian-Indian folk story. Without rereading the story, the participants had to repeatedly describe the events in the story. The story originated from a culture foreign to that of his English research participants. The research procedure was a *study, test, test, ..., test* experimental design. On successive retelling of the story, the participants shifted in their description of the story. There was a steady drift in the language away from the North American Indian terminology and closer toward their native English manner of speech and expression. Bartlett introduced the concept that memory is *reconstructed* by the participants to reflect their culture and reconstructed to incorporate posttraining suggestions. This study, as well as many other experiments, points out the fact that the memory for an event is not stable; it can also change according to postencoding experiences. Later in chapter 11, evidence is provided that the testing of memory is in itself a learning event. Consequently, if the person is mistaken on one occasion, then that mistake can become learned (i.e., the person becomes well practiced in making his or her mistake). Importantly, this phenomenon also shows us that the memory of an event is not static.

5.2.2   Myth of the Recovery of Lost Information via Hypnosis

Simon and Chabris (2011) also exposed another memory myth; this misconception was about the effectiveness of hypnosis for retrieving forgotten information of an event such as a crime. A total of 54.6 percent of the general public thought hypnosis was an effective memory retrieval method—more so than ordinary cuing. Again, *none* of the professional memory experts agreed with that statement. Yet it should be pointed out that in clinical psychology, there has been a longstanding acceptance of the idea that hypnosis can augment lost memories, and this belief within clinical psychology has led police to use hypnosis in an effort to recover "lost" or "repressed" memories of traumatic events (Kihlstrom, 1997). The origin of this idea goes back to Freud, who used hypnosis on the famous case of Anna O, whose real name was Bertha Pappen-heim (Breuer & Freud, 1895). This case is the original case for psychoanalysis, and it featured the use of hypnosis as a method of treatment and memory retrieval. The case involved a woman, who under hypnosis revealed a repressed resentment of the illness of her deceased father (Freud, 1910). The belief in repressed and recovered memories is a myth, but it is an idea that is difficult to subdue because of the deep faith of the devoted followers of Sigmund Freud in the psychoanalytic program. Nonetheless, this misconception is flawed on at least three separate grounds. One, the idea of repressed memories for traumatic experiences is itself a myth. Two, the idea that hypnosis can recover a "lost" or "repressed" memory is the central misconception addressed here in this section. Third, it is yet another myth that a memory from a hypnosis session is accurate and clinically therapeutic. Let us debunk these myths in order.

In regard to the belief in repressed memory of traumatic experiences, it is now clear from the study of patients who are known to have experienced trauma that their horrific experiences are difficult to forget (McNally, 2003, 2005). Rather than repressing these memories, the patient more typically dwells on them. Traumatic memories can lead to posttraumatic stress disorder as opposed to repression. For both animals and humans, the learning of an emotional event (particularly negative emotion) is quicker and stronger than the learning of emotionally neutral information (Cahill & McGaugh, 1998; Labar & Cabeza, 2006). Animal studies have demonstrated that higher negative emotionality enhances the memory retention even after a lengthy delay (McGaugh, 2004). Rather than repressing the memory of a negative event, the opposite seems to be the case; negative experiences are better remembered. Usually after a traumatic experience, there is a persistent reviewing of the event. The effect of this internal reviewing usually results in the event being well learned. Consequently, it is likely that the memory of a traumatic event is not forgotten, although the person might choose to not discuss the trauma for a long time (McNally, 2005). Thus, there is no support for the hypothesis of Freud that traumatic experiences are repressed from consciousness. Consequently, if memories are not repressed from consciousness (as assumed by Freud), then there is no reason to believe that the ideas that emerge from a hypnotic session are the recovery of true experiences.

In regard to the Freudian idea that hypnosis can unlock forgotten or repressed memories, it is now clear that hypnosis does not produce this type of an effect (Kihlstrom, 1997). In general, hypnosis is a condition of increased suggestibility, and under this condition, the person can exhibit a wide range of alternations in his or her normal behavior and beliefs (Hilgard, 1975; Kihlstrom, 1985; Kirsch & Lynn, 1995). But hypnosis does not increase the ability of a person to remember events that otherwise are not accessible. In fact, hypnosis results in distortions of memory. For example, Dywan (1988) showed that under hypnosis, participants produced more recognition false alarms without an increase in the hit rate. Hypnotized people might be more confident in their memory performance, but their actual accuracy is not enhanced

(Whitehouse et al., 1988).[1] To make matters substantially worse, under the influence of a hypnotist, people are very vulnerable to misinformation and deceptive postevent suggestions (Putnam, 1979). Constructed false memories created under hypnosis also result in confident "memory," even when the person is no longer in a hypnotic state (Laurence & Perry, 1983). Due to this type of distortion effect on memory, the U.S. courts in general do not accept testimony that resulted from hypnosis.[2] Consequently, it is clear that the Freudian notion that hypnosis can unlock repressed or forgotten information is false. Rather than unlocking a repressed memory, it is more probable under hypnosis that a confident false memory has been produced.

Finally, there is no evidence that a false belief that was generated under the influence of psychotherapy is nonetheless beneficial to the client. The belief that the client has a repressed memory of a traumatic event (or events) and the use of hypnosis by the therapist is a particularly dangerous combination. Suppose for example that a therapist believes that a high percentage of people who exhibit an eating disorder were sexually abused as children. Moreover, suppose that a patient, who is very thin, comes to the therapist complaining that she is unhappy with her life and her job. If this patient were hypnotized by the therapist, then it would not be surprising that the patient comes to "recover" a memory of early childhood sexual abuse that fits the therapist's preconception. It is also very possible that the patient's real source of unhappiness was not

1. In part II of this book, cognitive-psychometric methods are introduced that provide separate parameter estimates for memory storage and memory retrieval. As a developer of a number of these models, I was interested to see if participants under hypnosis could remember more than what the model estimated as the proportion stored. This question was assessed in an experiment. The purpose of an unpublished study was to check on the validity of the psychometric method. For example, if the psychometric model indicated that only 50 percent of the items were stored in memory, and hypnosis recovered all of the items, then the model must be in error. To test the research question, people first learned the information in a normal alert state, and they were then tested with the psychometric procedure to enable storage and retrieval estimates. Next the participants were hypnotized and given another memory test of the items. It was determined that the rate of memory recovery with hypnosis was actually lower than the estimated proportion of stored items. Thus, the rate of memory recovery with hypnosis was lower than the rate obtained with standard recognition-testing procedures.
2. In 1987, the Supreme Court of the United States ruled that it is acceptable to include the testimony from a person who had undergone hypnosis, provided that the person was the defendant. Essentially, the defendant's legal right to testify cannot be set aside because the defendant had undergone hypnosis. Most states prohibit testimony from someone who has undergone hypnosis. The Supreme Court did not state that hypnosis was acceptable in general; their ruling only applied to the defendant. In essence, for everyone except the defendant, the testimony about information recovered from hypnosis is not acceptable in most courts.

discovered or addressed. Instead, the patient is left with the belief that she is a victim. To make matters worse, this belief in victimization can result in injury to others. If an innocent third party is falsely accused of a crime, then someone else has a serious problem while the client is still left with an unresolved psychological issue. But even if others are not harmed by false accusations, the false belief that emerged from the therapeutic interaction can still leave the client with a lasting bizarre understanding of the real world. For example, in 1997, PBS aired a *NOVA* television documentary about a group of people who have come to believe that they were abducted and abused by aliens (April, 1997 *NOVA*). In the documentary, the viewers see how, in a hypnosis session, individuals come to "remember" being abducted and sexually abused by aliens from another planet. The hypnotist is a believer in alien abductions, and he regularly induces the false "memories" of alien abductions on the part of his "clients." Consequently, the Freudian belief that the unlocking of a repressed memory is therapeutic is in itself a flawed notion because (1) there is no repressed memory in the first place, (2) hypnosis cannot recover forgotten information, and (3) the real source of the patient's psychological problem has not been addressed. Yet the psychoanalytic program of unlocking supposedly repressed memories has been deeply promoted by devoted followers who continue to advance the ideas of Freud. Freud has been one of the most influential thinkers of the twentieth century, but he also was not correct in his model of memory.

5.2.3 The Myth of Videotape Memory

Simon and Chabris (2011) report that 54.0 percent of the general public mistakenly believe that memory is like a videotape—accurately recording images and sounds for later retrieval. As with the other myths, *none* of the polled experts agreed with the videotape model for memory. We have already seen that only a small fraction of the total perceptual information content is encoded in memory, and an even smaller proportion remains in memory after a retention interval of a day. A case was also developed that propositional relationships are produced to encode novel visual information. Videotapes do not have propositions embedded within its structure. Although the videotape model does not receive any support among memory scholars, there is a possible caveat that needs to be addressed—*eidetic imagery.*

Allport (1924, 1928) reviewed an extensive research literature (done mostly in Germany) about people with a photographic-like memory. These individuals with vivid images were mostly children, and the images were called eidetic images. The study of eidetic imagery faded from attention until the 1960s, when this research area attracted the interest of Ralph Haber and his associates (Haber, 1979). Haber and Haber (1964)

delineated seven criteria designed to differentiate eidetic imagery from other perceptual phenomena. Their seven criteria were (1) the person had to scan his or her eyes on
the original stimulus, (2) the image produced must be reported as being seen, (3) the
image must be located in the stimulus scanning plane, (4) the duration for the image
needed to be substantially longer than a sensory afterimage, (5) the person had to be
able to scan his or her eyes over the image to inspect it, (6) the recall accuracy had
to be higher than expected by the normal memory recall rate, and (7) the image had to
be described in the present tense until it faded from view. The incidents of individuals
labeled as eidetikers was low (i.e., about 5 percent of preadolescent children and less
than 0.1 percent of adults). Although people identified as eidetikers were mainly children, there also are claims that eidetic imagery returned again in old age; for example,
it has been reported that 25 percent of 90-year-olds have eidetic imagery (Giray, Altkin,
Rodin, Yoon, & Flagg, 1978). But other researchers have stated that these incident rates
are probably high because many of the people labeled as an eidetiker did not satisfy all
the criteria that were listed above (Haber, 1979; Leask, Haber, & Haber, 1969). In addition, other researchers found that many eidetikers have fragmentary images, and as a
result, their accuracy is not better than normal memory (Gray & Gummerman, 1975).

It is not convincing that eidetic imagery is qualitatively different (beyond the vividness of the image) from ordinary images that people routinely generate. People can
usually produce an image when given a set of instructions in the form of propositions.
Imagine a cartoon elephant in a polka-dotted dress, holding a pink umbrella and wearing red shoes. Most individuals can generate an image (to some varying quality) when
given these instructions. To investigate properties of a standardized image, a technique
has been developed where the participant first views a scene and is asked to learn the
scene for a later imagery study (Kosslyn, 1976). Later the participant is asked to produce
an image of the scene that he or she first saw. People are able to do this task (Kosslyn,
1976). Many properties of the process of generating images have been explored as well
as properties of the images themselves (Kosslyn, 1976, 1994). In the Kosslyn model,
both continuous graphical information and discrete propositional information are a
part of the cognitive foundation for an image.[3] This theoretical position is similar to the

3. A vigorous debate occurred between psychologist Steven Kosslyn and philosopher Zenon
Pylyshyn over the nature of imagery. Pylyshyn (1973, 1979, 1981, 2003) argued for strictly a
propositional basis for an image. He argued that the picture-like quality of an image was only
a by-product or epiphenomenon; at the core the memory foundation for an image was purely
propositions. What was lost to many outside of this debate was that both parties accepted the idea
that propositions were involved with images, but Kosslyn (1976, 1994) argued that it was not only
propositions. Images also possessed properties of size, distance, and color—properties that a set

model that was described previously in the section in chapter 4 about syntactic-pattern recognition. For both the case of visual-pattern memory and the case of imagery, the memory representation has both propositional and pictorial features. Moreover, the standard testing protocols for eidetic imagers (which is described in detail by Leask et al., 1969) is roughly similar to the method used by Kosslyn for studying imagery based on an experimenter-provided picture. Some children and a few adults seem to have more vivid images than the rest of the population, but apart from this difference, they appear to be producing images in a standard way. Unlike afterimages, both normal images and the images produced by an eidetic imager are in positive colors as opposed to complementary colors. Both image types are produced in a fragmentary fashion as opposed to the whole image being generated immediately.

Although a case was just made that eidetic imagery shares common features with standard imagery, there are, nonetheless, some remarkable cases that warrant attention. None more remarkable than the case described by Stromeyer and Psotka (1970). These authors studied a woman who was a Harvard teacher and who had excellent artistic skills. The woman was 23 years old at the time of their study. She was able to create an image at will and combine it with ongoing perception. For example, she could see a picture of a tree without leaves and hallucinate leaves for the tree or she could see a beardless man and hallucinate a beard. She could encode a display of a poem in a language that she did not know, and later she could generate an image of the display and use the image to rewrite the poem. Most remarkable, however, is the experiment that Stromeyer and Psotka (1970) did using random-dot patterns like the ones employed with random-dot stereograms. Julesz (1971) extensively studied the creation of three-dimensional figures that emerged from the viewing of two nearly identical displays of random dots. Each of the two figures is a meaningless display of random dots; however, if one display is shown to the left eye and the other display is shown to the right eye, then a three-dimensional figure emerges from the combination of the two random-dot displays. The object is an illusion created from the integration of nearly identical random-dot displays. In the Stromeyer and Psotka (1970) experiment, each display

---

of propositions could not possess. Propositions have information content, and propositions can have a truth value, but propositions do not have relative size, distance, or color. Moreover, these perceptual properties of an image corresponded to brain activation in the cortical regions that are also used for the processing of visual information (Kosslyn, 1994). The evidence that Kosslyn has brought to bear to support his thesis that images involve both propositional and depictive components is convincing. Although the relentless opposition of an extreme skeptic has improved the research in this area, the position that images are purely propositional is an untenable hypothesis. The memory for visual patterns requires both propositions and graphical/pictorial components.

was a 100-by-100 array of elements that were either black or white. The woman in the Stromeyer and Psotka (1970) experiment viewed one of the two random-dot displays to encode the random pattern. She studied the pattern with only her right eye for a total of 12 min. The next day, she was asked to image the random display that she previously encoded, and she was asked to localize the image on a particular spot. The researchers then presented the other random-dot display to her left eye in the same region as the image. The woman was easily able to detect correctly the object in depth. The two displays of the Julesz stereogram were never shown at the same time; nonetheless, the woman was able to produce the figure with the combination of a perceived display and a generated image.

The woman investigated by Stromeyer and Psotka (1970) had a remarkable skill for generating accurate and vivid images. She could also remember these images. Vivid images that are distinctive are events that are most easily remembered. The woman in the Stromeyer and Psotka study also could use her imagery talents as an effective tool for remembering information. Imagery in general can improve memory retention, especially if the image is linked to other information that also needs to be retained (Marschark & Surian, 1989). Imagery is potentially effective as a tool for increasing memory retention because the image promotes the distinctiveness of the item and reduces interitem interference. Solomon Shereshevsky, the famous Russian mnemonist who was documented by Luria (1968), was extremely skillful in creating images and associations. He too would qualify as having eidetic imagery. Despite the extraordinary memory skills of these few skilled mnemonists, there is no reason to believe that their imagery and memory retention are but quantitative differences in the skill level exhibited by the larger group of normal individuals.[4] Some researchers in the area of eidetic imagery have argued that the phenomenon is qualitatively different, and thus it warrants having a different name and its own theory (Haber, 1979). However, Gray and Gummerman (1975) argue that eidetic imagery is different only in a matter of degree rather than being an entirely different process. This view of eidetic imagery avoids the

4. Solomon Shereshevsky was also a synesthete, for example, he pictured sounds (Luria, 1968). Synesthesia is a qualitative difference in perception where the usual inhibition between the sensory channels is breached. For example, a synesthete might experience the number 7 as being purple or the sound of C sharp as being a sensation of yellow along with the sound. Shereshevsky was described as a five-way synesthete; each of the five primary senses also elicited an additional sensation, for example, music notes produced a color, touch sensations elicited a taste, and so on (Luria, 1968). Synesthesia is rare, but Shereshevsky might have been virtually unique. This condition is clearly qualitatively different from normal perception. See the review by Grossenbacher and Lovelace (2001) for a discussion of the possible neural basis for this condition. However, unlike synesthesia, eidetic imagery seems to be a matter of the degree of the vividness of the image.

mystery of a separate skill or ability that only a few people possess. Nonetheless, there is still an important question that requires an explanation as to why some individuals are so skillful in producing images, whereas others are more pedestrian in their imagery. With both standard imagery and eidetic imagery, the image is not purely pictorial but has important propositional information. Images in general are not videotaped scenes. Much of memory does not have a pictorial component whatsoever, and when the memory is of a visual scene, then the memory representation is fragmentary and linked with propositions. Videotapes do not have embedded propositions.

Another argument against the videotape model of memory comes from some philosophers who have argued in general that there are no pictures in the mind whatsoever (Pylyshyn, 2003; Ryle, 1960). Perhaps the best illustration of their point is with the perception of color. Visible light is within a narrow wavelength band, which is approximately between 400 and 700 nanometers (Purves & Beau Lotto, 2003). Color is a qualitative sensation correlated to the wavelength of light, but it is entirely a brain construction. *In the physical world, there is no such thing as blue, orange, or yellow.* There is only electromagnetic radiation of varying wavelengths. Yet humans with normal vision experience the sensation of seeing colors. The experiential phenomenon is a product of how the eye and brain process light.[5] Thus, the perceptual experience is different from the external world. But the perception of color is nonetheless a real neural-cognitive event, so it does not mean that there can never be depictive information in memory as claimed by Pylyshyn (2003). Moreover, philosopher Michael Tye (1988) has vigorously refuted the philosophical objections raised by Pylyshyn. More important, psychologist Steven Kosslyn (1994) has demonstrated that mental images share common neurological activation similar to actual visual perception, and the images exhibit properties that cannot be represented by propositions alone. But propositions and stored graphic primitives in memory can be used to reconstruct a perceptual-like experience when creating an image. Nonetheless, the constructed image might not have anything to do with an actual experience. To illustrate this point, consider again the example of imaging a cartoon elephant in a polka-dotted dress, holding a pink umbrella and wearing red shoes. The memory associated with this image was generated by propositional statements; the image was not a stimulus in the physical world. Thus, taken together,

---

5. There are three different types of cone sensory cells in the human retina (Blake & Sekuler, 2006). These cells have different detection sensitivity in terms of the wavelength of the light - short, medium, and long. As information is neurologically processed from the retinal level to the occipital cortex, there is further organization of the pattern of wavelength information. Ultimately at the cortical level, the neural firing results in the sensation of color (Blake & Sekuler, 2006). Color is merely the brain's way of representing wavelength information.

the common belief that memory is an accurate videotape, like a security camera that is recording actual visual events, is a false notion on many grounds; it is simply a myth.

5.2.4 The Penfield Hypothesis of Recovered Memories

Penfield, along with his associates, studied the effect of stimulating cortical cells (Penfield, 1958; Penfield & Jasper, 1954; Penfield & Perot, 1963). There are no pain receptors in the brain, so it is possible and even advantageous to keep the patient conscious during neurosurgery. Of course, a local anesthetic is required to prepare the brain for surgery. Penfield is reported to have performed neurosurgery on 1,132 patients who had epilepsy (Squire, 1987). The surgical treatment involved the removal of the site responsible for the epileptic seizure. But prior to performing surgery, Penfield electrically stimulated cortical cells to ascertain the function of those cells. For some of the cortical sites, the electrical stimulation elicited the perception of a meaningful event. Penfield argued that the probe recovered a memory. Because in some cases the event was not previously remembered, the notion that lost memories were still resident in the brain was a popular hypothesis advanced by Penfield. However, Squire (1987) has extensively critiqued the Penfield hypothesis of recovered memories. Squire makes an excellent case that the Penfield hypothesis is yet another myth about memory.

Penfield found that a triggered sensation/experience only occurred occasionally when either the left or right temporal lobe was electrically stimulated (Squire, 1987). No other cortical region resulted in a meaningful sense of an event. Of a total of 520 patients who had temporal-lobe stimulation, only 40 patients claimed they experienced a meaningful event (i.e., 7.7 percent of the temporal-lobe probes). It is interesting how such an infrequent occurrence has resulted in such strong claims about hidden memories being unlocked. There seems to be a strong theoretical bias toward believing that memory can be recovered—a bias in favor of the idea that forgetting is only a retrieval problem. Importantly Squire pointed out that 24 of the 40 patients who reported a memory-like event were also known to hallucinate when having an epileptic seizure. The electrical stimulation, like a triggering event of a seizure, behaves as a nonnormal cortical firing of a set of cells. If the seizures activate hallucinations, then it is possible that these patients were merely hallucinating to the probes. Halgren, Walter, Cherlow, and Crandall (1978) also electrically stimulated temporal lobe sites; their probes were in the hippocampus and amygdala. These researchers also found about the same percentage as Penfield (i.e., 7.6 percent) of the patients reported a memory-like experience, but the experiences were characterized as hallucinations rather than as actual memories. Moreover, about 50 percent of the experiences occurred when there were also afterdischarges triggered by the electrical stimulation. With afterdischarges, there is more extensive activation of other brain regions. A similar finding was also reported by Gloor, Olivier, Quesney, Andermann, and Horowitz (1982). Moreover, the reported experience is not reliably linked to a stimulation site, that is, the same experience occurred from different sites as well as different experiences from the same site (Balwin, 1960; Penfield, 1958). Squire (1987) also pointed out that there were cases where the triggered experience remained after that site was surgically removed. Consequently, there is not an objective establishment that the reported experience is a veridical event that would not otherwise be remembered.

In short, the hypothesis of Penfield that the reports of his patients were real memories is unsubstantiated. The few probes that appear to trigger a memory seem to be rather hallucinations. Moreover, our previous discussion of neural activations during memory encoding stressed the fact that the neural representation of the memory is widely distributed. The Penfield hypothesis of a localized memory is inconsistent with the idea of distributed memory. The evidence is simply not there to support Penfield's localized memory hypothesis. His hypothesis is yet another memory myth.

5.2.5 Myth of Permanent Memory

There is a common theme for each of the myths discussed previously, namely, a belief that forgetting is strictly due to a retrieval failure. It is curious why this idea of a permanent memory is so appealing. Perhaps it is because we have all experienced cases of forgetting some piece of information but later were able to remember the previously forgotten information. Such occurrences are temporary retrieval lapses. But suppose we do not remember later; why do people still want to believe that the target information is still stored in the memory system?

It is easy to show that the storage of information is not perfect (i.e., the availability of the target representation is less than 100 percent, and the rate of nonstorage can change between experimental conditions). My research associates and I have employed a contingent four-alternative forced-choice testing procedure that provides a method for showing that not all is stored, although that objective was not our primary goal (Chechile, 1998; Chechile & Sloboda, 2014; Chechile, Sloboda, & Chamberland, 2012; Chechile & Soraci, 1999). Initially, the research participants learned some material for a later memory test. At the time of memory assessment, there is a series of forced-choice recognition tests. The assessment procedure begins with a four-alternative forced-choice test (4-AFC), that is, there is a line-up with four items and a single target that was on the learning list along with three novel foil items that were not on the list. If the participant correctly selects the target item, then testing for that piece of information is over and another test trial is initiated. But if the participant is incorrect on the 4-AFC test, then the incorrectly selected foil is removed from the list, feedback is given to the participant, and the participant is asked to do a three-alternative forced-choice (3-AFC) test (i.e., another lineup with the same target item along with the two foils that were not selected). If the participant correctly picks the target on the 3-AFC test, then testing for that item is over and a new test trial for a different target is started. But if the participant is again incorrect, then the selected foil is removed from the test list, and the participant is asked to do a two-alternative forced-choice (2-AFC) test, which consists of the target item and the one foil not selected on the prior two forced-choice tests. Finally, the participant is either correct or incorrect on the 2-AFC test. Thus, it is possible for a participant to exhibit his or her total lack of knowledge of the target item by always being incorrect on the contingent series of forced-choice tests. Even if the participants had a fragmentary knowledge of the target, then they should have eliminated at least one of the foils as being incorrect and thus avoid being wrong on the final 2-AFC test.

Experiment 2 from the Chechile et al. (2012) paper provides evidence that the nonstorage rate is greater than zero and varies between experimental conditions. Participants studied word lists as paired items; the first word was the cue word and the second word was the response. There were two types of learning conditions. For one learning condition, called the READ condition, the response item was displayed intact (e.g., LAMP - COFFEE). For the other learning condition, called the GENERATION condition, a single letter was missing from the respond item (e.g., LAMP - CO _ FEE). The missing letter is not a difficult problem for the participant to solve, but the consequence of this increased demand during learning results in better memory retention. The effect is called the generation effect (Roenker, Wenger, Thompson, & Watkins, 1978; Slamecka & Graf, 1978). The generation effect is a robust result that has attracted considerable interest as a method of improving memory retention (Hirschman & Bjork, 1988; Soraci et al., 1994, 1999). The effect applies not just to stimuli such as words but also to numerical operations (McNamara & Healy, 1995), sentences (Graf, 1982), and pictorial information (Carlin, Soraci, Dennis, Chechile, & Loiselle, 2001; Kinjo & Snodgrass, 2000; Wills, Soraci, Chechile, & Taylor, 2000). For experiment 2 in the Chechile et al. (2012) study, there were 11.38 percent of the words in the READ condition where the participants were always incorrect on the contingent forced-choice testing procedure. Any fragmentary knowledge would have helped the participant to avoid always being wrong. Thus, for this 11.38 percent of the word list, there is no evidence of target storage. This rate is also significantly different from the corresponding rate of

3.62 percent in the GENERATE condition.[6] Consequently, there is evidence that on some occasions, the targets are not stored whatsoever, and the rate of nonstorage can vary between experimental conditions. Clearly, storage as well as retrieval is subject to failure.[7] Thus, the hypothesis that all-forgetting-is-retrieval is incorrect.

Although the hypothesis that all-forgetting-is-retrieval is not valid, there is a modified form of the hypothesis that is widely accepted in several major memory theories. For example, storage losses can occur either in the Atkinson and Shiffrin (1968) model or in the standard form of consolidation theory. But for both theories, there is also a permanent long-term memory. For information in this state of permanent memory, all subsequent forgetting is assumed to be caused by a retrieval failure. Thus, a modified version of the hypothesis that all-forgetting-is-retrieval is to limit the claim to some of the items. Yet there are both theoretical and empirical psychometric reasons why this modified version of the all-forgetting-is-retrieval hypothesis is also mistaken.

At a theoretical level, there is no known biological mechanism to stop the adaptive changes in memory from continuing and thereby possibly destroying the prior learning record. Given the two types of plasticity mechanisms previously discussed (i.e., receptor trafficking within the dendritic spines and the structural changes to the axion terminals), there is no biological or chemical mechanism known to stop those processes from continuing to operate. If the learning environment changes, then the same adaptive mechanisms that created the memory in the first place will continue to rearrange the structure of memory and potentially destroy the former memory representation. Viewed from this perspective, memory storage is a dynamic process. Moreover, from a consolidation theory perspective, there is a theoretical problem caused by the reintroduction of additional training prior to a disruptive event like an electroconvulsive

---

6. A classical chi-squared assessment of the difference in frequencies between the two learning conditions is statistically significant, $\chi^2(1) = 59.8$, $p < 10^{-13}$.

7. Chechile and Soraci (1999) studied the generation effect using cognitive psychometric models that enabled estimates of both storage and retrieval. That study concluded that generative encoding resulted in an improvement in *both* storage and retrieval processes. But the role of retrieval was limited to conditions where the participants had to produce their response to the generation puzzle in the context of multiple cues (i.e., the participants had to consider several candidate solutions before determining the correct response). The Chechile et al. (2012) study reconfirms the role of storage changes with generative encoding conditions, but it did not require the consideration of a solution to the generation puzzle in a fashion where there were multiple possible solutions—a condition known to result in retrieval differences (Soraci et al., 1994). In general, generative encoding instructions are complex and can result in changes in both storage and retrieval processes.

shock (ECS). To explain the memory loss, researchers concluded either (1) that the further training weakened the memory or (2) that the disruptive event only caused a temporary retrieval change. In the case of the first option, it is a tacit acknowledgment that the memory is not permanent—all one needs to do to disrupt a memory is to provide practice. This option is not theoretically creditable. If the trace were truly hardened, then further training should only augment the memory representation (i.e., it should continue to be a highly tuned and practiced memory pattern). For the second theoretical option, there is also a problem. The second option assumes that a disruptive event only causes a retrieval change. However, the recovery from retrograde amnesia is generally not complete (Barbizet, 1970), so the retrieval-only hypothesis is problematic. For both options, the problem is caused by the assumption that the memory was hardened and not susceptible to further change. A way to resolve this theoretical conundrum of consolidation theory is to reject the idea that memory is permanent in the first place. For example, Chechile (2006) assumed that memory is always vulnerable to storage loss, but the rate of the loss is a dynamic property. Memory can be highly unlikely to change under some circumstances, but it never is totally protected from the possibility of a storage loss. From this viewpoint, further practice of old learning continues to improve the quality of the stored memories, but it also resets the temporal clock and puts the target information back in the time zone of high-storage interference as would be predicted from Ribot's law and Jost's law. As was discussed previously in the section on retrograde amnesia, destructive interference has its maximum effect on the most recent information in memory. Hence, the essential problem of consolidation theory stems from the assumption of a permanent memory. The other features of consolidation pertaining to an indexing theory and the multiple time scales for memory loss for the two complementary systems are not being challenged here. It is only the idea that memory becomes hardened into a consolidated trace that is the theoretical problem.

The Atkinson and Shiffrin (1968) model (or the modal model of memory) is another theory that supposes that memories are permanent if they are transferred to the long-term store. The general consensus from this perspective is that long-term memory is the information that has been retained after a delay of about 20 sec. That is, if a memory survives a 20-sec delay in which other events are occurring, then it must be in the long-term store.[8] It is possible to test this assumption of the modal model by using advanced

8. In addition to the Atkinson-Shiffrin model, the SAM and REM models discussed in chapter 3 assume that items transferred to the long-term store are permanent. This assumption is also made for the crystallized long-term memory component of the Baddeley (2000) working-memory model.

psychometric measurement techniques that enable estimates of storage and retrieval components of the recall rate. The Chechile and Ehrensbeck (1983) experiment is a particularly good test of the hypothesis about the permanence of long-term memory. In their experiment, participants learned a paired-associate list to differing levels of mastery. In the partial learning condition, the training on the list was at an 83 percent accuracy level, whereas in the overlearning condition, the participants reached a 100 percent mastery level plus they had two overlearning training trials. With the paired-associate training method, the participants were first presented with each pair for study; the whole list consisted of eighteen pairs for learning. Following the study phase of the trial, the participants had to do a rehearsal preventing digit-shadowing task of repeating presented digits. The shadowing period was 20 sec in duration. The delay between the learning phase and the test phase of each trial was designed to ensure that the performance was from only the long-term store. Although there was a minimum of 20 sec between the study and the testing of a pair, the actual average lag between study and test was 88 sec; the extra time was caused by the testing of other pairs. Hence, at the end of the training on the initial list, there is a level of performance established for the long-term store. Following the testing on the criterion test trial, the participants next had to learn a second competing list of eighteen different associates. The training on the second list used the same experimental protocols as in the first list. After six trials on the second list, which took about 18 min, the participants were tested again on the first-list associates. If the modal model were correct, then the only forgetting caused by the study of the second list should be a change in retrieval.

In the Chechile and Ehrensbeck (1983) experiment, the test phase for each trial consisted of a random intermixing of recall, old, and foil recognition tests. For example, if one of the eighteen pairs was VAJ-clothing, then a recall test probe would be "VAJ-recall," an old recognition test probe would be "VAJ-clothing," and a foil recognition test might be "VAJ-pilot" where the response "pilot" was a response item for another nonsense syllable on the list. All stimulus items were constant-vowel-constant nonsense syllables, whereas all the responses were words. All materials were presented via audiotape. For old and foil recognition tests, the participants were asked to give a yes or no response along with a confidence rating. The participants did not know how any pair was to be tested on any particular trial. These testing protocols were designed to provide the information needed to measure the storage and retrieval components of the correct recall rate. Chechile and Ehrensbeck (1983) used the Chechile and Meyer (1976) psychometric model for those data, but since that time, Chechile (2004) developed an improved psychometric model for the same testing protocols. See table 5.1 for the results using the 6P model from Chechile (2004) for the analysis of Chechile

Table 5.1
Chechile (2004) 6P model analysis of the storage and retrieval components of recall (adapted from Chechile & Ehrensbeck, 1983)

Condition	Process	Criterion	Final
Overlearning	Storage	.993	.917
Overlearning	Retrieval	.999	.944
Partial	Storage	.870	.645
Partial	Retrieval	.997	.924

and Ehrensbeck (1983) data. The psychometric model enables separate probability estimates for storage and retrieval processes without assuming a specific cognitive model for memory. This model and others will be discussed in more detail in part II. For each condition and for each process parameter (i.e., storage or retrieval), a full probability distribution is developed for the parameter, and these probability distributions are the basis for contrasting the parameter between conditions. Of particular importance is the statistical assessment of the change in storage between the criterion trial and the final test trial. For the overlearning condition, there is a highly reliable decrease from .993 to .917 caused by the 18 min of learning a second paired-associate list.[9] In the partial-learning condition, there is also a highly reliable decrease in the storage caused by the 18 min of learning a competing association (i.e., the change from .870 to .645).[10] Thus, contrary to the modal model of memory, long-term storage is not permanent. It is both theoretically and empirically clear that the modified form of all-forgetting-is-retrieval hypothesis is incorrect. Like the retrieval process, storage is also susceptible to failure.[11]

The Sloboda (2012) study also provides compelling evidence against the idea of the permanence of items in the presumed long-term store. This study examined memory with the contingent forced-choice recognition test that was described at the beginning of this section. The details for the testing protocol are also described in the glossary under the label of **IES Task**, which refers to the implicit-explicit separation task. The goal of this task is the decomposition of memory into four separate components:

9. The probability that there is a decrease in storage for the overlearning condition is directly computed by means of a Bayesian analysis. The probability that the storage level is higher on the criterion trial is greater than .997, so there is a highly reliable decrease in the storage parameter.
10. Based on the Bayesian posterior distributions, the probability of a storage decline between these conditions is greater than .99997.
11. Although the decrease in storage over the 18-min retention interval is a problem for the multiple-store framework, this result is nonetheless consistent with the two-trace hazard theory, which was described in chapter 3. Recall that in the two-trace hazard model, both traces undergo risk of a loss, and there is not a time when old memories become risk free of a loss.

(1) explicit storage, (2) implicit storage, (3) fractional storage, and (4) nonstorage. The details about how this goal is accomplished are described in chapter 10, but for the present discussion, we need only focus on how the proportion of nonstored items (denoted as θ_N) can be estimated. The initial recognition test is a four-alternative forced-choice (4-AFC) task where there is one old target item presented in a lineup along with three foil items that were not previously presented. The participant has to pick the target from the lineup. If the participant is correct, then testing is finished for that item. But if the participant incorrectly selected a foil item, then the participant is given feedback, and that item is eliminated from the lineup. In this case, the participant is asked to select again from the three-item lineup (i.e., a 3-AFC task). If the participant is correct on the 3-AFC test, then testing for that item is completed. But if the participant is again incorrect, then the task reduces to a two-item lineup (i.e., a 2-AFC task). Consequently, if there is any knowledge of the target item whatsoever, then the participant should be able to avoid the case of being incorrect on a final 2-AFC test. If all three foils on the initial lineup were selected instead of selecting the target item, then there cannot be anything still stored in memory about that target. The proportion of the original list of targets that are in this nonstored state can be estimated strictly by knowing the proportion of items that were incorrect on all the forced-choice tests.[12]

In experiment 3 in the Sloboda (2012) study, participants were tested at ten different retention intervals. The target items were words that were presented for 3 sec each. For one condition, the participants were examined after a delay of 3 h. The mean proportion of the nonstored items is estimated as .573; therefore, 42.7 percent of the items have some type of memory after a 3-h delay. In the Atkinson-Shiffrin model, the correct performance for a temporal lag as long as 3 h could only be due to the information being in the long-term store. Thus, there should be no further decline from the 42.7 percent performance level set from the 3-h condition. Sloboda also tested participants after a memory lag of 144 h (i.e., 6 days). The mean proportion of the nonstored items for that condition is .755; therefore, only 24.5 percent of the items survived in some form after 6 days. Importantly, the decrease from 42.7 percent to 24.5 percent is statistically significant.[13,14] Consequently, there are different methods for rejecting the idea of a permanent memory system—an assumption employed by

12. Chechile, Sloboda, and Chamberland (2012) showed that a maximum likelihood estimate of the proportion of items that are not stored (θ_N) is equal to four times the proportion of the items that are incorrect on the eventual 2-AFC task.

13. The difference is significant via classical t test (i.e., $t(8) = 4.852$, $p < .0007$). A Bayesian analysis also ascertained that the difference between the conditions is statistically robust.

14. Sloboda (2012) also evaluated the performance of an explicit storage measure across all ten retention intervals and found that it was well described by the two-trace hazard model retention

a number of memory models. These above experiments instead strongly demonstrate that memory storage is a dynamically varying process. Thus, the belief in a permanent memory is mistaken.

5.3 Content-Addressable Memory

Previously, the term *content-addressable memory* (CAM) has been used to describe the structure of human memory. The term itself came from Dudley Allen Buck, a wunderkind graduate student inventor at MIT, who invented and built a number of advanced devices for miniaturized computer components long before the silicon chip but unfortunately died at an early age (Brock, 2014). In 1955, Buck created a system that used the information content itself to instruct the computer to go automatically to a memory address for the storage of that information along with other associated data. This type of memory avoids the need to search locations one-at-a-time to find the desired match. Even today most computers are not CAM machines but rather use random-access memory (RAM). To store some information in a RAM system, the next available computer memory address is used to store the data; there is no constraint on the location based on the information content, but there is a constraint on the memory address—it has to be free to accept a new piece of information. If we later need to find the information that we previously stored, then a search process is required to locate the information from all the possible items in memory. In contrast to RAM computers, a CAM arrangement uses the properties of the present information to localize a storage address. Of course, this means that new information can interact with older data previously located at that address. There are both advantages and disadvantages with the CAM arrangement. The advantage is the speed of access. Without a search process, there is a direct connection to a storage location.[15] The disadvantage is that

---

equation described in chapter 3 (i.e., equation 3.1). For the two-trace hazard model, all memories are vulnerable to storage loss.

15. A *hash function* from computer science is a software simulation of a CAM system (Konheim, 2010). A hash function maps data to a number that determines the address for storing the data in a *hash table*. As an example, suppose the information being stored is a string of letters. Moreover, suppose that the hash function is designed to store strings into a 336-element *hash array*. One way to accomplish this objective is to construct an algorithm that converts the string to a high-valued integer. For example, suppose the algorithm maps the string "World peace" to the number 51,877,123. The location for storing the letter string in the hash array might be determined by expressing the number in modulus-336 arithmetic. For example, 51,877,123 is $67_{mod(336)}$. This mathematical operation will always be able to map a large positive integer to an integer from 0 to 335. Yet it is possible that a different string might be mapped to the number 21,346,483, but notice

there might be some distortion caused by the storage of new data into an address where other information was previously stored.

In psychology, the term *direct access* is used more often than the term *content-addressable*, but the two terms are essentially the same. There are many examples of direct access models in theoretical psychology. The previously discussed feature-space models for recognition (e.g., Ashby, 1992; Biederman, 1990; Estes, 1994; Kruschke, 1992; Nosofsky, 1986) as well as neural-network model (e.g., McClelland & Rumelhart, 1981; Rumelhart & McClelland, 1982) are but a few of the models that endorse a CAM framework. Shiffrin and Atkinson (1969) also proposed that the long-term store of their multiple-store model is organized as a content-addressable system or, as they called it, a "self addressing" system. For all of these psychological models, the addressing function is a bottom-up automatic addressing mechanism where the sensory and perceptual processes involved in the detection of a stimulus automatically drive the "location" in the memory system where the information about the current event is stored.

One rationale for the CAM organization of memory is the slowness of a serial search to locate an item. From the Sternberg (1966) study, it is clear that the memory search time for items within the immediate memory span is about 40 msec per item. Although this time might seem rapid at first, it is far too slow to search the mental lexicon exhaustively. For example, people can quickly ascertain if a letter string is a word or not in a lexical decision task in 350 to 750 msec. Moreover, the size of the receptive lexicon (i.e., the words that one can recognize as words) is between 150,000 and 250,000 words for adults (Szubko-Sitarek, 2015). If, on average, half the lexicon must be searched, at the rate of 40 msec per word, and if the size of the lexicon is at least 150,000 words, then it would require 50 min for a serial-search process to locate the word in a random-access memory. This time is clearly too long by a factor of more than 5,000. Thus, from a theoretical perspective, a CAM system or a hybrid CAM/serial-search model is needed to speed up memory access time. Nonetheless, some experimental psycholinguists did not endorse this idea initially but instead suggested that the process of making a lexical decision involved an exhaustive search of the mental lexicon (Forster & Bednall, 1976). This early model soon gave way to modified search models where direct access is assumed to narrow the search set to a more manageable short list that was then searched serially. The Becker (1980) model is an example of a short-list lexical access model. In a short-list model, the size of the search set is about

---

that 21,346,483 is also $67_{mod(336)}$. Hence, different strings might have the same storage location in the hash array, so for hash functions, it is possible to overwrite previously stored information. Thus, both hash functions and biological CAM systems can quickly store information, but they can also disrupt previously stored information.

ten words, so direct access essentially narrows the lexical access by 99.993 percent. It is curious why some psycholinguists would find it desirable to stress a serial search process when it only accounts for a 0.007 percent narrowing of the possible candidates in the lexicon. It is not parsimonious to assume two mechanisms for the narrowing of the number of possible lexical candidates when one mechanism (direct access) is sufficient. The McClelland and Rumelhart (1981) interaction activation model is a good example of a pure direct-access model where there is no serial search whatsoever.[16]

Fleischman (1987) provided experimental support for the direct-access class of lexical decision models by using a speed-accuracy trade-off methodology. She presented participants with three types of lexical units: (1) words, (2) legal nonwords, and (3) legal pseudowords. The legal descriptor means that none of the letter strings violated the letter constraints for English. Pseudowords are nonwords that are one letter different from actual words (e.g., SWEEN, which is one letter different from the word SWEET). Nonwords are three letters different from actual words (e.g., SWULB, which is three letters off from the parent word of SWEET). With a speed-accuracy trade-off experiment, the participants are trained to respond within a narrow time window. For example, in an extremely short condition, the letter string was displayed for only 2 msec and the allowed response time band was between 130 and 430 msec after the letter string termination. If participants responded before the time window, then they were told that their response was too early and to please respond within the time window. If the participants responded beyond the 430-msec limit, then they were told that their response was too slow, and they were told to respond within the acceptable time band. The participants had nineteen shaping or training trials to learn to respond within the 130-msec to 430-msec time band after the 2-msec display. Next the participants were tested on twenty-eight additional trials to ascertain their accuracy on the lexical decision task given that they were responding in the specified time band. Fleischman also examined nine other display times. For example in an extremely long condition, the participants saw the letter string for 375 msec before the beginning of the 130-msec to 430-msec time band. Thus, with a speed-accuracy shaping procedure, the participants are trained to give a response within a narrow temporal range. In essence, the response time is really not an unconstrained dependent measure because it has been shaped. Rather, the key dependent variable is the percentage correct within the shaped time band. If there is a speed-accuracy trade-off, then it is expected that as time for processing is reduced, then the accuracy will decrease.

16. This model is described in more detail in the glossary. The key property that concerns us here is its bottom-up activation process based on the features of the stimulus. It is a connectionistic model with multiple levels (i.e. a feature level, a letter level, and a word level).

Fleischman reasoned that short-list models would predict the same speed-accuracy trade-off function for pseudowords and nonwords. Fleischman argued that there were two possible strategies for any short-list model for dealing with the case of an interrupted scanning of the list. One strategy was called the 100 percent rule, and for this assumed strategy, the participants would give a "no" response if the scan is interrupted. But notice that the "no" response is actually the correct decision for nonwords and pseudowords. If the participants complete the scanning of the list, then the response would again be a correct "no" decision. Consequently, with the 100 percent rule, there should be a 100 percent accuracy for all display times for both nonwords and pseudowords (i.e., a flat speed-accuracy trade-off function). Besides the 100 percent rule, Fleischman also considered another short-list default strategy if the scan process for a short-list model is interrupted. In the alternative strategy, the participant is assumed to guess at a fixed rate of 50 percent whenever the scan is incomplete. Now there would be a decrease in accuracy if the participant did not complete the scan of the short list. Assuming the same size of the short list for nonwords and pseudowords, there would again be the same guessing rate and the same trade-off function. Consequently, for either the 100 percent rule or the 50 percent rule, the short-list class of lexical-access models would predict the same speed-accuracy trade-off function for nonwords and pseudowords. Fleischman further reasoned that the interaction-activation model would predict a different trade-off function for pseudowords and nonwords. This prediction follows because pseudowords (relative to nonwords) are closer to actual words, so they should elicit more word-like activation, which in turn requires more processing time to inhibit. She found that there was clearly a different trade-off function for pseudowords and nonwords—a result contrary to the short-list class of lexical-access models where there is a serial search in addition to a direct activation process to produce the short candidate list. The observed trade-off functions were exactly the pattern expected by the interactive-activation model. Thus, there is no reason to assume that a memory search is required to do a lexical decision. The Fleischman experiment thus supports a CAM framework for lexical memory.[17] Consequently, there are both theoretical and experimental reasons for supporting the idea that memory is organized differently than a RAM memory.

Although it is clear that a CAM cognitive architecture of some type is required, it is not established what informational content is needed for unlocking previously stored

---

17. Corbett and Wickelgren (1978) also employed a speed-accuracy trade-off methodology to test models of semantic memory activation that required a search process. For example, the models by Rosch (1973) and Anderson and Bower (1973) utilize a serial search process of the exemplars of a category. The predictions of these models for a speed-accuracy trade-off were not supported.

memories. One possible answer, used in feature-space models, is the set of features of the stimulus. Here the perceptual processes elicited by the stimulus result in a reactivation of the feature set that was originally activated. Neither a top-down engagement on the part of the person nor a memory search are required for memory access. There are many examples in theoretical psychology for this type of memory framework. The McClelland and Rumelhart (1981) interaction-activation model fits within this class of bottom-up CAM architectures, as do the feature-space models for pattern recognition (e.g., Biederman, 1990). Yet we have already seen in the section in chapter 4 about syntactic-pattern recognition that the activation of the features were not sufficient for the person to recognize a complex visual pattern. Rather, a case was made that the pattern required a top-down analysis and construction of the propositions prior to gaining access to the stored knowledge about the pattern. Given this evidence for the necessity of elaborative encoding produced by an active participant, it is clear that an automatic direct-access model is not correct in general, but it might be applicable for some highly practiced stimuli where the elaborative encoding itself might have become automated.

In figure 5.1, there are illustrations of some possible CAM systems in different test contexts. The configuration in panel A illustrates an automatic CAM system for the recognition of an item. The stimulus (denoted as S_T) automatically elicits a set of features $\{f_i\}$, which in turn triggers the stored information $S_T past$. In the case of episodic memory, the current event triggers a past memory trace by means of the activation of the features of the stimulus. If the current event is an old recognition test, then the person produces the correct "yes" response. If the occasion is a test trial in a lexical decision task, then the content recovered is the word meaning and other associations as opposed to event memory. In that case, the response would be a correct word decision. The arrangement depicted in panel A is expected for stimuli that are within the memory span capacity or highly practiced stimuli that have become chunked into a meaningful unit. For example, in a delayed matching-to-sample test, if a musical note is either the same or different, then it is likely that a top-down process is not required, and a recognition decision can be made based only on a feature-space representation.[18]

---

18. It is important to stress that decisions for the panel A arrangement are based on the information recovered. Later there will be discussion in detail about the strength theory of recognition memory where old recognition decisions are assumed to be based on a strength associated with the set of features (e.g., Egan, 1958; Malmberg, 2008; Yonelinas, 2002). I will argue later that these strength models of memory are flawed and are not the way human episodic memory functions. It might work for sensitization, but it is problematic even for habituation because habituation is the learning that results in a decrease in strength rather than an increase.

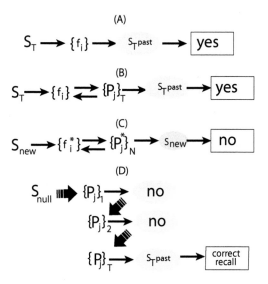

Figure 5.1

Panel (A) illustrates a CAM old recognition test for a feature-space model. Panel (B) is a display of modified CAM system with propositional elaborations as part of the addressing code, and panel (C) is a corresponding foil recognition test. Panel (D) illustrates a successful memory search for a recall task.

In light of the research on visual pattern recognition, the panel A arrangement is not suitable as a general CAM architecture. For more complex, supraspan targets, like the stimulus shown in figure 4.2, the localization of the stored information requires the reactivation of the propositions generated in the original encoding. This modified CAM model is depicted in figure 5.1B. To produce these propositions requires the top-down engagement of the person. The set of propositions along with the set of perceptual features elicits the memory from the previous event. This type of addressing is a modified CAM system involving both bottom-up activation of features and top-down proposition generation by a conscious, engaged person. The set of generated propositions is denoted as $\{P_j\}_T$. The stored trace $S_T past$ for panel B is likely to also include propositions. The activation of the stored information leads to a "yes" decision for old recognition testing. In the case of a foil recognition trial, a new stimulus (denoted as S_{novel}) is presented at the time of test.[19] This case is illustrated in figure 5.1C. The foil item elicits its own set of features $\{f_i^*\}$, and as demonstrated in

---

19. A proper foil is a stimulus to control for the base rate of guessing. In yes/no recognition, a person can be always correct by saying "yes" to each test probe. To detect guessing, the researcher

the Chechile et al. (1996) study, the foil stimulus also requires a propositional elaboration. These propositions are denoted as $\{P_j^*\}_N$. But this activation does not trigger any prior event memory, so the participant decides that it is a new item. But the novel item does establish its own memory representation $S\,new$, which can be remembered later. Perhaps the novel foil can become the basis of falsely recalling information from memory (i.e., a memory from a foil test trial might mistakenly be confused later as an event memory).

Memory representations undergo dynamic changes, and these changes can result in memory errors. In recognition testing, there are two types of errors—false alarms and misses. Let us consider two forms of a degraded memory trace. The first case is when there are missing aspects of the trace (i.e., a fractional or fragmented representation). The second case is the extreme case when the information about the stimulus is lost altogether (i.e., the nonstorage case). If either of these cases occurred, then errors are possible in recognition memory testing. For old recognition tests, the features of the test probe along with any elaborative encoding evoked by the old recognition probe will address the memory system, but because of the degradation of the memory trace, the person does not necessarily give the correct "yes" response. In the case of nonstorage, the person is likely to give an incorrect "no" response because of the absence of stored information at the localized memory address. In the case of fractional storage, the person is likely to guess because the match is only partial. When tested with a novel foil, the memory address of the foil stimulus is activated. It is unlikely that anything is stored there, so the person is likely to correctly give a "no" response. But the address triggered by the foil might have some information available from some other past event. In this case, the person might guess "yes" incorrectly.

A key feature of a CAM organization is its bottom-up activation. Yet there is a role for top-down memory search of the information in the CAM system. In a recall test trial, the person is only given directions to produce some item from the past. For example, what is the name of the person who spoke first on Wednesday afternoon? In the weak cue context, the person has to try to recover the name. Perhaps by trying some candidate answers. The candidate answers are like propositions. These candidates will either contain the stored information (if it still is available in the memory system) or not. If no match occurs, then there is an implicit "no" reaction, and another candidate is tried. This sequence of activity is portrayed in figure 5.1D. A recall failure to find information that is stored occurs when the person fails to localize the correct candidate within the

---

presents a novel item. A foil should not be highly similar to an actual target because the goal is simply to ascertain the base rate for giving a "yes" response.

time constraints of testing. The search can also lead to a correct localization and thus to a correct recall response.

Between the two extremes of an old recognition test and a recall test, there is the case of partial cuing. It is not surprising that the recovery of the target information is more effective with a greater degree of cuing. This property of recovery of the full target information with partial cues is indicative that not all the features of the stimulus are needed to narrow the memory address. As with the Teyler and DiScenna (1986) indexing theory that was illustrated in figure 3.5B, the partial cue is able to recover the entire target presumedly by having sufficient neocortical activation to trigger a hippocampal representation. Our understanding of the details of how a CAM system functions in order that addressing can occur with only a partial cue is still an open question, although there are some models for this process. For example, connectionist models, like the McClelland and Rumelhart (1981) interactive-activation model, do enable memory access on the basis of partial cues. In general, the process of reconstructing the whole encoding from a fragment is call redintegration (Schweickert, 1993).

5.4 Overview of Part I

In chapters 1 through 5, we have seen that biological systems possess a remarkable ability to model their experiences for future use. Habituation, sensitization, Pavlovian conditioning, instrumental conditioning, observational learning, procedure learning, semantic fact learning, and propositional encoding of relationships are all valuable records of the past that help us understand the past and enable us to function in the world. Unaided, we would not survive for long without the information stored in memory. The process of building a memory is one of the most fascinating processes in all of nature. For habituation and sensitization, there is an overall system-model representation in terms of a strength for a response to the stimulus. This system model captures the complex set of neural changes that occur without dealing with the biological details. However, with habituation and sensitization, the linkage between the stimulus and response is based on preexisting hardwiring. Without neurological damage, the connections that underlie habituation and sensitization are not lost; rather, they vary in the degree of activation. For all other forms of learning and memory, the newly acquired memory is based on novel connections that occur due to environmental experience. These connections were not part of the hardwiring of the animal. Processes such as long-term potentiation and long-term depression create novel linkages that occur due to experience. The storage of this information is a dynamic process. Although memories can quickly be developed, they can also become eroded. One of the myths about associative memory is the belief that it is permanent. Long-term potentiation and long-term depression are not permanent neural processes. Rather, there is a vigorous forgetting process that new memories undergo. Moreover, only a small fraction of the total information detected from the environment becomes stored in memory. Thus, there is a pruning of information to a streamlined structure that enables ready access to past information with a minimal amount of searching. The storage interference process occurs due to the fact that memory is a direct-access, modified content-addressable memory system in which the current sensory input activates previously stored representations. In fact, even a similar fragment of a previous encoded event can activate the complete encoding. Content-addressable memory is ideal for rapid access because memory search is not required provided that the sensory cues are available. But a content-addressable memory can also result in altering previously stored information. Although there are memory errors, it is more impressive that information is so quickly learned and retrieved.

II The Measurement of Memory

Introduction to Part II

There is a natural developmental process for any scientific topic. It begins with crude intuitions about the topic and later obtains some crude observations. But in order for the topic to fully blossom, measurements need to improve to the point where the fundamental processes have been recorded by some excellent measurement tools. Sophisticated theories eventually merge from the regularities and puzzles that are observed from our empirical studies with measurement tools. For example, in physics, the theory of gravitation had to wait for a genius of the caliber of Sir Isaac Newton to come along, and it also had to wait until an extensive body of data about planetary orbits was measured. The movements of the sun and the planets were perhaps the best scientific measurements available in any science at the time period in which Newton lived. It is very likely that Newton would not have discovered his laws of motion and the theory of gravitation without the availability of the astronomical measurements. Other topics in physics, such as thermodynamics, had to wait for good measurement gauges to be constructed for temperature and pressure. It also had to wait for a body of empirical findings to be generated from using these new measurement tools. With the basic groundwork established, another generation of physicists developed a sophisticated theory of thermodynamics. This pattern that emerged in physics is quite general. Scientific theory and measurement are tightly interlocked. Consequently, in part II, the topic of memory measurement is the central focus. More specifically, the focus is on memory for associative events. The chapters in part II reveal a surprising set of insights that occur from the development and utilization of good memory measurement tools.

Psychologists have an enthusiasm for theory, but a case can be made that theory without good measurement is premature and quite possibly wrong. Without resolving the memory measurement problem, we cannot know why there was a change in behavior on a task related to memory. For example, if there was a change in the memory recall rate between two experimental conditions, then was the change due to a storage difference or a retrieval difference? If we cannot answer this question, then it is difficult to see how we can make progress toward understanding the changes that occur as a function of the many factors that impact memory.

Although the measurement problem is central for advancing the study of memory, a number of theoretical roadblocks need to be resolved first. These roadblocks are examined in chapters 6 and 8. In chapter 6, there is a brief history of the measurement of an important physical property in order to see what good measurement is like in physics. Also in chapter 6, there is a critique of some of the misguided approaches toward measurement that have emerged in the psychological literature. Given this broad description of the chapter, some readers, who have a primary interest in memory, might be inclined to skip chapter 6, but I strongly urge all to carefully read this chapter because the critique includes some widely used methods in cognitive neuroscience. An even more important reason for carefully studying chapter 6 is that there is an outline in the chapter for any successful measurement method. The approach is called model-based measurement, and a case is made that model-based measurement is essential to get access to important cognitive processes for which there are no direct measures available.

In chapter 7, a successful memory measurement method is described in detail. This method exemplifies the model-based measurement approach. The measurement model is developed for a specially constructed task that enables the measurement of the underlying storage and retrieval processes that are in effect for a single experimental condition. The resulting storage and retrieval measures can then be studied across different experimental conditions to learn how the conditions altered the underlying memory processes. Evidence is advanced that demonstrates the validity of the storage and retrieval measures. Some of the key findings that were obtained with this measurement tool are also described in this chapter.

Yet not all model-based measurement approaches toward memory measurement are valid. If either the underlying theory or the task model is flawed, then the theoretical defect will result in a problematic measurement. In chapter 8, an extensively used method for measuring memory strength is described and critiqued. This measurement method emerged from a valid measurement tool in sensory psychology, but it has inappropriately been exported to the field of memory. A reader who is primarily interested in memory might be inclined to skip chapter 8, but the critique of strength measurement in psychology is in fact a major theoretical debate that is important to understand.

The method used in chapter 7 is designed for a specific task that can only be obtained with conscious humans who understand and follow verbal instructions. Yet memory is quite general, so the issue of the measurement of storage and retrieval in other animals requires a different measurement tool that does not depend on verbal instructions. In chapter 9, a novel method of obtaining storage and retrieval probability measures for pigeons is described along with some validation results.

If information in memory is not sufficiently stored, then it might be partially stored or it could be implicitly stored. It might not even be stored at all. For any given experimental condition, the information pertaining to previous events will be in one of four states: explicitly stored, implicitly stored, fractionally stored, or totally lost. Over a number of trials, there is a proportional breakdown for memory being in these four states. In chapter 10, a task and a method for estimating the proportions in these four states are described in detail. As with the other measurement methods, this measurement tool provides valuable insights about the structure of memory.

6   Historical Lessons

6.1   Chapter Overview

In this chapter, we see how physicists go about measuring quantities. From this brief and highly selective historical analysis, some general ideas emerge about the process of measurement. In contrast to this general approach, several misguided attempts toward measurement that have emerged from the psychological and neuroscience literature are described and critiqued. Finally, a case is made that measurement in cognitive psychology requires an approach called model-based measurement to overcome an entanglement problem that exists with all available dependent variables.

6.2   Introduction to Physical Measurement

Accurate and minute measurement seems to the non-scientific imagination, a less lofty and digni-fied work than looking for something new. But nearly all the grandest discoveries of science have been but the rewards of accurate measurement and patient long-continued labour in the minute sifting of numerical results. (William Thomson Kelvin, 1871, p. xci)

When you can measure what you are speaking about, and express it in numbers, you know something about it, when you cannot express it in numbers, your knowledge is of a meager and unsatisfactory kind; it may be the beginning of knowledge, but you have scarcely, in your thoughts advanced to the stage of science. (William Thomson Kelvin, 1889, p. 73)

Lord Kelvin, whose name is linked to a temperature measurement scale that is based on absolute zero, beautifully captures a fundamental feature of science—theoretical development and a deep understanding of nature are interlocked with scientific measurement. It is not an accident in physics that theoretical advances occurred in precisely the areas where there were rich databases built from excellent measurements.[1] But it is also the case that good measurement requires some theoretical foundation. In essence, theoretical constructs and scientific regularities are turned on their head to form the basis of a measurement system. To measure something well requires an understanding of some scientific regularity that is used to construct the measurement tool. Also in physics, excellent measurement often leads to new inventions that can open up new fields of study. For example, the measurements and inventions in the fields of electricity and magnetism led to devices that were used for the detection and understanding of the structure of atoms. Measurement, in general, is a pillar in the building of scientific understanding.

Historically, experimental physicists have been particularly good at measurement. Without an instruction manual, experimental physicists have learned to be creative and skillful in order to measure physical quantities. Although it might be valuable to discuss this process for some simple dimensions like weight or length that most cultures solved thousands of years ago, it is far more instructive to pick another simple dimension such as temperature that was more challenging for physics to measure. The struggle to measure temperature is more typical of the processes involved in experimental physics.

6.3   Brief History of Temperature Measurement

Prior to the first temperature gauge or thermometer, it was obvious there was variability with the property that has come to be called temperature. There were cold days when water froze and hot days when liquid water evaporated quickly. Fire was hot, and ice was cold relative to the human body. As objects were heated in fires, different substances changed from a solid state to a liquid state in a particular order. For example, ice melted before copper, which in turn melted before silver, which melted before gold. Although the rough idea of temperature is prehistoric, it took scientists a long time to distinguish between the concepts of heat and temperature. It is beyond the scope of this book to discuss these ideas in detail, but many excellent sources are available (e.g., Resnick, Halliday, & Krane, 2002). Eventually, physicists came to think of heat as energy and temperature as a property that predicted the directional flow of energy. For example, ice is at a colder temperature than the human hand. Ultimately, after a long intellectual struggle, temperature was linked theoretically to the average speed of atomic particles within the substance. The road to a deep understanding of the concept

1. I came to the field of psychology after initially working as an experimental physicist. From this experience, I came to marvel at the inventiveness of other experimental physicists. This background has also grounded me as to how psychological measurement should be conducted.

of temperature only occurred with the assistance of having accurate measurements of temperature.

While most physicists regard Newton as the intellectual giant who established the way to do theoretical physics, physicists regard Galileo as the creative father of experimental physics. In 1593, Galileo was credited with the construction of the first gauge for measuring temperature (Allaby, 2004). He cleverly constructed a glass tube of air that was placed upright in a water tank, so that there was water in the tube above the level of the water reservoir. The tube was sealed in the upright end and open to the water in the reservoir at the other end. The air pressure inside the glass tube was balanced with the atmospheric pressure on the water reservoir. But as the air in the tube was heated up, the pressure and volume of the air increased, so there was a decrease in the height of the column of water. Conversely, as the temperature of the air decreased, there was a corresponding increase in the height of the column of water. Galileo had invented a crude temperature measurement tool. It clearly changed in an appropriate way with the prehistoric and primitive understanding of the concept of temperature. Importantly, the manner in which the gauge worked relied on the pressure and volume relationship that is described in the ideal gas law, that is, $PV = KT$ where P, V, and T are respectively pressure, volume, and temperature and where K is a constant. But Galileo's device was built 241 years before the ideal gas law was published (Clapeyron, 1834)! Although Galileo's air thermometer is not a pure indicator of temperature because it was also affected by pressure changes, it did nonetheless capture the concept of temperature. Importantly, Galileo created a measurement device because he found a way by which his understanding of temperature varied systematically with what were obvious differences in temperature. He used the regularity of a physical relationship as a means for developing a measurement tool.[2]

2. Galileo's temperature gauge was different from a popular commercial product called a Galilean thermometer. These devices are more of a curiosity item. I have one such device in my office. The device contains a number of spheres where each sphere is filled with different colored liquid, and a metal tag is attached to each sphere with a temperature value. The spheres are equal in volume and weight, but the metal tags vary in weight. All of the small spheres are contained within a fluid column. The density of the fluid is a function of the temperature of the fluid. The weight of the tag is calibrated in order that the buoyancy force on the sphere balances its weight at a particular temperature (i.e., the sphere neither sinks nor floats to the top at a given temperature). As the temperature increases, some of the spheres float upward in the column and the rest remain at the bottom. The temperature is somewhere between the gap between the spheres in the fluid column. This very limited and crude device is affected by temperature in a different way than the original air thermometer of Galileo. This device is based on the effect of temperature on the density of the fluid, and it is very different from the air thermometer that Galileo actually built.

Had Galileo's air thermometer been placed inside a chamber where the air pressure of the water reservoir could be increased, then the height of the water in the column would also change because the instrument was also a pressure gauge. Galileo's instrument had an entanglement problem because the height of the water column was affected by both temperature and pressure. The familiar enclosed thermometer, which purely responded to temperature and not pressure, was invented by Ferdinand II, the grand duke of Tuscany, in 1654 (Allaby, 2004). Now a glass tube was sealed on both ends. This device worked on a different principle (i.e., the volume of alcohol in the tube increased as a function of temperature). Here again the scientist used a physical regularity to build an instrument for measuring temperature. Today there are many different sophisticated ways by which temperature can be measured. Included in the methods are thermocouples, thermistors, Langmuir probes, infrared thermometers, and Coulomb-blockade thermometers (McGee, 1988). With multiple measurement methods, one temperature gauge can also be used to calibrate and validate the readings on another temperature instrument. For each instrument, the science of studying the properties of matter was turned on its head to build a measurement gauge. Measurement in physics requires working in experimental physics. Sometimes the science surrounding the regularity upon which the scale is constructed is not fully understood initially. Nonetheless, the regularity of the relationship is enough to develop a method for measurement. Measurement is thus embedded within a research context.

Usually measurement in physics requires a model connecting the latent property to other measured quantities. In the area of temperature measurement, some methods are so simple that it does not seem that there is a model at all. For example, the fact that the volume of mercury varies with temperature means that a tube of the substance with a constant radius will increase linearly with temperature. This conclusion follows because the volume in a tube is equal to Ah, where A is the cross-sectional area and h is the height within the tube. But other temperature gauges have a more complex underlying model that only emerged after extensive research. For example, the infrared thermometer is based on Planck's theory of black-body radiation where the energy at a particular wavelength λ of infrared radiation is related to the temperature T by the formula $E = \dfrac{\frac{hc}{\lambda}}{\exp\left(\frac{hc}{k_b \lambda T}\right) - 1}$, where k_b, h, and c respectively are the Boltzman constant, Planck's constant, and the constant for the speed of light (Biró, 2011).[3] It

3. Historically, quantum theory in physics is linked to Planck's theory of black-body radiation. This conclusion follows because the characteristic spectrum for each element only has discrete wavelengths and because the energy at a fixed temperature is only a function of the wavelength; hence, energy also can only be at discrete levels. This remarkable conclusion about a fundamental

follows from this relationship that the temperature of infrared radiation at wavelength λ is $T = \dfrac{hc}{k_b \lambda \ln\left(1 + \frac{hc}{\lambda E}\right)}$. Thus, the energy recorded at a measured wavelength leads to a corresponding value of the temperature. In general, the measurement of a latent property, such as temperature, is dependent on a model that relates observable quantities to the latent quality.

6.4   Measurement Approaches Not Taken

Measurement of latent cognitive processes is not straightforward. Many measurement methods are not suitable for this task. Before discussing a method that does successfully achieve the goal of measuring underlying memory operations, it is instructive to examine why some alternative approaches are unsuitable.

6.4.1   Multiple-Item Test Theory

In physics, there is no major specialization area dealing exclusively with physical measurement. As stressed in the brief look at the history of temperature measurement, the process of measurement in physics is linked with the "nuts-and-bolts" work of being an experimental physicist within some experimental branch. There is no course in the physics curriculum that is solely limited to physical measurement in general. Measurement is fully nested within specialized experimental subfields.[4] But in psychology, not all of the approaches toward measurement have been deeply engaged within a subfield of experimental psychology. Historically, psychometrics, which is a standard course in the psychology curriculum, is not focused on experiments at all, but it stresses instead the study of complex psychological and educational tests. Psychological constructs,

---

truth of the physical world came about in part because temperature could be measured and controlled. Notice that quantum theory did not originate from a philosophical or pure mathematical analysis. It did not emerge from the testing of a preplanned hypothesis, like what is routinely done in experimental psychology. Rather, it emerged because of the discovery that each element only has a finite number of discrete wavelengths in its spectrum and because of the linkage of wavelength to energy. Measurement and observations led physicists to quantum mechanics. The mathematical details and the theory came later. This example also shows the wisdom of Kelvin's statements at the beginning of this chapter.

4. The physics curriculum does include training in computer programming and in electronics, which are general skills. There are also laboratory courses where students learn to conduct experiments in more established areas to sharpen their skills in experimentation and to learn more about substantive areas of physics. But those courses are not theory-based courses about generic measurement. Rather, students learn about measurement as it emerges from the various topic areas within physics.

such as intelligence, are ill defined and are not easily encapsulated by any one task or response measure. Consequently, psychometricians design a test with many different questions to assess the individual, where each question is roughly related to the trait being investigated. Psychometrics has invented many statistical tools for the analysis of tests such as the intelligence test (Nunnally & Bernstein, 1994; van der Linden & Hambleton, 1996). Although tests are a tool in current practice in clinical psychology and in educational psychology, multiple-item psychometric tests do not fit well within an experimental topic area where there are clear response measures that tap the cognitive process under investigation. There are a number of reasons why multiple-item psychometric exams are neither a necessary nor a suitable methodology for measuring memory processes such as the storage and retrieval rates of information. First, there are relevant response variables for memory experiments, whereas the whole idea of multiple-item test theory is that no response measure captures the psychological construct. Second, unlike test theory, which designs tests to have items of varying difficulty, memory experiments use stimuli as equal in difficulty as possible. Third, it is challenging in a nonexperimental testing context to validate the measures obtained from test theory, whereas psychological measures of basic cognitive processes obtained for a specific experimental context can be validated by studying how the measures vary across experimental conditions. Fourth, the latent processes in test theory are assumed to combine in a linear fashion, whereas key memory processes are nonlinear as will be shown later.[5] These four points are enough to refrain from using the psychometric approach of multiple-item test theory as a tool for measuring basic memory processes, but others (Michell, 2000, 2004; Trendler, 2009) are even more severe in their critique of psychometrics and have characterized the field as an area with fundamental flaws. It is outside the scope of this section to generally critique psychometrics beyond the suitability of multiple-item test theory from psychometrics for memory measurement. But I disagree with those aforementioned global critics of psychometrics because many of the statistical tools developed in psychometrics can have utility in many other applications both inside and outside of psychology.

5. Although the general approach of multiple-item test theory from psychometrics is eschewed here as a method for memory measurement, there have been contributions nonetheless from psychometrics to the statistical analysis of data that are generally useful (De Boeck, 2008; Lord & Novick, 1986). For example, memory studies have used psychometric methods to pull out the effects of individuals and stimulus items so as to remove their contaminating effect on memory data (Baayen, 2008; Matzke, Dolan, Batchelder, & Wagenmakers, 2015; Rouder, Lu, Morey, Sun, & Speckman, 2008). Also the development of multidimensional scaling (Kruskal & Wish, 1977; Torgerson, 1952) has been used as a theoretical model for conceptual knowledge, although that approach was critiqued in chapter 4.

6.4.2 Magnitude Estimation

In my opinion, model-based measurement within an experimental context is an established methodology for the physical sciences for measuring underlying properties, and it is unwise to depart from this approach because of some general, abstract philosophical principle. Yet some major figures in psychology have adopted a different framework for psychological measurement. Rather than take inspiration from the iconically successful scientific discipline of physics, which has demonstrated centuries of scientific progress, they have instead approached psychological measurement from a more mathematical and philosophical perspective.

One of the most influential theorists who set the stage for a purely mathematical approach toward measurement was Stevens (1936, 1951, 1957). The context for the work of Stevens came from a defense of psychological measurement from a vigorous attack from N. R. Campbell (Campbell, 1920, 1928; Campbell & Jeffreys, 1938). Campbell was a physicist who also engaged in philosophical theorizing about the nature of measurement. For Campbell, a scientific measurement had to satisfy a set of rules that can be roughly summarized as a verifiable method for assigning numbers to support the conclusion that the property under study is truly varying absolutely with the numbers. Campbell also argued that meaningful psychological measurement was fundamentally not possible because it could not satisfy his rules. These issues were debated at the British Association for the Advancement of Science, so psychological measurement was under a serious attack that was more than an isolated critical remark. Stevens ultimately provided a defense for psychological measurement.

Stevens had a different concept of measurement. He argued that measurement simply was an assignment of numbers to a property. The numbers can form one of three types of scales. The scales were (1) an ordinal scale where a higher number indicates a higher property value, (2) an interval scale where equal differences on the scale corresponded to equal differences for the property, and (3) a ratio scale where ratios of scale values reflect the true ratios for the property.[6] To have a ratio scale, a true zero for the property has to be established. For example, time, length, and weight have established true zeros, so it is true that a 4-min duration is twice as long as a 2-min duration. But the Celsius scale for temperature has an arbitrary zero when water begins to freeze. With an arbitrary zero, ratios are not meaningful for the Celsius scale. For example, the average kinetic speed of atoms within a substance at 20 degrees Celsius is not twice as fast as the average kinetic movement of atoms of a substance at 10 degrees Celsius.

---

6. Stevens also discussed nominal (or categorical) observations as a measurement, but here measurement scales are restricted to ones having at least the property of ordinality. Nominal information is treated here as a data type rather than a measurement scale.

Physical scientists eventually established that the true zero for temperature was at −273 degrees Celsius (i.e., the kinetic movement of particles would go to zero at that temperature). Hence, the Kelvin temperature scale (with a true zero) could be created by adding 273 degrees to the Celsius value. Now water froze at 273 degrees Kelvin. Moreover, it is true that the average kinetic speed of atoms at 20 degrees Kelvin is twice the value of the average kinetic speed of a substance at 10 degrees Kelvin.

In the language of the Stevens system of scales, measurement to N. R. Campbell requires a verifiable ratio scale or it was merely an ordinal relationship. In my opinion, Campbell's requirements are too restrictive even for physics. A mercury Celsius thermometer is a proper gauge in the physical sciences within a range of temperature values. Without instruments like the mercury thermometer, experimental physics would have been relegated to a primitive status.[7] Fortunately, Campbell's philosophical theorizing had little to no influence on his own discipline of physics; experimental physicists continued to have the good sense to know what was a proper measurement without the necessity of conferring with a philosophical or mathematical handbook.

While it is reasonable to reject Campbell's rigid requirements for measurement, it is nonetheless important to acknowledge the importance of the verification aspect of Campbell's system. Verification can take many forms. In some cases, verification can be for the model used for measurement. For example, the objective of verification for the mercury thermometer can be achieved because we can verify time measurements and use those values to show that the volume of mercury varies linearly with time as we heat up mercury. Because it is reasonable to hypothesize that the temperature of the substance should increase linearly with heating time, the experimental physicist correctly concludes that the volume increases with temperature after observing the relationship between heating time and volume. Hence, the model connecting the volume of mercury and temperature has received empirical verification. Verification can also be achieved by relating two different measurements for the same property. For example, a thermocouple thermometer can be used to validate a different temperature gauge that

---

7. Campbell, as a physicist, was well aware of the fact that the Kelvin scale of temperature was a ratio scale measure for temperature. Nonetheless, Campbell argued that temperature was not a measurable property. His requirements for a measurement were not even satisfied by all ratio scale quantities. For Campbell, a proper measurement had to satisfy an additivity property. For example, concatenating an object of 1 cm length with another object of 1 cm length results in a system that is 2 cm long. Consequently, length is a proper measurable property because it satisfies this additivity principle. However, Campbell deliberately singled out temperature as an unmeasurable property because when two objects of the same temperature are combined, it would not yield a system with twice the temperature (see Campbell & Jeffreys, 1938, p. 126). This absurd position is sufficient reason to reject Campbell's additivity principle for measurement.

operates on another physical regularity. Verification is a routine aspect of experimental physics, and Campbell as a former physicist turned philosopher also stressed the importance of verification. But Stevens rejected Campbell's view of measurement altogether (i.e., he rejected both the rigid rules along with the requirement of verification). Stevens argued that measurement was merely the assignment of numbers to a construct. Any intensive dimensional psychological construct can be measured according to Stevens. Consequently, in his magnitude estimation research, Stevens would ask individuals to assign numbers to their sensation of a stimulus. For example, the observer might be given a stimulus and be told that this standard should be given the number 10 for the magnitude of its brightness. The observer was further instructed to give a value of 20 if another stimulus was judged to be twice as bright and instructed to give the value of 5 if a stimulus was judged to be a half as bright.[8] For Stevens, it was not necessary to show that the psychological sensation of brightness for a stimulus with a magnitude estimation of 20 was in fact four times greater than the brightness sensation of another stimulus that received a magnitude estimation value of 5. Of course, there could be error in the judgments of 20 and 5, but is there a way to verify the linkage of the magnitude estimates to the psychological sensation? The failure to provide verification troubled Campbell as well as many others.

6.4.3 Abstract Measurement Theory

Michell (2000, 2004) is among the critics of psychological measurement in general and psychometrics in particular. A key concern for Michell is the testing of psychological measures. He is a critic of psychometrics, but he applauds the research in formal measurement theory in mathematical psychology as a positive development. There are two subfields for quantitative psychology: psychometrics and mathematical psychology. These subfields are quite different, and they have been frequently working at cross-purposes. Psychometrics has focused on nonexperimental assessment methods needed for applications arising in clinical practice or in education. But mathematical psychology has focused on theoretical models that emerge from experimental studies. Within mathematical psychology, there is a specialization area dealing with axiomatic

8. Stevens (1961) distinguished between prothetic continua like brightness versus metathetic continua like color. Brightness was a psychological continuum where magnitude estimation was possible. In fact, magnitude estimates were only recommended for prothetic continua. Stevens observed that brightness values combined physiologically in an additive fashion, whereas colors did not summate physiologically. Colors are different qualitatively because they elicited a different mixture of color receptors. Thus, prothetic continua varied quantitatively, whereas metathetic continua varied qualitatively.

or formal, representational measurement (e.g., Krantz, Luce, Suppes, & Tversky, 1971, 1989; Luce, Krantz, Suppes, & Tversky, 1990; Luce & Suppes, 2002; Narens, 1985; Narens & Luce, 1986). This work is quite abstract and general. In fact, it deals with measurement for any scientific discipline. It is also largely couched in a set of abstract conditional statements. Yet it is difficult to know if these conditions apply for any real measurement in any science. Cliff (1992) correctly observed that abstract measurement theory is the revolution that never happened in psychology. It has also not been embraced by any other scientific discipline. It is a method for testing properties of a scale under some limited conditions rather than a method for measuring a specific scientific property. In my opinion, successful real measurement requires using empirical properties from a scientific field as part of the methodology for measurement. Measurement in physics is not an abstract algorithm for assigning numbers. Physical measurement emerges from the details of the laboratory science. In my opinion, measurement of memory processes must also be grounded on experimental memory protocols.

6.4.4 Unsuitability of Experimental Double Disassociation

Researchers in psychology have used a number of response variables to explore cognitive processes, for example, response time (RT), percentage correctly recalled, percentage correctly recognized, average event-related potential (ERP), functional magnetic resonance image (fMRI), and positron emission tomography (PET). In the language of experimental psychology, these measures are called *dependent variables*. I have argued previously that each of these dependent variables is not a pure measure of any one cognitive process (Chechile, 2007; Chechile & Roder, 1998). For example, cognitive processes related to attention, memory encoding, storage, retrieval, guessing, and decision making can cause changes with any of the observable dependent variables. Currently, there is no known behavioral or physiological dependent measure that only yields a response signal when a single cognitive process is active and is dormant when the process is absent. All the dependent variables are functions of more than a single cognitive process. There is thus a fundamental entanglement problem that needs to be resolved to measure underlying processes. Nonetheless, there is a common (but misinformed) belief among many researchers in the cognitive neurosciences in the process-purity approach. Process purity is the hypothesis that a simple observable measure is a pure measure of an underlying cognitive process. Recall in part I that some researchers, who used the process-purity approach, argued that implicit memory had a different neurological representation than that of explicit memory (Cohen & Squire,

Table 6.1
A demonstration of double experimental dissociation

Experiments	I			II	
Independent Variable	A_1		A_2	B_1		B_2
Measure/task Y	.08	<	.20	.09	=	.09
Measure/task Z	.90	=	.90	.60	<	.65

1980; Parkin, 1993; Schacter, Chin, & Ochsner, 1993; Schacter & Graf, 1986; Squire, 1987; Tulving, 1985a; Weiskrantz, 1987, 1989). These researchers argued that their conclusion is valid because of an experimental double dissociation. So what is a double experimental dissociation, and what can be concluded if it occurs?

See table 6.1 for an idealized example of a double experimental dissociation. Suppose that there are two dependent variables for a specific task, and let us denote these variables as Y and Z. Both Y and Z are observable dependent variables. Moreover, suppose we can manipulate some independent variables. An independent variable, in the language of experimental psychology, is an experimental condition that can be altered by the experimenter. For example, the amount of time that the participant is allowed to study each item can be manipulated as an independent variable. Let us denote the various conditions for an independent variable with a subscripted Latin letter and call these conditions "levels" (e.g., at level A_1, the participant is allowed 2 sec to study each item, whereas at level A_2, the participant is allowed 8 sec per item). Suppose further that there is a different independent variable that is also manipulated over two levels, which are denoted as B_1 and B_2. An experimental double dissociation occurs if measure Y changes as a function of an independent variable, but measure Z does not change, whereas for another independent variable, measure Y does not change, but measure Z does change across levels. Note in table 6.1 that the Y variable changes in experiment I but it is a constant in experiment II, whereas the opposite pattern holds for variable Z.

Some researchers believe that the finding of an experimental double dissociation implies that measures Y and Z are each pure measures of a psychological process. Without a formal proof of the correctness of the process-purity conclusion, these researchers have used dissociations as a rationale for their theorizing about the cognitive neuroscience of memory (e.g., Bruce & Young, 1986; Coltheart, 1985; Crowder, 1976; Glanzer & Cunitz, 1966; Shallice, 1988; Tulving, 1983; Vallar, 1999). The double experimental dissociation approach has even been advocated in handbooks for

Table 6.2
Double experimental dissociation confound $Y = C \cdot U$ and $Z = C + U$

Experiments		I			II	
Independent Variable	A_1		A_2	B_1		B_2
Latent variable C	.80	>	.40	.30	>	.20
Latent variable U	.10	<	.50	.30	<	.45
Measure/task Y	.08	<	.20	.09	=	.09
Measure/task Z	.90	=	.90	.60	<	.65

the cognitive neurosciences (e.g., Ellis & Young, 1988; Rapp, 2001; Vallar, 1999). Despite the widespread use of this approach, the double experimental dissociation idea is hopelessly flawed, as will be shown from the discussion of tables 6.2 and 6.3.[9]

Table 6.2 demonstrates the error of concluding the process purity of measures on the basis of finding a double experimental dissociation. Suppose both experiment I and experiment II cause two latent and unseen cognitive factors (C and U) to change. Moreover, suppose that direct measure Y is dependent on these two latent processes via a product rule (i.e., $Y = C \cdot U$), whereas direct measure Z is dependent by means of an additive rule (i.e., $Z = C + U$). It is possible that there are values for the latent C and U parameters that combined to account for the observed double experimental dissociation (see table 6.2). Notice that for experiment I, which manipulates the independent variable A, there is a difference for *both* the C and U latent processes. Yet the dependent measure Z does not show a difference, but there is a detectable difference for the Y dependent variable. Moreover, for experiment II, which manipulates the independent variable B, there is again a difference for *both* the C and U latent processes. But for this experiment, dependent variable Y does not show a difference, but dependent variable Z does detect a different. Hence, the Y and Z measures exhibit a double experimental dissociation, but neither measures tapped only one latent process. For both experiments, the levels of the independent variable caused changes for both latent processes. If we concluded that measures Y and Z were each measuring

9. The studies that use the double experimental dissociation approach typically claim that there is no difference under some conditions because of the failure to detect a significant difference on a classical statistical test. But as noted earlier, this conclusion is not valid because the null hypothesis is assumed in the first place. Evidence can be provided for the null hypothesis from a Bayesian analysis, but that statistical method was not typically used in the dissociation studies. Despite the importance of this comment, there are more serious concerns about the double experimental dissociation approach as will be discussed.

Table 6.3
Other C and U combinations (see table 6.2)

	A_1	A_2	B_1	B_2
C	.80	.50	.30	.20
U	.10	.40	.30	.45
C	.10	.40	.30	.20
U	.80	.50	.30	.45
C	.10	.50	.30	.20
U	.80	.40	.30	.45
C	.80	.50	.30	.45
U	.10	.40	.30	.20
C	.80	.40	.30	.45
U	.10	.50	.30	.20
C	.80	.50	.30	.45
U	.10	.40	.30	.20
C	.10	.40	.30	.45
U	.80	.50	.30	.20

separate cognitive processes, then we would be mistaken. We would not be correctly understanding the latent processes, and we would be mistaken in our interpretation of the two experiments.

If one were inclined to think that the example in table 6.2 is an isolated outcome, then consider table 6.3 for seven more examples of double dissociations that result from the same two confounded direct measures Y and Z. In each of these examples, the two experiments caused both the C and U latent parameters to change but still result in a double experimental dissociation. In addition, there could be many other functions that might describe the observable Y and Z measures in terms of two or more latent processes. Thus, there is an unlimited number of ways by which confounded direct measures combine to yield a double experimental dissociation. Consequently, the finding of a double experimental dissociation is meaningless and does not support the assumption that the observable measures are pure measures of cognitive processes.[10] The belief in the process-purity hypothesis is naive wishful thinking.

10. If measure $Y = aC + b$ and $Z = cU + d$, where a, b, c and d are constants, then it would not be surprising to find a double dissociation. But there are no dependent measures Y and Z that are known to be a function of only a single underlying cognitive process. Consequently, the finding of a double dissociation does not imply process-pure dependent variables. A double dissociation can occur either with or without process-pure dependent variables.

Memory measurement is a tougher nut to crack than simply being a behaviorist and assuming that the observed behavior from experiments is directly measuring a single cognitive process.

6.5   Rationale for Model-Based Measurement

Some quantities in science are so fundamental that there is an obvious direct method for their measurement. Count, length, and mass are basic, and each has a direct measurement method. For these fundamental measurements, a model is not needed, and there is not a problem with obtaining agreement among observers. For most other quantities both inside and outside of psychology, a model is required to measure a property. In some cases, the property is defined as a relationship between fundamental measures. For example, speed is defined as the distance traveled divided by the time of travel. In this case, the definition functions as a simple model.[11] But for the case of temperature with a mercury thermometer, the measurement model is not based on a definition; rather, it is based on an empirical relationship that is supported by the regular volume changes that take place when a thermometer is heated.

Memory is complex and involves many processes. These processes are not fundamental properties like length that lead to an easy direct, model-free measurement procedure. Moreover, the memory processes cannot be measured by means of a simple definitional recipe in terms of other established measures (like the way the concept of speed is measured in physics). In addition, physiological studies of simple animals like *Aplysia*, which were used to study habituation, fail to capture the vitally important cognitive aspects of memory. To study human memory, it will be necessary to study intact people.

Some might look to physiological dependent variables as a way to solve the measurement problem for memory. But there is no physiological dependent variable

11. Actually, time is a quantity that has proven to be more difficult to measure directly without a more complex model. There are many clocks in physics, but they all function on the basis of an empirical process that has a fixed duration and where counts are possible. For example, a pendulum clock works by relating the number of swings of a pendulum with the elapsed time. But the strength of gravity will affect the period of the pendulum, so there are challenges for the experimental physicist to resolve to obtain an accurate clock. So even in the case of speed, some empirical scientific work is needed to get the elapsed time part of the speed definition. Once the physics is well understood and incorporated into the technology, the clocks are effectively a well-calibrated gauge. Hence, speed today is an example of a simple definitional model. This example also illustrates an old adage among experimental physicists—yesterday's sensation is today's calibration.

that is a pure measure of a cognitive process (i.e., where the dependent variable is in an "on" setting if the memory process is active and in an "off" setting when the memory process is inactive). An on-or-off response can occur for a feature neuron in the primary sensory cortex, but a typical memory stimulus is complex, variable, and not regulated by a single neuron. A stimulus activates a widely distributed pattern of neural activation that can vary in terms of the actual sensory input. For example, a visual stimulus can be perceived as the same event despite substantial variation in size and orientation (i.e., there is a many-to-one mapping where different neural patterns elicit the same event memory). What might be unifying across the set of different activations is the propositional relationships among the feature components. Propositional relationships are not controlled by single neurons. Thus, there is currently no physiological signal that only occurs when one memory process is active. In fact, it is difficult to invent a task that only requires a single process. Consequently, arguments in favor of neurological signals of a single cognitive process are difficult to establish, especially given the aforementioned problem of reasoning based on double experimental dissociation.

Despite the above argument against the process-purity hypothesis for physiological measurement, there are nonetheless many experimentalists who believed that there are signals of isolated cognitive processes. For example, there was a belief among some researchers that the N400 component of the event-related potential (ERP) measures an effort toward semantic integration (Kutas & Hillyard, 1980). However, since the original paper, there is still an ongoing debate about when the N400 ERP response occurs and what it means (Kutas & Federmeier, 2011). Perhaps some day there will be a broad consensus, but the ultimate conclusion is likely to be that the N400 is not a pure measure of a single cognitive operation. As another example from neuropsychology, consider the claim that the increased neural activation in the "facial fusiform" region occurs only when there is a facial stimulus (Kanwisher, McDermott, & Chun, 1997; Sergent, Ohta, & MacDonald, 1992). But this interpretation of a neural signal was later shown to be an oversimplification (Gauthier, Skudlarski, Gore, & Anderson, 2000). These two examples are not unusual; rather, they reflect a general pattern. There is initial enthusiasm about a simple behavioral or physiological dependent variable that is followed by the realization that the dependent variable also responds to other processes or stimuli.

It is theoretically naive to invest considerable research effort in the search for a process-pure neural or behavioral signal. The finding of a signal under some limited experimental condition is easy, but the understanding of the signal is more difficult. Also, it is unlikely that evidence can ever show the exclusivity of the signal. There are an unlimited number of experimental conditions, so how can it be known that the signal cannot occur for any of these other conditions? Thus, it is better to assume that an intact human when measured with any dependent variable will exhibit a number of different memorial processes rather than assume that there is a directly observable process-pure dependent variable.

Memory and other cognitive processes are entangled for any practical task. To disentangle the underlying processes, it will be necessary to model the important memory tasks in a fashion that will enable the contributions of the various processes to be numerically estimated. The task models should incorporate our knowledge about the tasks. For the right set of tasks and conditions, it is possible to separate the entangled processes by estimating the values of the individual components of the model. Thus, model-based measurement provides a method for measuring the underlying memory processes. This inescapable conclusion about the necessity for model-based measurement is likely to trouble many in cognitive psychology who are trained to be experimenters rather than measurement theoreticians, but it is not unusual in science for researchers to use model-based measurement. Most scientific measurement requires a model built from earlier empirical discoveries.

In terms of memory measurement, there are four critical questions to address for model-based measurement.

• Question 1: What memory constructs should we measure?
• Question 2: What information from empirical research links the latent psychological constructs to observable outcomes?
• Question 3: Is the model of the cognitive processes statistically identifiable, and does it result in numerical estimates of the model parameters?
• Question 4: Is there empirical validation evidence that supports key features of the measurement model?

Each of the above questions must be carefully addressed to build a successful measurement tool. Although the scientific burden of proof for developing and validating a model is high, the failure to use a measurement model leads to limited findings. At best, a new effect for a difference between experimental conditions can be discovered without a measurement model, but the reasons for the effect cannot be unambiguously linked to fundamental memory processes. For example, memory performance on a recall task changes over the life span of the individual, but how much of this change is due to a difference in the ability to store information (Belmont & Butterfield, 1969), and how much of the change is due to a difference in the ability to retrieve information (Kobasigawa, 1974)? Without a measurement model, there is a seemingly endless debate about the causes for the effect. Researchers might adopt a particular interpretation as a working hypothesis and look to see if that hypothesis is supported. But it is possible that a rival working hypothesis might also be a creditable account for the effect. Thus, progress toward a deeper understanding requires model-based memory measurement as an essential component of the research.[12]

6.6 Chapter Summary

In this chapter, we saw how physicists turn the process of measurement on its head. By measuring some quantities and by using a model for the observed regularities among those quantities, a new property is measured. This approach is called model-based measurement. A case is made that cognition similarly requires a model-based measurement approach because the underlying cognitive processes are entangled with the available dependent variables. Many others have used a process-purity approach instead, but the only support for that approach is the finding of experimental dissociations. But arguments are advanced that single and double dissociations are meaningless, so the process-purity approach is invalid because confounded measures can nonetheless exhibit a pattern of experimental dissociation. The belief in a process-purity approach is thus shown to be based on wishful thinking and on a misunderstanding about experimental dissociations. Consequently, the measuring of underlying memory processes requires model-based measurement in a fashion that is similar to measurement in the physical sciences.

12. There is still some value for experiments that do not employ model-based measurement. For example, by studying how the correct recall rate varies across a wide range of conditions, we can discover conditions that affect the recall rate. This knowledge is valuable. But in the next chapter, it will be shown that the correct recall rate is the product of a storage probability and a retrieval probability. Without model-based measurement, it is impossible to know how the underlying storage and retrieval processes vary across the conditions.

7 Measuring Storage and Retrieval

7.1 Chapter Overview

This chapter explores model-based measurement that is designed to obtain separate probability estimates for storage and retrieval processes. Each measurement model is linked to a specific set of test protocols. The Chechile-Meyer test procedure is discussed in considerable detail. Evidence is provided for the validity of the measurements obtained from using the Chechile-Meyer experimental procedure. Additional research findings are also examined; the focus for these studies is the detection of underlying storage and retrieval changes that occur between groups that differ in regard to factors such as age, cognitive disability, and drug state. Several recall-based models for separating storage and retrieval are also reviewed. The chapter further provides a gentle introduction to the topic of multinomial processing trees (MPTs).

7.2 Introduction to Storage-Retrieval Measurement

It was noted in part I that only some information from the environment is successfully stored in memory, and this information can undergo dynamic change as the individual encodes other information. So it is important to measure the amount of information that is stored for any experimental condition. But even when memory is stored, it is also possible that there is a temporary failure to retrieve the desired information. Thus, the concepts of storage and retrieval are fundamental for any information-processing system, regardless if it is a machine, a human, or an animal. Consequently, the central issue of memory measurement is obtaining metrics for these fundamental processes and examining how these metrics change as a function of experimental conditions.

Interestingly, the constructs of storage and retrieval, while compelling since the mid-twentieth century, were not terms used by earlier scholars. Why is it that storage and retrieval, which are unavoidable constructs from an information-processing framework, were not foreshadowed in earlier work in either psychology or philosophy? It is difficult to know the answer to this question. But it is interesting to observe that Aristotle discussed memory in terms of associations that have the property of strength, but he did not discuss either storage or retrieval. Perhaps his focus on the concept of association strength influenced other scholars to continue thinking in a similar fashion. But in information science, the ideas of storage and retrieval are essential while the concept of strength plays little or no role.

7.3   Chechile and Meyer (1976) Task

Prior to 1973, the existing experimental protocols or procedures lacked the informational richness to disentangle storage and retrieval. This methodological gap was circumvented by a procedure that has come to be called the Chechile-Meyer task.[1] The goal of the task is to provide the data needed to obtain separate estimates for target storage and target retrieval.

The experimental procedure is one where the participant is provided with a random intermixing of recall test trials along with old- and new-recognition tests. In the recall tests, the participant has to reproduce the memory target without the benefit of any specific information to assist in the retrieval of the target. In an old-recognition test, the target is re-presented, and in a new-recognition test, a novel item is presented. The participants on recognition testing are asked to respond either YES or NO. The recognition tests also required the participants to give a three-point confidence judgment. They are instructed to use a "3" rating if they are certain of the response and did not guess, to use a "2" rating if their recognition decision is an "educated" guess, and to use the "1" rating if their recognition decision is a "blind" guess.

Figure 7.1 illustrates the outcome categories for the Chechile-Meyer task. Notice that the two low-confidence categories are combined. A recognition response with a confidence rating less than 3 is assumed to be a guess because the participants are instructed to use those rating when guessing. The processing-tree model that will be described later makes a correction for guessing. But it is possible that even some of the 3-rating

1. Actually experiment 1 in the Chechile and Meyer (1976) paper was a part of my 1973 PhD dissertation. But the first published paper that used these experimental protocols was the Chechile and Butler (1975) paper—a project that was started after the initial Chechile and Meyer (1976) study was submitted for publication. But due to an unusually long difference in the processing rates between the two journals, the Chechile-Butler follow-up study was published earlier. In Chechile (2004), I labeled the experimental protocols as the Chechile-Meyer task because it was described more completely in the 1976 paper.

RECALL		RECOGNITION				
correct	incorrect		no 3	no 1 or 2	yes 1 or 2	yes 3
n_1	n_2	old	n_3	n_4	n_5	n_6
		new	n_7	n_8	n_9	n_{10}

Figure 7.1
The categories for the responses for the Chechile-Meyer task. The n_i for $i = 1, ..., 10$ are the outcome frequencies in the various cell categories.

responses are also guesses despite the fact that they were instructed to only use a 3 rating when they are certain. For example, responses in either cell 3 or cell 10 are high-confidence errors; these errors must be due to guessing. In the later models for this task, there will be a different correction for high-confidence guesses.

The random intermixing of recall, old recognition, and new recognition and the use of the 3-point confidence rating are unique defining features of the Chechile-Meyer experimental task. But some might find the recognition-testing procedure similar to the remember-know task suggested in the Tulving (1985b) paper. Tulving asked participants to use the REMEMBER response if they cognitively travel through time to the original encoding event and to use the KNOW response if they believe the item is an old item but cannot recall the encoding event. In some experiments, the participants are asked to give a NEW response if they believe the item is a foil, but other researchers did not provide that option and the participants have to give either a REMEMBER or KNOW response (Dunn, 2004). In other experiments, the participants are provided with an option of giving a GUESS response (Donaldson, 1996). Thus, it is not surprising that the remember-know instructions have been found to be confusing to participants and to produce questionable data (Geraci, McCabe, & Guillory, 2009; McCabe & Geraci, 2009). Some participants failed to grasp the difference between these verbal categories. The experimental protocols for the remember-know task were designed to obtain a pure experimental dependent variable that directly taps a cognitive state of interest. More specifically, the task was constructed as an attempt to measure explicit versus implicit memory directly by asking participants to evaluate the state of their knowledge. In my opinion, this goal is better achieved with model-based measurement and a different set of experimental protocols. In chapter 10, this topic is directly addressed. Hence, despite the considerable use of the remember-know task, I consider it to be a questionable procedure and to be quite different from the experimental protocols of the Chechile-Meyer task.

7.4   Recall Tests: Defining Storage and Retrieval

The recall task helps in the definition of storage and retrieval measures. Let us first examine how a storage measure can be defined in light of the requirement to recall previously experienced information.

7.4.1   Defining Sufficient Storage as a Probability Mixture Rate

There is considerable variability in the storage status of target events. At the time of test, some targets might have degraded to a point where there is no information about the target event whatsoever. Other targets might be only partially encoded (i.e., essential fragments of the target event are missing). Some items might be fully encoded in a vivid and elaborate memory representation, whereas other items might be fully encoded but with a less elaborate representation. Among the items stored, some might be encoded as a visual image, whereas others might have an auditory encoding. I found in my PhD dissertation that there were enormous differences among the participants in their successful encoding, even for items within the narrow class of consonant trigrams (Chechile, 1973). All the targets in the Chechile (1973) study were auditorily presented. Yet one participant reported that she imagined the letters being written on a blackboard. Another participant, who worked for a company that had a coded inventory, reported that some of the targets were similar to items in his company's inventory. Yet another participant reported that he encoded LRH as "Linda Rides Horses" along with an image of a person named Linda on a horse. Other participants simply tried to remember and encode the sound of the target letters. People also reported that they changed their strategy for encoding targets as they became more experienced.

Given all this variability in the nature of the representation, how can storage be defined and measured? The decision that I made was to classify a memory target as having a storage status at the time of memory testing as either (1) sufficiently stored or (2) insufficiently stored. There is a grouping of the items in both categories. Among the items in the sufficiently stored category, there are items based on different encoding properties (as noted above). Despite these differences, all the items in the sufficiently stored state would be able to support the correct recall of the whole item if that information were retrieved at the time of test, even if the cue was simply the phrase "please recall all the items presented earlier." Similarly for insufficiently stored targets, there can be variability among the items where some were deficient in only a few essential features, whereas other targets might be totally lost. Despite these differences among the insufficiently stored targets, the person would not be able to correctly recall the full target item. He or she might recall some aspect of the target but miss other information or might simply not be able to recall anything about the target. Given these two categories, it is possible to define the proportion of sufficiently stored items, which is denoted as θ_S. In essence, the requirements of the recall task shape the definition of sufficient versus insufficient storage. The θ_S parameter is a probability measure. It is also a measure of a mixture rate because all the targets are considered a mixture of items in two nonoverlapping categories. If all the items were stored sufficiently, then the value of θ_S would be 1, whereas if none of the items were sufficiently stored, then θ_S would be 0. In general, for a specific experimental condition, there will be a value for θ_S between 0 and 1. It is an empirical issue to determine the value of θ_S for that condition.

Although θ_S is a probability, it is not thought of as a random process where the same item is sometimes stored and sometimes not stored. Rather, the θ_S parameter is the proportion of all the targets that have sufficient storage to support recall.[2]

7.4.2 Defining Retrieval for the Recall Task

Let us now consider the retrieval process that must occur for a properly designed recall task. It makes little sense to discuss retrieval of information that is insufficiently stored. Clearly, if there is no memory representation, then there is nothing that can be retrieved unless the experimenter uses a trivial recall task. For example, some researchers have presented memory targets that have a strong pre-experimental association (e.g., BLACK-NIGHT). If given a later cued-recall test (BLACK-?), then there is good chance that the participant can correctly guess the item even in the case of a total memory loss. This type of a recall task is a poor instrument for studying retrieval

2. The mixture rate interpretation of the probability of sufficient storage is in sharp contrast to earlier sensory-threshold theories (Blackwell, 1953, 1963; Luce, 1963). Those models were not measurement models of memory, but rather they were models of sensory detection. Moreover, those models treated the response to the same sensory stimulus as having stochastic variation (i.e., probabilistic variability) Sometimes the response was above a sensory threshold for detection and sometimes it was below threshold. But with the advent of signal-detection theory, the idea of a sensory threshold was challenged (Green & Swets, 1966; Tanner & Swets, 1954). This issue will be discussed later in more detail, but it is mentioned here because some memory researchers, who advocate a signal-detection approach for the measurement of memory strength, might mistakenly think that the division of storage into categories for sufficient and insufficient storage implies a threshold model. Although I believe that the signal-detection approach to memory is questionable (as will be discussed later in chapter 8), it is important to stress here that the definition of sufficient storage as a mixture is not based on a strength interpretation at all, and it does not assume a threshold. Rather, the memory trace is regarded as a complex set of features that can be encoded in many ways. Storage does not stochastically waver at any given instant in time. However, some of the memory traces are lost after encoding, and the sufficient-storage parameter is simply the proportion of targets that are encoded sufficiently at the time of test.

processes, and it should be avoided. Tulving and his associates designed cued-recall items of this type in order to produce high performance levels for a supposed recall task (e.g., Flexser & Tulving, 1978; Tulving, 1974; Tulving & Thomson, 1973; Tulving & Watkins, 1977; Wiseman & Tulving, 1976). These researchers argued that recall rates can be higher than recognition rates when there is a cue-target preexperimental association. But they did not correct the cued-recall data by examining pairs that were not presented. In other words, participants can guess the response to a cue like BLACK, so corrections are required to take into account the rate of guessing for recall trials. Note that a properly designed cued-recall test should have unrelated pairs (e.g., BLACK- PRISM or BLACK-741) or simply have a nonsense syllable response like LRH. With unassociated pairs, the guessing rate in the absence of target storage would be effectively zero. If a researcher uses memory targets that are virtually impossible to guess when there is insufficient storage, then it will be possible to obtain valid measures of storage and retrieval processes because a correct recall will mean that the target is both stored and retrieved rather than guessed.

Retrieval is conditional on the fact that the information is sufficiently stored in the first place, so the two fundamental memory processes have an asymmetrical relationship. The storage status of the target is not dependent on the retrieval success rate, but retrieval can only be meaningfully understood in terms of the pool of targets that are sufficiently stored. Also, unlike storage, the retrieval of the target can exhibit stochastic variation even for the same target (i.e., there is randomness involved with a memory search process). There are factors that can influence the success rate of retrieval, such as the time allowed for recovering a stored item. The retrieval rate will also vary with respect to the cues provided. But if there are lots of cues, then it is challenging to know how to impose a correction rate for guessing. Consequently, the best way to study retrieval is with either an unrelated cue like in the example of BLACK-741 or with a generic prompt to recall all the previously presented targets.

In figure 7.2, a tree model for the recall test trial is shown in terms of the two fundamental information-processing measures for memory—the probability of sufficient storage, θ_S, and the conditional probability of successful retrieval given sufficient storage, θ_r. The probability of correct recall for a properly designed recall test must therefore be $\theta_S \cdot \theta_r$ because successful recall of a target can only occur if the target is both stored and retrieved. Incomplete or incorrect recall occurs either (1) because there is insufficient storage with probability $1 - \theta_S$ or (2) because there is sufficient storage with unsuccessful retrieval with probability $\theta_S(1 - \theta_r)$. Because either type of failure would result in a failure to recall, these two probabilities combine additively; thus, $(1 - \theta_S) + (\theta_S[1 - \theta_r]) = 1 - \theta_S \cdot \theta_r$. Clearly, if there were only recall data, then it would

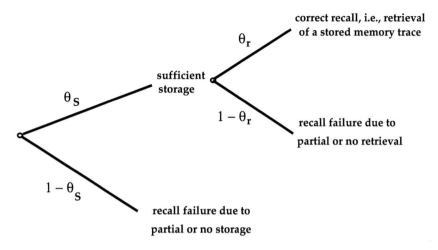

Figure 7.2
Chechile and Meyer (1976) probability tree representation for a recall test trial in terms of the probability of sufficient storage θ_S and the probability of successful retrieval θ_r conditional on sufficient target storage.

not be possible to discover the reason for a memory loss. More information is required to separate storage and retrieval processes, and that additional information is available in the old- and new-recognition data. The resulting storage-retrieval separation process is described in the next section.

7.5   Storage-Retrieval Separation

In this section, a task analysis will be provided for the old- and new-recognition test trials. This analysis will yield probability estimates of storage as well as several other factors that are part of the recognition testing. Importantly, it will not involve the retrieval parameter, so it will be possible to obtain separate estimates of storage and retrieval when all the information is combined. Yet to have the proper conditions to ensure a valid model, some methodological issues first need to be stressed about the recognition-testing protocols.

7.5.1   Issues for Proper Recognition Protocols
In part I, a case was made that old recognition was a data-driven process because memory had the feature of being a modified content-addressable system (see figure 5.1B). Thus, if a target is sufficiently stored at the time of test, then the person would not

need to search memory because the re-presentation of the target would automatically activate the stored encoding and would result in a YES response. But a YES response can occur even without the target being sufficiently stored, so there must be a correction for the tendency to just give an affirmative response. This correction is done in part by the experimental procedure and in part by a statistical correction for guessing. The experimental control is running the new-recognition test trials along with the old-recognition test trials. A foil is a new item presented in the same way as the old-recognition test. The participants come to know, through the instructions as well as from their experience, that some recognition probes match a memory target whereas other probes are just foils.[3] This knowledge means that the participant cannot just respond YES all the time and hope to be accurate. Thus, foil-test trials help make the interpretation of a YES response clearer. In addition, it enables a mathematical correction for guessing.

Another procedural issue deals with the homogeneity of encoding processes between occasions when the person has a recall test and when the person has a recognition test. If people are forewarned when they will be tested with a recall test, then it is likely that they will encode the items differently than when they are forewarned that they will be tested with a recognition test. Consequently, a key requirement of the Chechile-Meyer protocols is the lack of prediction about the manner of test. In fact, they are instructed that there will be a random intermixing (from trial-to-trial) of the recall, old, and new recognition tests.[4] This random intermixing thus eliminates the possibility of any differences in encoding and retention strategies between the recall and recognition trials. Thus, the value for the sufficient storage probability that is in effect for recall-test trials is the same value in effect for old-recognition testing.

The last part of the test protocols for the Chechile-Meyer task concerns the 3-point confidence ratings for the YES versus NO recognition decision. Remember that the participants were told to use the "3" rating for a certain non-guessed response, a "2" rating for an educated guess, and a rating of "1" for a random guess. In some experiments, children were told instead to say "sure," "maybe," or "I don't know," which were coded

3. The terms *lure*, *foil*, *distractor*, and *new recognition* are all used interchangeably.
4. A number of researchers examined memory retention as a function of test-type expectancy (Balota & Neely, 1980; Hall, Grossman, & Elwood, 1976; Neely & Balota, 1981). Sometimes the participants were told that they would be examined with a recall test, but in fact they were given a recognition test. Other times the participants were correctly told that they would be given a recognition test. The participants who incorrectly expected a recall test did better on a later recognition test than the participants who correctly expected a recognition test. This research indicates that it is important to remove predictability about the manner of test because people will encode targets differently based on the type of test. The random intermixing of the test trials is thus an important feature of the Chechile-Meyer experimental protocol.

respectively as 3, 2, or 1 (Chechile, Richman, Topinka, & Ehrensbeck, 1981). There is a correction for guessing for low-confidence responses, but there might also be a guessing correction even if the participant gives a high-confidence rating because there might be high-confidence recognition errors.[5]

7.5.2   Generic Recognition Trees for the Chechile-Meyer Task

There have been a number of correction-for-guessing models for the Chechile-Meyer task (Chechile & Meyer, 1976; Chechile, 2004, 2010a, 2013). For all the models, the representation for the recall trials is as shown in figure 7.2. The models for old/new-recognition test trials are different, but all of the models have the generic structure as shown in figure 7.3. Notice in panel A that the upper branch has the same branch probability for sufficient storage as the upper branch in figure 7.2 (i.e., the same value for θ_S applies for both the recall tests and for the old-recognition tests because of the random intermixing of the trials). Because memory is considered a content-addressable system, the stimulus in old recognition reactivates the information of the original encoding and results in a YES decision, provided that the target is stored. The reactivation triggers the additional encoded context information; thus, the participants are also confident that their response is not a guess. But if the target is insufficiently stored at the time of the old-recognition test, then there can be any of the six combinations for the yes versus no response and the three confidence ratings. The corrections for guessing in the aforementioned models will be described shortly.

To keep the participants from always guessing "yes," it is necessary to use foil items in the random mix of test trials. The presentation of a new item is thus a catch trial. A properly designed foil should be a nonpresented item from the common pool of stimuli from which the targets were selected. The generic model for foil test trials is shown in figure 7.3B. Because the foil is novel, it does not automatically elicit a memory target. But it is also possible on some occasions that the participant knows what the memory targets are and can reject the foil item because it mismatches the targets. In

5. In line with my earlier skepticism about the use of psychological measurement where the participant gives a numeric response without a calibration, it should be pointed out how the confidence ratings with the Chechile-Meyer task were really metacognitive responses about guessing. We assume if the participants say they are only guessing that they are accurately reflecting on their cognitive recognition decision. Yet if they are claiming not to be guessing, then there still is a correction for guessing to the extent of their accuracy when giving a "3" rating. For example, a person who never has a false alarm when he or she used a "3" rating will not have a correction of the old recognition "3" response. But a person with a substantial false alarm rate when he or she used a "3" rating will have a correspondingly large correction for the old recognition YES 3 response.

A.

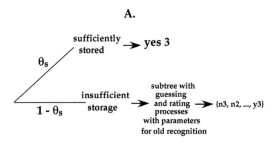

B.

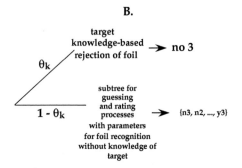

Figure 7.3
Generic tree representation for the Chechile-Meyer task in panel A for old recognition and a generic tree in panel B for foil recognition.

such cases, the participant will give a high-confidence NO response. The probability for this knowledge-based rejection of the foil is denoted as θ_k in figure 7.3. For the original Chechile and Meyer (1976) models for foil recognition, it was assumed that $\theta_k = \theta_S$, but later Chechile (2004, 2010a, 2013) argued against this simplification because it was neither necessary nor correct.[6] For example, Chechile (2010a) performed a statistical test of the data from Gerrein (1976) that showed $\theta_S \neq \theta_k$. One reason for this result is because foil items are not stimuli that automatically activate the memory target.

6. In the original Chechile and Meyer (1976) paper, there were four models developed, but models 1 and 2 were mainly developed for illustrative purposes. Perhaps the most important model discussed in that paper was model 3. For this model, there is no simplification in terms of confidence rating parameters (i.e., all of these parameters can be different). Instead in model 3, it was assumed that $\theta_k = \theta_S$—an assumption that was later challenged in the Chechile (2004) paper. Model 4 introduced a retrieval parameter (θ_{dr}) for foil recognition, so $\theta_k = \theta_S\theta_{dr}$. This model has problems when knowledge-base foil rejection is at a higher rate than θ_S. The Chechile and Meyer (1976) models are superseded by the Chechile (2004) models.

Another reason is because it might not be necessary to remember very much about the targets in order to reject some foils with high confidence. For example, if there were only two city names among the memory targets, and the foil was a different city name, then the participant can reject a city name foil with high confidence even though only some of the list is stored. The participant might only have 60 percent of the targets stored sufficiently, but the two city names might be among the stored targets, so in that case, the participant could reject any city name that is different from the two target cities.

For the subset of all the foil recognition test trials where the participant does not have a knowledge-based reason for rejecting the foil, there will be default guessing and rating processes. The proportion for such trials is $1 - \theta_k$, and this case is illustrated in the lower branch of the tree for figure 7.3B. Thus, for both old- and new-recognition tests, there are corrections for guessing.

If there is sufficient target storage and an old-recognition re-presentation of the target, then there is a bottom-up activation of the memory that results in a high-confidence YES response. But this same response can also occur when there is insufficient storage and a high-confidence guess. Consequently, the rate of a high-confidence YES response is a larger value than the probability of sufficient storage because having sufficient storage is only one way by which that response occurs. But correct recall requires both sufficient storage and successful retrieval, so the rate for sufficient storage must be greater than or equal to the recall rate. Thus, the estimate of sufficient storage is constrained on both the high and the low side. Taking into account all of the recognition responses, the model results in unique estimates for the storage and retrieval parameters.

There is a subtree for the case when there is insufficient storage for an old-recognition test (i.e., figure 7.4A). This subtree can be attached to the tree shown in figure 7.3A after the $1 - \theta_S$ path. The corresponding subtree for foil trials is shown in figure 7.4B. For these two subtrees plus the primary parameters shown in figures 7.2 and 7.3, there are a total of nine parameters. It is necessary to make some simplifying assumptions because there are too many parameters for model identifiability. In the Chechile (2004) paper, several identifiable models were examined. These models are discussed in the next subsection.

7.5.3   Chechile (2004) Models

There are seven statistical degrees of freedom associated with the categories shown in figure 7.1. The statistical degrees of freedom for multinomial data are one less than the number of category cells. Thus, there is one degree of freedom for the two-cell recall binomial and three degrees of freedom for each of the four-cell recognition

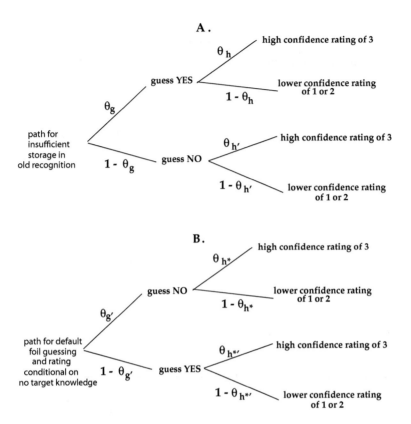

Figure 7.4
Generic subtrees for (A) old-recognition guessing and rating when there is insufficient storage and for (B) default guessing and rating in foil recognition when there is no knowledge-based foil rejection. These subtrees are a continuation and elaboration of the old- and new-recognition trees shown in figure 7.3.

multinomials. These statistical degrees of freedom combine for a total of seven. For model identification purposes, it is necessary to reduce the number of model parameters to be equal to or less than the statistical degrees of freedom.[7] This goal is achieved by making model-simplifying assumptions about the four parameters for high-confidence guessing (i.e., θ_h, $\theta_{h'}$, θ_{h*}, and $\theta_{h*'}$). These are nuisance parameters because they are only needed for the statistical correction of the use for

7. The identifiability of a model is a more complex topic than just having the number of model parameters being less than or equal to the statistical degrees of freedom. See the glossary for a more extensive discussion of the topic of model identifiability.

the high-confidence rating when in fact the participant was really guessing. Chechile (2004) considered three alternatives for model simplification (i.e., model 6P, model 7A, and model 7B).[8] The simplifying assumptions for these models are listed below.

Simplifying Assumptions for the 6P Model (See Figure 7.4)

$\theta_h = \theta_{h*} = \theta_{h'} = \theta_{h*'}$

Simplifying Assumptions for the Chechile (2004) 7A Model (See Figure 7.4)

$\theta_h = \theta_{h'}$

$\theta_{h*} = \theta_{h*'}$

Simplifying Assumptions for the Chechile (2004) 7B Model (See Figure 7.4)

$\theta_h = \theta_{h*'}$

$\theta_{h'} = \theta_{h*}$

It is convenient to discuss first the two more general models (i.e., model 7A and model 7B). Model 6P is a special case of these models. Of these two seven parameter models, there is a good reason to expect that model 7A is not a plausible model relative to model 7B. For model 7A, it is assumed that the rate for using the 3 rating for a guess in old recognition is the same regardless if the person responds either YES or NO. Also for model 7A, it is assumed that the rate for the high-confidence rating in new-recognition guessing is the same regardless of the YES or NO decision. Both assumptions are not creditable. Given that the person is guessing, then his or her state of knowledge about the target event is either fractionally stored or there is no storage whatsoever. If all insufficiently stored events were not stored at all, then there would be no need for either the 7B or 7A models. For this unrealistic case, only a five-parameter model would be needed, and that model would be a special case of model 6P where there is the further simplification of $\theta_g = 1 - \theta_{g'}$. However, from the Chechile (2004) paper, there is evidence from a study that demonstrates (1) that model 6P is a suitable method of analysis for the experiment and (2) that θ_g is very much larger than $1 - \theta_{g'}$. Hence, if information is insufficiently stored, then there is likely a mixture of events that are partially stored along with events where the information is totally lost from storage. For the occasions when the target is partially stored and the person is tested in old recognition, it seems that a YES guess might be more likely followed by a high-confidence rating than a NO guess. Also for the occasions when there is a partially stored target and the person is given a new-recognition test, it seems that a NO guess might be more likely followed

8. The 6 and 7 in the model names are in reference to the number of parameters. P is a reference to the word *parameter*.

by a high-confidence rating than a YES guess. Thus, both of the two assumptions for model 7A are unlikely to be valid.

It is instructive to do a similar analysis of the plausibility of model 7B in light of partial storage. Model 7B is based on two assumptions. First, it is assumed that the rate of a high-confidence guess of YES is the same regardless if the person is given either an old- or a new-recognition test. Second, it is assumed that the rate of a high-confidence guess of NO is the same regardless if the person were given either an old- or a new-recognition test. If there is partial information available at the time of test, and the person is inclined to judge that this information matches the test probe, then it is likely that the person will give a confident YES response regardless of the type of recognition test probe. Similarly, if there is partial information, and the person judges this to mismatch the test probe, then it is likely that the person would give a confident NO response regardless of the type of test probe. Thus, the two assumptions of model 7B are reasonable.

One might wonder why model 7A was even proposed at all in light of my skepticism about that model. In the Chechile (2004) paper, an advanced statistical method for model comparison was discussed. Consequently, I wanted to have an alternative seven-parameter model for the Chechile-Meyer task in order to assess the ability of the model-comparison statistic to select the correct model. Based on artificial data sets in which either of the models was selected to be "correct" and simulated data were sampled, evidence was presented that the model-comparison statistic did select (almost always) the correct model.[9] Data from a condition of the Chechile and Meyer (1976) experiment 1 showed clearly that the model comparison statistics selected model 7B over model 7A. Consequently, there is empirical support to justify the skepticism

---

9. Chechile (1998) developed and used a parameter estimation method called population parameter mapping (PPM). This method has a probability metric (called $P(coh)$) for the purpose of assessing the coherence of the model itself. For greater details about the PPM method and the $P(coh)$ metric, see the glossary. Theoretical Monte Carlo sampling studies were examined in Chechile (2004) to see if the coherent probability metric would be successful in selecting the correct seven-parameter model. Model 7A was useful simply as an alternative seven-parameter model; the model was not proposed as a credible psychological model. Those studies showed that even when the sample size was only about fifty simulated observations of old- and new-recognition data, the coherence probability almost always picked the correct model. In Monte Carlo studies, data are sampled from an idealized known population. Over repeated Monte Carlo data samples, the success rate for deciding between the 7A and 7B models was based on selecting the model with the higher coherence probability. The Monte Carlo studies showed that the rule of picking the model with the higher $P(coh)$ value had a 97 percent correct-decision rate even when there were only fifty old and fifty new recognition responses.

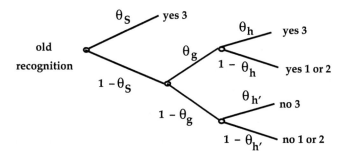

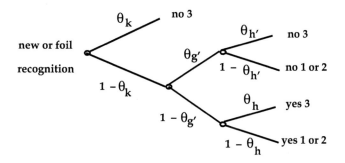

Figure 7.5
The full 7B process-tree model for old and new recognition.

about model 7A. Figure 7.5 illustrates the most general identifiable model for the Chechile-Meyer task (i.e., model 7B).

The seven parameters of model 7B are thus (1) the storage probability (θ_S), (2) the retrieval probability (θ_r), (3) the target-knowledge parameter for foil trials (θ_k), (4) the rate for correctly guessing YES on old-recognition tests (θ_g), (5) the rate for correctly guessing NO on foil-recognition tests ($\theta_{g'}$), (6) the rate for the use of the high-confidence rating when guessing YES (θ_h), and (7) the rate for the use of the high-confidence rating when guessing NO ($\theta_{h'}$). The statistical properties of the model have been developed in detail, and there is a formal proof of the identifiability of the model (Chechile, 2004).

The 6P model is the special case of the 7B model when $\theta_h = \theta_{h'}$. There is a simple test to assess if the assumptions of the 6P model are justified. If the conditional probability $P(confident\ miss|miss)$ is statistically different from the conditional probability $P(confident\ FA|FA)$, then the assumption that $\theta_h = \theta_{h'}$ is incorrect, and model 7B should be used instead. If the assumption of the 6P model is supported for a particular data set, then it is the preferred choice because there is no reason to use two parameters for the confidence rating when only one parameter will suffice. But if θ_h is substantially different from $\theta_{h'}$, then model 7B should be used.

Models 6P and 7B are examples of a multinomial-processing tree model. Additional general information about this class of measurement model is provided in the Appendix at the end of this chapter. Readers interested in a discussion of the history and the philosophy behind this class of models will find useful information in the Appendix. Also in the Appendix, there is a more technical discussion about statistical methods for estimating parameters as well as available software tools to assist in parameter estimation.

Finally, an example that illustrates the storage-retrieval measurement method is provided in the next section. The example is based on an actual experiment, and it highlights a number of properties of the storage-retrieval separation model as well as some important observations about memory obtained from the measurements.

7.6   Example: Letter-Shadowing Experiment

7.6.1   Introduction and Method of Experiment

Usually experiments have different conditions that enable the assessment of some hypotheses about how the independent variable affects the dependent variable. But as discussed previously, the dependent variable for a single experimental condition is a result of an entangled set of latent cognitive processes. Consequently, let us examine how the dependent variables in a single experimental condition can be analyzed in terms of the latent processes via model-based measurement. For this example, let us consider one of the conditions of experiment 1 from the Chechile and Meyer (1976) study. The condition used the Brown-Peterson experimental task for presentation but used the Chechile-Meyer procedure for testing. On each trial, a single, subspan target item (like LRH) was articulated in a female voice that was played through one channel of a headphone. The rate of presentation was a half second per letter. Immediately following the target presentation, the participants had to repeat (i.e., shadow) a string of auditory letters articulated in a male voice on the other channel of the headphone. The interpolated letters were presented at the rate of one letter per half-second. Following the interpolated task, the participant was tested randomly with either a *recall* cue (the word *recall*) or an old/new-recognition cue (like either LRH or VKW). The test probe was again presented in the female voice on the same channel as the target. The participants had 2 sec to respond to the test probe. On recognition trials, the participants also had to provide a 3-point confidence judgment as described earlier. After a 1-sec rest interval, another trial began. The participants were paid on the basis of the number of correct responses. In total, there were 240 tests for a 12-sec interpolated retention interval (80 recall, 80 old recognition, and 80 new recognition). There were three other retention intervals also tested, but in this example, only the 12-sec condition will be discussed. Twelve participants were tested over 4 days with this letter-shadowing task.

7.6.2 Overall Categorical Data and Confidence Analysis

For the overall experiment, there were 270 correct recalls of the target and 690 incorrect recalls. For the old recognition, there were 410 high-confidence hits, 258 lower-confidence hits, 107 high-confidence misses, and 185 lower-confidence misses. For the new recognition tests, there were 532 high-confidence correct rejections, 309 lower-confidence correct rejections, 17 high-confidence false alarms, and 102 lower-confidence false alarms.

It is assumed that participants can use the confidence rating scale to indicate when they are guessing, although there might still be guessing even when they give a 3 rating. To test this assumed skill, it is instructive to examine the probability of correct recognition, conditioned by the confidence rating. For old recognition, the conditional probabilities of a correct response given a 3, 2, and 1 rating are respectively .793, .587, and .543. In foil recognition, the corresponding conditional probabilities given a 3, 2, and 1 rating are respectively .969, .755, and .729. Pooling the two lower-confidence rating categories for old recognition, there is a significant decrease in the probability of a correct response with lower-confidence.[10] The corresponding test for foil items also indicates a significant decrease in the correct recognition rate with a lower-confidence rating.[11] Consequently, these statistical assessments support the assumption that a lower confidence rating is indicative of guessing and thus requires a correction. It is also important to observe that when there was a 3 rating, the conditional probability of a correct recognition decision was not perfect, so some correction for guessing must be made for even the high-confidence cases. If in some other experiment, which also used the Chechile-Meyer task, the data did not support the above monotonicity test for recognition accuracy, then that outcome would call into question the suitability of the storage-retrieval model for that experiment. In those cases, the participants either might not understand or might not be able to use the confidence scale given the current set of instructions and training. But this concern is not a problem for the letter-shadowing experiment.

10. A chi-squared test to assess this decline is significant, $\chi^2(1) = 50.01$, $p < 1.6 \cdot 10^{-12}$.
11. This decline was assessed by a chi-squared test that demonstrated a significant effect, $\chi^2(1) = 102.12$, $p < 10^{-16}$.

Table 7.1
Means (standard errors) for Chechile (2004) model 7B and Chechile and Meyer (1976) model 3
for the condition of letter shadowing for 12 sec

Parameter	Model 7B	Model 3
θ_S	.390 (.053)	.385 (.061)
θ_r	.592 (.057)	.614 (.058)
θ_k	.480 (.074)	$----^*$
θ_g	.539 (.057)	.553 (.060)
$\theta_{g'}$	.723 (.055)	.754 (.028)
θ_h	.124 (.055)	$----^+$
$\theta_{h'}$	.311 (.063)	$----^+$

*Parameter not defined for this model.
+Parameter not computed for Chechile and Meyer (1976) model 3.

7.6.3   Is the 6P Model Applicable?

For model 6P, there is a simple statistical test to see if that model is appropriate (i.e., the proportion of high-confidence misses over all the misses should be equal to the proportion of high-confidence false alarms over all the false alarms). That assumption was not supported for this experiment.[12] Consequently, we need to use model 7B to analyze the data.

7.6.4   Model 7B versus Chechile and Meyer (1976) Model 3

Model 7B was fit for each participant via a Bayesian analysis. This statistical method results in a probability distribution for each model parameter. The mean of the parameter distribution is taken as the optimal point estimate for the parameter.[13] The mean and standard errors for each parameter across the group of twelve participants are shown in table 7.1.

Before discussing the results from the letter-shadowing experiment, it is valuable to compare the results from fitting model 7B with the results originally reported in the Chechile and Meyer (1976) paper, which used their model 3. Model 3 differs from model 7B in a number of ways. First, with model 3, there is not a θ_k parameter; instead, the assumption was made that knowledge-based foil rejection is equal to the storage

---

12. A statistically significant chi-squared value would indicate that the 6P model is not appropriate for the experiment. For the letter-shadowing condition, this test is statistically significant ($\chi^2(1) = 20.06$, $p < .000008$).

13. The Bayesian analysis used a Markov chain Monte Carlo (MCMC) method (Hastings, 1970; Metropolis, Rosenbluth, Rosenbluth, Teller, & Teller, 1953).

parameter. Second, the assumptions that $\theta_h = \theta_{h*'}$ and $\theta_{h'} = \theta_{h*}$, which were made for model 7B, are not required in model 3.[14] Despite these different assumptions, the means were remarkably close; see the means and standard errors for the two models in table 7.1. Although model 3 has now been replaced by model 7B, the older model still produces very similar results for this condition.

7.6.5   Individual Differences and Further Model 7B Analyses

Not all the parameters of model 7B are of the same status. Four parameters θ_g, $\theta_{g'}$, θ_h, and $\theta_{h'}$ are merely correction parameters. These parameters were introduced to control for processes other than the key memory processes of interest (i.e., the storage and retrieval parameters). A fifth parameter θ_k is also of secondary interest because it deals with the ability of people to remember enough about a target to reject foil items, but foils are only introduced as a control so that participants cannot just respond YES all the time in recognition testing.

It is instructive to see how the model provides separate estimates of these parameters for just a single participant. Prior to testing a participant, each of the seven model parameters can have any value on the [0, 1] interval. Thus, for each parameter, a uniform prior probability distribution is assumed in a standard Bayesian analysis. A uniform prior probability has a probability density of 1.0 for all points on the [0, 1] interval.[15] Given the 7B model and the observations obtained from testing the participant, there is a posterior probability distribution for each model parameter. These distributions are called marginal distributions.[16] The probability density plots for the

14. With model 3 it was not feasible to estimate the high-confidence rating parameters. Chechile and Meyer (1976) showed how to obtain Bayesian probability distributions for four key parameters by integrating the posterior joint distribution with respect to all the parameters of the confidence ratings. The four parameters that could be computed in this fashion were θ_S, θ_r, θ_g, and $\theta_{g'}$.

15. All continuous distributions have a probability density for each possible value for the parameter. Probability between any two values for the variable is the area under the probability density curve between the two point values. The probability density for a uniform distribution on the [0, 1] interval is 1.0.

16. The statistical terms *prior* and *posterior* in a Bayesian analysis refer respectively to the probability distributions before and after the collection of information. The statistical jargon term *marginal distribution* originated from bivariate tables where one variable was studied by summing across the other variable. These numbers appeared on the margins of a 2×2 table. This idea of a marginal distribution generalizes to more than two dimensions and generalizes from discrete to continuous variables. Hence, a marginal probability distribution is a probability distribution for a single parameter after the joint probability distribution has been integrated across all the other parameters.

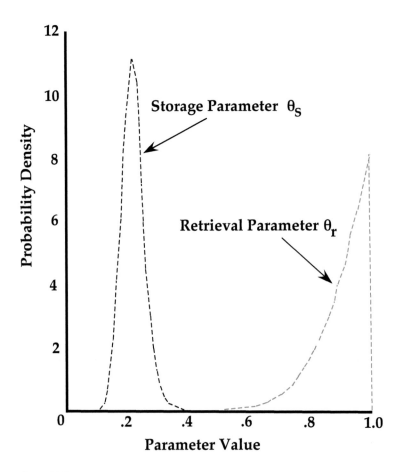

Figure 7.6
Probability density plots for the storage and retrieval parameters for participant P$_2$.

storage and retrieval parameters are provided in figure 7.6 for participant P$_2$.[17] Note that there is still uncertainty about the location of the parameters due to the sample size. Yet the uncertainty has been considerably reduced to a relatively small range, and within that range, some values are far more probable than others. The storage mean for that participant is .228, and the retrieval mean is .899. This person had difficulty

17. This participant recalled twenty-two targets correctly and was incorrect on fifty-eight recall tests. In old recognition, there were thirteen high-confidence hits, forty-five low-confidence hits, zero high-confidence misses, and twenty-two low-confidence misses. In new recognition for this participant, there were sixty-seven high-confidence correct rejections, four low-confidence correct rejections, zero high-confidence false alarms, and nine low-confidence false alarms.

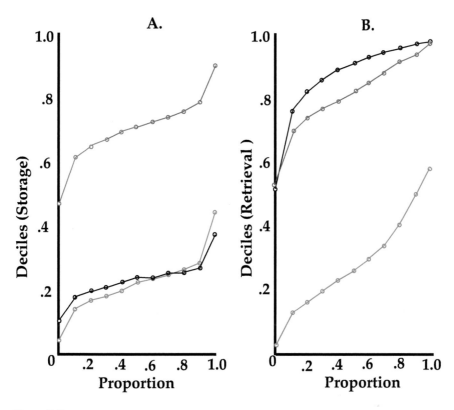

Figure 7.7
Decile plots for three participants (P_1 in red, P_2 in black, and P_3 in blue). Panel A is for the storage probability θ_S, and Panel B is for the retrieval probability θ_r. Decile points are connected, and the extreme points are also plotted.

maintaining the target information in storage. But if an item was stored sufficiently, then the participant had a relatively high probability of retrieving the item.

Not all participants exhibit the same characteristics as participant P_2. Participant P_2 had the highest rate for retrieval (.899), whereas participant P_1 had the lowest rate (.298). Participant P_3 had the highest rate for the storage parameter (.714), whereas participant P_1 had the lowest storage rate (.135). The differences among the participants for any given process were not just random fluctuations, but rather the variability reflected real differences in memory abilities. To illustrate some of the individual differences, it is convenient to examine decile plots for the key model parameters. See figure 7.7 for decile plots for three participants. Panel A shows decile plots for the sufficient storage parameter θ_S, whereas panel B shows the corresponding decile plots for the retrieval probability θ_r. The deciles are the values that divide the probability distribution into ten equal proportions. The deciles are connected in the plots for the participant. Also the extreme values for the distribution are added in the plot. Note that for the storage measure, participant P_1 is similar to participant P_2, whereas participant P_3 has considerably improved skill in storing information. But notice that for the retrieval parameter, participant P_3 is similar to participant P_2, and participant P_1 has a substantially poorer ability to retrieve information from memory.

7.6.6   Statistical Issues for Model Estimation

Although individual differences are important in clinical and educational psychology, in experimental psychology, interest typically is centered on contrasting the overall performance for a condition with the overall performance for other conditions. For this example, there were originally three other retention intervals that each participant experienced, and there were two other groups of participants who had different tasks to execute during the retention interval. The question arises as to the best way to represent the overall performance for a condition. There are two straightforward methods that can be used, but these methods do not necessarily yield the same results. One method (called the averaging method) estimates the values for a model parameter separately for each individual in the group and then averages these estimates. The alternative method (called the pooling method) combines the categorical observations over all the members of the group and then estimates the model parameters once for the pooled data. When there is a difference between these methods, how can a choice be made between these two approaches? The criterion for deciding is clear; the method that has the least error with respect to the true population mean for the model parameter should be used. Chechile (2009) explored this statistical question in exhaustive detail for a number of general models similar to the 6P and 7B models, and the results were very clear. First, there was a difference in the average absolute-value error for the two methods, and the size of the error was not small. Second, there is a sizable benefit for the pooling method. On the basis of the Chechile (2009) paper, it can be estimated that the difference in the error between the averaging and the pooling methods for a primary parameter like θ_S is approximately .03 for the letter-shadowing condition, and the most accurate estimate would be obtained via the pooling method.[18] There can be an even larger advantage for the pooling method for conditional parameters like θ_r

---

18. The reason for the pooling advantage is because the pooled frequencies in the various observational categories are the sufficient statistics for the Bayesian analysis. A sufficient statistic has the special feature of providing all the information needed to do an entire statistical inference. The parameter estimates for the individuals are not sufficient statistics. Chechile (2009) also provided

Table 7.2

Overall point and 95 percent highest density intervals (HDI) for model 7B using population parameter mapping (PPM) and MCMC methods for the letter-shadowing condition

Parameter	PPM Mean	95% HDI	MCMC Mean	95% HDI
θ_S	.379	(.33, .41)	.378	(.33, .41)
θ_r	.745	(.63, .85)	.748	(.64, .86)
θ_k	.365	(.30, .42)	.367	(.29, .42)
θ_g	.509	(.46, .55)	.510	(.46, .55)
$\theta_{g'}$	.802	(.76, .83)	.803	(.76, .83)
θ_h	.149	(.09, .21)	.151	(.09, .21)
$\theta_{h'}$	.367	(.31, .42)	.367	(.31, .42)

and the guessing and rating parameters. Consequently, the best estimates for the overall storage and retrieval probabilities for the letter-shadowing condition, after a 12-sec retention interval, should be based on the pooling method, rather than the averaging method. The averaging-method means are shown in table 7.1, and the pooling-method means are in table 7.2. The estimate of the storage parameter for the pooling method is within .011 of the estimate based on the averaging method; however, the estimates for the retrieval parameter exhibit a more sizable difference between the two methods of analysis. The greater accuracy for the pooling method is demonstrated in this experiment by the fact that the product of the storage and retrieval parameters matches the observed proportion for the correct recall rate, whereas the product of these parameters (based on the averaging method) underestimates the recall rate.

Besides a point estimate, it is important to also report an interval for each point that has a set level of probability for localizing the parameter. It is common to use a 95 percent probability interval. In a Bayesian statistical analysis, this is called the 95 percent highest density interval (HDI). The mean and 95 percent HDI for each parameter are provided in table 7.2; see the section of the table labeled MCMC. These values are based on a standard Bayesian analysis of the pooled frequencies computed by means of a Markov chain Monte Carlo (MCMC) sampling system. The MCMC method approximately converges to the exact posterior distribution. Chechile (1998, 2010a) also developed an alternative Bayesian analysis for models like the 6P and 7B models; this method is called population parameter mapping (PPM). The PPM approach also involves Monte Carlo sampling, but this method is based on sampling from the exact posterior distribution associated with the observational categories. With

---

several theorems about pooling that apply for both Bayesian statistics as well as for the classical maximum likelihood method.

the PPM method, there is a two-step process of first sampling a vector of population proportion values for the observational categories and then mapping that vector to a corresponding vector of values for the scientific model parameters (i.e., the 7B model). The collection of mapped points for the 7B parameters is the basis for obtaining point and interval estimates. In table 7.2, the results of the PPM analysis are provided. The PPM and MCMC analyses are remarkably close.

The PPM approach has a method for assessing the quality of the model because not all sampled vectors from the observational categories are consistent with the scientific model. For example, it is possible to sample a vector where the proportion for the category of an old recognition YES 3 response is smaller than the proportion for correct recall, but in this case, there are no values for the model parameters that can satisfy this inequality. In a case where the sampled observational proportion vector is inconsistent with the model, there is no mapping attempted. But the proportion of these samples that imply an inconsistent mapping is an estimate of the incoherence of the model. One minus this proportion is an estimate of model coherence, which is denoted as $P(coh)$. The value of $P(coh)$ is useful for assessing the model. When there are no data, each category has a frequency count of zero. In the no-data case, the $P(coh)$ value for model 7B is .10923. This small value illustrates that it is improbable when there are no data that model 7B is coherent. However, the posterior $P(coh)$ value can either go up or down from the initial no-data value. If the model is flawed, then the $P(coh)$ value will decline as sufficient data are observed. But if the model is a good representation of the underlying cognitive processes, then the posterior value for $P(coh)$ will increase with the sample size. For the letter-shadowing experiment, there were 30,000 samples from the posterior distribution for the proportions in the various observational categories; all of these samples were consistent with the 7B model. Thus, there is a posterior-to-prior odds ratio for the model of 9.15 (i.e., the data have increased our belief in the model by a factor greater than 9). In a standard Bayesian analysis, the 7B model is assumed to be correct, so the standard Bayesian analysis does not have a built-in test of the model itself as does the PPM method. Despite these differences, the two Bayesian methods are not distinguishable in terms of the point and interval estimates for the letter-shadowing condition.

Another statistical topic is the evaluation of hypotheses. In classical statistics, a null hypothesis is assumed in an analysis to see if that hypothesis can be rejected. But with Bayesian statistics, hypotheses are assessed directly by computing the probability for the hypothesis. If the probability for the hypothesis is large, say for example .95 or greater, then we can state that there is a "reliable" difference. The term *statistical significance* has a technical meaning in the framework of classical null-hypothesis testing;

hence, a different term is used to indicate a highly probable effect within the Bayesian framework. In some cases, it is obvious that there is a reliable difference. For example, it is clear that the values for storage and retrieval processes are different when the 95 percent intervals for these parameters do not overlap. But even when it is less obvious, it is possible to compute the probability that one parameter is greater than another parameter. If the probability for a difference about the parameters is greater than .95, then the difference is considered reliable.

7.6.7   Discussion of the Letter-Shadowing Condition

The Brown-Peterson procedure causes a rapid memory loss. In just 12 sec after the target encoding, the correct recall dropped from an initial 98.85 percent rate down to a 28.13 percent rate. The storage and retrieval probability were initially almost perfect, but both parameters underwent a change. As noted earlier, the nonoverlapping 95 percent intervals for the storage and retrieval parameters indicate that storage is reliably less than the retrieval parameter. The decrease of the retrieval parameter with the retention interval is demonstrated by the fact that the 95 percent HDI interval for θ_r is below .87. Thus, both storage and retrieval are affected by the retention interval of shadowing letters.

Another noteworthy finding is the nonoverlapping of the 95 percent intervals for the θ_h and $\theta_{h'}$ parameters; the difference in these parameters explains why the 6P model was not suitable for this experiment and why the 7B model was used instead. Finally, it is also noteworthy that θ_k was not reliably different from θ_S, which explains why the earlier model 3 from Chechile and Meyer (1976) is also a good model for this condition as noted earlier. Nonetheless, it is risky in general to assume that θ_k is equal to θ_S.

Overall, the participants in the 12-sec letter-shadowing condition were able to correctly recall the target items on 270 of the 960 recall trials to yield a correct recall proportion of .281. The correct recall of a difficult item requires both sufficient storage of the target and the successful retrieval of the stored target (i.e., the recall rate is the product of θ_S and θ_r). With the 7B model, the correct recall rate is uniquely decomposed into an estimate of .379 for storage and .745 for retrieval. Thus, this example illustrates how model-based measurement provides a method to uncover the fundamental, latent memory processes of storage and retrieval.

The 6P and 7B models are MPT models that satisfy the first three requirements listed in section 6.5 for any successful model-based measurement system. The probability of sufficient storage and the conditional probability of successful retrieval of stored information are defined as the two critical latent memory constructs that need to be measured. These constructs are linked to observable data by means of the

Chechile-Meyer experimental procedure. The models are identifiable and the parameters of the models can be readily estimated. The last remaining question about the models deals with the evidence for the validity of the models. What evidence do we have that the storage and retrieval parameters relate to those specific processes? There are actually now a number of studies that support the validity of the models; some of these studies will be described in the succeeding subsections.

7.7   Validation Studies for the 6P and 7B Models

7.7.1   Independent Latent Measures

An essential characteristic of the Chechile-Meyer procedure is the utilization of recognition test trials along with recall test trials in order to obtain measures for storage and retrieval. But there are two separate research traditions that view forgetting (at least the forgetting that occurs subsequent to a short temporal delay) as a result of a common retrieval mechanism. In chapter 5, two experiments were described that refuted this theoretical bias. Nonetheless, a number of theorists believe that forgetting for either a recall or a recognition test is due to a retrieval impairment. For example, Tulving and his associates (Flexser & Tulving, 1978; Tulving, 1974; Tulving & Thomson, 1973; Tulving & Watkins, 1977; Wiseman & Tulving, 1976) found cases where cued recall was at a higher rate than the level of correct recognition. But as mentioned in section 7.4.2, these investigators did not design a suitable recall task (i.e., the cues were highly likely to elicit a correct response even if the pair of words were not studied at all). Also for recognition, the participants had to pick words from a long list of responses that were intermixed with a large set of novel words. Importantly, the cue words were not presented for the recognition test. Consequently, the context was changed between study and test for recognition, but it was not changed for recall. In chapter 12, there is an extensive examination of context changes between the time of encoding and the time of test. We will see that changing the context can degrade the performance on a memory assessment. Thus, the selective changing of context is an experimental-design problem. In essence, the above experiments were not constructed to measure separate memory processes, but rather they were built to study the role of cuing.

Also from a different research tradition, some memory theorists constructed "global" models of memory that applied to both recall and recognition. Examples of these models would include the SAM, REM, TODAM, and CHARM models described in chapter 3 and in the glossary. From this research tradition, one might suspect that a change across experimental conditions for the recall rate should be correlated with a corresponding change in recognition. In fact, Kahana, Rizzuto, and Schneider (2005) found a consistent correlation of about .5 between recall and recognition. These investigators interpreted this finding as support for the global modeling approach.

Given the findings from the above two research traditions, some memory theorists might be skeptical about the Chechile-Meyer task because it employs recall and recognition to extract independent latent storage and retrieval measures. Yet according to models 6P and 7B, there really ought to be a correlation between a change in correct recall and the corresponding change in correct recognition because there is a shared storage parameter (θ_S) for both recall and recognition. This parameter in common should induce a correlation between the observable rates for recall and recognition. But are the resulting model-based measures of storage and retrieval nonetheless independent? Experiment 1 from Chechile and Meyer (1976) provides an answer to this question. The experimental procedure is described in subsection 7.6.1. In addition to the letter-shadowing condition, there were also two other interpolated tasks for the participants to perform in the retention interval of the Brown-Peterson task—a digit-shadowing condition and a counting-backward-by-3 condition. In total, thirty-six participants received 240 trials with a 0-sec retention interval as well as 240 trials with a 12-sec retention interval.[19] There was a sizable decrease in memory retention that occurred for the correct recall rate and correct recognition hits between the 0 and 12 retention interval. The correlation between these recall and recognition changes was .458, which is remarkably close to the value reported in the Kahana et al. (2005) study. Moreover, the 7B model estimates for storage had a significant decline of .474 for the θ_S parameter and a decline of .201 for the θ_r parameter.[20,21] Nonetheless, the correlation between the storage and retrieval changes is equal to −.0115, which is clearly not significant. Thus, model 7B is able to extract two statistically independent components of the correct recall rate despite the fact that both storage and retrieval changed across the 12-sec retention interval. Hence, the finding of independent components for recall and the definition of the recall components in figure 7.2 make for a strong case for valid measurement of storage and retrieval.

19. Each participant also had extensive testing for a 4-sec retention interval as well as an 8-sec interval. However, these intermediate retention-interval conditions are not needed for the correlational analyses.

20. The standard deviations for the change in storage and the change in retrieval across the retention intervals are respectively .202 and .225. Moreover, the decline in storage is significant as demonstrated by a classical t test, $t(35) = 14.07$, $p < 10^{-15}$, and the decline in retrieval is also significant, $t(35) = 5.36$, $p < .000003$.

21. Although the pooling method discussed in subsection 7.6.6 is more accurate, for this correlational analysis, model 7B was used separately for each participant.

7.7.2 Search Time Study

Experiment 2 from Chechile and Meyer (1976) explored the effect of greater search time at the time of test. The experiment was conducted in a similar fashion as with the letter-shadowing study described earlier. The participants had an interpolated task for the Brown-Peterson task of letter shadowing, digit shadowing, or counting backward-by-3. But these three interpolated tasks are grouped for the purpose of this analysis. There was a total of ninety volunteers in this experiment: forty-five participants in a condition with a short search-time deadline of 1.5 sec and forty-five participants in a condition with a longer search-time deadline of 3.5 sec. The time deadline was for the recall-test trials. In the 1.5-sec condition, the participants had to state aloud the target item completely within 1.5 sec or else the response was scored as an incorrect recall because it exceeded the time deadline. The participants in the 3.5-sec condition had an extra 2 sec to say the target item. The participants in the 1.5-sec condition were given an extra 2 sec for the rest interval between trials, so the total spacing of the targets was the same between the two groups of participants. Having 2 extra seconds to recall the target cannot meaningfully increase the probability of target storage, but the extra 2 sec might make it easier to retrieve a stored target. Thus, this experiment provides a method to assess the validity of the model because only one of the two fundamental memory components of recall should change. If there is a higher recall rate for the 3.5-sec condition, then only the retrieval parameter should be reliably improved with search time. Our understanding of the concept of retrieval and our everyday experience with the benefit of more search time on the retrieval success rate provides a way to see if the model parameters change in a right way.

In the 1.5-sec deadline condition, there were 443 correct recall responses and 907 incorrect or late responses, but in the 3.5-sec condition, there were 589 correct recalls and 761 incorrect recall responses. The extra 2 sec thus resulted in better recall.[22] In the 1.5-sec condition, the frequencies for the {high-confident NO, low-confident NO, low-confident YES, and high-confident YES} are respectively {101, 195, 228, 826} for old recognition, and the corresponding frequencies for new recognition are {1035, 249, 48, 18}. In the 3.5-sec condition, the old-recognition frequencies are {85, 175, 221, 869} and the new-recognition frequencies are {1067, 211, 54, 18}.[23] The resulting estimates

---

22. The recall rate significantly improved as demonstrated by a significant chi-squared test, $\chi^2(1) = 33.43$, $p < 10^{-8}$.

23. As with the letter-shadowing experiment, the confidence rating demonstrates a strong mono-tonicity effect. Combining across the two conditions, the probability of a correct response in old recognition when there is a high-confidence rating is .901, whereas a correct recognition response is only .548 when there is a low-confidence rating. The probability of a correct response in new

Table 7.3
Overall point and 95 percent highest density intervals (HDI) for the 1.5-sec and 3.5-sec search times conditions for both the 7B and 6P models

Parameter	Mean(1.5)	95% HDI(1.5)	Mean(3.5)	95% HDI(3.5)
θ_S(7B)	.544	(.49, .58)	.584	(.54, .62)
θ_r(7B)	.605	(.53, .67)	.758	(.68, .82)
θ_S(6P)	.526	(.49, .56)	.567	(.53, .60)
θ_r(6P)	.625	(.56, .69)	.780	(.71, .84)

for both the 6P and 7B models are shown in table 7.3. Both models are included because the 6P model is suitable for this experiment, but the 7B model is always acceptable.[24]

For both the 7B and 6P models, the θ_r 95 percent highest density intervals are nonoverlapping, so retrieval is reliably different between the two deadline conditions. Also for both models, the storage parameter does not reliably change between the two deadline conditions.[25] Thus, the search time study illustrates that an independent variable that should only affect retrieval did in fact only change that parameter.

7.7.3 Acoustic Similarity Effect

A number of investigators using short-term memory procedures like the Brown-Peterson task have found that phonemically similar targets are more likely to result in errors relative to phonemically dissimilar targets, and this effect can occur for both visually presented items and for aurally presented items (Chechile, 1977b; Conrad, 1967; Murray, 1968; Peterson & Johnson, 1971; Tell, 1972). Yet the phonemic similarity effect for visual targets disappears if the participants must articulate a neutral word (e.g., a word like *the*) while viewing the targets (Murray, 1968; Peterson & Johnson, 1971). Without a silent (subvocal) auditory representation of the visual targets, the phonemically similar targets were not different from visually presented targets that were phonemically dissimilar. Thus, it seems like the intraitem phonemic similarity effect should be a storage effect in short-term memory. If an acoustically similar target is sufficiently stored (despite any difficulties in achieving sufficient storage), then

recognition is .983 when there is a high-confidence rating but reduces to .819 when the confidence is low. The monotonicity effect demonstrates that the participants understood how to use the confidence judgment rating, and it demonstrates the value for obtaining this information.
24. The combined chi-squared test for the acceptability of the 6P model is $\chi^2(2) = 2.71$, which is not statistically significant. Consequently, the 6P model can be used for this experiment.
25. The hypothesis that storage is greater in the 3.5-sec deadline condition has a Bayesian probability that is below the .95 threshold for a reliable effect.

there is no reason why participants would have any extra trouble retrieving that item. Consequently, this effect provides another means for assessing the validity of the memory measurement model; this time the validation study focuses on an experimental manipulation that should only affect storage. In fact, Chechile (1977b) used model 4 from Chechile and Meyer (1976) and found that the acoustic similarity effect was only a storage effect. But model 4 has been subsequently upgraded to the models in the Chechile (2004) paper (i.e., either model 6P or model 7B). If a reanalysis for the Chechile (1977b) study with the new models also shows that the acoustic similarity effect is a pure storage effect, then that outcome would provide further support for the validation of the models.

All the stimuli in the Chechile (1977b) experiment were presented via stereo headphones. The acoustically similar targets were letter triads that shared a common phoneme (e.g., PGT), whereas the acoustically dissimilar targets were triads with no phonemes in common (e.g., RZM). The Brown-Peterson experimental design, along with the Chechile-Meyer test protocols, was used. During the interpolated task, the participants had to count backward by threes aloud from a four-digit number. Each participant was tested after retention intervals of 0, 3, or 9 sec.

The recall rates were significantly higher for dissimilar targets relative to similar targets at each retention interval.[26] Consequently, the similar intraitem targets were more difficult to initially encode and remember. Plots of the storage and retrieval parameters from the 7B model analysis are shown in figure 7.8. The retrieval probability (shown in black) is not different between the two levels of intraitem acoustic similarity for any of the retention intervals. In fact, the overall point estimate for the θ_r parameter was slightly (but not reliably) bigger in the similar condition (.750 versus .713). Because the retrieval mean for the similar condition is not lower, it is clear that the acoustic similarity effect found for the recall rate cannot be explained by the retrieval component of recall. The storage component (shown in red in figure 7.8) is consistently lower for the acoustically similar targets. The 95 percent highest density intervals are not overlapping in the 3-sec retention interval condition, so there is a clear storage difference at that retention interval. Also there is a probability of .964 that θ_S is lower in the similar condition for the 0-sec condition, and there is a probability of .981 that θ_S is also lower in the 9-sec condition.[27] Hence, the acoustic similarity effect (found

26. The observed χ^2 values for a test of a difference between the dissimilar and similar recall rates were respectively 13.4, 8.2, and 3.8 for the 0-, 3-, and 9-sec retention intervals.
27. The 6P model is suitable for this experiment, and the 6P model analysis also supported the same conclusions found for the 7B model.

for the recall rate at each retention interval) is attributable to solely a storage decrement. Targets that share a common phoneme are more difficult to initially encode correctly and are more vulnerable to storage losses relative to targets with no phonemes in common.

Finally, it should be stressed that there were changes over the 9-sec retention interval for both storage and retrieval. Thus, the same measurement model is able to detect (1) that the increase in retention interval affects both storage and retrieval and (2) that the acoustic similarity effect is purely due to a greater difficulty for storage. Thus, this experiment is further support for the validity of the measurement model for separating processes that are entangled at the level of any one dependent variable.

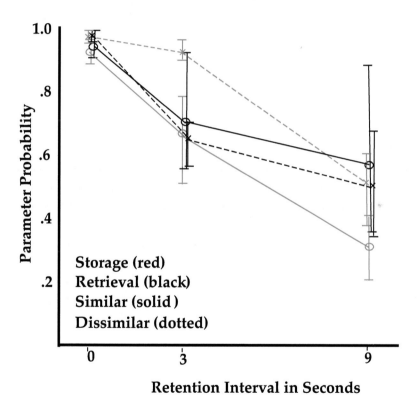

Figure 7.8
Storage and retrieval probabilities from model 7B are shown as a function of retention interval and intraitem acoustic similarity. The error bars show the 95 percent highest density interval for the parameter.

7.7.4 Paired-Associate Learning

Up to now we have only looked at studies that used the Chechile-Meyer procedure for testing the memory of a single target on each trial (i.e., the Brown-Peterson experimental procedure). With the Brown-Peterson procedure, the target for a trial is within the memory-span capacity. But many researchers are also interested in cases where the individual must learn a number of associations (i.e., a list of associations that is considerably beyond the memory-span capacity). To learn the list, individuals are given repeated training trials. Prior to training, the participants would not know any of the associations, but with training they will be able to store some of the associations. With information available in storage, it is possible that the association can be produced on a recall test. It is important to study the storage and retrieval characteristics of the learning process. Both processes should improve with training, but the rate of learning might be very different for the storage and retrieval components of correct recall. If the storage and retrieval components change in a reasonable fashion, then that would provide additional support for the validity of the storage-retrieval separation procedure.

Chechile and Ehrensbeck (1983) used the Chechile-Meyer test protocols in conjunction with a paired-associate list learning study. In that study, participants had to learn a set of eighteen associations of the type "VAJ-pilot." The participants studied each of the eighteen pairs for 2 sec, and next they had to verbally shadow digits for 20 sec before the beginning of the test phase. The order of presentation of pairs was random for each trial, and the testing order was also random. Six of the pairs were tested with a recall probe (e.g., "VAJ-recall"). The other twelve pairs were examined by a yes-no recognition test (e.g., "VAJ-pilot") for old recognition and "VAJ-clothing" for new recognition. The new-recognition word (e.g., *clothing*) was another item that was paired with a different nonsense triad. Over a block of three successive test trials, each pair was tested once for each of the three types of test trial. After a recognition probe, the participant gave a 3-point confidence judgment with the usual instructions for the Chechile-Meyer test protocols. All participants had six training trials.

With only six training trials, none of the participants reached 100 percent mastery of the eighteen-item list. Correct recall improved with each trial from trial 1 to trial 3 with the proportions being respectively .050, .273, and .377. Recall continued to improve to a value of .500 on trial 4.[28] However, the correct recall proportion did not statistically vary over trials 4 to 6; the rates were respectively, .500, .583, and .513.[29]

28. The change with each trial number from 1 to 4 is statistically significant as indicated by the large χ^2 values for the change between successive trials; the smallest $\chi^2(1) = 7.3$, $p < .007$.
29. The test of these values results in a $\chi^2(2) = 3.75$, which is not statistically significant at an alpha level of .05.

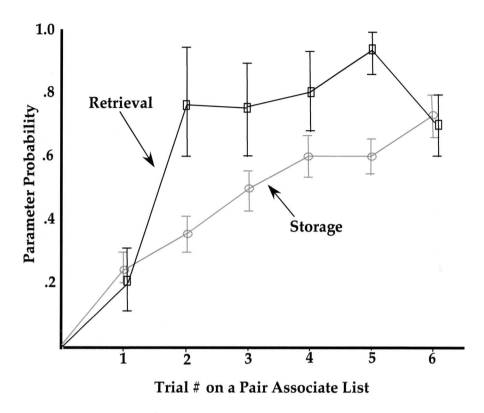

Figure 7.9
Storage and retrieval probabilities as a function of the number of trials of paired-associate learning.

See figure 7.9 for plots of the storage and retrieval components of correct recall. As expected, both the storage and retrieval probabilities improved with training, but the rates of the improvement are different. Except for trial 5, storage improves with each training trial. On trial 5, storage is virtually the same as on trial 4. Retrieval shows a more abrupt improvement with training. The θ_r parameter is the conditional probability of successful retrieval given that the target association is stored. This parameter does not reliably change from trial 2 to 5. Despite the fact that the correct recall rate does not improve on trial 6, there are nonetheless two reliable but opposite effects happening for the storage and retrieval components of recall. Storage is reliably improved on trial 6 relative to trial 5, as can be seen by the fact that the 95 percent highest density intervals are not overlapping. Consequently, on trial 6, there are more stored associations than on the previous trial, but there is not an improvement in the retrieval rate for these associations. As more associations are stored, more items have to be retrieved.

Apparently, the more recently acquired associations were not yet retrievable on the recall test, so the retrieval rate parameter decreased on trial 6. Yet with more training, both the storage and retrieval parameters would eventually get to 100 percent mastery level. This experiment illustrates that the storage-retrieval separation model is capable of detecting subtle effects that would not otherwise be found if there were only the recall response measure.

7.7.5 Monetary Incentives and Guessing

Although additional studies also point to the validity of the 6P and 7B models, it is more convenient to discuss these findings later in the book in the context of other memory issues. As in the physical sciences, model-based measurement needs to be used as a tool to tackle open research questions. With more experience, the properties of the measurement tool become better understood. If along the way, a study is done that challenges the validity of the measurement model itself, then progress has been achieved by learning about the faults of the tool. This process has resulted in the 6P and 7B models because the faults of the original Chechile and Meyer (1976) models became more apparent. The absence of contradictory evidence is itself informative. Currently, there are no problems found for the 6P and 7B models. Nonetheless, one additional validation study should be discussed here because (1) it would not otherwise be discussed in the book, and (2) it provides especially strong validation evidence for the models.

At the core of the 6P and 7B models for recognition test trials is the idea that there is the possibility of guessing, so model-based corrections for guessing are required. Given that there is guessing, it is likely that the guessing values can be altered by monetary incentives. Monetary reward might influence storage if the payoff structure were known prior to the study of the memory targets. But if the reward structure were only introduced after learning and prior to recognition memory testing, then the incentives should not affect the value of the storage probability. Chamberland and Chechile (2005) studied this hypothesis with an experiment that used only recognition test trials.

Participants in the Chamberland and Chechile (2005) experiment studied two separate lists of fifty words. A total of fifty-six participants were studied. The participants were informed that there would be points awarded for each correct response and points deducted for each memory error. Moreover, they were informed that there were five cash prizes for the participants who earned the most points; there was one $20 prize, one $15 prize, and three $10 prizes. The participants did not know the point payoff system before the list was presented, but they were informed of the payoffs just before recognition memory testing. For one of the lists, the point reward structure favored responding YES, whereas for the other list, the reward system favored responding NO.

Table 7.4
Chamberland and Chechile (2005) incentives study

Parameter	Risk Mean	Taking 95 percent HDI	Risk Mean	Adverse 95 percent HDI
θ_S	.342	(.30, .37)	.306	(.27, .34)
θ_g	.660	(.61, .69)	.406	(.36, .44)
$\theta_{g'}$	.520	(.48, .56)	.790	(.75, .81)
θ_k	.152	(.11, .19)	.159	(.12, .19)
θ_h	.096	(.06, .12)	.141	(.08, .19)
$\theta_{h'}$	.175	(.12, .22)	.103	(.07, .13)

The YES bias was achieved by a reward structure that gave more points for a correct YES in old recognition than the points earned by responding NO in new recognition. Also in terms of errors, there were more points lost by responding NO in old recognition (i.e., a memory miss) than were the points lost by responding YES in new recognition (i.e., a false alarm). This YES-bias condition is called the risk-taking condition. For the other list, the reward structure favored a NO response by a point system that was opposite of the risk-taking condition. Half the participants had the risk-taking list first, and the other half of the participants had the risk-adverse condition first. The words on the test list included both targets and foils, and the usual Chechile-Meyer instructions were also given to the participants for doing the 3-point confidence judgment.

Model 7B without the retrieval parameter was used for the analysis because there were no recall test trials. The means and 95 percent highest density intervals for the other six model parameters in the two incentive conditions are shown in table 7.4. The incentives induced major changes in the guessing parameters. The θ_g parameter is related to the tendency for a YES guess in old recognition. The $\theta_{g'}$ parameter is the tendency for a NO guess in new recognition. Both of these guessing parameters changed as a function of the incentives. The θ_g parameter is reliably larger in the risk-taking condition, whereas the $\theta_{g'}$ parameter is larger in the risk-adverse condition. These changes in the parameters are exactly what would be expected given the bias introduced by the payoff systems. There was also a change in one of the two confidence judgment parameters (i.e., the $\theta_{h'}$ parameter). The $\theta_{h'}$ parameter is the probability for giving the high-confidence rating following a NO guess in old recognition. That parameter is reliably larger in the risk-taking condition.[30] This effect is reasonable because the reward structure in that condition favors responding YES, so if the participants, nonetheless, responded NO, then they might think (incorrectly) that they did

30. The probability that $\theta_{h'}$ is greater in the risk-taking condition is .985.

not guess. Most important, there were no reliable differences between the incentive conditions for either the storage parameter θ_S or the knowledge-based foil rejection parameter θ_k. This study thus provides validation support for the 7B model because it shows (1) that the guessing parameters are behaving as expected and (2) that the θ_S and θ_k parameters were not altered by monetary incentives.

7.8   Population Differences for Storage and Retrieval

The studies in the previous section provide strong validation evidence for the models developed for the Chechile-Meyer task. Thus, this task provides a means for examining memory effects in terms of the fundamental storage and retrieval processes (e.g., Butler & Chechile, 1976; Chechile, 1987; Chechile & Butler, 1975; Skoff & Chechile, 1977). One useful application of this task is to examine the storage and retrieval processes that vary between different populations. For example, younger and older children might differ in terms of their success rates for either storage or retrieval.

7.8.1   Developmental Changes in Memory

It is obvious as children mature that there are changes in their memory performance (Gathercole, 1998; Howe, 2000). The question is: how do the storage and retrieval components of memory change with child development? Chechile, Richman, Topinka, and Ehrensbeck (1981) used the Chechile-Meyer test protocols to assess this question over the school-age timeframe. This time period is especially important because the children are engaged in formal education and thus are in the process of substantially building their semantic memory and propositional world knowledge on a wide range of topics. In essence, each age is a qualitatively different population.

In the Chechile et al. (1981) experiment, the children had to learn an association between a picture (along with its name) and the position of the picture on a list. They saw each picture for .5 sec, and they were allowed to rehearse the picture-position information for 2 sec. Following the presentation of a five-item list, the children had to shadow digits for 20 sec before they were tested. First- and sixth-grade children along with college students (who had a mean age of 19.3 years) were studied in this experiment. The storage parameter was different between the first graders (.716) and sixth graders (.836), but the storage probability for the sixth graders did not statistically differ from the storage value for the college students (.828). The failure to find a storage change between the sixth graders and college students is especially noteworthy because the college students were from a private university (Tufts University) that has rigorous selection criteria for acceptance, whereas the sixth-grade children were from a

Boston suburban public school with no selection criteria. Despite this substantial selection bias in favor of the college students, the sixth graders were still able to exhibit the same value of storage proficiency. This finding supports the idea that memory storage properties have matured by the time the child is 11 years old. Perhaps other changes might still occur but were not tapped by the storage requirements of the task of remembering an association between a picture and list position. But at this time, there is no convincing case for further age-related improvements in storage beyond the level of proficiency that is already reached by 11 years of age.

The story about memory improvements with age is different for retrieval. The retrieval parameter demonstrated a statistical difference between each age with the mean values being .347 (first grade), .498 (sixth grade), and .646 (college). Consequently, the improvement in memory skill between the sixth-grade level and the level of college students was strictly due to their enhanced retrieval success. A reasonable hypothesis why retrieval improves with age is due to the benefit of the more extensive semantic memory network of information for the older students. As the child develops a richer semantic memory, there are more potential retrieval routes to recover the target information. Chechile and Richman (1982) provided support for this hypothesis. Usually in developmental memory research, the same target items are used across age levels. But Richman, Nida, and Pittman (1976) showed that the number of associations for words increased with age. Thus, by using the same words, the older students have more associations for the words compared to younger children. Chechile and Richman (1982) and Richman et al. (1976) demonstrated that if different words are used for the various ages in order to control for the number of associations, then the recall rates did not vary between age levels. This matching of the number of associations effectively equates the semantic memory between the age levels and eliminates the change in memory retention rates.

7.8.2 Memory Deficits and Developmental Dyslexia

The storage-retrieval testing method was used to study different populations of children that differ profoundly in their reading ability (Chechile, 2007; Chechile & Roder, 1998). This research focused on fourth-grade children who were identified as poor readers by the Metropolitan Reading Achievement Test because they were more than one year behind in their reading level. Two comparison groups were also studied (i.e., an average reading group and an above-average reading group). Each child in the average and above-average groups was matched with a corresponding child in the dyslexic group on the basis of age, gender, and IQ, so all the groups were matched on those predictive variables. All the children came from the same Boston suburban school system.

Hence, the three groups of children were highly similar except for their substantial differences in reading ability.

The memory targets were separate words displayed on cards. The word stimuli did not form a sentence, and all the words were selected from the Entwisle (1966) word-association norm for first-grade children. All the children were able to read and to comprehend the words on the list. In fact, after being shown a word, each child was required to say the word aloud. The children were asked to learn the position of the words on the list. Any trial where the target word was not pronounced correctly at encoding was eliminated in the subsequent data analysis. Consequently, the memory test was a comparison of the ability to *remember* the word-position association for words that were *correctly read*. There were six words on each list. The lists consisted of four types: (1) visually similar, (2) phonologically similar, (3) semantically similar, and (4) dissimilar words.

Although the groups differ on the basis of their reading level, this does not mean that the groups need to be different in their memory for word associations. In fact, the poor readers are actually better in their ability to retrieve word-position associations, given that the associations are formed in the first place. The retrieval parameter is at a 75.3 percent level for the poor readers, whereas retrieval is only at a 56.3 percent level for the average readers and a 48 percent level for the above-average readers. The high level of performance for retrieving information demonstrates that the poor readers do not have problems in retrieving and using information that is successfully encoded in memory. But the story about the memory for the poor readers is quite different when we examine storage. If the words on the list are similar, then the poor readers have a severe storage problem in preserving a representation of the word-position association in memory. The poor readers maintain only 28.1 percent of the word-position associations when the list consists of similar words, whereas the storage percentage for similar words is 47.2 percent for the average readers and 40.2 percent for the above-average readers. Poor readers have a particularly difficult time remembering word-position associations for visually similar words (e.g., *pals, slaps, spill, lips*) with a mean storage probability of only 8.6 percent relative to the 40.0 percent rate for the average readers and the 38.0 percent rate for the above-average readers. Because the recall rate is the product of the storage and retrieval parameters, the impaired storage retention results in worse recall for the poor readers on similar lists despite the high level of retrieval for these readers. Although the poor readers have profound difficulty remembering word-position associations for words that share similarity, the poor readers have normal storage when the words are dissimilar. The storage rate for poor readers is 39.9 percent on dissimilar lists relative to the 40.7 percent rate for the average readers and 45.4 percent rate for the above-average readers. Consequently, poor readers have a different pattern in terms of their storage and retrieval of information.

Poor readers exhibit very profound storage problems for visually similar lists. One straightforward hypothesis to account for this storage impairment is to propose that these children have a distorted perceptual encoding of visual information (Bryan, 1974; Vellutino, 1977). However, Chechile and Roder (1998) argued that the perceptual distortion hypothesis is deficient for a number of reasons. First, the children could correctly pronounce each stimulus. Second, studies of the iconic imagery of poor readers provide evidence that the initial encoding is not degraded in terms of either the duration or accuracy of the iconic image (Doering & Rabinovitch, 1969; McGrady & Olson, 1970; Morrison, Giordani, & Nagy, 1977). Chechile and Roder proposed instead an alternative hypothesis. They argued that poor readers are not spontaneously encoding visual word forms effectively. A competent encoding strategy would stress the distinctive differences among the words rather than stress the commonality among the words. If the words are not distinctive, then the word-position associations are not as likely to remain stored. High similarity promotes a greater degree of storage interference of the word-position association. But if the words are dissimilar, then there is less of a problem with storage interference. In fact, the poor readers are quite normal when the words on the list are dissimilar. Based on the Chechile and Roder hypothesis for the memory impairment of poor readers, there is no reason why poor readers cannot be taught a more effective strategy for encoding words in a fashion that promotes the distinctive features of each word.

Reading skill depends on a host of cognitive abilities, and any process that is inefficient can impair reading. Eye tracking, attention, visual processing, lexical access, and phonological and semantic integration are some of the cognitive operations that are needed to support successful reading. Of course, remembering information is also an important skill for proficient reading. Poor readers have a problem recalling information about recently presented words. The fact that this impairment is localized to storage indicates that poor readers would benefit from training that promotes improved memory of recent information. Moreover, their problem of remembering items is further localized to similar lists. Chechile and Roder (1998) suggest that poor readers need to learn to encode information flexibly (i.e., adopt an appropriate memory encoding that stresses the distinctive properties of each word).

The assessment of developmental dyslexia is advanced by the use of model-based measurement. With separate measures for storage and retrieval processes, a more complex understanding is gained about how poor readers differ from average and above-average readers. In turn, the new insights from model-based measurement lead to the possibility of a more targeted reading training program that is designed to circumvent the learning disability.

7.8.3 Memory Assessment of Intellectually Disabled Adults

Although reading-disadvantaged children have a relatively limited cognitive difference between themselves and normal readers, there might be another group difference that reflects a major deficit for both storage and retrieval. For example, IQ is a measure that should be related to the ability to learn and remember new information. Consequently, it is reasonable to expect intellectually disabled individuals to show a deficit for both storage and retrieval (Detterman, 1979). To explore this possibility, Gutowski and Chechile (1987) used the storage-retrieval measurement methodology to study memory processing differences for a group of intellectually disabled adults. The intellectually disabled group had a mean IQ of 60.38 and had a mean chronological age of 33.12 years. The comparison group was an age-matched set of adults.

The experimental task consisted of a continuous paired-association procedure where there was an intermixing of study and test trials (i.e., the retention interval consisted of either the studying or testing of another paired associate). The duration for either the testing or studying of a pair was 8 sec. All the stimuli were pictures selected from the Peabody Picture Vocabulary Test. The response that needed to be linked with each picture was a concrete noun that was read aloud by the experimenter while the picture was displayed for 8 sec. There was no obvious semantic relationship between the pictures. The retention intervals of 0, 8, 16, 32, 64, and 192 sec were examined.

In other research with populations, the participants could execute the immediate encoding task at a 100 percent accuracy level. But it is possible with some populations that the immediate encoding might not be so accurate. In fact, the immediate encoding accuracy for the intellectually disabled adults was 87.8 percent, whereas the comparison group encoding accuracy was 100 percent. Presumedly inattention or distraction resulted in some failures to initially encode the picture-word pair. Gutowski and Chechile (1987) introduced an encoding probability parameter θ_e. The value for this parameter was 1.0 for the control group and .878 for the intellectually disabled adults. The model-dependent measure for sufficient storage was thus considered $\theta_e\theta_S$, where θ_S is the usual probability of sufficient storage given that the target was encoded. Thus, the model output for storage was divided by the θ_e to correct for differences between the groups in the initial encoding. For example, if for the intellectually disabled group the computer estimate for storage were .750 for a condition, then that estimate would be divided by .878 to obtain a sufficient storage value of .854. But if the estimate for storage were .75 for the control group, then the sufficient storage value

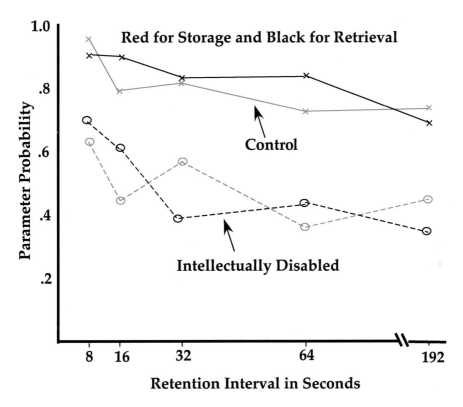

Figure 7.10
Storage and retrieval probabilities from model 7B are shown as a function of retention interval for intellectually disabled adults (dashed lines) along with a control group (solid). The storage parameter θ_S is displayed in red, and the retrieval parameter θ_r is shown in black.

would remain .750 because the encoding parameter is 1.0 for the control group. No correction is required for the retrieval parameter. The resulting storage and retrieval values for the 7B model are displayed in figure 7.10 for each group. Because the 7B model was published subsequent to the Gutowski and Chechile (1987) paper, the original data were reanalyzed. The reanalysis resulted in the same conclusions as the original paper. The intellectually disabled adults had problems with encoding, storage, and retrieval.

Although the parameters other than storage and retrieval probabilities are regarded as secondary or nuisance parameters, the values for these parameters still can reveal additional features about the cognitive processing of the intellectually disabled participants. The $\theta_{g'}$ parameter is the probability of correctly giving a NO response when tested with an incorrect picture-word association. The intellectually disabled adults were reliably less accurate on foil test trials (.469) versus the control group mean for $\theta_{g'}$ (.778). Yet when probed with an old-recognition test, the guessing mean θ_g (.624) for the intellectually disabled group is not reliably different than the control value (.614). The tendency to guess YES in old recognition minus the tendency to guess YES in foil recognition is $\theta_g - (1 - \theta_{g'})$, and this difference is .392 for the control group but is .093 for the intellectually disabled adults. The difference in correct guessing is an indication of partial information that informs guessing. The larger value for this difference for the control group is indicative of better utilization of partial information.

Another indication of a reduced ability to use stored information on foil-recognition test trials is found by the θ_k parameter, which is the probability to use knowledge of the target to reject a foil. The θ_k mean for the controls was .799, but the corresponding mean for the intellectually disabled adults was .250. Consequently, the intellectually disabled group shows a deficit in all aspects of memory processing. They have a reduced ability to encode the picture-word association as well as a reduced ability to maintain and retrieve information. They also have a reduced ability to use partial information in recognition memory tests. This general deficit has been called the "everything hypothesis" by Detterman (1979), but he suggested it was not possible to demonstrate this in a single experiment. Yet with model-based measurement, the "everything hypothesis" is supported by just one experiment.

7.8.4   Dynamic Memory Changes with Korsakoff Patients

Oliver Sacks (1985), in his famous and popular book about patients with profound neurological impairments, described a Korsakoff patient. The case study entitled "the lost mariner" provides a detailed description of a 49-year-old man who still only remembered himself as a 19-year-old. He had little memory of his life for several decades. Korsakoff syndrome is linked to bilateral brain damage to the medial thalamus and the mammillary bodies of the hypothalamus, although some patients also have damage in additional regions (Brion, 1969; Markowitsch, 2000; Whitty & Lewin, 1960). Initially, this syndrome was associated with chronic alcoholics, but now it is believed that the brain damage is caused by a severe vitamin B1 deficiency (Thomson, 2000). Presumedly, the poor diets of chronic alcoholics was the cause of the brain damage rather than the alcohol per se. Thus, this syndrome can occur just because of a chronic deficiency in the B1 vitamin.

Larner (1979) used the storage-retrieval separation method to study a group of ten Korsakoff patients along with ten carefully matched control patients. The targets were a triad of words that were followed by a digit-shadowing interpolated task. Larner tested the patients after delays of either 4 or 12 sec. Chechile (2010b) reexamined the data

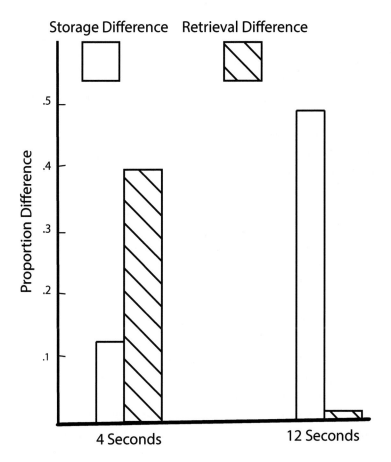

Figure 7.11
The difference between the control subjects and Korsakoff patients for storage $\Delta\theta_S = \theta_S$(control) $-\theta_S$(Korsakoff), and the corresponding difference between the groups for retrieval $\Delta\theta_r$ are displayed for two retention intervals.

from that study using the 6P model. Let us define $\Delta\theta_S$ as the difference in the storage probability between the two groups (the control patients minus the Korsakoff patients), and similarly define $\Delta\theta_r$ as the retrieval parameter difference. These differences are displayed as a function of the retention interval in figure 7.11.

The Korsakoff patients demonstrate a remarkably dramatic and complex pattern of storage and retrieval changes. After a brief delay of only 4 sec, the Korsakoff patients are only slightly impaired relative to the control group in terms of their storage of information. But there is a large and reliable impairment in the ability of the Korsakoff patients to retrieve their stored representation. Consequently, the substantial recall impairment that occurs after only a 4-sec delay is mostly due to a retrieval problem. But after another 8 sec of shadowing digits, the storage probability is only .377 for the Korsakoff patients compared to the storage probability for the control group of .819. Consequently, the two groups have very different rates for sufficient storage after a 12-sec retention interval. But the few memory traces that are stored for the Korsakoff patients after a 12-sec retention interval are retrieved at nearly the same rate as for the control patients. Thus, a retrieval impairment characterizes the difference between the groups after a 4-sec delay, but it is a storage impairment that differentiates the groups after a 12-sec delay. The near equivalence for retrieval after a 12-sec delay indicates that control and Korsakoff patients are well matched. The two groups were matched for age, gender, verbal IQ, occupational status, and educational level. Nonetheless, there is a profound storage difference between the control and Korsakoff patients after a 12-sec delay.

Chechile (2010b) explained the pattern of storage and retrieval differences between the two groups by a difference in the quality of the memory encodings. Unlike the intellectually disabled adults in the Gutowski and Chechile (1987) study, the Korsakoff patients were able to initially encode the target items. It should also be pointed out that Korsakoff patients are known to have normal memory span (Baddeley & Warrington, 1970). But having an initial encoding does not mean that the quality of the encoding is as effective as the encodings developed by the control patients. Cermack, Butters, and Gerrein (1973) reported that Korsakoff patients could encode targets on an acoustic, associative, or semantic basis when they were instructed to do so, but they did not consistently use a semantic code on their own. Furthermore, Talland (1965) observed that Korsakoff patients inaccurately grouped picture stimuli, so it is likely that the Korsakoff patients did not consistently generate high-quality semantic encodings. Low-quality encodings are likely to undergo a rapid memory loss, but this memory loss is likely to take more time than 4 sec. Retrieval is more challenging if there are many low-quality, nonsemantic encodings stored. So after only 4 sec, the Korsakoff patients are mainly different from the control patients because of their inferior memory retrieval. However, given 12 sec of digit shadowing, low-quality encodings are now likely to undergo a vigorous storage interference. Consequently, after 12 sec, the Korsakoff patients have far fewer memory traces available. Yet some traces still survived the 12-sec delay. Perhaps those few targets were the ones that the Korsakoff patients were able to encode in a fashion similar to that used by the control patients. As a result, the retrieval rate of the few surviving traces in the 12-sec condition is virtually the same as the retrieval rate for the control patients.

7.8.5 Drug Effects beyond State Dependency

Animal studies have shown a *state-dependency effect* for a number of drugs (Girden & Culler, 1937). These studies demonstrate that a drug can effectively produce an internal stimulus state that is a part of the original encoding of the learned event. For example, if an animal learns a task under the influence of sodium pentobarbital and is tested later when the drug is not in the animal's system, then there is poor memory retention. But if sodium pentobarbital is administered again without further training, then the memory retention is better (Overton, 1964). Overton (1964) also showed the opposite effect for the absence of sodium pentobarbital. In that research, the original learning is done without the drug, and later the animal is tested under the influence of sodium pentobarbital. The effect of the drug at the time of testing results in poor performance. But Overton also showed that if the nondrug state is reinstated without further training, then the performance improves. Thus, some drugs have stimulus qualities that can be mismatched later in testing and result in a retrieval impairment. This effect is called state-dependent learning.

It is also possible that a drug can alter memory, but to establish this effect, it is necessary to take into account the possibility of state-dependent memory. A straightforward way to assess a drug in light of the state-dependency effect is to have both learning and testing done under the same drug state. If the same drug state is present for both the learning and for the memory retention test, then the effect of the drug on memory cannot be due to state-dependent learning (i.e., there will be an effect beyond that of state dependency).

The study by Gerrein and Chechile (1977) illustrates how a legal substance in widespread use (or abuse) can be examined to evaluate how memory storage and retrieval processes are influenced by the substance. Since the learning and testing were conducted in the same drug state, any memory effects observed would be beyond that of state-dependent learning. The drug examined was alcohol. A number of researchers have previously demonstrated that alcohol results in state-dependent learning (Goodwin, Powell, Bremer, Hoine, & Stern, 1969; Ryback, 1971; Storm & Caird, 1967). The Gerrein and Chechile study was designed to see if alcohol has an effect on either storage or retrieval processes.

Volunteers in the experiment were examined in a double-blind procedure where neither the participant nor the experimenter knew if the substance was an alcohol–orange juice drink or a placebo drink of orange juice with an alcohol aroma (Gerrein & Chechile, 1977). But there was a research planner who knew the treatment conditions for each participant. The participants were presented with a triad of words for each trial. The memory targets were words either from the common category (e.g., GRAPE

CHERRY WATERMELON) or from different categories (e.g., CHERRY TIN DENIM). The 6P model estimates for the storage and retrieval probabilities were computed for both the LOW and HIGH target organization conditions for the two groups. Like in the study of Korsakoff patients, a difference score between the PLACEBO and ALCOHOL groups was computed for both storage [$\Delta\theta_S = \theta_S$(Placebo) $- \theta_S$(Alcohol)] and retrieval [$\Delta\theta_r = \theta_r$(Placebo) $- \theta_r$(Alcohol)]. When $\Delta\theta_S$ and $\Delta\theta_r$ are positive, it indicates that the placebo group is performing at a higher level, whereas a negative value indicates that the placebo group is not performing as well as the alcohol group. The values for $\Delta\theta_S$ and $\Delta\theta_r$ are displayed in figure 7.12 as a function of target organization. The statistical analysis demonstrates that there is a reliable group storage difference for both the high and low levels of target organization. Unlike storage, the retrieval parameter has an interesting interaction with the organizational level of the target. When the target consists of three words from different categories (LOW organization), the intoxicated participants have reliably poorer retrieval than that of the placebo group. But if the target has a strong organization (three words from a common category), then the alcohol group retrieval is not impaired relative to the retrieval for the placebo group.[31] Alcohol thus affects both storage and retrieval processes. Under the influence of alcohol, the ability to store a durable memory is impaired, and alcohol also affects the ability of participants to search memory, especially when the memory targets consist of items from three different categories. When the target consists of items from different categories and each of these words is stored, then there is an increased burden on the retrieval system to quickly find these items on a time-limited recall task. Alcohol impairs the search across categories, so there is an interaction for retrieval between the drug state and target organization.

Finally, it is noteworthy that any new pharmaceutical substance requires an assessment of the effectiveness of the drug along with an assessment of potential other "side effects." Thus, the new drug must undergo a rigorous clinical trial testing to achieve approval for use. The clinical trial tests assess many metrics, including a cognitive assessment, to understand the effects of the drug. Typically, a cognitive assessment might include tasks such as digit span, visual paired-associate learning, and verbal

---

31. The interaction of the group difference is reliable via a Bayesian analysis. The output from the 6P model for each group in each condition includes a full probability distribution for the model parameters. Therefore, an analysis was done where a random vector of values for the θ_r parameter was found for each condition. Moreover, a difference score related to the change in the group difference with organization was computed for each random vector of values, that is, $\Delta^* = [\theta_r$(Alcohol) $- \theta_r$(Placebo)]$_{low} - [\theta_r$(Alcohol) $- \theta_r$(Placebo)]$_{high}$. Based on 150,000 random vectors sampled, the mean value for (Δ^*) is .488. The probability that the interaction is positive is greater than .971. This analysis thus demonstrates that there is an interaction for the retrieval parameter as a function of target organization.

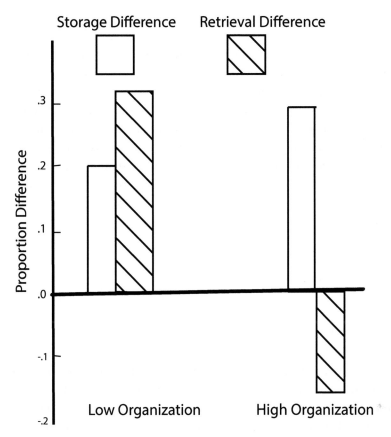

Figure 7.12
The difference between the control and the alcohol groups after a 12-sec delay for storage $\Delta\theta_S$, which is equal to $\theta_S(\text{Control}) - \theta_S(\text{Alcohol})$, and the corresponding difference between the groups for retrieval $\Delta\theta_r$. The differences are shown as a function of the target organization.

memory tasks such as the Rey auditory verbal learning test. Although these tasks are sensitive to aggregate cognitive functioning, the tests are nonetheless confounded tasks that do not isolate the specific cognitive process that might be affected by the substance. For example, a new drug proposed to treat high blood pressure will require a wide range of testing to see if the drug is effective and to see if it causes some unwanted side effect like causing people to be more forgetful. Cognitive model-based measurement could substantially improve the cognitive-assessment part of the clinical trials testing. It would be important to know if the drug affects storage or retrieval processes. Currently, model-based measurement of latent cognitive processes is not a part of the current assessment battery.

7.9 Storage and Retrieval for Recall-Based Tasks

The Chechile-Meyer task uses recall and recognition data to separate the storage and retrieval components of the correct recall proportion. Another model, which also uses recall and recognition, is presented in chapter 9; this model results in storage-retrieval measures for nonverbal animals. Furthermore, in chapter 10, there is an examination of some models that only use recognition testing, but those models currently were not employed to measure the retrieval process.[32] There are also models that use a combination of recall tests that ostensibly obtain values for storage and retrieval. Yet given the model for recall illustrated in figure 7.2, it would seem at first improbable that a combination of recall tests would result in a separation of storage and retrieval. But let us examine three models in this category of MPT models to see how they are suited as storage-retrieval measurement methods.

7.9.1 The Greeno et al. (1978) Model

The Greeno, James, DaPolito, and Polson (1978) model emerged from their interest in understanding the interference effects that occurred with paired-associate learning. This topic is reviewed in chapter 3. For the experimental condition in a study of proactive interference, the participants first study a set of paired associates **A-B**, and later the participants are given a second **A-C** list. At a still later time, the participants are cued with the **A** items, and they are asked to provide both the **B** and **C** responses. There are four observable outcome events for this procedure: event E_1 (both responses are recalled **BC**), event E_2 (the first response is recalled but not the second **B$\bar{\text{C}}$**), event E_3 (the second list response is recalled but not the first list response **$\bar{\text{B}}$C**), and event E_4 (neither list response is recalled **$\bar{\text{B}}\bar{\text{C}}$**). We only need to focus on the model-predicted proportions for the first three events because the sum of all four outcome proportions must be 1. Before presenting the Greeno et al. (1978) model, let us first express the above three events in terms of a model where the probability for recalling **B** is defined as $\theta_{sB}\theta_{rB}$, and the corresponding probability for recalling **C** is $\theta_{sC}\theta_{rC}$. These definitions are simply applying the general model for recall from figure 7.2. Let ϕ_i, $i = 1, 2, 3$, be the theoretical proportions for the first three outcome events. These proportions are

$$\phi_1 = \theta_{sB}\theta_{sC}\theta_{rB}\theta_{rC}, \tag{7.1}$$

$$\phi_2 = \theta_{sB}\theta_{sC}\theta_{rB}(1 - \theta_{rC}) + \theta_{sB}(1 - \theta_{sC})\theta_{rB}, \tag{7.2}$$

---

32. However, with a modification of the experimental task, it is possible to also measure retrieval for some of the models in chapter 10.

$$\phi_3 = \theta_{sB}\theta_{sC}(1 - \theta_{rB})\theta_{rC} + (1 - \theta_{sB})\theta_{sC}\theta_{rC}. \tag{7.3}$$

Equation (7.1) is based on the assumption that the only way to get both the **B** and **C** associates is for both items to be stored and retrieved. This assumption is reasonable provided that the responses are not already associated with the cue prior to training. Equation 7.2 is based on the two possible ways by which the participant can get the **B** item but not recall the **C** item. The first way is when both responses are stored but only the **B** item is retrieved. The second way corresponds to when only the **B** item is stored and retrieved. The rationale for equation (7.3) follows a similar reasoning process as for the E_3 outcomes.

In order for this model to be identifiable, there needs to be some assumptions to reduce the number of parameters from four to three or fewer parameters. The mathematically tractable model proposed by Greeno et al. has the following probabilities for the first three outcome cells:

$$\phi_1 = pqr, \tag{7.4}$$

$$\phi_2 = pq(1 - r), \tag{7.5}$$

$$\phi_3 = p(1 - q)r. \tag{7.6}$$

Clearly, the p parameter must be equal to $\theta_{sB}\theta_{sC}$.[33] Also, the r parameter in the Greeno et al. model is equal to θ_{rC}, and the q parameter is equal to θ_{rB}. Notice these equivalences result in equations (7.1) and (7.4) being matched. However, to match the two equations for event E_2, it is necessary to assume that the second term on the righthand side of equation (7.2) is 0. This term of $\theta_{sB}(1 - \theta_{sC})\theta_{rB}$ can only meaningfully be zero if we assume that $\theta_{sC} = 1$. In a similar fashion, to match the two equations for event E_3, the second term in equation (7.3) must also be zero. This result can only be accomplished if we assume that $\theta_{sB} = 1$. Thus, the Greeno et al. (1978) model is mathematically tractable only by assuming that both the **B** items and the **C** items are stored. In essence, these assumptions inadvertently make the model based on perfect storage. Hence, retrieval is overestimated and the model is not a method for obtaining separate measures for storage and retrieval.[34,35]

33. Greeno et al. defined the p parameter as the probability of reaching a node in memory from which both associations could possibly be retrieved (see p. 182 of the Greeno et al. book).
34. Bender, Wallstein, and Ornstein (1996) did essentially this same analysis to arrive at the same conclusion (see their Appendix A).
35. If perfect storage is assumed, then it follows that $p = \theta_{sB}\theta_{sC} = 1$. Interestingly, Riefer and Batchelder (1988) fit the Greeno et al. model to data from an experiment conducted by DaPolito and found that p was in fact approximately 1—a result predetermined by the assumptions of the model.

7.9.2 The Rouder and Batchelder (1998) Model

Riefer and Rouder (1992) developed a storage-retrieval measurement model for a free-then-cued recall task where the memory targets are words that have a natural pairing. Riefer and Rouder examined the memory of nouns that were the key words of an imagery instruction to the participants at the time of encoding, for example (the SNOWFLAKE fell on the MOUNTAIN). Alternatively, in a bizarre imagery condition, the participant would receive an instruction such as (the SNOWFLAKE climbed the MOUNTAIN). Later at the time of test, the participants were given a chance to free recall all the key nouns, and then they were given a cued-recall test where the first word is provided as a cue for the recall of the second word. There are six possible outcome events for this experimental procedure, and these outcomes are illustrated in figure 7.13A. For example, event E_1 is the case where both members of a pair are recalled on the free-recall test, and the participant is correct on the cued-recall test, whereas event E_5 is the case where one item is recalled on the free-recall test, but the person is incorrect on the cued-recall test.

Rouder and Batchelder (1998) revised the original model for this task. Model 4a from Rouder and Batchelder (1998) is shown in figure 7.13B.[36] The notation for this tree is the same as in the Rouder-Batchelder paper. The a parameter is the probability that both members of a pair are stored. The r parameter is the probability that the two members of a pair are retrieved together, and this would result in the outcome E_1. But if these items are not retrieved as a chunk (with probability $1 - r$), then there is an assumed secondary process of recovering two, one, or none of the items. These three conditional outcomes are modeled by a two-trial binomial distribution. If s is the success rate for the retrieval of an item on a binomial trial, then the outcome of both items being retrieved is s^2, whereas the probability for retrieving only one item is equal to $2s(1-s)$, and the probability of two retrieval failures is $(1-s)^2$. For all of these cases, the participant would be correct on the cued-recall test because both items are stored and linked to the cue word. However, if both of the words were not stored (with probability $1 - a$), then Rouder and Batchelder assume that it is still possible for the participants to free recall two, one, or none of the items. They assume that these three outcomes are the result of a different two-trial binomial distribution where the u parameter is the probability of successfully recovering an item. However, I find this assumption puzzling. For example, if both items are not stored with probability $1 - a$, then how is it possible to recall both items? Nonetheless, by (1) defining storage for a pair of items, (2) having both free

36. Rouder and Batchelder (1998) considered other highly similar models that were predicated on different assumptions, but model 4a was the version that fit the data the best.

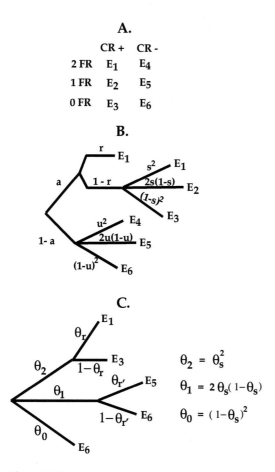

Figure 7.13
The data structure for the free-then-cued recall task is shown in panel A. Panel B is model 4a from Rouder and Batchelder (1998), and panel C is a new alternative model for this task.

and cued recall data, and (3) assuming binomial distributions for secondary processes, it is possible to obtain separate estimates for storage and retrieval. But there are other potential models for this task.

A new alternative tree model for the free-then-cued recall task is shown in figure 7.13C. The goal of this model is to see if it is possible to find a storage parameter for the case where only one member of the pair is encoded as well as for the case when neither item is stored. For the alternative model, a binomial process is employed for the storage of items; the binomial success rate is defined as θ_s. Consequently, the probability that both items are stored is $\theta_2 = \theta_s^2$, and the probability of one member of the pair being stored is $\theta_1 = 2\theta_s(1 - \theta_s)$. It follows that the probability of no storage is $\theta_0 = (1 - \theta_s)^2$. If both items are stored, then the probability of retrieval of both is defined to be θ_r. This outcome results in event E_1. If retrieval fails on the free-recall test, then it is assumed that no items are recalled, which is an E_3 outcome. Essentially, it is assumed that if one item were retrieved, then the organization of the pair would result in the other also being retrieved. For the case when one item is stored (with probability θ_1), there is a different retrieval parameter for the probability of retrieving the stored item; this parameter is $\theta_{r'}$. But in this case, the person would not be correct on cued recall (i.e., it is an E_5 event). Either the failure to retrieve a stored item or the failure to store anything results in outcome E_6.

It would be instructive to compare the above two models by examining the data from experiment 3 of the Riefer and Rouder (1992) study of common and bizarre imagery. For model 4a, the respective a, r, s, and u parameters in the bizarre imagery condition are .828, .675, .031, and .156. For the alternative model from figure 7.13C, the respective estimates for the θ_s, θ_r, and $\theta_{r'}$ parameters are .9098, .691, and .237. Both models fit the data well.[37] For the alternative model, the probability that both items are stored is $\theta_2 = \theta_s^2 = .828$, which is exactly the same value found for model 4a. The conditional retrieval parameters for the two models are very close in value (i.e., .674 for model 4a and .691 for the alternative model). But the alternative model predicts that the proportion of singleton storage is $\theta_1 = 2\theta_s(1 - \theta_s) = .164$, and the proportion where neither item is stored is .008. Comparable estimates of these proportions are not available for model 4a.

For the condition when the participants had to generate common images such as (the SNOWFLAKE fell on the MOUNTAIN), the model 4a a, r, s, and u parameter values respectively are .796, .547, .008, and .237. For the alternative model, the θ_s, θ_r, and $\theta_{r'}$ parameters respectively are .8975, .552, and .272. Both models fit the data acceptably.[38] The probability for storing both items is $\theta_2 = \theta_s^2 = .806$, which is remarkably close to the model 4a value (i.e., $a = .796$). Moreover, the θ_r parameter value of .552 is very close to the model 4a r parameter value of .547.

The above analyses indicate that the pair storage parameter a and the retrieval parameter r are values that demonstrate robust measurement because nearly the same

37. A χ^2 goodness-of-fit assessment resulted in a value of .189 for model 4a and a value of .206 for the alternative model). Neither model demonstrated a significant goodness-of-fit problem for fitting the data, although it is noteworthy that the alternative model has one less parameter.
38. The χ^2 assessment test for model 4a is 1.572, whereas the alternative model value is .485. Again the alternative model has one less parameter.

values were found by two different models. Also note that the alternative model adds new information about singleton storage and nonstorage. Moreover, it is interesting that the θ_r parameter is capturing the difference between the bizarre and common imagery conditions. Storage is remarkably high for both imagery conditions, and bizarre images are not stored at a higher rate. In chapter 13, we will return to the effect of imagery for enhancing memory. This study supports the benefits of imagery for encoding items and the additional advantage of bizarre images for having improved retrieval. It is also interesting to observe that the retrieval of a singleton $\theta_{r'}$ is substantially less than the retrieval of a stored pair of items, and it does not substantially vary across the two types of imagery instructions. When a pair is successfully stored, the quality of the encoding is substantially different from the encoding of a singleton. If a singleton is stored, then it functions like an isolated word rather than as an image. Thus, the retrieval of a singleton is substantially less successful compared to the retrieval of a vivid image.

Although the alternative model shown in of figure 7.13C appears to work well, there is a problematic aspect of the model. If two items are stored separately in memory, then it is reasonable to think that the retrieval of these items could be statistically independent. Hence, a two-trial binomial distribution of the type employed in model 4a in the Rouder-Batchelder paper is reasonable. However, does it make psychological sense to employ the idea of independent storage as was done for the alternative model? The use of a two-trial binomial distribution for storage is especially questionable for the case when the items are naturally linked via an image. Consequently, the singleton storage and the nonstorage estimates obtained with the alternative model are questionable. The alternative model really is most useful as evidence for the robustness of the storage and retrieval values from model 4a. But it would be still better for an experiment to be conducted that uses both the free-then-cued recall procedure and the Chechile-Meyer procedure. If the model 4a storage and retrieval values are still close to the corresponding estimates from model 7B, which have been extensively validated, then the measurements of these processes would be cross-validated.

Finally, although storage and retrieval can be measured with the free-then-cued recall procedure, the measurement is still dependent on the assumption of the two-trial binomial distribution and natural pairing of items of comparable meaningfulness. There are cases, such as for the Chechile and Ehrensbeck (1983) study that was reviewed in subsection 7.7.4, where the two members of the pair are not comparable in meaningfulness (e.g., a pair of items like VAJ-PILOT). The Rouder-Batchelder model would not be suitable for that type of experiment.

7.9.3   The Batchelder and Riefer (1980) Model

The Batchelder and Riefer (1980) model is designed to measure the storage and retrieval for a clustered pair of words. Suppose, for example, that a list of words is presented to a participant one at a time and where adjacent words on occasion are highly associated such as OXYGEN HYDROGEN or DOCTOR NURSE. On other occasions, the adjacent words are unrelated such as OXYGEN SLIPPER. This type of adjacency is the presentation of singleton items. These investigators were interested in the storage of a pair cluster and the retrieval of the pair. At the time of free recall, the output of the participant is examined in terms of both the recall of items and the adjacency of the output responses. For items that originated from a pair of highly associated items, there are four outcome events: E_1 (both members of the pair are recalled adjacently at the time of test), E_2 (both members of the pair are recalled but not adjacently), E_3 (only one member of the pair is recalled), and E_4 (neither member of the pair is recalled). These four outcome states form a four-cell multinomial. The singleton items are part of a separate two-cell multinomial, which we can denote as F_1 (correct recall of the item) and F_2 (incorrect recall).

The Batchelder-Riefer pair-clustering model has three parameters, which are c, r, and u. The c parameter is the probability that the pair is clustered in memory, that is, a cluster (**AB**). The r parameter is the conditional probability of retrieving the pair on the recall test given that the pair is stored. The u is the storage and retrieval of a singleton. The model for the four-cell multinomial is fully specified by the following three equations because the sum of the probability across all four outcomes must be 1. Let ϕ_i, $i = 1, 2, 3$ denote the model prediction for the first three outcome states. For the pair-cluster model, the predicted ϕ_i are

$$\phi_1 = cr,$$

$$\phi_2 = (1 - c)u^2,$$

$$\phi_3 = (1 - c)2u(1 - u).$$

If the cluster (**AB**) is stored with probability c, then there is a probability of r for an adjacent free recall of the pair; hence, that outcome has the probability of cr. In the case that the pair is not clustered with probability $1 - c$, there are four possible states of memory. These states are (1) two members stored separably (**A**) (**B**), (2) only the first item is stored (**A**), (3) only the second item is stored (**B**), and (4) neither item is stored (). It is assumed that adjacent output will not occur in any of those states of memory. Moreover, the model assumes that events E_2, E_3, and E_4 are determined by a two-trial binomial process with the success rate being the same as the u parameter associated with the storage and retrieval of a singleton. Hence, the probability event E_2 is equal to $(1 - c)u^2$, whereas the probability of event E_3 is equal to $(1 - c)2u(1 - u)$.

The paired-clustering model can estimate the storage and retrieval of a clustered pair of items by assuming a conditional two-trial binomial process and by using the output adjacency information. The singleton trials also help to provide an extra degree of freedom for model testing. Like the Rouder and Batchelder (1998) model, this model works only when the two members of the cluster are equally meaningful. Furthermore, this model is not capable of separately measuring the storage and retrieval of a singleton item. Finally, it would be informative to see how the pair-clustering model compares to model 7B, but to achieve that goal, a new experiment would be needed where both the paired-clustering methodology and the Chechile-Meyer procedure were employed.

7.10   General Discussion and Summary

Psychology has exhibited a strong affinity toward theory or, to be more accurate, a high regard for theory in the form of a working hypotheses. Theories are exciting, and their formulation and test have occupied a prominent place in the psychological literature. In contrast, the focus on measurement is often seen as an important but secondary issue. But facts in science are what matter the most, and the facts are only detected by using a measurement tool. If the measurement tool is flawed because of the entanglement of processes, then it is difficult to build a body of facts. Theories and hypotheses proliferate without achieving progress toward a consensus about conflicting hypotheses. In the words of the fictional character, Sherlock Holmes,

It is a capital mistake to theorize before one has the data. Insensibly one begins to twist facts to suit theories, instead of theories to suit facts. (Arthur Conan Doyle, *The Adventures of Sherlock Holmes: A Scandal in Bohemia*, 1892, p. 10)

Once good measures are developed, and data are collected by using the measurement tool, then preexisting hypotheses and theories are often shown to be overly simplistic or just plain wrong. Measurement is the foundation of experimental science.

A case is made here that storage and retrieval are fundamental for understanding memory, and probability metrics for these processes can be obtained by means of model-based measurement. The 6P and 7B models are shown to have high construct validation (i.e., the storage and retrieval measures from these models covary in a fashion that makes sense given the constructs of storage and retrieval). The 6P and 7B models are designed for a specific task that employs the random intermixing of recall test trials with old- and new-recognition test trials. Storage and retrieval probabilities are not assumed; rather, the values estimated for these processes are a function of the data. In general, both processes can exhibit a decrement. So far, we have only discussed a few experiments that have used these measurement models, but nonetheless, the memory story revealed by these studies is more complex and nuanced than the previous theorizing. For example, the developmental studies establish that by age 11, the ability to *store* memory targets has matured to the same level as for college-aged students, but the college students can use a richer and more elaborated semantic memory to enhance their retrieval of information. In another study, the measurement model showed that Korsakoff patients exhibit a previously undetected pattern of dynamic variation of their ability to store and retrieve information. The main difference between the Korsakoff and control patients after a short delay of 4 sec is with the retrieval rate, but after a longer delay of 12 sec, the storage parameter is the key discriminatory factor between the groups. In yet another study, the measurement model detected a novel memory impairment in dyslexic children in terms of their ability to preserve memory representations of words they could read, but this problem was limited to cases where the material had similar features. We also saw that the measurement model is able to detect drug effects beyond that of state-dependent memory. Moreover, a storage-retrieval separation model for the free-then-cued recall task resulted in new insights about the memory retention of items encoded as images. Finally, as discussed previously in chapter 5, the measurement model detected a storage loss for information that should have been in a permanent store according to the widely believed Atkinson and Shiffrin (1968) theory (Chechile & Ehrensbeck, 1983). The Atkinson-Shiffrin model hypothesized that information in their long-term store was permanent, but no evidence was directly provided for this assumption. The finding of a storage change for items presumed to be in the long-term store illustrates the problem of premature theories. Theories and hypotheses are important, but theories are best after good measurements have established a solid foundation.

Appendix: Multinomial-Processing Tree (MPT) Models

The models developed for the Chechile-Meyer task were among the first examples in psychology of a class of models that have come to be called multinomial-processing tree (MPT) models. The term *multinomial-processing tree* emerged after a paper by Riefer and Batchelder (1988), but this class of models already had a history both inside and outside of psychology. The book by Elandt-Johnson (1971) used MPT models in the field of genetics. The Chechile and Meyer (1976) paper was a formal model in this class, but it was based on the earlier Chechile (1973) paper. MPT models are measurement models based on (1) a specific experimental procedure; (2) categorical data with its characteristic likelihood function in terms of the probabilities for the various observational categories, which are denoted here as ϕ_i, $i = 1, \ldots, k$; and (3) a scientific rationale that describes each ϕ_i probability in terms of latent psychological process probabilities θ_j, for $j = 1, \ldots, m$. Because the observational outcomes are probabilities and the latent parameters are also probabilities, it follows that there is a probability-tree representation for the model. MPT models have now been employed on a wide variety of topics. For reviews of some of the research with MPT models in psychology, see Batchelder and Riefer (1999) and Erdfelder et al. (2009).[39]

A MPT model has a number of desirable properties. First, the model is based on categorical data, and typically there is not a problem with measurement or recording error for categorical data. For example, it is clear how to score a recall trial in terms of the observational categories of correct versus incorrect. We need not worry about scoring reliability, and we do not need to have different raters make judgments about the scoring of the data. The data from the experiment are fully captured by the frequencies in the various observational categories. The term in statistics for categorical data is *multinomial*. For example, for the Chechile-Meyer task, there are three separate multinomials. The recall data consist of a two-cell multinomial (or a binomial) data structure. The old- and new-recognition data each consist of a four-cell multinomial for the four scoring categories (high-confident NO, low-confident NO, low-confident YES, and high-confident YES).

Multinomial data also have the advantage of complete agreement about the statistical likelihood function. For a multinomial with $k + 1$ outcome categories, the likelihood function is the probability of obtaining the frequency data n_i, for $i = 1 \cdots k + 1$ given the population proportions ϕ_i for the outcome categories. For any multinomial, the likelihood is

$$P\left([n_1, \ldots, n_{k+1}] \mid [\phi_1, \ldots, \phi_{k+1}]\right) = \frac{n!}{n_1! \, n_2! \cdots n_{k+1}!} \phi_1^{n_1} \phi_2^{n_2} \cdots \phi_{k+1}^{n_{k+1}} \tag{7.7}$$

where $n = n_1 + \cdots + n_{k+1}$ and $\phi_{k+1} = 1 - \phi_1 - \phi_2 - \cdots - \phi_k$. When there are separate multinomials as with the Chechile-Meyer task, the combined likelihood function is the product of the individual likelihood functions. With MPT modeling, the psychologist proposes a tree structure to explain how the outcome cell probabilities ϕ_i are functions of the latent probabilities of interest. There might be alternative models for the latent

---

39. MPT models have also been expanded to account for more complex data that have categorical information and continuous information, for example, the response times for correct and incorrect data (Hu, 2001; Schweicker & Chen, 2008).

psychological processes, but there is total agreement about the statistical likelihood function shown in equation (7.7). It is unusual in statistics to have a consensus about the likelihood function. For example, the likelihood function for response-time data might be a wide number of possibilities. Consequently, a strength of MPT modeling is that it is grounded on the foundation of a universally accepted likelihood function.

The modeling philosophy for the MPT approach is different from other models in psychology. There are many models in cognitive psychology, but these models are designed to account for effects. Sometimes a model is more ambitious because it attempts to account for more than one effect. For the conventional models in cognition, the parameters are designed to relate to general processes, and the theorist tries to limit the number of parameters to a minimum. In conventional models, the predictions of the model apply across many conditions and experiments, and the fitting of the parameters also spans multiple conditions. With conventional models, if there are two competing models that are equally explanatory, but the models differ in terms of their theoretical complexity, then there is a preference for the more simple model. The goal of conventional models is to provide an understanding of the change in the dependent variable. But a key difficulty with conventional models is the entanglement problem of the dependent variable. For example, a conventional model that focuses exclusively on retrieval processes to account for behavioral changes would be flawed when the dependent variable actually is affected by processes other than retrieval. Conventional models work best after clean measures of underlying memory processes have first been obtained. A conventional model is thus likely to be premature if the model is fit to confounded dependent variables that tap multiple latent processes.

In contrast to conventional cognitive models, MPT models are not designed to explain any psychological effect. Also, MPT models are not designed to make predictions across experimental conditions. The models are also not necessarily designed for standard tasks. Rather, the goal is the measurement of fundamental cognitive processes. A new experimental task might be constructed to obtain the necessary data for measuring the latent processes. The MPT model is an analysis of the factors that are involved on specific tasks. A good MPT model should be based on solidly established facts or principles that have been gleaned from prior research and based on careful reasoning about the experimental task. Unlike conventional models, there is no effort to have only a few parameters. If there are many potential confounding factors that compromise the interpretation of the data, then these factors are included in the model. If there are several alternative ways to model the confounding factors, then the alternative models are also examined. Because the MPT model must be identifiable in each experimental condition, it might be necessary to make some model simplifications to achieve this goal. We saw examples of model simplification in the 6P and 7B models. Both of those models were proven to be identifiable. Model identifiability deals with the unique linking of the likelihood function to a unique configuration of model parameters (Chechile, 1977a, 1998). If there were two potential configurations of the model parameters that result in the same likelihood function, then those two configurations would be *observationally equivalent*. There are no observationally equivalent points for an identifiable model. Chechile (1998) provided a formal method that can be used to assess if a proposed model is identifiable.[40]

The parameters of the model are a mixed bag of important process parameters along with many nuisance parameters. The nuisance parameters are still important because they provide the basis to correct for factors that would otherwise confound the dependent measures. When the key parameters of the MPT model are estimated, the latent processes have been unambiguously measured. With such measurement, it is possible to explain effects such as the loss of memory with retention interval in terms of changes in the key latent parameters.

Although the calculations for estimating the parameters of the MPT model are elaborate, there are many software tools available. Examples of general packages include MBT (Hu, 1999), GPT (Hu & Phillips, 1999), MultiTree (Moshagen, 2010), AppleTree (Rothkegel, 1999), MPTinR (Singmann & Kellen, 2013), and HMMTree (Stahl & Klauer, 2007). The 6P and 7B models have been implemented so that both MCMC and PPM fits are produced along with a host of other useful point and distributional information, but this software is designed for these specific models.[41]

40. The identifiability issue is complex and not as simple as just having fewer parameters than degrees of freedom. It is not as simple as demonstrating a unique parameter estimate for the parameters. Yet there is a straightforward procedure for determining if a model is identifiable. The method begins by assuming that the model is not identifiable and that there are two observationally equivalent points. But in the subsequent analysis, if it is established that these points are actually the same point, then it is proved that the model is identifiable. However, if the model is not identifiable, then the procedure discussed in Chechile (1998) will find examples of observationally equivalent points.

41. Classical maximum likelihood estimates are also computed for the 6P and 7B models, along with a sophisticated overall goodness-of-fit metric using the power divergence principles from Read and Cressie (1988).

8   What Does Memory Strength Mean?

8.1   Chapter Overview

Not all model-based measurements of memory are valid. If either the model or the theoretical construct is flawed, then the subsequent measurement is questionable. A case is made in this chapter that the measurement of the theoretical construct of memory strength is problematic in many applications: (1) because the concept of a single memory strength value is difficult to justify for associative memory of realistic events and (2) because of problems with the measurement model itself. However, the critique in this chapter is limited to the area of associative-event memory because we have previously shown in chapter 1 that strength is an important construct underlying habituation and sensitization. Also, strength is likely a valid construct for long-term semantic memory.

8.2   The Problem of Measuring Memory Strength

Prior to measuring any quantity, it is important to have a rough premeasurement understanding of the property. For example, a vague understanding of temperature predated the early temperature gauges by thousands of years. Certainly, an informal case can be made that memory strength is a meaningful concept. We all have some vivid memories that are unlikely to be forgotten over our entire life, whereas we also have other memories that seem less vivid and might be eventually lost. On a more technical basis, studies in the area of lexical decisions have found a robust priming effect that is called *repetition priming* (Forbach, Stanners, & Hochhaus, 1974: Kirsner & Smith, 1974; Scarborough, Cortese, & Scarborough, 1977). In a lexical decision task, the participants must decide if a letter string is a word or not. If a letter string is repeated later, then the decision tends to be faster. A plausible hypothesis for repetition priming is the idea that the presentation of the stimulus results in a temporary strengthening of the representation for the word in memory. The repetition-priming effect has also been reported for a component of the event-related potential (ERP).[1] A repeated stimulus tends to elicit a change in the magnitude of the N400 component of the ERP relative to that of the first presentation of the stimulus (Rugg, 1985, 1987). But repetition priming is a complex effect that could be caused by a number of other processes. Experimental measures are not tapping a single process; hence, there must be a model for isolating a strength measure.

Although there are reasons for the concept of memory strength, it is also important to point out that strength is a slippery notion as well. There are many questions about the notion of memory strength. For example, are there different types of strength in a similar fashion as the distinction between explicit and implicit memories? How does the value of memory strength reflect the possibility that a piece of the memory might be lost, but the rest of the memory is intact? How does a strength arise for a memory trace with qualitatively different features? Later in chapter 10, a measurement model called the IES model will be described that addresses these questions. The IES model is an MPT model, and it was not developed for the purpose of measuring a unitary value for strength.

There is a standard strength measurement method in widespread use in experimental psychology. This tool is based on a model-based measurement method that was developed in another area of psychological research (i.e., the signal detection theory from sensory psychology). But there is now considerable evidence that demonstrates that this measurement method is problematic for some applications. Consequently, in this chapter, the signal-detection approach for measuring memory strength is discussed and critiqued. The critique also highlights additional reasons for the storage-retrieval approach discussed in the previous chapter. But before critiquing the signal-detection approach, it is important to first describe the historical framework of strength measurement from sensory psychology and psychophysics. This background is discussed in the next section.

---

1. The ERP is obtained by averaging the raw EEG electrical potentials at a particular site on the scalp. For example, at a particular scalp site and after a fixed time from the onset of the critical event, the various EEG potentials are averaged. The signal averaging is assumed to cancel all electrical potentials that are not consistent across the collection of EEG waveforms. The result of this averaging is the ERP signal at each time unit and recording site. The N400 component is an important negative going peak that tends to occur 400 msec after the stimulus onset. See Kutas and Federmeier (2011) for a review of the ERP literature. For additional information about the EEG, see the glossary.

8.3 The Basic Signal-Detection Representation

A classic research topic in the early days of experimental psychology dealt with the functional relationship between the strength of a physical quantity and the human's ability for detecting the strength of the quantity. This research question was first examined by Fechner (1860). Psychophysics is the discipline within experimental psychology concerned with the functional linkage between physical measurement and psychological measurement. Examples of physical dimensions that result in a perceptual strength would include the luminosity of light and the amplitude of sound. For each of these physical dimensions, there is a perceived strength dimension (e.g., brightness for luminosity and loudness for sound amplitude). For example, the intensity of a sound is measured by the amplitude of the sound wave, whereas the detection of loudness is measured by the human's response to the tone. This linkage can be examined in an absolute magnitude context, as in the case of a soundproof room where the sound amplitude is increased slowly until a person can detect the tone. Alternatively, the relationship can also be done in a difference context where there is a noise background of a particular physical intensity, and the noise intensity is slightly increased until the person can detect the difference. Based on Fechner's work and the work of subsequent researchers, it is now known that there is a nonlinear relationship between the physical strength and the perceived strength. Yet it should be stressed here that not all physical quantities result in a perceived psychological strength. For example, the wavelength (or alternatively the frequency) of light can be varied in physics along a continuum, but animals do not detect a corresponding strength continuum for wavelength. Rather, different wavelengths of light are detected psychologically as qualitative differences (i.e., different colors). The difference between the two cases when there is a psychological strength continuum (such as brightness) and when there is not a strength continuum (such as color) is best understood by the human's ability to order stimuli. If a dimension has a sensory strength continuum, then the person can easily order any two different stimuli in terms of that property. For example, for the brightness dimension, people can detect the brighter of two stimuli on that dimension. If a person is given two stimuli that differ in color, then the person cannot order the two colors in terms of their strength. Dimensions that result in a psychological strength continuum must be properties that can be ordered, and not all physical properties result in a psychological response that can be ordered. This important distinction between types of continua is not based on physics, but it is due to the physiological properties of the sensory system (Stevens, 1961). For example, in the Békésy (1960) place theory for hearing, the frequency of a sound, which is a qualitative property rather than a strength property, corresponds to the specific hair cells being activated on the basilar membrane of the ear. Different hair cells result in the detection of different pitches. But the loudness of a sound (which has a strength property) is due to the number of hair cells activated and by the firing rate of those hair cells. Thus, loudness has the property of being ordered in terms of intensity, but pitch, like color, is a sense that is not ordinal. If people evolved in a fashion where pitch caused a physiological variability in a firing rate, then pitch would be ordered, and there would be a sensory strength dimension for pitch. But we did not evolve in that fashion, so when it comes to the dimension of pitch, there is not a sensory strength continuum. Stevens (1961) called continua that could be ordered on a sensory strength dimension as *prothetic* and called continua that could not be ordered on a sensory strength dimension as *metathetic*. For metathetic continua, Stevens stressed that there is a substitutive property in terms of the physiological response. For example, different specific hair cells are frequency detectors in the Békésy (1960) place theory for hearing. A qualitative difference is detected when one detector is substituted for another detector. In psychophysical scaling, the focus was on the prothetic properties that possessed a magnitude such as the light intensity, the weight of objects, and the sound-loudness level. But not all sensory information is prothetic. Many physical properties are metathetic, and strength is not a meaningful construct for such properties. The anatomy of the sensory system determines which attributes are qualitative and which attributes are quantitative. The human sensory system has both qualitative and quantitative properties depending on the sensory attribute. Strength is a concept limited to quantitative properties.

When it comes to the measurement of the strength for a prothetic sensory attribute, there is an entanglement problem for the behavioral measures. Two latent processes influence the detection of the signal (i.e., the magnitude of sensory response and the observer's decision criterion). A risky observer is a person who has a tendency for responding in the affirmative, even when no stimulus is present. Such a person will make more false-alarm errors relative to a person that uses a more stringent decision criterion for an affirmative response. Signal-detection theory was invented to deal with the entanglement of sensory sensitivity with the criterion for making a response (Green & Swets, 1966; Tanner & Swets, 1954). Signal-detection theory (SDT) was one of the first model-based measurement methods developed in mathematical psychology.

Figure 8.1 illustrates the basic psychological representation for SDT. It is assumed that there is an underlying psychological strength distribution for the background "noise" condition. The term *noise* refers to the strength distribution for the trials when the target stimulus is absent. In the initial application, the stimulus-absent condition was in fact a white noise (static-sounding) background. For the stimulus-present

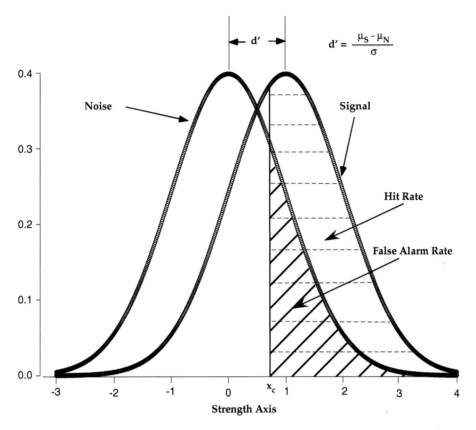

Figure 8.1

The hypothesized latent representation for two sensory conditions: (1) a signal condition where the target is present and (2) a noise condition where the target is absent. Also shown are the hit and the false-alarm rates when the decision criterion is located at the point x_c on the strength axis. The measure of signal sensitivity is $d' = \frac{\mu_S - \mu_N}{\sigma}$ where μ_S is the mean of the signal distribution, μ_N is the mean of the noise distribution, and σ is the standard deviation of the distributions. Each distribution is a Gaussian or "normal" distribution.

condition, there was a slightly louder white noise background. The observer's task was to detect the trials where there was a louder static sound. Now the noise condition has come to have a generic meaning for the stimulus-absent condition. Thus, the noise condition could be the background luminosity of a screen or the saltiness of a solution. The strength representation for the noise condition is assumed to be characterized by a Gaussian (or normal) probability distribution on the underlying psychological strength continuum. In other words, there is variability assumed for the response to the noise stimuli. The underlying normal distribution is assumed and is not directly observed. The mean of the noise distribution is denoted as μ_N and σ is the standard deviation.[2] The other condition in a sensory-detection experiment is for the test trials when the stimulus is present. For example, the stimulus-present condition might be a slightly louder white noise sound, a brighter light, or a saltier solution. The generic term used for any of these examples is the *signal* condition. It is assumed that there is another normal probability distribution for the signal condition. The mean of the signal distribution is μ_S and the standard deviation is again σ. In the more general formulation of SDT, the standard deviations of the two distributions are different (i.e., $\sigma_N \neq \sigma_S$). It is also assumed that the person adopts (perhaps temporarily) a decision criterion x_c for making decisions about whether a test stimulus is a target or not. If the test stimulus elicits a strength $x \geq x_c$, then the person is assumed to respond YES. But if the strength is less than x_c, then the person is assumed to respond NO. The area under the probability density curve for the signal to the right of the decision point x_c (the horizontal dashed shaded area) is the hit rate (i.e., the proportion of YES responses on signal-present tests). The area to the right of x_c under the probability density curve for the noise distribution (the diagonally shaded area) corresponds to the false-alarm rate (i.e., the proportion for responding YES on noise trials). In SDT, the measure of the sensitivity to the sensory stimulus is $d' = \frac{\mu_S - \mu_N}{\sigma}$. The value for d' is strictly a function of the separation between the two hypothesized distributions in units of the standard deviation. The parameter x_c is strictly a measure of the person's decision criterion. In STD, the quantities of d' and x_c can be extracted from the observed rates for hits and false alarms.

If the value for x_c changes because of a shift in the person's decision criterion, then the hit rate and false-alarm rate change, but the value of d' remains constant. The constant value for d' causes there to be a functional relationship between the hit rate and the false-alarm rate. As the value of x_c decreases, then the hit rate goes up (i.e., the area for $x \geq x_c$ under the signal distribution increases), but the false-alarm rate also goes up because the corresponding area under the noise distribution also increases. The curve that shows the trade-off of the hit and false-alarm rates for a fixed sensitivity or d' value is called the *receiver operator characteristic* (ROC). An ROC curve is a plot of the possible hit rate (on the y axis) versus the corresponding false-alarm rate (on the x axis) as x_c takes on all possible values. Since the ROC is for a fixed sensitivity or d', the ROC

---

2. For the normal distribution, the y axis corresponds to probability density $f(x)$, and $f(x) = \frac{1}{\sqrt{2\pi}\sigma} \exp\left(\frac{-(x-\mu)^2}{2\sigma^2}\right)$, where the mean and standard deviation of the bell-shaped distribution are respectively μ and σ. The probability for an outcome between any two points corresponds to the area under the probability density function between the two points.

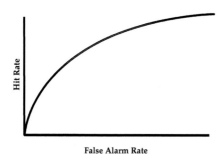

Figure 8.2
The receiver operator curve (ROC) for a standard normal model of signal detection.

curve is often called an isosensitivity curve. See figure 8.2 for a typical ROC curve. If the person has no detectability, then the signal and noise distributions are the same, and the ROC is a diagonal straight line between the (0, 0) point and the (1, 1) point. If d' is large, reflecting the fact that the two distributions have a sizable separation, then the ROC curve still begins at the (0, 0) point and ends at the (1, 1) point, but the curve is bowed like the one shown in the figure.

The developers of SDT argued against threshold models for sensory detection. For example, consider the Blackwell (1963) threshold model. In this model, the person is assumed to have two discrete states for detection. One state is below threshold, and the other state is above threshold. There is a probability (δ) that a target stimulus can trigger the above-threshold state, but it is assumed that the noise distribution has no chance for triggering the above-threshold state. There are two possible *pure* strategies for this model. Strategy A is one of always responding YES regardless of the detection state elicited, and strategy B is the one of responding YES only when the above-threshold state is triggered by the test stimulus. Strategy A corresponds to the (1, 1) point on the ROC plot, whereas strategy B corresponds to the point (0, δ). Although the two strategies are extreme cases, an ROC function results by considering all the possible mixtures of these two pure strategies. The person is assumed to adopt a mixture of the two pure strategies. For example, consider the case when strategy B is followed on p_B proportion of the test trials, and strategy A is followed on $1 - p_B$ of the test trials. The person with this mixture would have a hit rate of $p_B\delta + (1 - p_B)$ and a false-alarm rate of $1 - p_B$. As the value of p_B varies over all possible cases, the resulting ROC function is the straight line between the two extreme points. This ROC is illustrated in figure 8.3. Consequently, the ROC for the Blackwell model is a straight line, whereas the standard SDT model has a curved ROC. Like the signal-detection model of Green and Swets (1966), the Blackwell

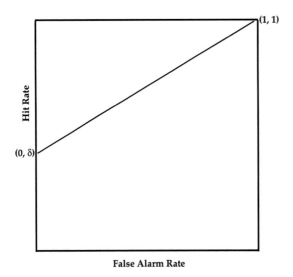

Figure 8.3
The receiver operator curve (ROC) for the Blackwell (1963) threshold model.

model treats the behavioral responses of the hit and false-alarm rates as a function of two latent parameters. One parameter (δ) deals strictly with sensory detectability (i.e., it is the proportion of times that the stimulus triggers the above-threshold state). The other parameter (p_B) is strictly a function of the observer's decision strategy (i.e., the proportion of times that strategy B is adopted). The Blackwell threshold model is a different form for a strength model. As the stimulus gets stronger, the value of δ increases, whereas in the classic signal-detection model, a stronger stimulus has a larger value for d'.

The ROC functions for the classic SDT model and the Blackwell threshold model are different. Consequently, a method for testing these two strength theories is to examine empirical ROC functions. To obtain an empirical ROC, the experimenter needs to plot the observed hit and false-alarm rates for several different conditions that vary in the observer's decision criterion. This assessment is challenging because (1) there are statistical errors in both the vertical and horizontal directions for each point, and (2) there are only a few points for the assessment. Nonetheless, empirical ROCs do not support the prediction that all the points lie on a *single* straight line (Swets, Tanner, & Birdsall, 1961). Yet Laming (1973) observed that the empirical ROCs do not rule out a theory where there are two or more different straight-line segments for the function. Nonetheless, the rejection of the prediction of the Blackwell (1963) model has been widely interpreted (1) as an indication that the classic SDT is correct and (2) as an indication that the concept of a sensory threshold is incorrect. But these conclusions are an overreach on both accounts. Clearly, as any clinical audiologist routinely demonstrates on a daily basis, there are stimuli that are below a detection threshold, and this value is a function of the age and of the hearing health of the patient. Nonetheless, given the reasonableness of the basic signal-detection model and the problem of the Blackwell (1963) sensory-threshold theory, the signal-detection framework became an accepted method in sensory psychophysics, although it is important to remember that it is limited to cases where the stimulus could vary in an intensity (like brightness) as opposed to varying in terms of a qualitative property (like color).

Strength theory for memory began with the paper by Egan (1958). Egan observed that in yes/no recognition testing, there are target-present and target-absent trials that parallel the signal-present and signal-absent trials in sensory-based SDT. In the memory application, it is assumed that there is a strength axis associated with the strength of the memory trace. Foil items are also assumed to have a distribution associated with their strength like that of the noise distribution, and old items presented for learning are assumed to have an augmented strength distribution like that of the signal distribution. In memory strength theory, strength is equated to the construct of "familiarity." Even novel foils might have a sense of familiarity, but old items should have a greater sense of familiarity.[3] Consequently, in the strength theory of memory, the general technology of sensory detection is borrowed and transformed to a psychological strength/familiarity latent dimension for representing memories. Murdock (1974) provided some modest support for the SDT application to memory strength by finding a nonlinear ROC curve as would be predicted from the SDT framework. Murdock (1974) also tested and rejected the Blackwell (1963) threshold model. But the rejection of the Blackwell threshold model is a long way from providing validating evidence for

3. Some might argue that the x axis represents "discriminability" rather than strength. However, the term *discriminability* is a relative construct (i.e., discriminating between two light sources). A difference in the strength between two stimuli predicts discriminability, but the opposite statement is not true. For example, suppose Light 1 has a strength value of 18 and Light 2 has a strength value of 10. The difference between these two stimuli is 8 units, and this number predicts the discriminability between the stimuli. Moreover, suppose further that there is a Light 3 that has 17 units of strength. The discriminability between Light 1 and Light 3 is 1. Thus, distinguishing between Light 1 and Light 2 is easier than distinguishing between Light 1 and Light 3. Given any two stimuli, we can compute a discriminability score based on the strength values of the two stimuli. However, we cannot replace the strength values with a discriminability index. Light 1 has many different values for discriminability depending on the intensity of the other light source. Therefore, the x axis is a strength axis rather than a discriminability axis.

the memory strength measurement model obtained by borrowing the signal-detection framework of sensory psychology.

A number of additional features of the SDT approach to memory strength have been subsequently advanced, but only two ideas will be discussed here.[4] First, memory researchers have found it necessary to modify the equal-variance SDT model as shown in figure 8.1 with the unequal-variance version of the theory (i.e., $\sigma_S \neq \sigma_N$). Empirical fits to data have found that the ratio $\frac{\sigma_S}{\sigma_N}$ is often greater than 1 (Ratcliff, McKoon, & Tindall, 1994; Ratcliff, Sheu, & Gronlund, 1992).[5] Second, some researchers have advanced a dual-process variation on the SDT strength model for memory (Wixted & Stretch, 2004; Yonelinas, 1994, 1997, 2002). The two processes discussed in these models are familiarity and recollection. For example, in the Yonelinas (1994) model, the representation of foil items is that of a unimodal normal distribution like for noise in figure 8.1. But the corresponding model for old items is a mixture of two processes. It is assumed that on R_0 proportion of the old- recognition test trials, the person is able to recall the target information and produce the YES response. If the person can produce the target when probed with the old item, then clearly the person would always respond YES on a recognition test. But for all the other old recognition test trials where the target is not recollected, the person has a familiarity process just like that of the signal-detection model with equal variance for targets and foils.[6] The resulting ROC curve for the dual-process model is shown in figure 8.4. The ROC has extreme points of $(0, R_0)$ and $(1, 1)$ and between these extremes there is a curved path.

8.4   The Problem with Defining Strength

There is a substantial body of research in psychology based on the SDT approach for memory measurement. Nonetheless, I believe that this approach is questionable. To see why, let us begin the critique with perhaps the most subtle point. This point concerns the problem of meaningfully defining strength in a fashion that is consistent with our

4. Readers who would like additional information about SDT can find valuable information in a number of texts (Green & Swets, 1966; Macmillan & Creelman, 2005; Wickens, 2002). Also, a brief paper by Stanislaw and Todorov (1999) is a practical guide about calculating a variety of signal-detection measures.

5. This finding has been studied in terms of models such the SAM model reviewed in chapter 3. However, other nonstrength theories can also account for this feature.

6. This second characteristic of a mixture of processes will also be shown for foils. This finding combined with the assumption of unequal variances between the two distributions results in an identifiability problem for the dual-process model as will be noted in section 8.7. Thus, for the dual-process model, the standard deviations of the two distributions are assumed to be equal.

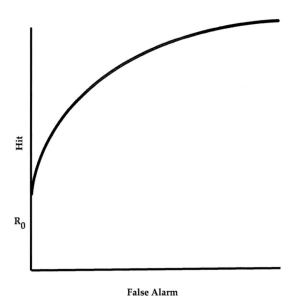

Figure 8.4
The receiver operator curve (ROC) for the Yonelinas (1994) dual-process model where R_0 is the proportion of old-recognition test trials that result in a YES response because of a recollection of the target.

contemporary understanding of the nature of the memory trace. This section is thus focused on theoretical problems with the SDT approach for strength measurement. Later sections of the critique will focus on empirical studies that result in problems for the SDT approach for memory strength measurement.

The concept of strength is an old idea, dating back to Aristotle, and it is a basic aspect of classical association theory. In this framework, there is a single bond or association between the mental representation of two environmental events, and that bond has a strength value. For recognition memory, the stimulus is assumed to have a strength for the bond between the stimulus **A** and all other stimuli that are part of the learning context **X**. In association theory, the stimulus representation does not have a strength value, but rather it is the connection between the stimulus and the experimental context that is assumed to have a strength. Yet the idea of a single bond or association in contemporary neuroscience has been replaced with a complex neural network. For example, in part I, a case was made that the memory trace is a widely distributed neural web of features and propositions. Where is the strength in such a network? One possible answer to this question is to claim that the connections between nodes activated by the stimulus have a gradient of strength values. Yet this answer is not the whole story because there would be many strength values since the stimulus activates a whole network of links and nodes. Even for the simple case of habituation in the *Aplysia*, memory required a vector of strength values, and it was necessary to have a mathematical operation to obtain a single value for the strength of the habituation response to the target stimulus. The sum of the component strength values was used for the habituation model. With habituation, the memory is manifested by a *reduction* in strength and not by an increase in strength because habituation is a reduction of a hardwired habit. Sensitization deals with the increased strength of a different hardwired habit. Thus, the role of strength is complex even for hardwired habits in animals such as the *Aplysia*. But the overall strength in human recognition memory is vastly more complex than habituation/sensitization because the information content is not a simple habit and because the trace has qualitative attributes and propositional elaborations within its distributed representation. It is easy to claim that there is a familiarity response that has an overall strength value, but it is not clear how this aggregated strength is produced. There are global memory models, like the SAM and REM models discussed in chapter 3, where the memory trace has features, each with strength linkage to the features of a context. Any particular test probe (either a target or a foil) elicits an overall activation, which is mapped onto a single continuum. But can all knowledge be represented in such a representation?

Some psychologists have proposed a feature-space theory that might serve as a model for the generation of an overall strength value for the stimulus (Ashby, 1992; Biederman, 1986, 1987, 1990; Estes, 1994; Kruschke, 1992; Nosofsky, 1986). A feature-space model assumes that all existing memories are stored in a vector space. Thus, a memory is a point in a vector space with many dimensions. A key concept in these feature-space theories is the similarity between memories. These theories typically use the mathematical operation of taking a scalar product of two vectors as a measure of the similarity between the test stimulus and any other memory.[7] The overall strength in these feature-space theories is a function of the elicited similarity, which is the sum of

---

7. The scalar product of two vectors is also called either an inner product or dot product of vectors. The operation is the sum of the product of the corresponding components of the two vectors. For example, for vector $\mathbf{a} = (a_1, a_2, \ldots, a_m)$ and vector $\mathbf{b} = (b_1, b_2, \ldots, b_m)$, the scalar product is equal to $\sum_{i=1}^{m} a_i b_i$. It is well known that the scalar product of two vectors is also equal to the product of the length of the two vectors times the cosine of the angle between the vectors. For vectors at right angles, the scalar product is 0, whereas for vectors that are identical, the cosine is 1. Also, the scalar product of vectors is related to the correlation between the two vectors. For these reasons, the scalar product has been used as a measure of the similarity between two vectors.

the scalar product of the test stimulus vector with *all of the other vectors in memory*. This operation will yield a value for novel test stimuli as well as for old items. For a memory study with word stimuli, the presentation of a word earlier is assumed to increase the component strength values from that of the preexperimental representation. But it should be pointed out that many feature-space papers discuss the recognition process in general without supplying a model for realistic stimuli. Instead, the papers discuss a general outline of the feature-space approach, and the only implementations developed are for a circumscribed stimulus set.[8] While this theoretical position does provide a possible answer for how an overall strength can be produced for any test stimulus, it also raises far more questions. It also rises and falls with the fate of the feature-space model itself. In chapter 4, there is a lengthy critique of the feature-space approach for understanding concepts and categories. For example, the experiment by Chechile et al. (1996) on syntactic-pattern recognition raises concerns about how similarity itself is measured in light of the evidence for propositional information. (See the discussion about syntactic-pattern recognition in chapter 4.) Stimuli typically are encoded in a propositional context, which is not captured by a feature space. A vector space can represent some information, but it falls short as a construct for representing relational information and propositions. Hence, the strength theory assumption of a strength axis is itself built on a huge set of assumptions.[9] These issues are not as troubling for the original sensory application of SDT because the experimenter actually presented a stimulus that varied in intensity or strength value (i.e., the observer was presented with either a noise sound or a slightly louder noise sound). But in the memory application, the axis is not clearly linked to properties of a neurally distributed pattern of activation.

It was argued in part I that memory is a modified content-addressable system. See figure 5.1 for the various arrangements for a content-addressable memory (CAM). Only

8. This observation is admittedly extreme. There are some exceptions (e.g., see Lacroix, Murre, Postma, & van den Herik, 2006).

9. There are alternative modeling approaches to that of a simple feature space. For example, computer scientists have constructed realistic models for encoding propositional knowledge (e.g., Fu, 1985; Gonzalez & Thomason, 1978; Winston, 1975). Jones and Mewhort (2007) developed a holographic representation for knowledge of language comprehension. Jones and Love (2007) studied how events with relational structure can be linked to other events that have relational similarity. Landauer and Dumais (1997) developed an approach of multidimensional *latent semantic analysis* for representing world knowledge. All of these approaches are interesting representational systems, but unfortunately these topics are outside of the scope of this volume to review and thoughtfully analyze. Importantly, none of these modeling approaches are simple feature spaces, and none of the models attempt to represent information as being distributed along a single strength axis.

the representation in panel A is consistent with a strength theory approach. Importantly, it was previously argued that the panel A arrangement is not an accurate representation. Instead, it is more reasonable to argue that we recognize a stimulus as an old item because it results in the recovery of other temporal and elaborative information that is not part of the current test stimulus. Panels B, C, and D illustrate modified CAM arrangements where it is possible to recover more than the presented information. For example, suppose that the person sees and encodes the stimulus LRH as "Lynda Rides Horses" along with a decoding strategy. The presentation of LRH later elicits a YES response, not because the LRH stimulus is reactivated but because some content is recovered about the original encoding event. If the person does not have any other temporal or elaborative information stored in memory, then the activation of LRH is not in itself likely to elicit a confident YES response.

Stated another way, it is problematic that the memory of a real-world event is modeled in terms of a mathematical structure that is not designed to handle qualitative differences, relational information, and propositions. Upon hearing the opening few bars of a symphony, the individual experiences a complex pattern of information. A memory for that event is likely to encode many tonal, rhythmic, relational, and temporal features. We also have prior knowledge and expectations that can be part of the encoding of this experience. This event is not at all like a sensory psychological experiment with featureless white noise at two slightly different intensities.

In addition, real-world events often have different informational attributes that can have different fates in terms of their retention in memory. For example, consider a brief bit of dialogue articulated in a play. The sentence stated by the actor has (1) explicit message content, (2) identifying information about who stated the message, (3) a temporal context for the statement, (4) a thematic context for the message, and (5) the dramatic reaction to the statement. Some of these components of memory might be retained while other components become lost. How can this complex event be measured by a single strength value?[10] Thus, memory traces typically have a complex structure that is not conducive for a simple strength value. Memory measurement needs to focus on constructs that are important for the preservation of the original encoding. Previously,

10. In contradistinction to the measurement problems posed by this type of an event, researchers, who have employed MPT models like the ones discussed in chapter 7, have been able to obtain separate memory measures for the item content (*the what message*) and the source content (*the who information*) (Batchelder & Riefer, 1990; Bayen, Murnane, & Erdfelder, 1996; Chechile, 2016). Unfortunately, this topic is outside the scope of the present volume. But the point here is that the MPT modeling approach can measure the separate attributes of a target event for a case that clearly eludes representation in SDT theory.

a case was built for the importance of knowing about the storage and retrieval of encoded information. Those constructs were defined in terms of mixture rates. In contrast, it is difficult to link the concept of strength to the rich informational structure of a memory encoding.

As a further illustration of the representational impoverishment of a strength value, consider the problem of interpreting d' in terms of information processing. Suppose that the d' value in one condition is 1.0, and in another condition is 0.8, then how many fewer memory traces would be remembered in a recall test in the second condition relative to the first? The value of strength as measured by d' does not map onto a prediction for the proportion of the traces that are too weak to be recovered on a recall test regardless of the length of time in attempting to recover the target. So the question remains: what does a d' value mean? It is clear from the theory what the definition of the d' is, but we do not have a strength axis in the brain. Instead, there is a distributed memory representation of features and propositions, and this structure does not map to a single strength value.

The problem with interpreting d' is not limited to just memory applications. The perceptual detection of complex stimuli differs from the simple sensory-intensity framework in which SDT was designed. For example, Hanley and McNeil (1982) studied a radiologist who examined CT scans for a group of actual patients who either had a medical abnormality or were normal. The radiologist was unaware of these classifications and examined each patient's CT scan. The radiologist gave a normal or abnormal classification along with a 3-point confidence rating. Chechile (2014) analyzed these data in terms of both a classic SDT model as well as with a new perceptual detection process-tree model that is nearly the same as the 6P model for memory minus the retrieval parameter. In the perceptual detection model, there is a θ_d parameter for the identification of a certain abnormality when in fact the patient has an abnormality, and there is a θ_{nt} parameter for the identification of a normal scan when in fact the patient is normal. These parameters are the perceptual counterpart to the θ_S and θ_k parameters in the 6P model. The usual SDT analysis resulted in a d' value of 2.33 along with the estimate of $\frac{\sigma_S}{\sigma_N} = 1.41$. But what do these numbers mean? Is the doctor doing well or is there room for improving his diagnoses? How many misses and false alarms are there? Is the doctor guessing? The SDT model does not answer these pertinent questions. However, these same data can also be analyzed in terms of the perceptual detection model. The analysis of the doctor's CT scan judgments resulted in the finding that the doctor correctly identified 55.2 percent of the abnormal patients with high confidence (i.e., $\theta_d = .552$). For 44.8 percent of the patients, the CT scans were more challenging, and the doctor had to give an educated guess. The θ_g parameter was .734, so 73.4 percent of those difficult cases resulted in the doctor correctly guessing that the patient had an abnormality. But this result also means that for 11.9 percent of the patients, the doctor missed detecting an abnormality (i.e, .266 times .448). Consequently, there is room for improvement in either the quality of the CT scans or in the ability of the radiologist to interpret the scan data. Moreover, the radiologist correctly detected 49.6 percent of the normal patients as being normal, so 50.4 percent of the scans of normal patients were not seen as clearly interpretable. In fact, the doctor incorrectly guessed that 59.5 percent of these cases had an abnormality. Thus, 30 percent of the normal patients were judged incorrectly to have a medical abnormality. The doctor's decision making is very good in general, but there is clearly room for improvement. Given a complex perceptual event, such as interpreting a medical CT scan, the signal-detection model is not the most appropriate tool to use to understand the underlying cognitive processes. Perceptual categories are different from sensory intensity levels. Perception is different from sensation because with perception, the observer has a rich knowledge structure to draw upon to analyze the qualitative pattern of information. Memory data are also complex and more similar to the detection of perceptual categories. In short, memory data are not well suited for a simple SDT representation on a single strength axis.

8.5   The Problem of Target-Foil Similarity

The similarity between targets and foils is a way to contrast the SDT approach for strength measurement with the previously described storage-retrieval approach for the Chechile-Meyer procedure. The contrast between these measurement methods produces a paradoxical finding for the SDT approach but results in providing additional validation evidence for the 6P model. The two measurement approaches can be compared for the same data because there is a way to obtain the SDT measures for the recognition test trials that are part of the Chechile-Meyer procedure. This experimental task obtains a YES versus a NO response followed by a 3-point confidence rating. The most signal-like response is a YES 3 and the most noise-like response is a NO 3. In figure 8.5, a strength-theory representation for this type of data is illustrated. There are five decision points. These are denoted as $x_1, \ldots, x_5$, and they divide the responses into six categories. From the SDT framework, if the test stimulus results in a strength of value s, then that strength results in a response on the task according to the following set of decision rules:

$$s > x_5 \Rightarrow YES\,3$$

$$x_4 < s < x_5 \Rightarrow YES\,2$$

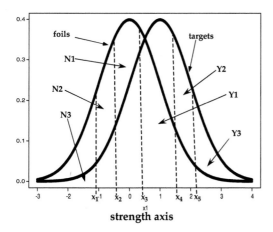

Figure 8.5
The SDT strength representation of the responses possible for the recognition data that also includes a 3-point confidence rating. YES and NO responses are denoted by Y and N respectively, and the decision points $x_1, \ldots, x_5$ are dividing the responses into six possible categories.

$$x_3 < s < x_4 \Rightarrow YES\,1$$

$$x_2 < s < x_3 \Rightarrow NO\,1$$

$$x_1 < s < x_2 \Rightarrow NO\,2$$

$$s < x_1 \Rightarrow NO\,3$$

The values for x_i for $i = 1, \ldots, 5$ correspond to five different decision operating points on an ROC curve. Standard signal-detection theory can thus be used to estimate d' and the ratio of the standard deviations. Consequently, it is possible to obtain a signal-detection strength measure from the same data used for estimating probabilities for storage and retrieval. The comparison between these two approaches for measurement is particularly valuable when considering the issue of the similarity between targets and foils.

Foils in recognition testing are there as catch trials to keep the participant from always responding YES. There is a good reason to suspect that the similarity between the foils and targets should influence a signal-detection measure because strength is defined as a separation between the target and foil strength distributions. But if this occurs, it is a paradoxical outcome. Why should the strength of items studied and remembered take on different values depending on the properties of novel test probes that were *not* learned?

Table 8.1
Gerrein (1976) study of target-foil similarity

Exp.	Measure	Dissimilar	Similar
	False Alarm%	0.0%	8.3%
	θ_S	.651	.624
Exp. 1 and 2	θ_k	.988	.788
	d'	3.36	2.36
	False Alarm%	0.2%	18.8%
	θ_S	.464	.424
Exp. 3	θ_k	.964	.575
	d'	4.00	1.53

Gerrein (1976) examined target-foil similarity with the Chechile-Meyer task. The memory targets were word triads presented via the Brown-Peterson experimental paradigm. There were two separate groups of thirty participants (i.e., thirty in experiment 1 and thirty in experiment 2). The target items were identical between the two experiments. The only systematic difference between the two experiments was the similarity of the foils to the target. The foils were either highly similar to the target or dissimilar to the target. For example, in the case of the target of BUTTERFLY WOOL GLACIER, a highly similar foil would be either BUTTERFLY GLACIER WOOL or BUTTERFLY OCEAN GLACIER, whereas a dissimilar foil would be an item like TOE LETTER VEST. Clearly, the participants in the similar condition knew that the foils were very close to the target, and the participants knew in the dissimilar condition that the foils were all different words because the similarity manipulation was done between the groups. In experiment 3, the similarity of the foils to the targets was examined by means of a within-subjects design (i.e., the same person had experience with both similarity conditions). The key model parameter estimates for both the 6P model and the signal-detection model are provided in table 8.1.

The manipulation of similarity did have a significant effect on the false-alarm rate. The false alarms were near zero in the dissimilar condition, and they were significantly higher in the similar condition. Despite this difference, the storage parameter (θ_S) from the 6P model was not reliably different. But the parameter for the knowledge of the target in foil recognition (θ_k) was substantially larger in the dissimilar foil condition. This result is reasonable because when dissimilar foils are used, only a little bit of information about the target needs to be recovered to lead the participant to a confident rejection of the foil item, whereas in the similar foil condition, a great deal of information needs to be recovered to support a confident rejection. Consequently, the probability of successfully recovering knowledge of the target during foil recognition testing should be greater in the dissimilar condition. But it is important to stress again that the target storage parameter (θ_S) is invariant with respect to the similarity between the foils and targets.

In contradistinction to the 6P model description of the Gerrein (1976) data, the signal-detection d' values are dramatically altered by the degree of the similarity of the foils to the targets. In experiments 1 to 2, the value of d' is more than 40 percent larger in the dissimilar condition. In experiment 3, the value of d' is 260 percent larger in the dissimilar condition. This dependency on foil similarity is a very serious problem for the SDT approach to recognition memory, and it illustrates that the numeric value of d' is meaningless without the knowledge of the target-foil similarity. This problem with the signal-detection approach is clearly due to representing the strength of the targets relative to the strength of the foils. If the foils become more similar to the targets, then the signal-detection measure of strength must change. But why should the strength of the items that were presented be altered by the nature of items not previously presented to the participants? In contrast to the paradoxical result for the signal-detection analysis, the invariance of the storage parameter from the 6P model with respect to target-foil similarity is yet another source of validation for that model.

8.6 The Problem of the Mirror Effect

The mirror effect is a robust outcome in recognition memory that has been explored by Glanzer and his colleagues (Glanzer & Adams, 1985; Glanzer, Adams, & Iverson, 1991; Glanzer, Adams, Iverson, & Kim, 1993; Glanzer, Kim, & Adams, 1998). The mirror effect refers to the ordering of the means for the distributions on the signal-detection strength axis for cases where there are experimental conditions that result in different levels of memory retention. Let us label one condition as "strong" and the other condition as "weak." Not surprisingly, the strength-axis representation of the targets is ordered such that the strong condition mean is shifted to the right of the weak condition. But Glanzer and colleagues showed that the corresponding foils have the reversed order on the strength axis (i.e., the foil mean for the strong condition is shifted to the left of the mean for the weak condition). See figure 8.6 for a display of a typical mirror effect that occurs when there are two types of encoding conditions, and the data are analyzed in the context of the signal-detection framework. This effect is problematic for the standard SDT model of recognition memory because none of the foils have been shown to the participant. There is no reason why the strength

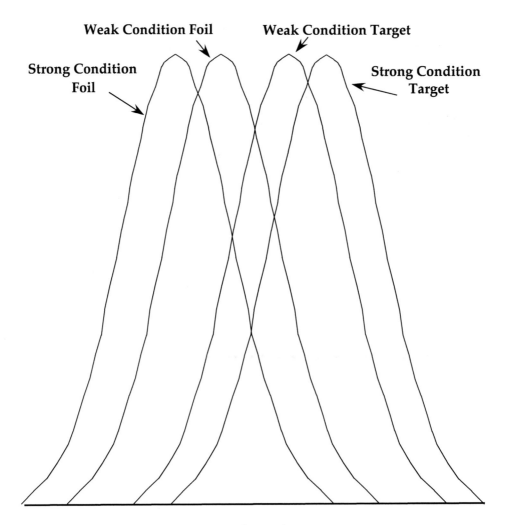

Strength Axis

Figure 8.6
This figure illustrates the mirror effect in recognition memory that arises in the signal-detection analysis when there a two conditions with different memory retention (i.e., a "strong" and a "weak" condition).

distributions for the two foil distributions should be different. It would seem more plausible in the SDT framework to assume that all the nonpresented items originate from a common strength distribution.

For the 6P and 7B models, the foil trials are divided between occasions (with probability θ_k) where there is *target knowledge* and occasions (with probability $1 - \theta_k$) *without any target knowledge*. If the value for the θ_k parameter is different between the strong and weak conditions, then that difference is enough to account for the mirror effect. For example, in the 3-sec retention interval for the acoustically dissimilar targets discussed in section 7.7.3 of the previous chapter, the θ_k parameter was .892, whereas in the 9-sec retention interval, the θ_k parameter was .542. If one were to analyze these two conditions with a signal-detection strength model, then it would result in the foil distribution for the "strong" 3-sec condition being shifted to the left by .87 of the foil distribution for the "weak" 9-sec condition. The difference in θ_k itself is not surprising. But imposing the signal-detection framework on the data results in a paradoxical shift to the left for the strong-condition foils on the assumed strength axis. Thus, the mirror effect illustrates a problem with using a signal-detection framework for examining recognition memory, whereas the changes in the parameters for the storage-retrieval model result in a meaningful pattern. Thus, strength theory has a fundamental problem accounting for the mirror effect. Shifts in the location of the items not presented to the person is yet another indicator of the lack of validity for the signal-detection method for measuring memory strength.

8.7 The Problem of Foil Mixtures

Homogeneous normal distributions are assumed for both targets and foils in the standard form of the signal-detection strength model. An alternative to a homogeneous normal distribution is the case where the items come from two or more different probability distributions; such cases are called stochastic mixtures. An example of a stochastic mixture would be the case where 20 percent of the items came from a normal distribution that has a mean of 5 and a standard deviation of 1, and 80 percent of the items come from a normal distribution that has a mean of 0 and a standard deviation of 1. A mixture for the targets can be explained by the previously discussed dual-process model. The dual-process model is a form of the strength model where there is a stochastic mixture because R_0 proportion of the targets are super strong and the rest of the targets have a strength distribution characterized by a normal distribution with a mean of μ_s and standard deviation of σ. But the Yonelinas (1994) dual-process model for foils still assumes a single-process normal distribution (i.e., the dual-process model does not predict a mixture for foils).[11] Consequently, a critical assumption of the signal-detection strength model is a homogeneous normal distribution for the foil items. Yet as shown earlier, the 6P and 7B models are mixture models for both the targets and the foils. Thus, a critical test of both the dual-process model and the standard signal-detection strength model is to ascertain if the foils originate from a homogeneous distribution or not. Evidence of a stochastic mixture for foils would be inconsistent with signal-detection strength models.

In general, mixtures are challenging to detect. To illustrate this point, consider the probability density display shown in figure 8.7 where there is a mixture of two normal distributions where each component contributes 50 percent. The components are separated by 1.5 units, and each component has a standard deviation of 1. A multimodal density display is a clear sign of a stochastic mixture. Moreover, a 50/50 split between the components of a mixture is the optimal mixing rate for detecting bimodal distributions. Yet despite a substantial separation between the components and an optimal mixing rate, there is still no indication of a bimodal distribution. Thus, a unimodal distribution does not mean that there is a homogeneous distribution (as in the case of figure 8.7), but a multimodal distribution does signal the presence of a mixture. With a greater separation between the components, it is possible to detect the mixture by finding a bimodal density function. For example, a 50/50 split of two normal components that are separated by 3 units is clearly detected, as can be seen in figure 8.8. But notice even if the separation between the components is 3 units, a density display fails to detect a bimodal profile for the mixing rates of 95 percent versus 5 percent, as shown in the figure. Consequently, the detection of a mixture from a probability density display is not a particularly sensitive method.

11. It is not an accident that the foil distribution of the dual-process model is a homogeneous normal distribution. If there is a mixture of two normal distributions, then the foil model would not be identifiable in the sense that more than two model configurations would yield the same likelihood distribution for the data. For example, if it is assumed that R_N of the foils originate from a normal distribution with a large negative mean, and the rest of the foils come from a normal with a mean of 0 and a standard deviation of 1, then any number of possible values for R_N and the decision points x_i for $i = 1, \ldots, 5$ would predict the same likelihood function. For example, suppose $R_N = .1$ and the decision points are $x_1 = -1$, $x_2 = -.7$, $x_3 = -.5$, $x_4 = 0$, and $x_5 = .5$. That configuration is indistinguishable from another model configuration where $R_N = .2$ and the decision points are $x_1 = -1.611944$, $x_2 = -1.0484785$, $x_3 = -.7651045$, $x_4 = -.157311$, and $x_5 = .393149$. The model is unidentifiable when more than one configuration of the model parameters produces the same likelihood function. If it is assumed that the foils only originate from a single normal distribution, then the model is identifiable. See the glossary for more detail about the topic of model identifiability.

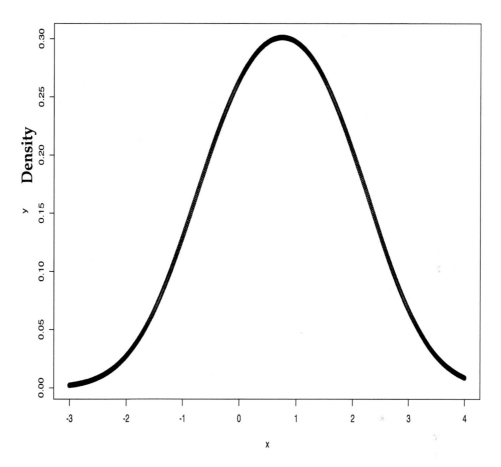

Figure 8.7
A 50 percent–50 percent mixture of two normal probability components where there is 1.5-unit separation between the components. The standard deviation of each component is 1.

Chechile (2003, 2011, 2013) showed that mixtures can be detected by using a tool called the hazard function.[12] To understand what a hazard function is, it is necessary to define several terms for a continuous random variable x (i.e., a variable that has a probability distribution). The height of the probability density display like that of a normal distribution is denoted as $f(x)$. The cumulative probability is denoted as $F(x)$

12. In chapter 3, hazard was used to assess forgetting functions, whereas in this chapter, hazard is used to examine the properties of the strength continuum that is assumed in the signal-detection strength model.

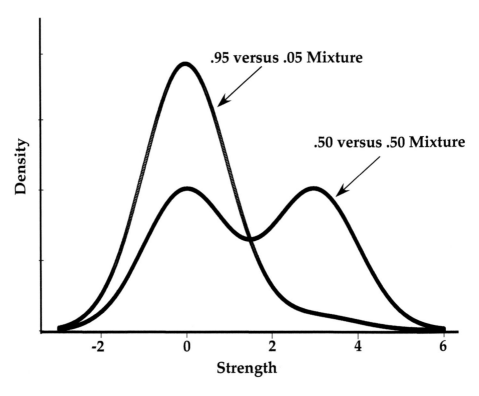

Figure 8.8
Two illustrations of stochastic mixtures of normal distributions. The two components of the mixture are separated by 3 units and the standard deviation of each component is 1.

and is obtained from the integration of the density for all values less than or equal to x. In terms of these constructs, hazard $h(x)$ is defined as

$$h(x) = \frac{f(x)}{1 - F(x)}.$$ (8.1)

Hazard typically deals with a state change, such as a task going from being uncompleted to completed, or a trace going from sufficiently stored to insufficiently stored. Originally, hazard was developed from an insurance industry context to examine the state change from living to deceased. The term $1 - F(x)$ is called the survivor function, and it is the area under the density curve to the right of point x. In essence, at the point x, $F(x)$ is the proportion of the items that have already transitioned to state 2, whereas $1 - F(x)$ is the proportion of the items that are still in state 1. The survivor function is thus the proportion of the traces that have not yet transitioned. Hazard is thus the instantaneous risk of a transition taking place at the point x. The division by the survivor function reconditions the probability of a state change given that the change has yet to occur.

Hazard functions have many well-known properties that have been proven mathematically. For example, a mixture is known to have decreasing hazard if it is made up of component distributions that have constant or decreasing hazard (Chechile, 2003). But the normal distribution is known to have monotonically increasing hazard (Chechile, 2003). Hazard functions are more complex for mixtures where the components of the mixture have increasing hazard. If the separation between the components is below a critical distance, then the mixture also has increasing hazard. But if the separation is greater than a critical distance, then there is a dramatically different hazard profile. This profile can be called a IPDVI structure where I stands for increasing, P for a local peak, D for decreasing, and V for a local valley. Notice in figure 8.9 the IPDVI profile occurs for the hazard function for a mixture of two normal distributions that have a separation of 3 and where the mixture rates are 90 percent and 10 percent. Chechile (2013) showed that this mixture could not be detected via a density display (i.e., it did not result in a bimodal distribution). Consequently, the hazard function can be a more sensitive tool for detecting stochastic mixtures.

If the hazard function for foil items was not monotonically increasing, then there would be evidence for a mixture process for foils. This finding would be inconsistent with a signal-detection strength model of memory. But the detection of the mixture could be explained by the 6P or 7B models because these models stipulate the possibility of mixtures of foil trials (i.e., θ_k proportion of the trials result in a confident knowledge-based foil rejection, whereas the rest of the foil tests would be a default guessing process).

Chechile (2013) developed several mathematical theorems that enable the estimation of the mean hazard for four of the intervals shown in figure 8.5. In eight experiments, the mean hazard was found for these intervals. The result of this hazard analysis indicated that a profile of mixture was present in seven of the eight experiments. This analysis assumed a normal distribution for the target trials in order to find the decision points $x_1, \ldots, x_5$. Based on these values, the mean hazard was found for each of the those intervals. See figure 8.10 for a median (across the eight experiments) of the hazard in the four internal intervals on the presumed strength axis. If the foil distribution were a single normal distribution, then hazard must be monotonically increasing. The results from these eight experiments demonstrate that the foil distribution must be a stochastic mixture because the hazard in the interval between x_4 and x_5 is significantly less than the previous two intervals.

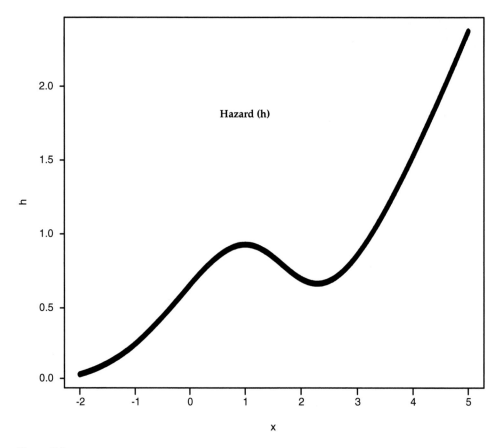

Figure 8.9
The hazard for a 90 percent–10 percent mixture of two normal components that are separated by 3 units and where each component has a standard deviation of 1.

One might challenge the assumption of that analysis of a normal distribution for the targets because the dual-process model treats targets as a mixture. But if one were to assume instead the dual-process model for the targets rather than a normal distribution, then the evidence for a foil mixture is *even stronger* (Chechile, 2013). Thus, the hazard analysis is able to detect a foil mixture that is contrary to the predictions of the signal-detection strength model for foils.[13]

13. Kellen and Klauer (2015) used a different critical test of the SDT model for foils. These investigators used a function that is similar to a hazard function to reject a hypothesis from the SDT framework. The hypothesis deals with the assumption of continuous variability of the confidence rating for foil tests. They showed instead evidence for the two-high-threshold model. The

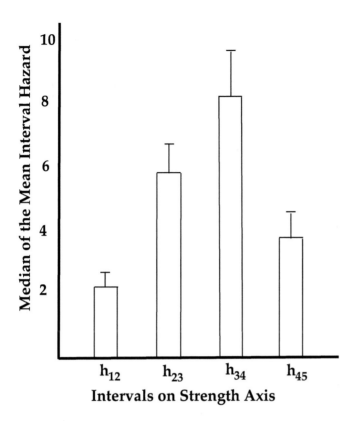

Figure 8.10
The figure shows the median across eight experiments for the mean hazard of the foil distribution between the intervals on a presumed strength axis. For each experiment, the distribution of the targets was assumed to be a standard normal. This assumption resulted in five decision points $x_1, \ldots, x_5$. See figure 8.5 for the definition of the decision points.

The hazard analysis relies on new mathematical theorems to establish that there is a mixture among the foil items. But there is a simple experimental method for finding evidence for a mixture for the foil trials. In each of the eight experiments examined in Chechile (2013), the Chechile-Meyer task was employed. Consequently, there are recall trials intermixed with the target and foil trials. For each experiment, the recall rate is greater than zero (Chechile, 2013). Thus, it is established that on some test trials, the participant can reproduce the entire target from just the cue word RECALL. Clearly,

---

two-high-threshold model, which is described in the glossary, is an MPT model that is similar to the 6P and 7B models.

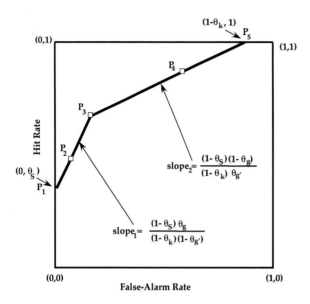

Figure 8.11
The ROC profile for the 6P and 7B models.

if the participant can reproduce the entire item, then the participant would reject a foil with certain knowledge that the item was wrong. Thus, we have an inescapable conclusion that there is a mixture process involved on foil trials, and this finding is further proof that the signal-detection model for measuring memory strength is not a valid measure.

8.8 The Problem of Ambiguous ROC Data

The signal-detection model for strength measurement fails to achieve validation support. Rejection of the Blackwell (1963) threshold model does not mean that there is evidence for the critical features of the signal-detection model. There are other theories besides the Blackwell model. Laming (1973) noted that the properties of the ROC are not the most sensitive method for comparing rival models. The 6P and 7B models do not predict a Blackwell threshold ROC, and those models do not predict the same ROC as the standard signal-detection ROC curve. Both the 6P and 7B models have the ROC shown in figure 8.11. There are five points of interest on the ROC path. Points P_1 and P_5 are extreme path end points. Point P_1 corresponds to the theoretical limit of a conservative decision bias where the false-alarm rate goes to 0. The hit rate for point P_1 is θ_S. Point P_5 is the other end point for the extreme limit for risk taking when the hit rate goes to 1. In this case, the corresponding false-alarm rate is not 1, but it is instead $1 - \theta_k$. Between the extreme end points, there are three operating points. Point P_2 is the most conservative *observed* point, and it corresponds to when an affirmative target decision is defined as a YES 3 response. The second most conservative observed point is P_3, and it corresponds to when an affirmative target decision is defined as any YES response regardless of the confidence rating. The slope of the ROC is a constant along the path from P_1 to P_3. The last observational point is the most risky observed case when an affirmative decision is considered as any response other than a NO 3. The path along the ROC from point P_3 to P_5 is also a constant, but it is not in general the same as the slope from P_1 to P_3. The equations for the slopes of the two straight line segments of the ROC are shown in figure 8.11. There is a special case when the two slopes are equal, and that case occurs when $\theta_g = 1 - \theta_{g'}$. This special case is not typically found, but it does have theoretical significance. If $\theta_g = 1 - \theta_{g'}$, then the 6P and 7B models become the same as the two-high-threshold (2HTM) model by Bröder and Schütz (2009).[14] But in most experiments, the value of θ_g is greater than $1 - \theta_{g'}$. The θ_g parameter is the tendency to guess YES on old recognition, whereas $1 - \theta_{g'}$ is the tendency to guess YES on new recognition. If there is partial information, then the person can use that information to be more effective in guessing. Improved guessing based on partial information results in $\theta_g > 1 - \theta_{g'}$. Moreover, when $\theta_g > 1 - \theta_{g'}$, the P_1 to P_3 slope is larger than the P_3 to P_5 slope. So in general, the ROC path for the 6P and 7B models differs from the path for the 2HTM model.

The 6P model is a special case of the 7B model, but the two models have the same ROC path. The only difference between these two models concerns the high-confidence rating parameter. In the special case when θ_h equals $\theta_{h'}$, the two models are equal. But when these two confidence rating parameters differ, then the location of the point P_2 on the P_1 to P_3 path is proportionally different compared to the location of point P_4 point on the P_3 to P_5 path. Apart from the relative locations of the P_2 and P_4 points, the 6P and 7B models have the same overall ROC path.

14. The two-high-threshold model (2HTM) is a process tree model for yes/no recognition without the confidence rating. It is a strength model that has separate thresholds for (1) when the strength is high enough to trigger a YES response and (2) when the strength is so low that the person gives a NO response. Strength values between the two thresholds result in guessing. Thus, the 2HTM model is another method for memory strength measurement. But the 6P and 7B models are not strength models at all; instead, they are mixture models dealing with the probabilities for sufficient storage in old recognition and for the knowledge of target information in new recognition.

Consequently, the vigorous debate in the literature about the signal-detection model versus the 2HTM model is not relevant for the critical assessment of the strength measurement approach itself. The shape of the ROC is a goodness-of-fit debate that is not the best way to evaluate rival measurement models. A number of ROC paths are different from the one in the Blackwell (1963) threshold model. Strength theory needs to develop positive validation evidence for the measurement method rather than arguing against a particular rival model. With only a few observed points for fitting the ROC path, the models are difficult to compare. Fitting the ROC function is not a central psychological problem; rather, interest should be focused on the validation of the model-based measurement of an essential memory process.

8.9   Summary Remarks about Memory Strength

There have been a number of enthusiastic supporters of the SDT approach for memory measurement. But instead of developing a strong set of validating experiments, the SDT approach has run into a wide array of paradoxical findings. Among the problems for the SDT approach are (1) the dependency of d' on the target-foil similarity, (2) the mirror effect, and (3) the finding of mixtures for both targets and foils. In addition to these troubling effects, there are other problems with the d' measure from SDT. For example, Chechile (1978) reported finding that d' was measuring at least two different properties, so it was not a pure strength measure in the first place. Given these failings, the question arises why the signal-detection approach still is used so widely. It is difficult to know for sure, but I suspect that it reflects a failure to appreciate the difference between a featureless sensory stimulus, which has an intensity value, and an informative stimulus that is encoded into memory along with relational and propositional information. In the original application of SDT, it was not an accident that white noise (static) was used as a stimulus. But real-world memories, such as the perception and understanding of social events, are complex patterns of perceptual features, prior knowledge, and elaborative information. Memory traces such as this do not fit the framework designed for the detection of a sensory intensity level.

Perhaps the high volume of papers using the SDT approach has also led to an implicit acceptance and general comfort with the SDT framework and thus a failure to critically think about the suitability of the framework. The basic idea of the SDT approach is modeled after a classical t test, which is a familiar analysis to all researchers. Furthermore, there are applications, such as the detection of the sound intensity level that does fit the SDT framework well. But one tool or method rarely fits all jobs. So critical thinking about the application of SDT to the area of memory must focus on basic questions about the approach in order to validate it. But researchers have replaced the quest for validating evidence with attempts to explain the empirical ROC or to patch a failing theory in order to deal with puzzling empirical findings. The empirical ROC is simply a plot of the data, and it does not constitute measurement of latent psychological properties of interest. The shape of the ROC is also not a fundamental problem in psychology. Moreover, the rejection of the Blackwell (1963) threshold model does not constitute validation support for the SDT model.

A question such as asking if target items are distributed on a strength axis (or in a multidimensional feature space) with a *single homogeneous distribution* would begin to address a critical assumption of the SDT approach. But instead of finding consistent support for this assumption, there is a consensus that the targets are not all distributed as a homogeneous distribution. Hence, the standard SDT model had to be replaced or patched in the form of the dual-process model. If there is a mixture of targets, then the most important measurements are the rate values for the separate components. In general, it is difficult to detect stochastic mixtures with density-display plots. Hazard analyses are a more sensitive method for detecting mixtures. Yet an even better way to detect mixtures is to assume that they exist and to employ an MPT model to measure the rates for the various mixture components. That approach is exactly what the MPT models discussed in chapter 7 do. An MPT model can detect a homogeneous process by finding that the proportions for the mixture components are at the extremes for probability (i.e., either close to 1 or 0). But if the mixture rates are different from the extremes, then a mixture has been detected. Using a hazard analysis, evidence was provided in this chapter that foil items also demonstrate a stochastic mixture. The combination of (1) mixtures for both targets and foils and (2) the finding of unequal variance for the targets and foils distributions is an especially lethal combination for the SDT approach because it results in the model being unidentifiable. A model that is unidentifiable exhibits a fundamental modeling pathology.

So why do models such as the 6P and 7B demonstrate a strong record of validation while the SDT measurement of memory struggles to find validation support? Well unlike the SDT approach, those two MPT models do not attempt to specify the details of the memory encoding beyond the targets being sufficiently encoded or insufficiently encoded. As discussed in chapter 7, the format of the encoding can vary widely, and it still can be sufficient for the recall of the target, provided that it is recovered at the time of test. The encoding can have propositional information or not for the categorization of sufficient versus insufficient storage. Moreover, the proportion of sufficient storage is a value with a true zero (i.e., $\theta_S = 0$ is a true zero representing the loss of all the targets). Hence, the storage measure is a ratio-scale metric in terms of the Stevens category of measurement scales. Thus, the storage measure is not dependent on the similarity between the foils and the targets. In contradistinction to this theoretical approach, the SDT approach (1) makes detailed assumptions about the nature of the memory representation, which does not take into account propositional encodings, and (2) it is, at best, an interval scale in the Stevens measurement classification because targets are scaled relative to the scaling of foils. Consequently, as the foils are more similar to the target, there is a corresponding decrease in d', whereas the θ_S metric is invariant to this change. Also in chapter 10, a case will be made that targets that are insufficiently stored can be decomposed further (by a different MPT model) into three categories of (1) implicit storage, (2) fractional storage, and (3) nonstorage. In contrast to this decomposition, the SDT model is silent about the relationship between explicit and implicit memory. Thus, the notion of single strength value will be challenged to account for the multiple attributes and types of memory. In addition, in this chapter, a case was made that there are events that have multiple attributes related to the (what), (who), (where), and (when) components of the target event. These attributes can be separately measured with MPT models, and these different attributes can have different fates for memory retention. But in contrast to this fine-grained measurement within the MPT framework, the SDT framework is still laboring with a single strength value.

Finally, we have already seen that the theoretical concept of strength played a central role in the system models of habituation and sensitization, which were discussed in chapter 1. But those system models did not require the SDT measurement approach. It is also likely that a model of semantic memory activation will need the concept of memory strength. This topic is outside the scope of this book, but it is likely that it too can be modeled without the use of the SDT measurement framework.

9 Storage-Retrieval: Animal Model

9.1 Chapter Overview

Memory is clearly not a uniquely human ability. Consequently, it is important to measure latent cognitive memory processes, such as storage and retrieval, for animals. In this chapter, the Cook, Chechile, and Roberts (2003) model for obtaining separate measures for storage and retrieval will be described for pigeons. The chapter also contains evidence for the validation of the model estimates of the underlying processes. Finally, there is a technical appendix for those readers with a particular interest in the statistical estimation of the model parameters.

9.2 Storage-Retrieval Measurement for Animals

Research with human adults has an enormous advantage—the participants can follow complex verbal instructions. For example, with the Chechile-Meyer task, the participants quickly can be taught that there will be recall, old-recognition, or new-recognition test trials. Moreover, on recognition trials, they learn to give a 3-point rating and learn the guidelines for each rating. With this method, it is possible to obtain measures for storage and retrieval. But this type of experimental protocol does not transfer to memory measurement for nonhuman species. So how can storage and retrieval measures be obtained for animals?

It is straightforward to train animals to select the stimulus that was presented earlier in a trial. For example, it is possible for pigeons to select on a forced-choice test the stimulus that was previously presented in the learning phase. Pigeons have excellent color vision, so stimuli such as the ones shown in figure 9.1 can be used as memory targets. In the Cook et al. (2003) research, eight different color patches were used as stimulus components. Memory targets could be a singleton (one color), dyads (two colors), or triads (a combination of three colors). The total number of such targets is shown in

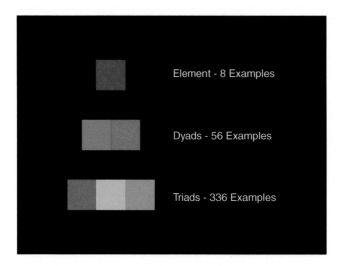

Figure 9.1
Examples of singleton, dyadic, and triadic color stimuli.

figure 9.1. The birds had difficulty encoding the triad targets, and the one-element targets were too simple, so the dyadic targets were the best combination for studying the storage and retrieval of information. Consequently, the model for separating storage and retrieval is designed for dyadic targets.

There are fifty-six unique combinations of two different colors, that is, $n*(n-1) = 8 \cdot 7 = 56$. Because the same eight colors are used for all cases, a given color can be both correct and incorrect depending on the trial. At the time of test, the exact target pair of colors can be presented either with two other color pairs for a three-alternative forced-choice test or with one other color pair for a two-alternative forced-choice test. If we denote the target item for the current trial as AB, then there are two different types of two-alternative forced-choice tests: (1) a test with no shared components of the target (i.e., a choice between AB and YX) and (2) a test with one shared component (i.e., either a choice between AB and YB or a choice between AB and AX). In a similar fashion, there are three different types of three-alternative forced-choice tests: (1) a test with no shared components (i.e., a choice among AB, YX, and ZW), (2) a test with one shared component (i.e., a choice among AB, YB, and YX or a choice among AB, AX, and YX), and (3) a test with two shared components (i.e., a choice among AB, AX, and YB).

A novel feature of the test protocols is the use of a color wheel for the pigeons to type in the pair of colors that they experienced during the learning phase of the experiment. The color wheel is shown in figure 9.2. The color wheel is effectively the

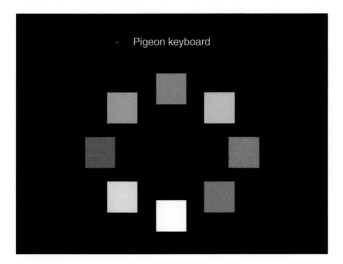

Figure 9.2
The pigeons' keyboard for typing (or actually pecking) in their answer on recall test trials.

pigeons' typewriter for inputting the colors for the current test trial. Consequently, the storage-retrieval separation model uses a novel type of recall test trial to supplement the more traditional forced-choice recognition test trials.

The combination of recall and forced-choice recognition test trials enables the estimation of the probability of sufficient storage, θ_S; the probability of fractional storage, θ_F; and the probability of nonstorage, θ_N, where $\theta_S + \theta_F + \theta_N = 1$. Sufficient storage is the memory for the two colors of the current trial, whereas fractional storage corresponds to the case where the animal knows one of the two colors. The nonstorage case is when the animal does not know either color of the current trial. In addition, it is also possible to estimate the probability of retrieval θ_r. The model-based measurement method is based on six binomials. The six binomials are illustrated in figure 9.3. Cells 1 and 2 correspond to the outcomes of incorrect recall and correct recall, respectively. For the two-alternative forced-choice tests, there are separate binomials for the none-shared test (cells 3 and 4) and the one-shared test (cells 5 and 6). Finally, the last three binomials correspond to the forced-choice tests with three alternatives: (1) the case for no shared components (cells 7 and 8), (2) the case with one shared component (cells 9 and 10), and (3) the case with two shared components (cells 11 and 12).

There are process trees for each of the six binomial test-trial types. The trees describe how the observed outcomes are modeled in terms of the latent memory processes associated with sufficient storage, fractional storage, nonstorage, and retrieval. The process

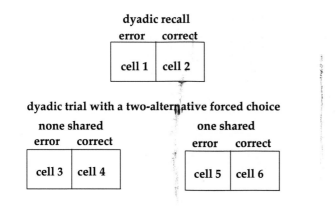

Figure 9.3
The six binomial test trials for testing dyadic stimuli.

trees also account for guessing. The process-tree model for the recall test trials is shown in figure 9.4. The upper branch of the process tree corresponds to the case when the current pair of colors are stored and the bird can retrieve that information. This outcome has a probability of $\theta_S \theta_r$ and results in the bird correctly pecking only the correct two colors on the color wheel. The lower branch of the tree corresponds to the case either when the target information is not stored or when the colors are stored but not retrieved. The combined probability for these two cases is $1 - \theta_S \theta_r$. There is also a nonnegligible chance of $\frac{1}{56}$ that the bird might type the correct two colors by random guessing. Consequently, the probability for incorrect recall or cell 1 (which is denoted as ϕ_1) is

$$\phi_1 = (1 - \theta_S \theta_r) \frac{55}{56}, \tag{9.1}$$

and the probability of correct recall is $\phi_2 = 1 - \phi_1$. Clearly, from equation (9.1) alone, there are more latent memory parameters than statistical degrees of freedom, but there is additional information available from the other binomial cells that enables the model to be fully identified.

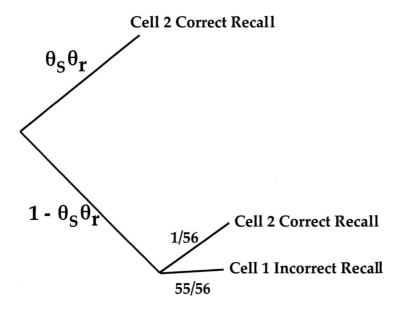

Figure 9.4
Tree model for recall test trials.

The two binomials where the forced-choice alternatives do not share any of the colors of the current target are particularly informative tests. The process trees for these two binomials are illustrated in figure 9.5. Panel A corresponds to the case of two-alternative testing (i.e., choosing between AB YX), and panel B shows the tree for three-alternative testing (i.e., choosing among AB YX ZW). The non-storage branch for both trees leads to a pure guessing case. For the two-alternative case, the correct guessing rate is $\frac{1}{2}$, whereas in the three-alternative case, the correct guessing rate is $\frac{1}{3}$. In the case of any target knowledge (which occurs with probability $1 - \theta_N$), the animal can select the correct AB choice because the alternatives do not share any of the target component colors. Consequently, the probability of an error for the two-alternative forced-choice test (denoted as ϕ_3) and the error probability on the three-alternative forced-choice test (ϕ_7) are, respectively,

$$\phi_3 = \frac{\theta_N}{2}, \tag{9.2}$$

$$\phi_7 = \frac{2\theta_N}{3}, \tag{9.3}$$

(A)

None Shared: Two_Alternative Forced Choice, e.g. AB YX

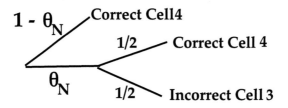

(B)

None Shared: Three-Alternative Forced Choice, e.g. AB YX ZW

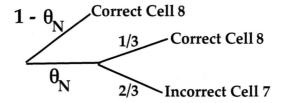

Figure 9.5
Process trees for the case where the components are not shared for (A) the two-alternative forced-choice test trials and (B) the three-alternative forced-choice tests. The pathway labeled with the probability θ_N denotes the case when the animal does not remember either of the two colors for the current trial.

where the corresponding correct recognition rates for the case of no shared features are $\phi_4 = 1 - \phi_3$ and $\phi_8 = 1 - \phi_7$. Note that equations (9.2) and (9.3) are strictly a function of θ_N, so it is possible to learn the value of the nonstorage parameter from these binomials cells.

In figure 9.6, the process trees are provided for the cases when the alternatives in the forced-choice testing share a feature in common with the target. Panel A is for the case of one shared feature and when there is a two-alternative forced-choice (2-AFC) test. Clearly, when both colors are stored, the animal will definitely be correct on the test (i.e., cell 6), and when neither color is stored, the correct 2-AFC recognition rate is $\frac{1}{2}$. But in the fractional storage case, when the animal knows one of the two colors, the rate of correct recognition on the 2-AFC test is more complex. Suppose the animal knows

(A)
One Shared Feature 2 AFC, e.g. (AB YB) or (AB AX)

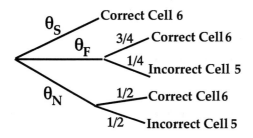

(B)
One Shared Feature 3 AFC, e.g. (AB YB YX) or (AB AX YX)

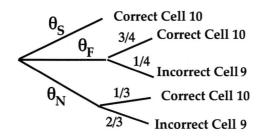

(C)
Two Shared Feature 3 AFC, e.g. (AB AX YB)

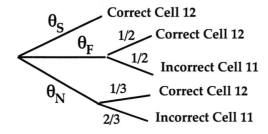

Figure 9.6
Shared components process-tree models for (A) one shared component for the two-alternative forced-choice test trials (2-AFC), (B) one shared component for three-alternative forced-choice (3-AFC) tests, and (C) two shared components for 3-AFC tests. The pathway labeled with the probability θ_N denotes the case when the animal does not remember either of the two colors for the current trial, whereas θ_F has one remembered component, and θ_S is the case of sufficient storage for both target colors.

only the B member of the target pair, then the pigeon would be 100 percent correct if the 2-AFC test pair were AB AX, but the animal would be at a 50 percent correct rate if the test pair were AB YB. In the experiment, both types of pairing occurred equally as often. Consequently, for this case, the overall correct recognition rate is 75 percent. In the case when the animal is tested with a 3-AFC recognition test, and there is one shared feature (shown in panel B), then there is a similar set of outcomes as in the panel A analysis. For example, the pigeon would be definitely correct when there is sufficient storage, and the animal would have a chance rate of $\frac{1}{3}$ of being correct when there is no storage. Again notice in the case of fractional storage, if the one stored component of the target pair were B, then the pigeon would be 100 percent correct when given the alternatives of AB AX YX, but the animal would be at a 50 percent rate when given the test probe of AB YB YX. Consequently, in the case of fractional storage, the probability of correct recognition is again $\frac{3}{4}$. The final case corresponds to when there is a 3-AFC test and there are two shared features among the alternatives. The process tree for this case is shown in panel C. Once again, the sufficiently stored case and the nonstored case have outcome probabilities similar to the analysis described in the previous two panels, but the outcome success rate for fractional storage is reduced when there are two shared features (i.e., when the alternatives are AB AX YB). Regardless of which component is stored, the pigeon is only at a 50 percent correct rate. From these three trees for shared features, the proportions for the incorrect recognition rate are as follows:

$$\phi_5 = \frac{\theta_N}{2} + \frac{\theta_F}{4}, \tag{9.4}$$

$$\phi_9 = \frac{2\theta_N}{3} + \frac{\theta_F}{4}. \tag{9.5}$$

$$\phi_{11} = \frac{2\theta_N}{3} + \frac{\theta_F}{2}. \tag{9.6}$$

Of course, $\phi_6 = 1 - \phi_5$, $\phi_{10} = 1 - \phi_9$, and $\phi_{12} = 1 - \phi_{11}$.

Based on equations (9.2–9.3), an estimate of θ_N can be obtained, and with that value along with equations (9.4–9.6), there is an estimate for θ_F. Finally, since $\theta_S + \theta_F + \theta_N = 1$, it follows that an estimate of θ_S can be obtained by subtracting the θ_N and θ_F estimates from one. Consequently, all of the parameters of the storage-retrieval model can be estimated. The technical details about the estimation of the parameters are described in the Appendix at the end of the chapter.

Consequently, like the models in chapter 7, it is possible to measure the latent memory processes with model-based measurement. The model is different because we cannot use confidence rating judgments with animals to do the necessary corrections for guessing. Despite this difference, the storage and retrieval processes are nonetheless measurable provided that the animals can be trained to understand the task requirements so that there is a random intermixing of six types of test trials as illustrated in figure 9.3. As with any model-based measurement method, it is important to establish that the model is valid. In the next section, evidence is provided for model validation.

9.3 Validation Studies for the Animal Model

9.3.1 Encoding Level

It is essential for pigeons to encode the target in order to form a memory trace. But there are many things that might be distracting the pigeons while the target is being displayed. Thus, it is necessary that the animals demonstrate their attention to the target. Pigeons are trained in the Cook Laboratory at Tufts University to peck on a touch-sensitive screen. To show that they see a memory target, the birds must first peck on the screen in the location of the two displayed colors. But a single peck might not indicate a careful study of the target. Consequently, the bird might be required to peck the screen a number of times in the location of the target in order to trigger the beginning of a trial. The number of responses required is a depth-of-encoding variable. Up to some point, the number of pecks on the target is a measure of the time for *potential* memory encoding. Perhaps if there are too many pecks, then there might be a habituation effect to the visual array or a shift of attentional focus. The terminology for the number of required responses in the encoding period is called the fixed ratio (FR). A FR 5 condition indicates that five pecks are required to initiate a trial. Consequently, the FR factor is a way to examine the validity of the storage-retrieval measurement model. If the model is valid, then it is expected that the probability for the storage of both colors should increase with the value of the FR schedule, and the value of nonstorage should decrease. It is not clear how much memory retrieval should change, but the retrieval probability should not decrease markedly with increasing FR values.

Four birds were trained so that they could perform the combination of different test trial types shown in figure 9.3. The mean values for the model parameters are shown in table 9.1. With only a single peck on the target, there was a high probability (.739) that the birds did not encode either of the target colors. But when the birds were required to peck the target twenty times, the probability of a failure for encoding both colors dropped to a value of .182. Thus, the requirement to peck the target a multiple number of times helped the birds to store something. The probability for storing both colors increased as expected as the FR value increased; there was an improvement by more than a factor of 3. The fractional storage (i.e., the storage of a single color of the

Table 9.1

FR influence on storage quality and retrieval

FR Value	θ_S	θ_F	θ_N	θ_r
1	.117	.144	.739	.643
5	.235	.377	.388	.673
10	.286	.451	.263	.720
20	.365	.454	.182	.684

Table 9.2

Retention interval influence on storage quality and retrieval

Delay	θ_S	θ_F	θ_N	θ_r
0 sec	.352	.477	.176	.786
1 sec	.290	.456	.254	.827
2 sec	.257	.408	.335	.757
4 sec	.209	.341	.451	.818

target) also improved with the increase in FR value. Moreover, the fractional storage probability was larger than the probability for storing both target colors for each FR condition. Consequently, the storage of either one color (θ_F) or two colors (θ_S) increased markedly as expected with the FR value. Finally, the retrieval parameter (θ_r) corresponds to the probability of recovering both colors on the recall task when in fact the two current colors are stored. This parameter is $.680 \pm .04$, so the retrieval parameter is stable across the various values for the FR factor. Taken together, these results are consistent with what would be expected of a valid measurement system.

9.3.2 Retention Interval Effect

The same animals that were trained to perform the previous study were also tested to see the effects of the delay between study and test. There ought to be a decrease in the probability for having both colors stored as the retention interval is increased (i.e., θ_S should decrease with increased delay). Also, the nonstorage probability (θ_N) should increase with increasing memory delay. It is less clear a priori that either the fractional storage parameter or the retrieval parameter ought to change with a short increase in the retention interval.

The mean parameter values for the model are shown in table 9.2 as a function of the delay between presentation and testing. Four delay values of 0, 1, 2, and 4 sec were used. Notice that both the sufficient storage parameter (for two colors) and the fractional storage parameter (for one color) decreased as a function of the delay.

Moreover, the nonstorage value increased by almost a factor of 3 with delay. Again, the retrieval parameter was relatively stable and had a value of $.797 \pm .035$. As with the study with the level of encoding, the effects of delay are also consistent with expectations. Consequently, this experiment also provides validation support for the method of measuring storage and retrieval components with a nonhuman species.

9.4 Chapter Summary and Discussion

A key behavioral assessment procedure to use for any species is a matching-to-sample recognition test. Such tests are straightforward to implement on a wide range of species. The challenge is to design and train animals to do a recall task. The color wheel was the method used in the experiments with pigeons because the birds can detect color and understand the general requirement of matching the presentation stimuli to the test stimuli. But is the color wheel truly a recall task? Perhaps the pigeons are simply doing a matching to sample like in the forced-choice test trials; however, the recall display is always the same, but the dyadic pair of colors is a different-looking stimulus. Yet there is a statistical test to ascertain if the recall task is effectively another recognition test. If the correct recall rate is a product of two independent measures, then it is likely that the recall task involves more than a matching-to-sample recognition. The correlation between θ_S and θ_r for the retention interval study is $r = -.063$. This low and not statistically significant value would not be expected if the recall task and the forced-choice recognition test were tapping the same processes. Thus, there is evidence that the color wheel recall task involves different processes from the matching-to-sample task. That is, recall involves separate storage and retrieval processes.

Memory is not limited to human species, and in some cases, scientific questions are better assessed with nonhuman species. For example, research questions, which involve high risks, are not appropriate for humans, but some of these questions might be examined with animal studies. Also, it is an interesting and fascinating question in its own right to see how various animals are similar or different in terms of their memory. For example, there are enormous differences between avian and mammalian brains, so memory processes are expected to be very different. Nonetheless, processes such as storage and retrieval are pertinent for any species. The method of measurement of these processes is likely to be different, but importantly with model-based measurement, it is possible to measure storage and retrieval probabilities with potentially any species. Currently, we are still almost at the starting line in terms of understanding memory processes across the phylogenetic kingdom. Yet the work in this chapter is a beginning toward the development of a measurement program for studying animal memory in terms of storage and retrieval. Perhaps other species can be studied with the same procedure that was used in this chapter. Yet it is important to recognize that prior to this research, a great deal had been already learned about the capabilities of pigeons. The model was based on a task with which researchers had considerable experience, and the model focused on the memory of stimuli that were known to be detectable by pigeons. To measure storage and retrieval processes for another species, the investigators will also need to use a task and stimuli that are within the capabilities of that species.

Appendix: Parameter Estimates for the Animal Model

This appendix is designed to provide the technical details underlying the parameter estimation for the storage-retrieval model used for the animal model because the details were not published elsewhere. The statistical methods for the other models in part II can already be found in the cited papers about the models. The appendix can be skipped for readers who are willing to accept the fact that the parameters of the models can be estimated from the data of the six independent binomials shown in figure 9.3.

There are a number of different methods for determining the model parameters from data, for example, maximum likelihood estimation (MLE), standard Bayesian estimation, Bayesian Markov chain Monte Carlo estimation, and population-parameter mapping estimation (PPM). The MLE approach is the preferred method from a relative-frequency framework for statistical inference. Yet Bayesian statisticians have critiqued the reliance on the MLE because (1) the point estimates are not the best for realistic cost functions, and (2) point estimation itself needs to be considered within the larger context of the probability distributions for the model parameters (Bernardo & Smith, 1994; Chechile, 1977a, 2010a; Le Cam, 1986; Stuart & Ord, 1991). The last three methods listed above are based on Bayesian statistical theory and thus (in my opinion) are superior methods for estimating parameters.

The PPM method (Chechile, 1998, 2004, 2010a) provides a convenient way to generate both point estimates and probability distributions for the model parameters. This method employs an exact Bayesian Monte Carlo sampling of the population parameters for the six independent binomials. Let ϕ_1, ϕ_3, ϕ_5, ϕ_7, ϕ_9, and ϕ_{11} be the binomial population parameters, so it follows that $\phi_{2i} = 1 - \phi_{2i-1}$ for $i = 1, \ldots, 6$. It is a well-established fact from the statistical literature that the Bayesian posterior for each of these binomials is in the form of a beta distribution (Box & Tiao, 1973; Hartigan, 1983; Lad, 1996; Lee, 1989; Press, 1989). Given the data for the frequencies in each of the 12 outcome cells $D = (n_1, \ldots, n_{12})$, the set for six independent binomial distributions has a joint posterior probability density function for $\langle\Phi\rangle = (\phi_1, \phi_3, \phi_5, \phi_7, \phi_9, \phi_{11})$, which is denoted as $f(\langle\Phi\rangle)$ and is given as

$$f(\langle\Phi\rangle|D) = \prod_{i=1}^{6} \frac{(n_{2i-1} + n_{2i} + 1)!}{n_{2i-1}! \, n_{2i}!} \, \phi_{2i-1}^{n''_{2i-1}} (1 - \phi_{2i-1})^{n''_{2i}}, \tag{9.7}$$

where $n''_j = n'_j + n_j$ for $j = 1, \ldots, 12$ and where the n'_j values are the values for the prior Bayesian density function. For this research, a diffuse uniform distribution for each binomial parameter is used and that is equivalent to the case where $n'_j = 0$ for $j = 1, \ldots, 12$. Consequently, for the current application, $n''_j = n_j$. Equation (9.7) is a product of six beta distributions, and the parameters of the beta distribution are $a_{2i-1} = n_{2i-1} + 1$ and $a_{2i} = n_{2i} + 1$, for $i = 1, \ldots, 6$. An efficient algorithm for randomly sampling a value from a beta distribution was developed by Cheng (1978). Thus, a random vector $\langle\Phi\rangle_k$ can be readily generated. Notice that with this method, the sampling is from an exact posterior distribution, and the method does not require a Markov chain sampling algorithm to obtain an approximate posterior distribution.

Next with the PPM method, a random vector from the statistical population is mapped to a corresponding vector in the scientific model space, that is, the kth vector $\langle\Phi\rangle_k \longrightarrow \langle\Theta\rangle_k$, where $\langle\Theta\rangle_k$ is the corresponding vector of values $(\theta_S, \theta_F, \theta_N, \theta_r)_k$ and where $\theta_S + \theta_F + \theta_N = 1$. Due to the constraint on the three storage quality measures, there are only two free parameter values for storage quality (i.e., given θ_N and θ_F, it follows that $\theta_S = 1 - \theta_N - \theta_F$). The retrieval parameter is the third free parameter for the scientific model. There is some modeling error because the number of dimensions for the Φ-space representation is six, and the number of dimensions for the Θ-space representation is three. Modeling error is unavoidable when the dimensions of the space that is the origin of the mapping are larger than the dimensions of the space that is being mapped on. This fact does not mean that there is an identification problem. If more than one Θ-space point can be mapped to the same Φ-space point, then the model would not be identifiable. However, it is usually the case that the dimensions of the scientific model (i.e., Θ space) are less than the dimensions of the statistical data structure (i.e., Φ space). In this case, the model is still identifiable, but there is some modeling error. The mapping from Φ space to Θ space is designed to produce a representative point for the values of the scientific model.

A convenient mapping of the θ_N and θ_F parameters is the following:

$$\theta_N = \frac{12A - 20B}{19}, \tag{9.8}$$

$$\theta_F = \frac{4B - 5\hat{\theta}_N}{3}, \tag{9.9}$$

where $\hat{\theta}_N$ is the mapped value from equation (9.8) and where

$$A = \phi_3 + \phi_5 + \frac{4}{3}(\phi_7 + \phi_9 + \phi_{11}),$$

$$B = \phi_{11} + \frac{1}{2}(\phi_5 + \phi_9).$$

The value for θ_S is given as $\theta_S = 1 - \hat{\theta}_N - \hat{\theta}_F$, where $\hat{\theta}_N$ and $\hat{\theta}_F$ are determined from equations (9.8) and (9.9), respectively. Finally θ_r is given as

$$\theta_r = \frac{1 - \frac{56\phi_1}{55}}{\hat{\theta}_S}, \tag{9.10}$$

where $\hat{\theta}_S$ is the mapped θ_S value.

The mapping of a sampled vector from Φ space to a corresponding vector in Θ space is done repeatedly. Typically 30,000 random vectors from Φ space are sampled and mapped to Θ space. If any of the mappings result in a θ parameter being outside the (0, 1) interval, then that vector from the posterior distribution for $\langle \Phi \rangle$ is considered absurd. But if all the model parameters are on the (0, 1) interval, then the mapping is considered coherent. The overall proportion of coherent mappings is a measure of the quality of the model, and it is a unique feature of the PPM method. The subsequent coherent mappings are the basis for the posterior probability distributions for the model parameters. The mean of the posterior distribution is used as the point estimate for the parameter. The mapped vectors thus provide information for both point estimates and the probability distributions for the parameters.

10 Implicit-Explicit Separation

10.1 Chapter Overview

With the 7B and 6P models, an event was measured by means of estimating the probability of sufficient storage (denoted as θ_S). This metric is the proportion of the events that had an encoding in enough detail that it had the potential to support the recall of the target event. Thus, a proportion of the target events $(1 - \theta_S)$ are insufficiently stored, so even with perfect access to this information, the person cannot correctly recall the target event. Insufficient storage might be the total absence of any information about the target event, but it might also be some fragment or some faint residue that we can call implicit memory. The model-based measurement approaches discussed so far do not separate explicit storage (i.e., sufficient storage) from implicit memory storage. In this chapter, the focus is on further decomposition of the encodings that are not explicitly stored. We will examine two models for measuring implicit memory. The first method is based on an instruction to the participants about when they should respond YES depending on the source of the information. This method is called the *extended process-dissociation model*. After an examination of this method, a case is made for a different methodology that does not rely on the person's knowledge about the source of the target information. This second method is called the *implicit-explicit separation (IES) model*. Evidence will also be presented for the validity of the IES measurement method. Furthermore, some important findings are discussed about the relationship between explicit and implicit memory.

10.2 Implicit-Explicit Storage Measurement

A case was made in part I for a distinction between explicit and implicit memory. For example, Warrington and Weiskrantz (1974) found that if amnesic patients heard a word earlier, then later they were more likely to complete an ambiguous letter fragment with the solution of the word previously presented. That finding in itself is not compelling, but Warrington and Weiskrantz reported that the patient did not consciously remember experiencing the word earlier. The fragment completion task was said to be a task that was sensitive to implicit memory. There is an extensive experimental literature of reports similar to the research of Warrington and Weiskrantz (Roediger & McDermott, 1993; Schacter, 1987). But these studies were conducted within the questionable process-purity experimental approach. The evidence that the researchers offered for a different memory type was based on an experimental dissociation. But in chapter 6, we have seen (1) that experimental dissociations are meaningless and (2) that each of the experimental tasks can potentially tap a variety of latent processes. Moreover, several papers have argued (1) that a single memory system can account for both recognition and priming (Kinder & Shanks, 2003) and (2) that recognition, priming, and fluency are better modeled within a single-system framework rather than separate memory systems for explicit and implicit memory (Berry, Shanks, Speekenbrink, & Henson, 2012). But those studies did not employ model-based measurement to assess the hypothesis of separate systems for explicit and implicit memory. Consequently, it is imperative to use a model-based measurement approach to disentangle the underlying processes of explicit and implicit memory. Fortunately, implicit memory can be measured with model-based measurement. In fact, there are several models for separating explicit and implicit memory, so the question arises as to which model is most suitable.

10.3   The Extended Process-Dissociation Model

The extended process-dissociation model is an enhanced version of the earlier process-dissociation model by Jacoby (1991). As with other measurement models, the process-dissociation model was based on a novel experimental procedure. The key features in the experimental task were (1) the use of two information sources such as a first list (list 1) and a second list (list 2) and (2) to have two types of instructions for the participants. In a typical experiment, the participants are first presented with all the target information from both sources. At the time of test, some participants are given the *Exclusion* instructional rule, which directs the participants to only respond YES for items from list 2.[1] Other participants are given the *Inclusion* instructional rule, which

---

1. There are variations on this requirement. For example, there might be two types of memory items, and the participants are directed with the *Exclusion* instruction to only respond in the affirmative if the information is from one of the sources.

directs the participants to respond YES for any of the targets regardless of the source of the target. This task has come to be called the process-dissociation procedure. Buchner, Erdfelder and Vaterrodt-Plünnecke (1995) developed an improved version of the Jacoby process-dissociation model that they called the extended process-dissociation model. The key new feature with the extended process-dissociation model is a correction for guessing by means of an analysis of the foil trials when the participants are probed with novel items that were not from either of the two information sources.

The extended two-source process-dissociation model is based on four separate binomials. Each binomial corresponds to the two outcomes for a test trial (i.e., a YES versus a NO response). The four binomials are (1) list 1 items with the Exclusion instruction, (2) list 1 items with the Inclusion instruction, (3) foil items with the Exclusion instructions, and (4) foil items with Inclusion instructions. The Jacoby (1991) process-dissociation model does not utilize the information from foil trials, which is a questionable aspect of the model because the rates for the YES and NO responses for target items should be compared to their base rates for foil items. The extended process-dissociation model repairs this theoretical shortcoming of the Jacoby (1991) model. The process trees for the extended model are shown in figure 10.1 for the four binomial test trials. The model parameters (θ_C, θ_U, θ_{ge}, and θ_{gi}) correspond respectively to the rates associated with (explicit) conscious memory, unconscious (implicit) memory, guessing under the exclusion instructional condition, and guessing in the inclusion condition.[2] The Jacoby (1991) process-dissociation model is a special case of the extended model when $\theta_{ge} = \theta_{gi} = 0$, so there is no loss of generality by using the extended process-dissociation model where the guessing parameters are estimated from the foil test trials.

According to the extended process-dissociation model, if the participants can consciously remember a list 1 item (with probability θ_C), then the participants should respond NO when they are given the Exclusion instruction, but they should respond YES when they receive the Inclusion instruction. If the participants cannot consciously remember the list 1 item, then there still might be an implicit memory of the target (with probability θ_U), and it is assumed that this implicit trace leads participants to respond YES regardless of the instructional condition. Moreover, if the list 1 item is neither explicitly nor implicitly represented, then there still is a chance that the participant

2. Jacoby (1991) and Buchner et al. (1995) used U and C symbols to denote respectively the explicit and implicit probabilities. I have changed that notation here in order to comply with the standard in the field of statistics of using Greek symbols for population parameters and Latin letters for sample statistics. The model is a description of the population probabilities.

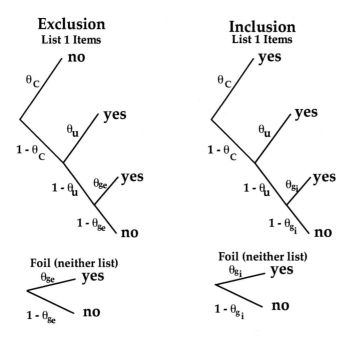

Figure 10.1
Process tree models for exclusion and inclusion instructional conditions when the target is present as well as for foil test trials. The upper left tree is for the list 1 items when the participant had the Exclusion instruction, whereas the upper right tree is for list 1 items for Inclusion instructions. The two lower trees are for foil test trials.

can guess YES. The guessing rate in the Exclusion condition is θ_{ge}, and the guessing rate in the Inclusion condition is θ_{go}.[3]

3. See Buchner et al. (1995) for the details of the parameter estimation; these details are not needed for the present discussion of the extended process-dissociation model. But it should be noted that in the special case of the Jacoby (1991) model when $\theta_{ge} = \theta_{gi} = 0$, the estimate for the explicit memory component $\hat{\theta}_C$ is equal to $P(NO|Exclusion) - P(NO|Inclusion)$, and the implicit component is equal to $P(YES|Exclusion)/[P(YES|Exclusion) + P(NO|Inclusion)]$. Clearly, the Jacoby (1991) model is suspect relative to the extended process-dissociation model because it assumes that the recognition false-alarm rate for list learning is zero. But it is common knowledge among memory researchers that people sometimes respond positively to foil items. Nonetheless, many other researchers have used the original Jacoby (1991) model for reporting their results rather than use the improved extended process-dissociation model. Although it is not known why researchers used an older and questionable model relative to a more general model that subsumes the original process-dissociation model as a special case, I suspect it is due in part to the preference for the remarkably simple formulae for computing the explicit and implicit memory

In addition to memory researchers, social psychologists have also been interested in the use of the process-dissociation model, presumedly because it can be used to study unconscious social biases (e.g., Anooshian & Seibert, 1996; Buchner & Wippich, 1996; Debner & Jacoby, 1994; Gaunt, Leyens, & Demoulin, 2002; Ott, Curio, & Scholz, 2000). Yet the term *unconscious* is debatable in either the memory or social psychological context. As we have seen in the case of Pavlovian conditioning, it is difficult to convincingly establish that the process is outside of consciousness. Regardless if implicit memory is conscious or not, we need only think about it as a nondeclarative memory as described in part I. But for the process tree in figure 10.1, the term *unconscious* is still used because it was used in the original process-dissociation papers.

10.4   Extended Process-Dissociation Critique

Despite the widespread use of the process-dissociation task in both memory research and social psychology, this model has been sharply criticized (Chechile, Sloboda, & Chamberland, 2012; Dodson & Johnson, 1996; Yu & Bellezza, 2002). All of the points raised in those papers will not be repeated here, but instead only a few essential criticisms will be readdressed.

As noted previously, the data for the process-dissociation task consists of four separate binomials. With this data structure, it is only possible that the resulting model have a maximum of four parameters. The extended process-dissociation model does have four parameters, but to achieve this condition, it was necessary to make two questionable assumptions. If either of these assumptions is incorrect, then the method is not valid.

One of the questionable assumptions concerns the equality of the rate for implicit storage across the two instructional conditions. In the original Jacoby (1991) paper, the unconscious or implicit process is described as a familiarity-based recognition response. Yet if the person detects a sense of familiarity with the test probe, then there should be

---

components rather than use the more complex estimation method recommended by Buchner et al. (1995). It is also curious that Yonelinas and Jacoby (1996) criticized the Buchner et al. (1995) model as to how response bias in foil recognition is modeled. Yonelinas and Jacoby (1996) provided a signal-detection variant of the process-dissociation model instead of the guessing model. Given the critique provided in chapter 8 of the use of the signal-detection approach to memory measurement, it is of questionable value to invoke a signal-detection framework onto the process-dissociation task. Moreover, in subsequent research by Jacoby and others, the original Jacoby (1991) model is used rather than the Yonelinas and Jacoby (1996) model. Consequently, the Yonelinas-Jacoby model did not gain traction even with the developers of the model.

differential consequences as a function of the response instructions. With the Inclusion instruction, the person should give the YES response when there is a sense of familiarity. But with the Exclusion instruction, it is not clear what the person should do. The item with a sense of familiarity might be from list 1, and therefore a NO response might be the appropriate decision. Confusion about the source of this sense of familiarity puts the person in a state of decisional ambiguity. Moreover, if the person is certain that the sense of familiarity is unquestionably from list 2, then the response would be YES; but how can a person be certain of the list source without being conscious of the item? Familiarity with total clarity about the list source is likely to be because the person has a conscious memory of the item. But the issue is about the role of familiarity in the conditional state when there is the absence of explicit memory. Note that even if the true implicit probability were the same in the two instructional conditions, the behavioral response would be different. Thus, the effective θ_U value would be different in the two instructional conditions. But if separate θ_U parameters are required for the two instructional conditions, then the extended model would have more parameters than is allowed for an identifiable model.

Another problematic issue with the process-dissociation model is the assumed equivalence of the explicit memory parameter θ_C across the two instructional conditions. As with the implicit memory parameter, there are reasons for doubting that this measure is the same in the two conditions. For example, suppose a person remembers the target item but does not remember if the item was on the first list or the second list. That person would still respond YES in the Inclusion condition, but in the Exclusion condition, it is not clear that the person would necessarily respond NO given the uncertainty about the list identification. This criticism about the process-dissociation procedure has also been raised by other investigators (Dodson & Johnson, 1996). If the rates of conscious memory are different in the two instructional conditions, then different parameters are required and that again makes the model unidentifiable.

In addition to the above two problems, the process-dissociation model also has the problem of not accounting for the case of partial information. We saw in the Cook et al. (2003) model for pigeons that fractional storage does in fact occur. Yet in the decomposition of states of knowledge, the process-dissociation model has only three configurations: (1) explicitly stored, (2) implicitly stored, and (3) neither explicit nor implicit storage. But the state of neither explicit nor implicit knowledge might be further decomposed into a fractional storage state versus a nonstored state. These two states would have different consequences under the two instructional conditions. For the case of nonstorage, the response would be NO regardless of the instructions, but in the case of fractional storage, there would be different responses depending on the instructions. Given partial storage and the Inclusion instruction, the person is likely to respond YES, but if the person is tested under the Exclusion instruction, then the person might respond NO because of the ambiguity about the source of the fragment.

Consequently, the extended process-dissociation model is questionable on multiple grounds. Fractional storage needs to be taken into account. It is also important to avoid using an experimental procedure that assumes that people's judgments about the *source* of their knowledge are free of errors.

10.5 The IES Model

The Chechile et al. (2012) model was designed to circumvent the problems with the process-dissociation approach for measuring explicit and implicit memory. The model is called the implicit-explicit separation (IES) model. The task for the IES model does not use the exclusion versus inclusion instructional manipulation. The IES model also takes into account the proportion of the targets that are partially stored. Thus, the IES model results in probability parameters for (1) explicit storage (θ_{se}), (2) implicit storage (θ_{si}), (3) fractional storage (θ_F), and (4) the null state of no storage (θ_N).

As with all the other model-based measurement methods, the IES model is based on a novel set of experimental protocols, which are illustrated in figure 10.2. After the initial learning, there is a test of the previously presented information. At the time of test, the participants see a lineup list of four alternatives and are asked to respond either yes or no with the instructions to respond YES if any of the items in the list were previously presented or otherwise to respond NO. There are two types of lineup test lists—target-present and target-absent. The target-present lineup has a single old item along with three novel foils, and the target-absent lineup consists of four foils. Participants are told in advance about the two types of lineup lists. After the initial recognition decision to the lineup, the participants are cued to give a 3-point confidence rating with the instructions to use the 3 rating for certain responses, the 2 rating for educated guesses, and the 1 rating for pure guesses. This is the same use of a 3-point rating as with the Chechile-Meyer task. On target-present trials, there is additional testing with a series of contingent forced-choice decisions. On the first forced-choice test, the participants are asked to pick out one of the items on the lineup list; this is a four-alternative forced-choice task (4-AFC). If the participant is correct, then the person is told that the choice is correct, and the testing for that item is done. But if the participant failed to select the target, then feedback is given that the choice is incorrect, and the person is asked to pick again from the remaining items (i.e., it becomes a 3-AFC task). If the response on the 3-AFC test is correct, then feedback is provided and the testing for this target item

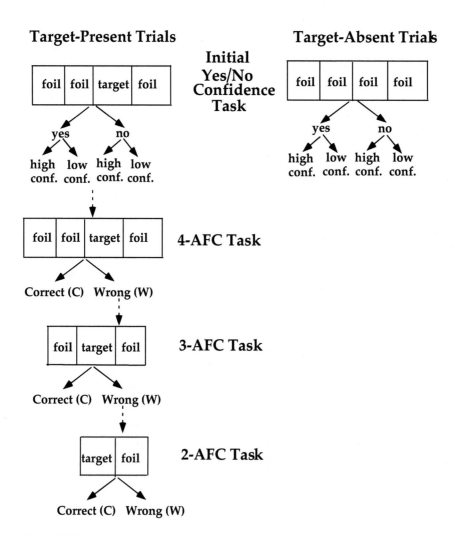

Figure 10.2
The sequence of test operations for target-present and target-absent trials for the IES procedure.

Target-Present Categories

	Yes Hi. Conf.	No Hi. Conf.	Yes Lo. Conf.	No Lo. Conf.
Cor. 4-AFC	Cell 1	Cell 2	Cell 3	Cell 4
Cor. 3-AFC	Cell 5	Cell 6	Cell 7	Cell 8
Cor. 2-AFC	Cell 9	Cell 10	Cell 11	Cell 12
Incor. 2-AFC	Cell 13	Cell 14	Cell 15	Cell 16

Target-Absent Categories

No High Conf.	No Low Conf.	Yes Low Conf.	Yes High Conf.
Cell 17	Cell 18	Cell 19	Cell 20

Figure 10.3

The sixteen response cells for target-present trials and the four response multinomial cells for target-absent trials (see figure 10.2 for task illustration). Note that the yes/no decision and the high- and low-confidence rating are to the initial presentation of the four-item test. The forced-choice testing follows this test only when the target is present.

is completed. But if the response on the 3-AFC task is again incorrect, then feedback is provided, the incorrect item is removed from the list, and the person is asked to do a 2-AFC task. For the target-absent tests, there is not a 4-AFC task because all four alternatives are novel. Thus, the participant is only asked to select an item when in fact there is a target in the lineup list.

The rationale for this complex experimental protocol might not be clear, but this procedure will be shown to enable the measurement of four mutually exclusive states for memory storage: explicit, implicit, fractional, and nonstorage. The experimental procedure results in two multinomials: a sixteen-cell multinomial for the response outcomes for target-present tests and a four-cell multinomial for the outcomes on target-absent tests. The response outcomes for both multinomials are illustrated in figure 10.3.

If information is explicitly stored with probability θ_{se}, then the participant will respond YES 3 on the lineup recognition test and will correctly select the target on the 4-AFC task (i.e., the participant will give a cell 1 response). But it will be also possible to produce a cell 1 response when there is either fractional storage with probability θ_F or nonstorage with probability θ_N. Consequently, corrections are needed to factor out the guesses from these other states from all the cell 1 responses. Thus, the rate for the cell 1 response is greater than the value for θ_{se}.

There is a different set of response expectations for information in the implicit-storage state. Now the participant might say either NO on the recognition test or say YES with low confidence, but the participant is nonetheless correct on the 4-AFC task. Information in the implicit-storage state cannot be correctly recalled and would result

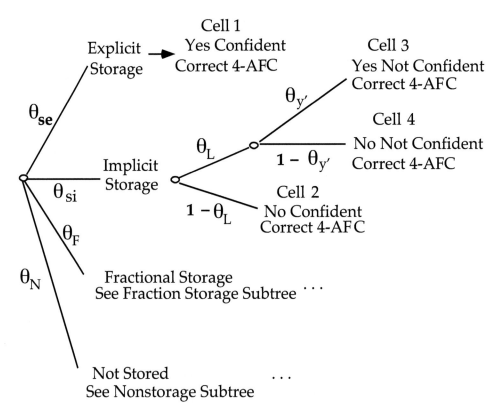

Figure 10.4
The four-part processing for the state of memory storage. For the cases of fractional storage and nonstorage, there are other subtrees not shown. See figure 10.3 for the cell definitions.

in either unconfident or incorrect recognition; yet the implicit representation is still good enough to result in a correct selection of the target on the 4-AFC test. Thus, the IES method for measuring implicit storage is based on the idea that this type of information results in the selection of the correct item despite the fact there is not a confident declarative memory for having experienced the target before. In other words, the person with implicit storage should produce a response in cell 2, 3, or 4, as shown in figure 10.3. Yet responses in these cells can also occur via guessing when the target is in either the fractional or nonstorage state. Thus, the sum of the probabilities for a response in cells 2, 3, and 4 will be greater than the implicit-storage probability θ_{si}.

Figure 10.4 is the basic tree for the IES experimental protocol. The subtree for when there is nonstorage is shown in figure 10.5, and the subtree for when there is fractional

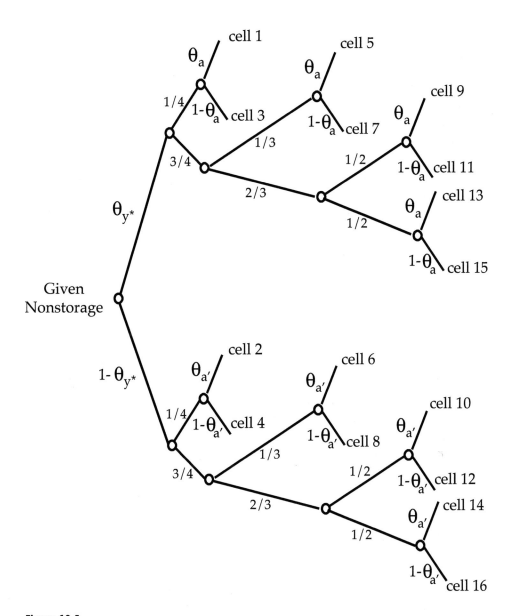

Figure 10.5
The subtree-processing model for the IES task when there is no target storage. See figure 10.3 for the cell definitions.

storage is shown in figure 10.6. There are many nuisance parameters in the IES model for dealing with response biases and for correcting for guessing. These parameters play a valuable role by correcting for nuisance factors, but beyond that role, these parameters are of little psychological interest. The important parameters of interest are the probabilities shown on the four paths emerging from the root of the tree shown in figure 10.4.

If there is either explicit or implicit storage of the entire target, then the person will select the target on the 4-AFC task (i.e., cells 1 through 4), but it is also possible that responses in those cell categories can occur with nonstorage or fractional storage. For example, notice in the subtree for nonstorage (shown in figure 10.5) that there are guessing responses for cells 1 to 4. Similarly for the fractional-storage subtree (shown in figure 10.6), there is the possibility of guessing responses for those cell categories. Hence, the IES model automatically corrects for the guessing rate.

When there is fractional storage, the person ought to eliminate either one or two of the foils on the 4-AFC task. If the person can eliminate all three foils on the 4-AFC task, then that would be either explicit or implicit storage. Also, if the person cannot eliminate any of the foils, then that is effectively nonstorage. Consequently, with fractional storage, the person will be able to eliminate either one or two of the three foils on the 4-AFC test. Thus, if the target is fractionally stored, then that target would never be wrong on a contingent 2-AFC task. For example, if the person can only eliminate one foil and guesses incorrectly both on the 4-AFC task and on the follow-up 3-AFC task, then the remaining foil on the 2-AFC task will be the one item that can be eliminated. Thus, with fractional storage, the person will never respond incorrectly on the 2-AFC task (cells 13 through 16). A response in those cells can only occur because there is nonstorage of the target (see figure 10.5).

The final processing tree for the IES model is for the target-absent case when all the items are foils. This tree is shown in figure 10.7. Like the 6P and 7B models, there is an important parameter for the probability of knowledge of the target in foil-recognition testing, and this parameter is labeled as in those other models (i.e., θ_k). Thus, when there is some target knowledge, the person is assumed to give a confident NO response. But when there is no knowledge of the target whatsoever, which occurs with probability $1 - \theta_k$, then the guessing is modeled the same as for nonstorage on the target-present test. Consequently, the θ_{y*}, θ_a and $\theta_{a'}$ parameters have the same values as in the nonstorage subtree shown in figure 10.5.

The trees shown in figures 10.5, 10.6, and 10.7 are elaborate and contain many nuisance parameters that are there only to correct for response biases and guessing. Chechile et al. (2012) discussed the full model in more detail and provided an analysis

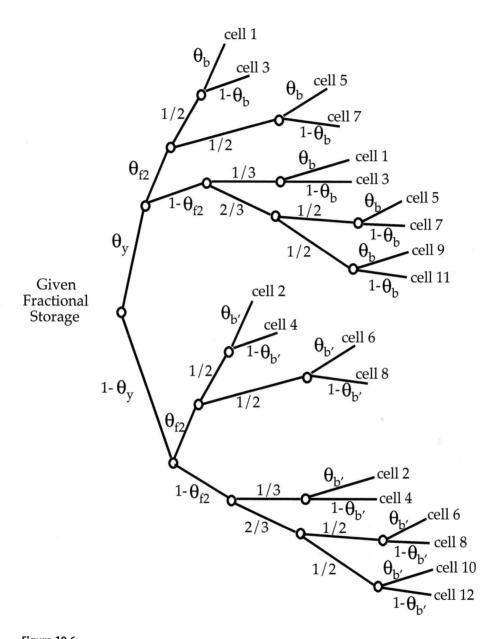

Figure 10.6
The subtree-processing model for the IES task when there is fractional or partial target storage. See figure 10.3 for the cell definitions.

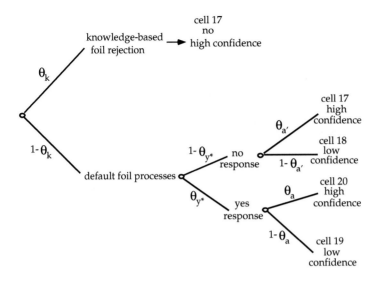

Figure 10.7
The processing-tree model for test trials where all four items are foils. See figure 10.3 for the cell
definitions.

of the statistical properties of the model. For this chapter, we need only focus on five
parameters for the probabilities of (1) explicit storage θ_{se}, (2) implicit storage θ_{si}, (3)
fractional storage θ_F, (4) nonstorage θ_N, and (5) knowledge-based foil rejection θ_k. Moreover, because storage is a four-part mixture of different representational states, it must
be the case that $\theta_{se} + \theta_{si} + \theta_F + \theta_N = 1$.

10.6   Validation Evidence for the IES Model

10.6.1   Encoding Time Study
Chechile et al. (2012) used the IES experimental procedure to study the memory of a
six-item lists of words. The lists had three different presentation times. Each list was
presented once and tested immediately after list presentation. The resulting estimates
for the key IES model parameters provide evidence for the validation of the IES measurement model. The time per word was either 45 msec, 100 msec, or 1,000 msec. The
first two conditions are extremely rapid presentation rates, and the last condition is
a more typical presentation rate. All words were visually presented. The experimental
details as well as the statistical details are described more completely in the Chechile
et al. (2012) paper. The IES parameter estimates for the key parameters are provided

Table 10.1
Effect of encoding time on IES parameters

Parameter	45 msec	100 msec	1,000 msec
Explicit θ_{se}	.197	.326	.608
Implicit θ_{si}	.094	.106	.101
Fractional θ_F	.062	.100	.154
Nonstorage θ_N	.647	.468	.137
Foil rejection θ_k	.009	.016	.278

in table 10.1. As would be expected, the increase in the encoding time per word had a reliable increase for the explicit storage (θ_{se}) parameter. Moreover, the nonstorage parameter (θ_N) decreased reliably with increased encoding time. The fractional-storage parameter (θ_F) also exhibited a reliable difference between the 45-msec and 1,000-msec conditions. The implicit-storage (θ_{si}) parameter was nonzero in each condition, but it did not vary with presentation time. The fact that there was a nonzero value for θ_{si} means that there is implicit memory that is detectable even after the correction for fractional storage. Finally, the parameter for target knowledge on foil recognition (θ_k) revealed particularly strong validation evidence. In the two rapid presentation conditions, it is nearly impossible for participants to encode all six words on the list. Consequently, in the case of the target-absent test condition, it is virtually impossible for the participants to reject the set of four foils because they do not know all of the words on the list. The value for θ_k is not reliably different from 0 in those two rapid presentation conditions. Yet in the encoding condition where each word is visual for 1 sec, there is a chance that all the words on the test lineup of four foils can be rejected with high confidence because all the target words are stored. Thus, the same model that shows virtually no recollective effects on foil trials when the presentation rate is extremely rapid is still able to have a 27.8 percent rate for knowledge-based foil rejection when the words are presented more slowly.

10.6.2 Divided Viewing

Just as people have hand dominance, there is also eye dominance.[4] The next experiment was done to examine explicit and implicit memory for simultaneously presented

---

4. A standard way to ascertain eye dominance can be done by a simple test. View a black circle on the wall and cover one eye. Next hold up a thumb so it is in the black circle. If the covered eye is the dominant eye, then the image of the thumb appears to shift immediately after the cover over the eye is removed.

Table 10.2
Divided viewing experiment

Parameter	Dominant	Nondominant
Explicit θ_{se}	.790	.547
Implicit θ_{si}	.104	.147
Fractional θ_F	.054	.096
Nonstorage θ_N	.052	.209
Foil rejection θ_k	.679	.575

figures.[5] The hypothesis for the study is to see if information presented on the non-dominant eye is less likely to be encoded initially with an explicit representation but more likely to have an implicit representation.

The participants in this experiment placed their chins on a resting stand and viewed a computer screen. There was a partition between the two eyes so that a figure on the left side of the computer display was only visible to the left eye, and likewise a figure on the right side of the display was only visible to the right eye. On each trial of the experiment, the participant saw two figures—one on the left side and one on the right side. The participant's task when viewing the patterns was to determine if the pattern on the dominant-eye side represented a living thing. The figures were two-dimensional drawings of common images taken from a standard corpus of experimental stimuli. Immediately after a brief 750-msec simultaneous presentation of the two figures, the participants were tested with the IES test procedure. Some target-present tests were for the figure on the dominant eye, and on other trials, the test was for the pattern on the nondominant eye.

This task is difficult. The two simultaneous figures were displayed only for 750 msec. In that short time, the participant is trying to encode two figures and make a judgment about the possibility that the dominant-eye figure represents a living thing. The results for this experiment are shown in table 10.2. It is thus not surprising that the explicit-storage rate for an immediate test for the figures presented to the dominant eye is not at a 100 percent rate but is instead at a 79 percent rate. Moreover, the nondominant-eye figure is encoded reliably at a lower 54.7 percent rate for explicit storage. In terms of nonstorage, the value for the nondominant eye was reliably larger than the value for the dominant eye. Importantly, the implicit-storage rate for the nondominant eye is reliably larger than the implicit-storage rate for the dominant eye.

---

5. This research is a pilot study conducted in collaboration with Holly Taylor, Erin Warren, and Daniel Barch.

This change in the immediate encoding condition supports the idea that some information in the perceptual field is not explicitly encoded, but it might nonetheless be implicitly encoded. This effect is, however, very small, and it would be a mistake to argue that all the information that is not explicitly stored is implicitly stored. Even when there are only two figures presented simultaneously, the combined nonstorage and fractional-storage rates are about twice the probability of an implicit memory representation.

10.6.3 Effect of Retention Interval

A number of researchers have hypothesized that explicit and implicit memory are different systems for storage (e.g., Cohen & Squire, 1980; Graf & Schacter, 1985; Johnson, 1983; Parkin, 1993; Schacter, 1989; Squire, 1986, 1987; Tulving, 1983, 1985a; Weiskrantz, 1987, 1989). In fact, some researchers even go so far as to suggest that explicit and implicit memory are contained within different brain regions (e.g., Schacter, Chin, & Ochsner, 1993). Yet the argument for a separate memory system is based on the rationale of experimental dissociation. But we have seen earlier in chapter 6 that even double experimental dissociation is a questionable methodology that cannot render a meaningful conclusion. Moreover, if an implicit memory is in a separate memory system based on the initial encoding, then increasing the retention interval should result in a potential loss of information from this system. But with the IES model, for any experimental condition, there can be a mixture of three different storage trace types (explicit, implicit, and fractional) along with some targets being lost altogether. It is possible that the rates for these four states can change with the length of the retention interval because items that were once explicitly stored might degrade to one of the other states. With the IES model, the rates for the four states can be measured as a function of the retention interval. If the rate of implicit memory increases with the length of the retention interval, then that finding would be a challenge to the multiple-systems model of memory. Experiment 3 from Chechile et al. (2012) examined the effect of the length of the retention interval on the parameters of the IES model.

The participants in the experiment briefly heard a triad of nonsense letters on each trial and then had to repeat an auditory string of digits. The digits were presented at the rate of three digits per second. The number of interpolated digits was 1, 4, 12, 36, or 90. Following the interpolated task of digit shadowing, the participants were tested by means of the IES experimental protocols. See the Chechile et al. (2012) paper for further details about the experiment.

In figure 10.8, the IES model parameters are plotted as a function of the retention interval. The explicit-storage parameter decreases with the retention interval, whereas

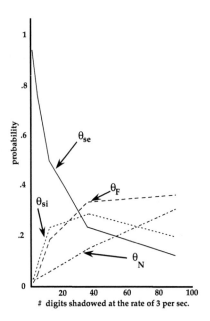

Figure 10.8
IES model parameters as a function of the retention interval. Explicit storage, implicit storage, fractional storage, and nonstorage are respectively θ_{se}, θ_{si}, θ_F, and θ_N.

the nonstorage parameter and the fractional-storage parameter increase with the length of the interval. Unlike these three parameters, the implicit-storage parameter has a biphasic profile. Implicit storage increases from the $\frac{1}{3}$-sec condition to the 12-sec condition. But implicit storage also has a reliable decrease from the 12-sec condition to the 30-sec condition. The pattern for the θ_{si} parameter is most likely caused by the loss of some items that were previously stored explicitly. If an item is initially encoded as an explicit trace, then it does not mean that the item rigidly remains as an explicit representation. A previously explicit memory trace can degrade in quality to the point that it becomes an implicit memory representation. Memories undergo dynamic changes. Consequently, the IES measurement model has detected evidence against the idea that implicit memory is a separate system as hypothesized by some researchers (Parkin, 1993; Weiskrantz, 1987, 1989). The increase of implicit memory with the length of the retention interval is difficult to explain from a separate-systems viewpoint for explicit and implicit memory. Note that from the current position that considers explicit, implicit, and fractional storage as different states for a memory representation, it is possible that items in the implicit state can also undergo change and degrade to either a fractional representation or become destroyed altogether.

10.7 Summary: Explicit-Implicit Memory Measures

The above validation experiments demonstrate that the problems of the process-dissociation task can be circumvented. The IES model parameters are a sophisticated way for characterizing four states for the storage of information. The focus in this research was on the idea that traces that are insufficiently stored in the 6P or 7B models are not necessarily in a state of complete memory loss. Some traces might have totally degraded, but not all insufficiently stored items are lost altogether. What is called sufficient storage in the 6P or 7B model is explicit storage in the IES model. The insufficient storage rate for the 6P or 7B model is a mixture of rates for implicitly stored, fractionally stored, and nonstored items.

While the focus of this section is on the measurement of explicit and implicit memory, it is possible to obtain a retrieval estimate as well. To get the retrieval measure, a modified IES task is needed. For this modified task, the participant is first examined with a recall test, which is then followed by the usual IES test procedure if the person could not recall the target. At the present time, no experiment has been done with this experimental protocol, but in principle, a retrieval measure could be found with the modified IES task.

Finally, it should be pointed out that the decision in the IES model for using a four-item lineup is not arbitrary. A simple 2-AFC task does not have sufficient information to enable estimates of fractional storage. It is possible to measure the four states of memory with a task that has a three-item lineup. But a pilot study was done by Erin Warren to compare the three-item lineup test with a four-item lineup. That study showed that the three-item lineup underestimated the fractional-storage rate. Warren also did a pilot study that used a five-item lineup test. The problem with the five-item lineup was it took more time to conduct the experiment and it required more foils. Moreover, the results based on the four-item lineup did not differ reliably from the results based on the five-item lineup. Consequently, the four-item lineup test was the best combination for a sufficiently elaborate test protocol to get good estimates of the four states of item storage, but it was not more complex than it needed to be.

10.8 Summary for Part II

In part II, a case is made that real progress in memory research is dependent on solving measurement problems. Without valid measures of underlying processes such as the probability of information storage and the probability of retrieval of memory traces from storage, it is impossible to develop an understanding of how memories develop and how they change as a function of subsequent events. The field of memory research has observed a number of effects, but the understanding of these effects is in most cases ambiguous because the behavioral measures do not directly measure fundamental storage and retrieval processes. Process-purity experimentation is predicated on the idea of experimental dissociations. But under close examination, the rationale for experimental dissociations is itself unjustified. Counterexamples are provided that show even double experimental dissociation is ambiguous and does not support claims that any direct dependent variable is a pure measure of target storage or a pure measure of item retrieval. Yet it is possible to measure fundamental memory processes with a method called model-based measurement. The 6P and 7B models are examples of measurement models that enable the estimation of quantities for the probability of target storage and the conditional probability of retrieval of stored targets. Model-based measurement treats the status of memory storage as a mixture of some items being sufficiently stored and the other items being insufficiently stored. With sufficient storage, the person can recall the target information if that information can be retrieved. Insufficient storage cannot support correct recall regardless of the subsequent efforts toward retrieval. But insufficiently stored information can still be the basis of correct recognition via a guessing process. In the implicit-explicit separation (IES) model, the insufficiently stored items are parsed further into implicitly stored items, fractionally stored items, and items that are lost altogether. Storage is seen from this perspective as a mixture of items in four possible states. The proportion in each state is a property of the overall quality of memory under a given condition. As conditions change, there are corresponding changes in the proportions for the various states of storage. The probability of retrieval can also vary across experimental conditions.

A key idea in model-based measurement is the development of a novel experimental procedure or protocol that enables the estimation of the fundamental memory components of interest and corrects for the factors that could compromise the measurement. There is also a method for determining if the measurement model is identifiable. Another key idea with model-based measurement is the process of validating the model itself (i.e., showing that the model is properly measuring what we want). Some models have run into problems with the validation stage. For example, the process-dissociation model has been shown to be problematic for several reasons. These difficulties led to the IES model as an improved model that did not suffer from the problems of the process-dissociation model.

Although much of the research on model-based measurement has focused on human experiments, a model was also developed that enables the measurement of storage and retrieval for animals. Several experiments with pigeons using this measurement model demonstrated that the method is both plausible and reasonable.

Finally, experiments with model-based measurement have provided novel and important insights about memory. For example, using a storage-retrieval separation method, Chechile and Ehrensbeck (1983) showed that a critical assumption of the Atkinson-Shiffrin model about permanent storage for items that should be in the long-term store was not supported because this information continued to show memory losses. Also, Chechile et al. (2012) showed that implicit memory is not a separate storage system. Information that was previously stored explicitly can degrade to become an implicit representation; this is an effect that is inconsistent with implicit memory as a separate storage system. Model-based measurement studies also showed new insights about developmental dyslexia, Korsakoff's amnesia, and other effects.

III Factors Affecting Encoding Quality

Introduction to Part III

> The stream of thought flows on; but most of its segments fall into the bottomless abyss of oblivion. Of some, no memory survives the instant of their passage. Of others, it is confined to a few moments, hours, or days. Others, again leave vestiges which are indestructible, and by means of which they may be recalled as long as life endures. Can we explain these differences?
> (William James, 1890a, p. 643)

Memory is a representation of previous learning. But some prior events are more memorable than others as noted in the above quote from William James. The focus of part III is on the factors that influence the effectiveness of learning. Some factors deal with properties of the stimuli and training, whereas other factors deal with the person's cognitive strategy.

The focus for chapter 11 is a set of issues concerning the repetition of training. There is an extensive research literature in psychology about this topic. Although the gradual learning curve has a long history, it will be shown in this chapter there is a richer and more complex set of processes involved than just a continuously improving learning function. There are dynamic changes to the memory representation as a function of repeated study. There are also effects for repeated testing. Chapter 11 also includes a novel dual-trace account for what is learned as the individual engages in repeated study and test trials.

The focus for chapter 12 is on factors that influence the encoding of events on a single trial. There is an exploration of how meaning is extracted and how it varies as a function of the encoding strategy. Unconscious information processing is shown to be momentarily produced, but in some cases it is quickly inhibited if it did not fit the encoding context. The chapter also addresses the role of changes in the context between the time of encoding and the time of test.

Chapter 13 deals with enhanced encoding effects for single-trial presentation. In particular, a number of encoding procedures induce the person to be more actively engaged. These methods also produce a more elaborated memory representation that improves both the retention of the encoding and the retrievability of the stored information.

In chapter 14, the spotlight is on the effect of emotion and memory. Emotional events trigger a number of physiological and neurological processes that further elaborate the encoding of information. Sometimes these processes improve the memory of some aspects of the event, but they can also produce inattention to other features of the experience.

Finally, throughout the four chapters in part III, there is a theoretical exploration of the problems with existing hypotheses in the research literature. A systematic account of the factors that alter the effectiveness of the memory encoding is developed and justified.

11 Repetition Effects

11.1 Chapter Overview

As the familiar joke goes: A self-absorbed musician is walking in Manhattan and is stopped by a lost person looking for directions. The person asks, "How do you get to Carnegie Hall?" The musician quickly answers, "Practice, practice, practice." Psychological studies have supported the truth of this joke, but the investigation of practice effects has also uncovered interesting properties about the learning process, the time spent studying, the spacing of practice, and the role of testing on the level of learning. The studies of these effects reveal insights about the process of establishing a memory as well as the structure of memory. These topics are the central focus for this chapter.

11.2 The Learning Curve

It was clear to everyone since the time of the foundational book by Ebbinghaus in 1885 that the learning of a list of items develops gradually because usually many trials are required before the list is mastered. The repetition of training is a standard educational method. For example, most students are required in some courses to memorize verbatim either a poem or some lines of a play. The mastery of these lines does not come automatically, but rather it is mastered after many trials of training. Yet prose material has an overall semantic organization that might have a prior familiarity. Consequently, the study of this type of realistic information is not the best way to examine learning for a tightly controlled experiment. Hence, many memory researchers prefer to investigate the learning of a novel set of words or a novel set of associations. In psychology, most memory studies are based on the study of a list of novel items. The list can vary in length, but it is usually longer than the immediate memory span.

The learning of a list is perhaps the most widely studied procedure in memory research. Sometimes the list is in the same order; this type of learning protocol is called *serial learning*. The ends of serial lists are learned more rapidly than the middle of the list. If participants are given enough training trials, then they eventually come to master the entire serial list. To avoid the confounding of list-position effects, some researchers use a paired-associate (PA) list-learning procedure. PA lists consist of paired target items (i.e., QEV-ACCOUNT, ZUQ-SECRET, BOJ-FAVOR etc). The first member of the pair is called the stimulus item, and the second member of the pair is called the response item. The stimulus and response items are carefully constructed to be novel. The order of the pairs is randomized for each presentation and each test trial in order to defeat the simple strategy of chaining the response items together. A recall test for the PA task consists of asking for the response member of an associative pair when provided with the stimulus member (i.e., QEV - ?). The PA learning of a list of about fifteen pairs takes a number of training trials before the list is mastered. A typical gradual learning curve is illustrated in figure 11.1. There is little dispute that the mastery of a PA list results in a gradual learning curve, like the one shown in figure 11.1. But it is less clear what the underlying mechanisms are for this learning. The next section explores this issue in more detail.

11.3   Is Learning Incremental or All-or-None?

11.3.1   The Problem

Learning a paired-associate list or learning a challenging poem are typically tasks that require slow, repeated practice. Yet in the opinion of some influential theorists, the apparent gradual improvement masks a sudden all-or-none learning process (Bower, 1961; Estes, 1956, 1960, 1961; Restle, 1962; Rock, 1957). The all-or-none proposal was in stark contrast to the prevailing opinion about learning, and it ignited a vigorous intellectual controversy in psychology. But before examining the all-or-none hypothesis, let us first examine the continuity theory for associative learning.

11.3.2   Continuity Theory or Incremental Strength Theory

The continuity model of learning is based on a number of different theoretical perspectives. For example, in classical association theory, units are held together in memory with an associative bond that has a strength magnitude. Association theory began in antiquity (Aristotle, 350 BCE), but it still was the basis for the behaviorial S-R learning theories of the early and mid- twentieth century. Within this general behaviorist framework of associationism, the repetition of training is assumed to increase the associative strength incrementally. Thus, continuity theory had its roots long before the cognitive

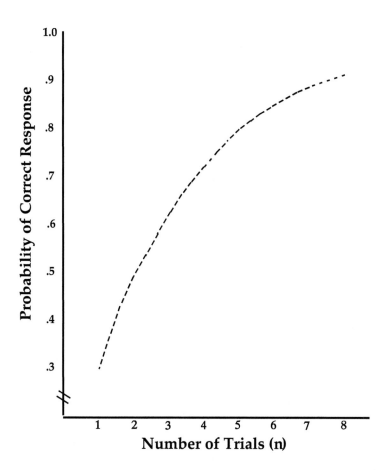

Figure 11.1
The weighted average theoretical learning curve for the hypothesized group from table 11.1.

revolution in psychology.[1] Yet there are still continuity models within the framework of cognitive psychology. For example neural network models are generally consistent with a continuity theory for learning. To see this point, let us consider the Rizzuto and Kahana (2001) model for paired-associate learning.

1. In the 1960s, it became apparent that the behaviorist program of ignoring cognitive processes was fundamentally flawed. In order to understand processes, such as language, attention, reasoning, memory, and problem solving, required psychologists to go beyond instrumental and classical conditioning. This shift in thinking within experimental psychology is referred to as the cognitive revolution.

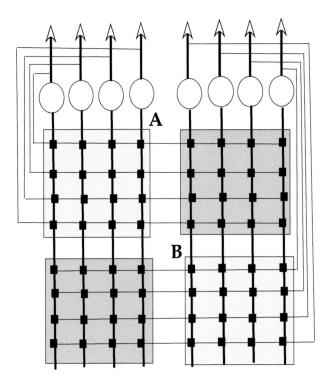

Figure 11.2
An illustration of the Rizzuto and Kahana (2001) model for paired-associate learning within a modified Hopfield (1982) neural framework. The **A** stimulus is represented by the four oval nodes on the left side, whereas the **B** member of the pair is represented by the four oval nodes on the right. The output of a node is denoted by a bold line with an arrowhead. The dark squares are synapse nodes on the various input sections of the nodes.

The Rizzuto and Kahana neural network model is illustrated in figure 11.2. Both the stimulus and response members of an associative pair are represented by an N-dimension feature vector of binary values (i.e., values of either $+1$ or -1). In figure 11.2, N is four, which is an unrealistically small value because the number of feature nodes needed to encode all the items for an entire paired-associate list would be considerably larger; nonetheless, $N = 4$ is sufficient for explanatory purposes. The stimulus item is denoted as **A**, and the response item is **B**. The N nodes for each component of the association are denoted in the figure with ovals. The output side of each node has a bold arrow, whereas the input side of the node has a regular bold line. Each node has a synaptic connection to each of the other nodes, and it has a synaptic connection to itself. The synapses are denoted by small black squares. There is thus a $2N \times 2N$

array of synapses. The same $2N \times 2N$ matrix of weight values is used for all the pairs on the paired-associate list. The Rizzuto and Kahana model thus combines two different types of Hopfield (1982) neural networks into a common cognitive representation. The light yellow sections of the array are called autoassociative networks. Autoassociative networks can be used for storing items in memory. The light purple sections shown in the figure are heteroassociative networks. Heteroassociative networks can be used for storing connections between separate stored memories (i.e., the association from **A** to **B** and the association from **B** to **A**). Most important for our current purposes, the synapses have numerical weights that vary as a function of training. Repeated practice causes the weights to change until the network settles on values for the entire list of associations. In essence, the weights are continuous positive and negative numbers that represent the level of learning for the entire list (i.e., the numerical weights represent all the **A** stimuli and all the **B** responses as well as the association between each stimulus-response pair). It is outside the scope of the current book to explain how the weight values change as a function of learning; those additional details are described in Kahana (2012). The key point here is the model is an example of the continuity theory of learning because the weights in the network are continuous values that gradually change with learning experience.

I have previously critiqued in part I the feature-space approach for representing memories because those models do not encode propositional and structural information that are essential for a complete encoding of the target event. Note that the Rizzuto and Kahana model is a feature-space approach that ignores relational information beyond that of the association between the **A** and **B** stimuli. Thus, I believe that this model has major shortcomings as a general memory representation. It is presented here simply as an example of a cognitive architecture that is consistent with the continuity framework. For another example of a more creditable cognitive representation, consider again figure 11.3 that was discussed in part I.[2] It is possible that the links shown in this figure are built and strengthened over repeated presentation. But a detailed theory of the learning processes for the model has yet to be advanced. It is also possible that the links in figure 11.3 are created abruptly.[3]

---

2. Gallistel and King (2009) have developed the thesis that connectionist models are fundamentally incomplete. They similarly argue for the necessity of symbolic representations in memory such those shown in figure 11.3.

3. One might wonder how it is possible to believe that the links in figure 11.3 could have strength values and also believe in the critique of strength theory in chapter 8. The two positions are not contradictory because the critique in chapter 8 was against a signal-detection model for representing the memory trace. Instead, it is argued that memory is a structured pattern of features and

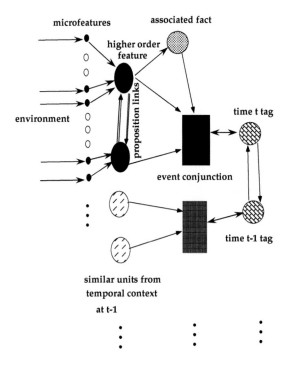

microfeatures   associated fact

higher order feature

environment

proposition links

time t tag

event conjunction

time t-1 tag

similar units from
temporal context
at t-1

Figure 11.3
A distributed representation of a memory representation within a temporal context. This figure
was also discussed in chapter 4.

11.3.3 All-or-None Theory

In an early paper, Estes (1956) argued that a smooth learning curve might possibly be
due to the averaging of abrupt step functions.[4] This idea is illustrated in table 11.1. The
assumption that underlies the example in the table is that each individual in a group
has a step-function learning "curve" (i.e., a function where the probability of a correct
response is 0 for $n < n_c$ and then it abruptly jumps to 1 for $n \geq n_c$, where n_c is the critical
trial number when the material is learned). For example, in table 11.1, a subgroup
of 30 percent of the total is assumed to know the memory list after only one trial (i.e.,

---

propositions, and it is not simply a value on a hypothetical strength axis. The representation in
figure 11.3 is not a signal-detection representation.

4. Before the Estes paper, Edwin Guthrie (1935) advocated a simple associative learning principle,
and he rejected the idea that the frequency of the pairing of a stimulus with a response increased
the strength of an association. For Guthrie, one trial was sufficient for learning. Guthrie thus
foreshadowed the later all-or-none learning hypothesis by a considerable period of time.

Table 11.1

Possible all-or-none learning profiles for a group

% of Group	$n = 1$	$n = 2$	Prop. $n = 3$	Corr. $n = 4$	$n = 5$	$n = 6$	$n = 7$	$n = 8$
30%	1	1	1	1	1	1	1	1
20%	0	1	1	1	1	1	1	1
12%	0	0	1	1	1	1	1	1
10%	0	0	0	1	1	1	1	1
8%	0	0	0	0	1	1	1	1
5%	0	0	0	0	0	1	1	1
4%	0	0	0	0	0	0	1	1
3%	0	0	0	0	0	0	0	1
8%	0	0	0	0	0	0	0	0
wt. sum	.30	.50	.62	.72	.80	.85	.89	.92

$n_c = 1$ for this subgroup). Another subgroup of 8 percent of the total is assumed to learn nothing after four trials but this subgroup masters the list on the fifth trial (i.e., $n_c = 5$). Finally there is another subgroup of 8 percent of the total that still has not learned anything after eight trials. The statistical averaging of these step functions results in the weighted sum shown on the last row of the table. The weighted average is plotted in figure 11.1, and it looks like a gradual learning function. Although this hypothetical example is focused on the group learning curve arising from the averaging of individual step functions, Estes also argued that this averaging artifact might hold for a single individual whenever the learning function involves the averaging over different memory targets (i.e., each item is assumed to undergo an abrupt learning process, but there is variability as to when this abrupt learning takes place). Consequently, when there is a long list of associates, many trials are needed before all the difficult associations make their step-function transition from the none state to the learned state.

11.3.4 The Debate

The all-or-none hypothesis versus gradualism became a heated debated in psychology. Conflict within a play or a movie is a key dramatic device to generate viewer interest. The scientific literature has also experienced its share of conflicts and controversies. Although conflict might be entertaining for a while in science, and it might help sharpen our thinking, unresolved controversies are not a good sign of a healthy science. A prolonged inability to resolve a controversy is a signal that there are limitations either with the available research tools or with the methods of analysis or with overly rigid thinking by key leaders within the discipline. The major leaders in learning theory in the 1950s and 1960s did not accept the notion of all-or-none learning and were quick to fault the Rock (1957) experiment that was seen as the triggering event for the controversy.[5] While there is some merit to the critical reaction to the Rock (1957) experimental paper, the theoretical and empirical arguments in the papers by Bower (1961), Estes (1960, 1961), and Restle (1962) were not as successfully challenged until more recently. I do believe that the all-or-none hypothesis is incomplete, but the hypothesis has helped sharpen our thinking about learning. Consequently, it is important to unpack this controversy carefully and to avoid the common bias of dismissing the all-or-none proposal out of hand.

First, let us review the Rock (1957) experimental paper. Rock used a dropout procedure to study learning. With the dropout procedure, an association that is incorrect on trial n is replaced by a new association for trial $n + 1$. If learning were all-or-none, then the incorrect association could be replaced by a new association, and there would be no effect on the overall learning of the list. Rock did not find a difference in the rate of list learning between the dropout method and the standard method for list learning. Williams (1961) replicated the Rock findings, but also argued that the effect is due to a statistical artifact. She pointed out that if an association is incorrect, then it is reasonable to assume that it is a more difficult association. With the dropout procedure, an incorrect and presumedly harder association is removed before the next trial and is replaced with another random association. If the new item is an easy association, then it is likely to be correct and therefore remain on the list. But if the new item is another hard association, then it is likely to also be incorrectly recalled and removed. The consequence of this selection procedure is thus to create a list of easy associations relative to the standard list-learning condition. By using two additional control conditions in her experiment, Williams found evidence to support the statistical-artifact interpretation of the Rock experiment. Yet her experiment itself is not without faults. For example, one of her two control conditions only had two participants, which is hardly enough people for drawing robust statistical conclusions. Nonetheless, other memory researchers quickly echoed the Williams's criticism and therefore rejected the possibility that learning is an all-or-none process. For example, Crowder (1976)

5. Irvin Rock was a psychologist working in the area of perception and did not publish a previous paper concerning either learning or memory. Moreover, his 1957 paper did not cite the prevailing ideas and papers in the memory literature. So it is not surprising that the established leaders in memory research reacted negatively to an outsider publishing a paper that challenged a fundamental tenet in their theory of associative learning. Yet their theories were in fact flawed and incomplete. A review of the hostile reaction to the Rock paper is described in detail by Roediger and Arnold (2012).

described the Williams (1961) criticism of the Rock experiment as "decisive," so he concluded that the Rock paper was, in his words, "irrelevant." Yet the rationale for the all-or-none position advanced by Bower (1961) and Estes (1960, 1961) was not based on the dropout procedure, and this different approach has received theoretical and experimental support from a number of researchers (Batchelder, 1975; Bjork, 1968; Brainerd & Howe, 1978). Thus, the all-or-none hypothesis does not solely rest on the merit of the Rock (1957) dropout procedure. Consequently, it is important to carefully examine the approach used in the Bower (1961) and Estes (1960, 1961) papers.

The Bower (1961) and the Estes (1960, 1961) papers were based on a particular mathematical form of the all-or-none learning model. These models employed a Markov chain for learning. A key feature of any Markov chain is a transition matrix. The entries in the matrix are transition probabilities between states. For example, in the Bower (1961) model, there were only two states: (1) the learned state (L) and (2) the unlearned state (U). The transition probabilities are conditional probabilities for a memory in one state on trial n being in each of the states on trial $n + 1$. For example, with the Bower model, the transition probability for a memory going from the unlearned state to the learned state is $p(L_{n+1} \mid U_n) = c$, and therefore the probability that the memory remains in the unlearned state is $p(U_{n+1} \mid U_n) = 1 - c$. The two transition probabilities for memory in the learned state on trial n are $p(L_{n+1} \mid L_n) = 1$ and $p(U_{n+1} \mid L_n) = 0$ and reflect the fact that the learned state is an absorbing state (i.e., once learned, it is assumed to remain learned on all subsequent trials). A key assumption of any Markov chain is that the transition probabilities between states are constants and hence do not depend on n. This condition is called *stationarity*. If the transition probabilities change with the trial number, then a Markov chain is not valid. If stationarity does not hold, but there are still discrete states, then the model is a Markov process but it cannot be a Markov chain.[6] Chechile and Sloboda (2014) reformulated Markov chain models and more general Markov process models in terms of a hazard function framework. Hazard functions were discussed briefly in chapter 8 and will be discussed in more detail later in this chapter. Chechile and Sloboda (2014) developed mathematical proofs that the transition probabilities in a Markov chain and in a more general Markov process are hazard values. In order for a Markov chain to be a valid description of learning, it is required that the hazard function is a constant. Chechile and Sloboda (2014) also established that the constant hazard condition is not possible if there is any variability

6. A Markov process has discrete states for memory, and it has transition probabilities between states that are dependent on the current state of the system. A Markov chain is a special case of a more general Markov process where the transition probabilities between the states are fixed values.

whatsoever in item difficulty. It is virtually certain that any list has some item variability. There is some slight variation in item difficulty even for the best controlled studies. Thus, it follows that without constant hazard, the Bower (1961) model and the Estes (1960, 1961) model cannot be correct. Later in this chapter, a hazard function analysis will be used to argue for a different learning equation. This equation is not an all-or-none model, and it is not a Markov chain model.

In light of the work from a hazard function perspective, the all-or-none model is not creditable for a general theory of paired-associate learning, but there still might be other learning events where an all-or-none model could be a valid characterization. One potential case for all-or-none learning is for concept-learning problems when insight is required for the solution. Prior to achieving insight and seeing a solution to the problem, the participant is incorrect. Restle (1962) was interested in this application of all-or-none concept learning. Yet even in this case, it is not entirely correct to assume that no learning occurs until some critical trial when insight occurs. Participants are very likely to remember prior failed attempts to solve a problem. This memory of failed attempts should be helpful in the process of hitting upon a proper solution. Concept-learning behavior might look like an all-or-none function because researchers are not measuring the participant's knowledge of what does not work. Because learning what does not work is an important step toward the discovery of an ultimate solution, the all-or-none hypothesis is not a complete process model for even the case of concept learning.

Another special case where all-or-none learning might be a creditable theory is for aversive conditioning. Typically, animals do not usually require repeated trials to associate a stimulus to a highly aversive event. For example, Maatsch (1959) showed that rats were able to learn to escape a platform on which they previously received a single-exposure trial to electrical shock. Demonstrations of single-trial learning are sufficient to show that learning on one trial is possible. But these demonstrations do not mean that all learning occurs in a single trial. So how does demonstrations of one-trial learning fit within a more broad learning context?

11.3.5 Dynamic Encoding-Forgetting (DEF) Model

There is a way to reconcile aspects of the all-or-none hypothesis with aspects of the incremental-learning position. The revised learning theory stresses the fact that a single item on a list is almost always initially encoded into memory, but that encoding undergoes dynamic changes. In chapter 10, the implicit-explicit separation (IES) measurement method has provided experimental support for the dynamic changes in the storage of a single target item as a result of the retention interval. Notice in figure 10.8 from chapter 10 that the probability of explicit storage is nearly perfect initially, but it rapidly decreases for longer retention intervals. For the Chechile, Sloboda, and Chamberland (2012) experiment, the probability for explicit memory storage (θ_{se}) was found to be .935 after a brief delay of 333 msec, but it fell to a very low value of .126 after a 30-sec retention interval. Yet the change in memory did not result in all memory traces being lost because there was, for the 30-sec retention interval, a probability of .197 for implicit memory storage (θ_{si}) and a probability of .363 for fractional storage (θ_F). In a paired-associate list-learning experiment, there is also a delay between the presentation of an item and the later test of the item. The retention interval is filled either with the learning of other associations or with the testing of other associations on the list. Thus, it is reasonable to assume that the storage of an association in a list-learning experiment also undergoes dynamic changes between the presentation of the association and the later test of the association. When the association is presented on the second trial of a list-learning experiment, then the explicit-storage probability again is assumed to spike back to 1. But after the second intratrial retention interval, there is again a dynamic decline in the storage representation. This theoretical idea of perfect storage initially followed by a within-trial storage loss is illustrated in figure 11.4. The figure provides two learning-retention functions. Relative to the blue curve, the red curve has a more rapid learning rate and a reduced loss due to the intratrial retention interval. The curves have a spike in the initial probability for explicit storage at the integer trial values of $n = 1, 2, ..., 8$, but they also show a decline as a function of the intratrial retention interval. The horizontal axis for the figure is the time after presentation plus $n - 1$.[7]

The explicit storage plotted in figure 11.4 is considered a mixture rate in the same way as for the IES model discussed in chapter 10. Each association item is (1) explicitly stored, (2) implicitly stored, (3) fractionally stored, or (4) not stored whatsoever. The value of θ_{se} is the estimated proportion of memory targets (averaged over participants

---

7. The plots shown in figure 11.4 are based on the Chechile (2006) two-trace hazard model discussed in chapter 3 and the learning model advanced by Sloboda and Chechile (2008) that will be discussed later in this chapter. The probability of explicit storage for this figure is a function of the trial number n and a proportional measure of the length of the most recent intratrial retention interval x. For this figure, the trial number n varies from 1 to 8, and x varies over the (0, .99) interval. The functional form for explicit storage is given as $\theta_{se} = 1 - exp(-bn^c) + exp(-bn^c - dx^2)$ where b and c are parameters associated with the learning rate, and d is a parameter associated with the speed of forgetting. For the blue curve, the values for b, c, and d parameters respectively are .4, .8, and 6, whereas the parameter values for the red curve are respectively .9, .8, and 2. When there is a new presentation, there is a step jump in the value for n, and the x value is reset to 0.

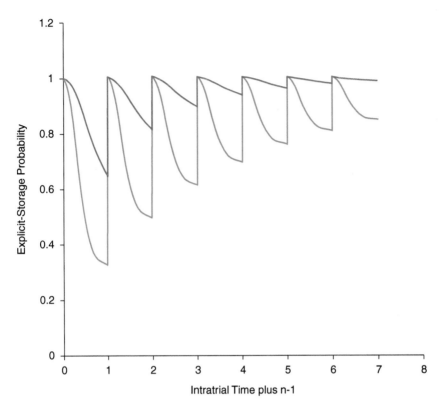

Figure 11.4
Two theoretical learning-retention functions for a list that is presented repeatedly. Both curves have a spike in the explicit-storage probability at the integer values n, which represents the number of presentations. Subsequent to the presentation, there is a loss of storage during the intratrial retention interval. Relative to the blue curve, the red curve has a more rapid learning rate and slower storage loss with the intratrial retention interval.

and targets) that are explicitly stored. If the probability for explicit storage changes, then those changes merely reflect the fact that the proportion of the explicitly stored traces varied. Thus, the figure is not plotting a strength for the association because each association is considered to be either explicitly stored or not. Let us coin the name "Dynamic Encoding-Forgetting" (or the DEF model) for this view of the learning and forgetting processes that are involved with the paired-association learning procedure. The DEF model differs from the all-or-none position in the case when the target is not explicitly stored after the intratrial retention interval. For the all-or-none hypothesis, a performance failure at the time of test is indicative of nothing being stored; this is the "none" feature of the all-or-none position. But with the DEF model, the associations that are not explicitly stored might still have either an implicit memory representation or a fractional memory representation.

It is reasonable to assume that the associations that survived the last intratrial retention interval will be able to also survive the next intratrial retention interval. Yet subsequent presentation trials provide an opportunity for new associations to also survive the vigorous memory loss caused by the retention interval; thus, the decline in storage should become less with each presentation trial.

For the all-or-none model, learning *can* occur on a single trial, but in the DEF model, learning typically only needs a single trial for developing a memory representation. It is not just a subset of items that is explicitly stored, but rather it is standard for humans to learn all the targets, provided that (1) the participants intend to encode the targets, (2) the targets are within the memory span capacity, and (3) the encoding time is adequate. But after the encoding event occurs, there is also a massive storage loss, so the rapid learning can be quickly lost. Support for this position is based on the storage-retrieval measurement models discussed in part II. Human studies show a near-perfect storage rate for targets that satisfy the above three conditions, and those studies show that the initial encoding degrades quickly as a function of the retention interval. Animals are not as diligent in their initial encoding of memory targets as demonstrated by the storage-retrieval model developed for pigeons in chapter 9, so the initial encoding level for the DEF model would need to be adjusted downward for nonhuman species.[8] The massive storage loss idea in the DEF model is not a feature in either the all-or-none position or the incremental-learning position.[9]

Like the all-or-none theory, the DEF model is not a strength-based theory of learning. Rather, the repetition of training increases the proportion of explicitly stored memory traces that survive the damaging effect of the retention interval. The DEF model is thus clearly different from a pure continuity theory of learning. But unlike the all-or-none

8. Pigeons were found to have an average sufficient storage value of only 36.5 percent, even after pecking twenty times directly on the target. But humans are quite good at encoding a target, provided that the target is within memory span and presented for about 1 to 2 sec.

9. Tulving (1964) developed a model that also stressed the combination of intratrial forgetting and a near-perfect initial encoding. However, the Tulving (1964) model differs from the present theoretical position because it did not stress the storage loss of information. Storage in the Tulving model was assumed to be invariant and near perfect. Forgetting in the Tulving model was assumed to be strictly a retrieval loss, but the Tulving paper was written prior to the advent of the storage-retrieval separation models that were discussed in part II, and it came before the numerous studies that demonstrated the dynamic loss of storage after the initial encoding.

theory, the DEF model does not result in the only alternative to explicit storage being nothing. Rather, if there is not explicit storage, then the memory representation can be implicit, fractional, nor nonstorage.[10] Consequently, the DEF model differs from both the continuity theory and the all-or-none theory; it is neither a strength-based incremental model nor an all-or-none model.

Beyond the improvement in the probability of storage that results from increased practice, it is also likely that the memory encoding becomes elaborated with repeated training. For example, in the context of memory representations like the one shown in figure 11.3, it is possible that further practice provides the opportunity for establishing additional nodes. Thus, the features, propositions, and associations that are stored in the DEF model are regarded as changeable. But what is not shown in figure 11.3 is the redundancy of the encoding in the DEF model. Recall from chapter 3 that the index theory of Teyler and DiScenna (1986) hypothesized two interrelated representations of a target event as shown in figure 11.5. Also remember that Buonomano (1999) found that in vitro hippocampal cells and neocortical cells differed in regard to their speed of responding to stimulation. In the complementary storage model discussed in chapter 3, the hippocampal representation is considered to be rapidly established, whereas the neocortical representation develops more gradually. Also in chapter 3, the Chechile (2006) two-trace hazard theory was a system-level model that assumed two traces are formed with the encoding of the target event. Thus, the DEF model borrows ideas from these models of memory. A key property of the DEF model is trace redundancy, *even for the initial presentation of a stimulus.*

In the DEF model, the repetition of a target stimulus involves a process of restoring and elaborating the existing memory representation. But there are still alternations of the memory representation after the repetition event due to storage interference. New memories are fragile, and they are easily disrupted. So there are dynamic changes to the information stored. This model is not a strength model, but instead it is a model

---

10. One might argue that the DEF model is merely a form of the Atkinson and Shiffrin (1968) model where there is a short-term store representation established at item presentation, but the short-term trace is lost for some of the items over the retention interval. But the multiple-store model of Atkinson and Shiffrin is a form of the all-or-none position because items are either stored forever if they are transferred to the long-term store or they are 100 percent lost if they are not in the long-term store. The multiple-store model does not have a state of implicit storage. The finding of implicit storage for some items that are not explicitly stored is thus contrary to the multiple-store model. Remember from chapter 5 that the Atkinson-Shiffrin model also had an additional problem of not being able to account for the Chechile and Ehrensbeck (1983) experiment that found storage loss for items presumed to be in the long-term store.

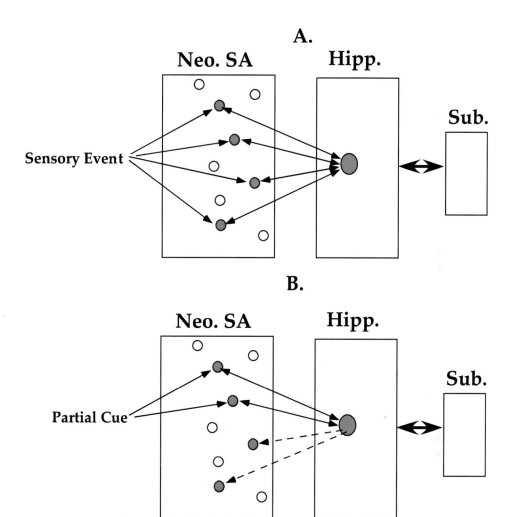

Figure 11.5
Illustration of the index theory of Teyler and DiScenna (1986). (A) Upon experiencing an event, the association areas of the neocortical sensory association areas (denoted as Neo. SA) receive activation of the features present and these in turn activate a conjunction binding representation in the hippocampus (denoted as Hipp.). The hippocampus also receives input from subcortical areas (denoted Sub.), which include the amygdala, medial septum, thalamus, hypothalamus, raphe nuclei, and locus coeruleus. The hippocampus activation in turn has an indexing property, so it can activate the neocortical cells that originally activated it. (B) A partial cue of the target cue can reactivate the hippocampal indexing cells, which can regenerate the missing features of the target event.

A.

$T_1 = \{S_1, C_1, P_{1,1}, E_{1,1}, E_{1,2}, t_1, S_{r1}\}$

$T'_1 = \{S_1, C_1, E'_{1,1}, E'_{1,2}, t_1, S_{r1}\}$

B.

$T_1 = \{--, C_1, P_{1,1}, --, E_{1,2}, t_1, S_{r1}\}$

$T'_1 = \{--, C_1, --, --, --, --, S_{r1}\}$

C.

$T_1 = \{S_1, C_1, P_{1,1}, P_{1,2}, P_{1,3}, E_{1,1}, E_{1,2}, A_{1,1}, t_1, t_n, S_{r1}, S_{rn}\}$

$T'_1 = \{S_1, C_1, E'_{1,1}, E'_{1,2}, t_n, S_{rn}\}$

D.

$T_1 = \{--, C_1, P_{1,1}, --, P_{1,3}, E_{1,1}, --, A_{1,1}, t_1, t_n, --, S_{r1}\}$

$T'_1 = \{--, C_1, --, E'_{1,2}, t_n, S_{rn}\}$

E.

$T_1 = \{S_1, C_1, C_2, P_{1,1}, P_{1,3}, P_{1,4}, E_{1,1}, E_{1,2}, A_{1,1}, t_1, t_n, t_{2n},$
$S_{r1}, S_{rn}, S_{r2n}\}$

$T'_1 = \{S_1, C_1, C_2, E'_{1,1}, E'_{1,2}, t_n, t_{2n}, S_{r2n}\}$

Figure 11.6

An illustration of the dynamic changes with a sequence of encodings (i.e., the stimulus S_1). Panel A is the initial encoding representation. Panel B is the resultant after $n - 1$ interpolated events, and panel C is the result of the second encoding of the stimulus. Panels D and E are the respective representations prior to and after the third presentation of the stimulus.

where the informational content is deleted and added as a function of experience. To see how these processes alter memory, we need a different notational system for representing the memory encoding. Graphical representations, like the one shown in figure 11.3, are not convenient to illustrate the dynamic processes of encoding and forgetting. Consequently, a new shorthand notation is introduced. This notation and the model for repetitive learning are illustrated in figure 11.6.

In panel A of figure 11.6, two traces are shown for stimulus S_1; these are denoted as T_1 and T'_1 and are shown as sets of labeled properties. Trace T_1 can be thought of as the slower developing neocortical representation, whereas trace T'_1 is the rapid hippocampal representation. The distributed set of perceptual features activated by the stimulus in figure 11.3 is represented by the vector of nodes S_1. The vector C_1 represents the contextual components of the encoding; this is not shown in the graphical representation in figure 11.3, but it is an important aspect of memory encoding. The context would include the goals and instructions as well as the incidental features that are not crucial for representing the target event but are nonetheless a part of the conditions of encoding. A propositional relationship is denoted by $P_{1,1}$. The event-binding features that are shown in figure 11.3 are now denoted as $E_{1,1}$ and $E_{1,2}$. The time-tag component is represented as t_1. Finally, any associated motor response, decision, or reaction to the stimulus is represented by the vector $\mathbf{S}_{r1}$. The $\mathbf{T}'_1$ trace also includes the stimulus vector $\mathbf{S}_1$, the context vector, a comparable set of event-binding components $E'_{1,1}$ and $E'_{1,2}$ as well as the time tag, and the response vector $\mathbf{S}_{r1}$. The rapidly developing $\mathbf{T}'_1$ trace does not have elaborated facts, propositions, and associations. These informational components will develop more slowly in the $\mathbf{T}_1$ representation because these features usually require more time to be encoded.

Suppose after the target stimulus is presented that there are $n-1$ other stimulus events. These intervening events can be either the learning of other items or the testing of previous items. The state of the target representation just prior to the second presentation of the target stimulus is shown in figure 11.6B. Note that both traces have been damaged by the encoding of the $n-1$ interpolated events. Both traces quickly lose the $\mathbf{S}_1$ sensory-feature components because people do not easily maintain these features without rehearsal.[11] The event-binding components are critical for the memory of the episodic event. Only one of these event components remains for the $\mathbf{T}_1$ trace, and all have been lost for the $\mathbf{T}'_1$ trace.

Overall, the survival of any component in the $\mathbf{T}_1$ trace is a probabilistic process that is likely to vary with the nature of the interpolated material. Similar but different interpolated items are more likely to cause a disruption effect. For example, an interpolated item, which shares some features in common with $\mathbf{S}_1$ but is associated with a different response, will have a greater potential (relative to a dissimilar stimulus) for causing a loss of an event-binding component in either of the traces for $\mathbf{S}_1$. Yet the survival for any component in the $\mathbf{T}_1$ trace is a separate probabilistic process from the interference process for the $\mathbf{T}'_1$ trace. Consequently, it is possible that a property that was initially redundant between the two traces is lost from one trace, but it is not lost from the other trace. Redundancy slows the rate of information loss. But it is also possible that a critical aspect of the memory representation might be lost altogether because it is lost from both traces.

In panel C, both trace representations are reinvigorated by the repetition of the target. Importantly, the $\mathbf{T}_1$ trace representation has developed some new encoding

11. Recall in chapter 5 the discussion of unusual people with a strong ability to generate and maintain eidetic images of prior stimuli (Stromeyer & Psotka, 1970). Yet most people do not maintain the prior stimulus without engaging in rehearsal. Rehearsal is difficult if the person is trying to study other targets.

components. Two additional relational propositions ($P_{1,2}$ and $P_{1,3}$) have been created along with a new association $A_{1,1}$. A second time tag and a new response vector are also elaborated. The T_1' trace is also augmented from its previously impoverished state. Yet the revised encoding is soon battered again by the subsequent $n-1$ presentations of the other intervening stimuli. The resulting state after this storage disruption is portrayed in panel D. Finally, in panel E, both traces are again augmented by a new learning trial in an altered context C_2. Notice that the memory representation is informationally developed and altered as a function of both repetitive learning and by the effects of interference from the other stimuli (i.e., the combination of two traces is different for each of the three presentations).

The repetition of training can also cause an improvement in the probability of successful retrieval on a recall test. For example, by the fifth training trial for the red curve in figure 11.4, the explicit storage is reaching a ceiling on the probability of the explicit storage. Yet the probability of correct recall might still not be at a mastery level for the list. Note the elaborations discussed above can be used for improving later retrieval. Although the DEF model is a storage-based theory, it is also possible that retrieval still benefits from the additional training by providing multiple contexts, temporal cues, and retrieval routes for recovering the target on a recall test.

The multiple-trace view of repetition advocated here for the DEF model differs from other multiple-trace models in psychology. The original multiple-trace proposal was suggested by Semon (1921). Each presentation of the stimulus in Semon's theory is assumed to result in a separate but related representation. These representations were called *engrams*—a term that Semon coined (Schacter, 1982). Semon also proposed that the later activation of memory involves the elicitation of the system of interdependent engrams; this process he described by the term *ecphory*. To Semon, ecphory refers to the simultaneous activation of memory representations that are produced by the triggering event. Ecphory is analogous to the vocal response of an entire chorus of singers rather than the sound of just a solo singer.

Although the Semon view of a multiple-trace model is in terms of a colorful musical metaphor, a similar idea has been advanced in a mathematical form with the Minerva 2 computer model (Hintzman, 1984, 1988). In this model, each memory event is treated as a vector of features; the values for these features are 1, 0, or -1. Importantly, the same target is assumed to have a separate feature vector for each of the presentations of the item.[12] For this model, there is an "echo" response to a probe, and this response

12. Another multiple-trace theory is the replica model (Bernbach, 1969, 1970). For this theory, as with the Minerva 2 model, each presentation of a stimulus results in another trace or replica being stored. If humans had a random-access memory (RAM) like most computers, then the creation

has a strength or intensity value. The echo intensity is equal to the sum of the activation strength values from all the vectors in the *entire* memory system.[13] The individual activation values for a memory trace are a function of the similarity value between a test probe and the memory trace. It is outside of the scope of the present discussion to fully describe the properties of the Minerva 2 model, but there are many problematic aspects of the model. For example, the model does not include relational and elaborative propositions as a property of the memory encoding. It is essentially a feature-space model, so it is subject to the same set of problems discussed in part I.[14] Also Minerva 2 uses a signal-detection framework, so it is subject to the same set of criticisms of that position delineated in chapter 8. Finally, experiments studying frequency judgments are not consistent with the idea in Minerva 2 that each event is separately represented (Greene, 1990; Tussing & Greene, 2001).

In contradistinction to the multiple-trace models of Semon (1921) and Hintzman (1984, 1988), the DEF model regards all memory encodings as having two related traces—even for the first presentation of a target. The subsequent presentations of other stimuli result in dynamic variation in the informational content of the two traces. Thus, the DEF model differs from the other multiple-trace models of memory. It would be more appropriate to regard the DEF model as an elaborated, two-trace representation

---

of multiple copies would be reasonable and in fact necessary. But it is not clear why a content-addressable memory (CAM) would layer another copy onto a previously established trace. I believe that humans have a modified CAM system, as discussed in chapter 5, so the idea of a separate trace for each presentation is not plausible.

13. Minerva 2 is an example of a global-memory matching model. Another example of this approach would be the Search of Associative Memory (SAM) model (Gillund & Shiffrin, 1984; Raaijmakers & Shiffrin, 1980, 1981). The idea that all of memory is probed in recognition and contributes to a strength value is, in my opinion, a physiologically implausible concept. Instead, I have argued for a content-addressable system where the sensory features of the specific target stimulus trigger the *specific* neural network in the neocortical sensory association areas that are linked with the processing of the stimulus rather than the broad activation of the whole of memory. How can it be neurologically possible to trigger the activation of all memories by the presentation of a single stimulus? The global-memory matching models in general and the Minerva 2 model in particular are thus implausible.

14. In the Chechile et al. (1996) study, visual patterns of varying degrees of relational complexity were learned. The more complex arrangements took more trials before people learned the patterns. Later in a speeded-recognition memory test, the more complex patterns required more processing time before a recognition decision was rendered. But in the Minerva 2 model, a memory with more repetitions should have a larger echo-intensity response and presumedly a faster response. See part I for other problems with a pure feature-space approach for representing the memory encoding.

that undergoes dynamic variation rather than a multiple-trace model where each presentation has a separate representation. Yet the DEF model does share with the Semon (1921) and the Hintzman (1984, 1988) models the idea that the repetition of learning leaves some record from the previous encodings. But rather than having each repetition event being a separate memory, in the DEF model, the history of the previous experiences is carried along with the time tags t_i, the context vectors $\mathbf{C}_i$, and the reactions to the individual events S_{ri}, as shown in the example in figure 11.6.

11.4   Is Total Study Time Sufficient?

The total-time hypothesis is the proposal by Cooper and Pantle (1967) that learning is mainly due to the total study time. In this section, we will carefully explore and critique this proposal. Importantly, the failure of the total-time hypothesis reveals important aspects of learning. But before directly addressing the total-time hypothesis, let us first examine the trade-off between the length of a list of associates and the presentation time per item. The trade-off function and two additional topics are important considerations that must be addressed before dealing directly with the total-time hypothesis.

11.4.1   Presentation Time and List Length Trade-off

By 1960, an enormous number of studies had been done to study learning with the paired-associate (PA) task. Nonetheless, Murdock (1960) still managed to detect a number of original observations about the learning of a set of unrelated stimuli. Murdock examined recall performance for word lists over a series of twelve experiments. The presentation time per word (PT) as well as the length of the lists (L) varied across the experimental conditions. Murdock observed the following relationship between the number of words recalled immediately after the first presentation of the word list (R_1) and the total learning time T_1:

$$R_1 = kT_1 + m, \tag{11.1}$$

where $T_1 = PT \times L$, and where k and m are constants. For his experiments, the median values for k and m were respectively 0.06 and 6.1. Equation (11.1) describes a trade-off between the list length and presentation time. As an example, consider the following two cases: (1) a list of thirty words presented at the rate of 2 sec per word and (2) a list of fifteen words presented at the rate of 4 sec per word. According to equation (11.1), the predicted number of correctly recalled words immediately after list presentation is 9.7 words (on average) for both cases because the total study time is

$T_1 = 60$ sec in both cases. Yet the proportion of correct recall $P_1 = \frac{R_1}{L}$ would be different in the two cases (i.e., .323 for the first case and .647 for the second case) because the lists differ in length. No theory was advanced to support equation (11.1), and Murdock realized that the equation only applied over a restricted range of values for the presentation time and for the list length. If the list is too long or too short, then the trade-off equation would not hold up, although his experiments spanned list lengths over the substantial range of 5 to 400 words. Also, he stated that extremes of presentation times can also invalidate the predicted relationship; the presentation times in his experiments ranged from .5 to 5 sec per word. Implicitly, one might further restrict the applicability of equation (11.1) to the stimulus class of a homogeneous set of words.[15] Within these stipulated limits, the total time for the first list presentation does accurately predict the average number of the words recalled. Yet this statement is *not* what has come to be called the total-time hypothesis. Murdock's paper did inspire a subsequent paper by Cooper and Pantle (1967) in which the total-time hypothesis was first advanced. But before addressing the total-time hypothesis, it is important to first discuss another regularity that Murdock observed in his 1960 paper—a regularity dealing with the effect of repetition. It is also important to discuss the most recent research on learning hazard functions before addressing the total-time hypothesis. Consequently, the learning function and the learning hazard function will be discussed below in the next two minor subsections before returning to the total-time hypothesis.

11.4.2 The Murdock (1960) Learning Function

In a standard list-learning procedure, all the items are tested after the presentation of the entire list. Murdock (1960) observed, across his twelve experiments, that a simple mathematical relationship appeared to accurately predict the number of words recalled after the nth presentation of the list. Murdock proposed the following exponential function linking the number of words recalled (R_n) as a function of the

---

15. One might wonder about the importance of Equation (11.1) in light of the previously discussed more general theory of the Dynamic Encoding-Forgetting (DEF) model. With the DEF model, the recall rate for any item on the PA list will be less than 1 because of the storage interference from other list items. Yet the DEF model does not make a specific numerical prediction for the recall rate after one study trial of the whole list. The recall rate is obviously a function of the amount of study time per item and the length of the retention interval. In the DEF model, more study time per item on the list will enable a more elaborate encoding that should enable better recall. The Murdock equation thus provides a numerical estimate, based on his experiments, of the recall rate after the first time of studying the list.

number of learning trials (n):

$$R_n = L(1 - e^{-bn}),\tag{11.2}$$

$$P_n = (1 - e^{-bn}),\tag{11.3}$$

where P_n is the proportion correct after the nth presentation, and it is obtained by dividing equation (11.2) by the list length (L). The b parameter is important in the learning equation because it affects the learning rate. For example, suppose that there is a list of ten words and $b = .4$, then from equation (11.3) the predicted performance would be at an 86.5 percent level after $n = 5$ trials. If instead $b = .7$, then the predicted performance level would be at a 97.0 percent level after five trials. Thus, the rate of learning increases as the value of b increases. Because the b parameter controls the rate of learning, it is a function of factors such as (1) the presentation time per word, (2) the participant's learning ability and learning strategy, and (3) the properties of the words. But for a homogeneous set of word stimuli that are presented at the same rate, b should be a constant for any given individual, provided that the individual has a fixed learning strategy and level of attention. Once again, Murdock did not offer a theory for his proposal of the exponential learning function.

The Murdock equation is a form of the continuity theory for learning. But earlier in this chapter, the continuity theory was rejected in favor of the DEF model. Yet it is still a fair question to ask if equation (11.3) still works as a good prediction model for paired-associate learning. Chechile and Sloboda (2014) addressed this question. We found that the empirical fits of the Murdock's learning function were reasonable for some experiments, but nonetheless, equation (11.3) could not be precisely true. Our argument was based on a hazard function analysis, and the rationale for rejecting the Murdock equation is described in the next subsection. Also, an alternative function is described.

11.4.3 Hazard Functions for Learning

As discussed in chapters 3 and 8 as well as earlier in this chapter, hazard is a general concept from probability theory, and it deals with the probability of a change of state occurring at a specific point, conditionalized by the fact that the change did not occur previously. Historically, the term *hazard* was first used in the insurance industry to characterize the risk of death, conditioned by the fact that the person lived up to some current age (Steffensen, 1930). Hazard can be defined for any probability distribution on an ordered dimension. The definition of hazard for continuous probability distributions is $h(x) = \frac{f(x)}{1 - F(x)}$, where x is a continuous value such as time,

Table 11.2

Example of discrete hazard calculations

Measure	$n=1$	$n=2$	$n=3$	$n=4$	$n=5$	$n=6$
Probability p_n	.570	.230	.105	.045	.020	.030
Cumulative prob. P_n	.570	.800	.905	.950	.970	1.00
Survivor function $S_n = 1 - P_n$	.430	.200	.095	.050	.030	.000
Hazard $h_n = \frac{p_n}{S_{n-1}}$	.570	.535	.525	.474	.400	1.00

$f(x)$ is the probability density at the value x, and $F(x)$ is the cumulative probability.[16] For additional details about hazard for continuous variables, see the discussion in part II about equation (8.1). There is also a corresponding definition for hazard for a discrete distribution such as the case when probability is restricted to integer trial numbers.

Chechile (2003) provided a general definition for discrete hazard (h_n) as

$$h_n = \begin{cases} P_1 & \text{for } n=1, \\ \dfrac{P_n - P_{n-1}}{1 - P_{n-1}} & \text{for } n > 1. \end{cases} \tag{11.4}$$

To illustrate the meaning of equation (11.4), let us consider an example of a lifetime distribution for a process completed by a specific integer trial number. For this example, suppose that the lifetime distribution values for trials $n = 1, \ldots, 6$ are respectively .570, .230, .105, .045, .020, and .030. That is, 57 percent of the people complete the task in one trial, whereas 4.5 percent of the people finished after the fourth trial. These values for the lifetime distribution are denoted as p_n. In table 11.2, the computations of the discrete hazard for this example are displayed in detail. To compute hazard, several other important concepts are also required, that is, the cumulative probability (P_n) and the survivor function ($S_n = 1 - P_n$). The cumulative probability is the sum of the lifetime distribution by the nth event; thus, $P_n = p_1 + \cdots + p_n$. After the nth trial, $1 - P_n$ proportion of the population remain as uncompleted. This proportion has a

---

16. For continuous probability distributions, there is a relationship between the probability density $f(x)$ and the cumulative probability $F(x)$. The density is the derivative of cumulative probability (i.e., $f(x) = \frac{dF(x)}{dx}$), and the cumulative probability is the integral of the density from the lower limit of the distribution (denoted as L) up to the value of x (i.e., $F(x) = \int_L^x f(x)\,dx$). For discrete probability distributions, there cannot be a probability density because the random variable can only take on discrete values such as in the example of integer values for the number of repetitions. Thus, there is a different definition of hazard for discrete probability distributions as discussed in Chechile (2003).

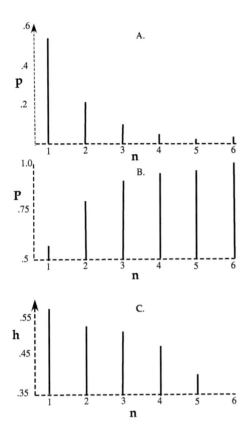

Figure 11.7
Example of hazard for discrete distributions. Panel A is a lifetime discrete probability distribution, and panel B is the corresponding cumulative probability distribution. Discrete hazard is displayed in panel C.

name in probability theory (i.e., the survivor function S_n). Thus, just prior to the nth trial, there is S_{n-1} proportion of the population that has yet to complete the process. The proportion of this remaining group that completes the process on precisely the nth trial is the ratio $\frac{p_n}{S_{n-1}}$. This conditionalized proportion is called the discrete hazard. Hazard for both the discrete and continuous case is a sensitive measure of a probability distribution, and it is a powerful tool for exploring psychological processes (Chechile, 2003, 2006, 2013). Displays of the lifetime distribution, the cumulative probability, and the hazard for this hypothetical example are shown in figure 11.7. The hazard for the final trial is omitted in the figure because the discrete hazard is always 1.0 for the final trial for all *finite* discrete distributions (Chechile, 2003).[17] Notice that the hazard display for this hypothetical example changes with the value of n (i.e., hazard is a function of n). In this case, the hazard monotonically decreases as a function of n for all but the last trial. Finally, note that the hazard in equation (11.4) is defined for $n=1$ as $h_1 = p_1 = P_1$ because it is assumed that the process is uncompleted for all individuals *before* the first trial.

The Weibull probability distribution is an especially important distribution in probability theory because it was invented as a way for studying hazard processes (Weibull, 1951). The cumulative probability of the Weibull distribution for the discrete case is

$$P_n = 1 - \exp(-bn^c), \tag{11.5}$$

where the b and c parameters are positive constants. The b parameter affects the rate of change with increasing n, whereas the c parameter affects the shape of the probability distribution. The c parameter also has a dramatic effect on the hazard function (Chechile, 2003). If the value for c is less than 1, then the discrete Weibull distribution has monotonically decreasing hazard, but if the c parameter is greater than 1, then the hazard function is monotonically increasing (Chechile, 2003). In the special case of $c=1$, the hazard is a constant. Hazard is a property of the probability distribution, so the discrete Weibull distribution is a very general representation that can approximate most other discrete distributions because with an appropriately selected c value, the Weibull distribution and many other distributions can look very similar. The Weibull model is also a method for studying the unknown hazard properties of a system. If a Weibull distribution is assumed and the c parameter is estimated empirically, then from the estimated value of c, we can learn if the hazard function of the system is (1) monotonically increasing, (2) monotonically decreasing, or (3) a constant.

Sloboda and Chechile (2008) defined a learning hazard function that corresponded to a change of an item from being incorrect on trial $n-1$ to being correct on trial n. Moreover, they assumed the general Weibull model as represented by equation (11.5) to develop a general learning equation. Note that the Murdock learning equation is the special case of the Weibull model for when the c parameter is 1.0. Thus, the Murdock learning equation (11.3) is the model that assumes the learning hazard function is a constant because the hazard is a constant when the c parameter is equal to 1.0. But the condition of constant hazard is a virtual impossibility to achieve in light of the variability among the items on the list and variability among individuals, as well as the

---

17. Theoretical models for the cumulative probability typically do not limit the value of n. For example, n is unbounded for the Murdock learning function (11.3), so there is never a finite n value. It is only when there is a fixed limit to n that the hazard for the last n is 1.0.

variability in the alertness level of the learner over training. Any experiment involves a mixture of some sort. Most often, researchers average the learning performance over individuals; for such cases, the proportion correct for a group is a mixture. Even in cases where a single person is studied, there is still a grouping over memory targets because the experiment constitutes learning a list of different items, and undoubtedly the items on the list have some variability in their ease of learning (i.e., the learning rate b parameters for these items are not all the same). In addition, it is also likely that for any given individual, there are shifts in attention. Momentary shifts in the alertness level result in a source of variability for the learning rate of the items on a list. Thus, some degree of variability is inevitable. Chechile (2003) proved a general theorem that a mixture of probability distributions, which individually have constant hazard, results in a mixture having a monotonically decreasing hazard function.[18] To clarify this point further, suppose for a single individual that the list consists of only two items that have item learning rates of b_1 and b_2. Even if the c parameter were 1.0 for each individual item, the mixture of two items would have a c parameter that is less than 1.0 whenever b_1 is not *exactly equal* to b_2. This property about mixtures is not an empirical finding that is subject to experimental error, but rather it is a mathematically proven property of mixtures. Thus, the Murdock learning equation (11.3) cannot be correct because a mixture of some sort is always the reality in learning experiments. Thus, the learning equation should be the general Weibull model as shown in equation (11.5) rather than the Murdock learning equation.[19]

18. In chapter 8, the case of a mixture of monotonically increasing hazard functions is discussed. If the separation between the mixture components is small enough, then the mixture is also an increasing hazard function, but if the components are separated widely enough, then the mixture has a complex biphasic profile like the one shown in figure 8.9. However, a mixture of either exponential distributions or a mixture of monotonically decreasing hazard functions results in the mixture having monotonically decreasing hazard (Chechile, 2003).

19. Earlier in this chapter, the DEF model for encoding and forgetting was discussed. The Weibull learning model is consistent with the learning component in the DEF model that survives the intratrial retention (i.e., the delay between presentation and test). The DEF model also includes a term for a second memory representation that is created at the time of encoding, but this trace is highly susceptible to memory loss. Mathematically, the DEF equation is $\theta_{se} = 1 - \exp(-bn^c) + \exp(-bn^c - dx^2)$, where x is proportional to the length of the intratrial retention interval. If we ignore the intratrial x value and only focus on the learning after discrete values for n, then x is approximately a constant, that is, $\exp(-dx^2) = \exp(-K)$ where K is a constant. In this case, the DEF equation becomes $\theta_{se} = 1 - (1 - \exp -K) \exp(-bn^c) = 1 - \exp -b^*n^c$, which is in the same mathematical form as equation (11.5). Note that the effect of retention interval has altered the rate of the learning parameter (i.e., it results in a new b^* value that is less than b). Consequently, the DEF model and equation (11.5) are compatible.

Given the general Weibull learning model, what is the value for the c parameter? To answer this question, Sloboda and Chechile (2008) fit equation (11.5) to the data from twenty-eight different conditions from many of the classic papers on list learning in the psychological literature (Anderson, 1981; Hellyer, 1962; Runquist & Freeman, 1960; Underwood, 1983; Underwood, Runquist, & Schultz, 1959; Underwood & Schultz, 1960, 1961). The mean value for the c parameter was .870 over these twenty-eight experimental conditions. The dependent variable for each of these experiments was the proportion correctly recalled, a measure that was shown in chapter 7 to be a function of both storage and retrieval probabilities. But Sloboda and Chechile (2008) also fit the Weibull learning equation to each of eighty individuals from the Ehrensbeck (1979) experiment. The Ehrensbeck experiment used the Chechile-Meyer test procedure to enable the estimation of a memory storage probability. Sloboda and Chechile (2008) reexamined these data by means of the Chechile (2004) 6P model to estimate the probability of the storage of an association as a function of learning trials. The mean value for the c parameter for the Ehrensbeck data was .73. Hence, it is clear that the c parameter for the learning equation should be less than 1.0, and thus the learning hazard is a monotonically decreasing function.[20]

11.4.4   Total-Time Hypothesis and Its Disconfirmation

Cooper and Pantle (1967) reviewed a wide number of learning experiments and concluded that learning is simply a function of the total study time. The claim by Cooper and Pantle is the total time to learn a list is only a function of the number of items on the list, the time of study per item, and the number of replications of the list. If one were to have less time per item and more training trials, then the total-time hypothesis claims the level of learning would be the same, provided that the total study time was equal. Underwood (1983) argued that the total-time hypothesis was so credible that it really should not be called a hypothesis at all. But the Cooper and Pantle paper was a nonquantitative review of the literature, and it did not subject the hypothesis to a

---

20. The entire 95 percent probability interval for the c parameter for the eighty participants in the Ehrensbeck (1979) experiment was less than 1.0, so the Weibull model for memory storage is highly likely to be one where the c parameter is less than 1.0 and thus reflective of a monotonically decreasing learning hazard function. Yet Sloboda and Chechile (2008) found that some of the twenty-eight conditions in their literature review had a c parameter estimate that was greater than 1.0 for the recall measure, although most experiments had a c estimate less than 1.0. But the recall measure is the multiplicative product of the storage and retrieval probabilities. Sloboda and Chechile further showed that it is possible for the product of a Weibull model for storage and a similar Weibull model for retrieval (both with the c parameter less than 1.0) to result in a Weibull model for recall that has a c parameter greater than 1.0.

rigorous test. Moreover, we can demonstrate that the total-time hypothesis cannot be correct because it is inconsistent with the well-established trade-off of presentation time and list length (Murdock, 1960) and the Weibull learning model (Sloboda & Chechile, 2008).

To demonstrate the problem with the total-time hypothesis, let us consider two learning conditions where the participants have to learn a set of forty paired associates. The participants in the first condition are given 1 sec of study time per pair, whereas the participants in the second condition are allowed 4 sec of study per pair. From the Murdock (1960) trade-off equation between presentation time and list length, the participants should have 21.25 percent of the list learned after one trial for the 1-sec condition, but the participants in the 4-sec condition should have 39.25 percent of the list learned after one trial.[21] From the Weibull learning model, shown in equation (11.5), it is possible to estimate the b learning-rate parameter from the proportion correct after one training trial. After one trial, the proportion correct from equation (11.5) is $P_1 = 1 - e^{-b}$. It thus follows that $b = -\ln(1 - P_1) = -\ln(1 - .2125) = .239$ for the 1-sec condition. A similar calculation results in the learning-rate parameter in the 4-sec condition being $b' = .498$. For the value of c, let us use the mean value based on the Ehrensbeck (1979) study (i.e., $c = .73$). Next let us find from equation (11.5) the trial numbers n_s and n'_s for the two conditions where the level of learning is the same. It is convenient to select the 63.2 percent level of mastery as the point of comparison because it corresponds to the trial where $P_n = 1 - e^{-1} = .632$. From equation (11.5), it can be shown that the value for n_s is 5.18 for the 1-sec presentation-time condition, and n'_s is 2.23 for the 4-sec presentation-time condition.[22] Finally, we can calculate the total study time to achieve a learning level of 63.2 percent mastery for the 1-sec condition by multiplying the value of n_s by the 40 sec of study per list presentation; this calculation results in the theoretical prediction that the total study time is 207 sec. For the 4-sec

21. The list length is equal to 40, so the total study time after one trial (T_{total}) in the 1-sec presentation is 40 sec, whereas the total study time in the 4-sec condition (T'_{total}) is 160 sec. Given the Murdock (1960) trade-off equation between presentation time and list length, the number of items correctly recalled for the 1-sec condition after one trial is $R_1 = (.06 \cdot 40) + 6.1 = 8.5$, whereas in the 4-sec condition, the number correct is $R'_1 = (.06 \cdot 160) + 6.1 = 15.7$. The values of .06 and 6.1 are the estimates of the slope and intercept that were determined by Murdock for the trade-off equation. Thus, the proportion correct after one trial is $P_1 = .2125$ in the 1-sec condition and is $P'_1 = .3925$ in the 4-sec condition.

22. The value of n_s can be found from equation (11.5) where $b = .239$ and $c = .73$. More specifically, it can be found from the relationship that $bn_s^c = 1$ because this condition coincides to the case where $P_n = .632$. Consequently, $n_s = \left(\frac{1}{b}\right)^{\frac{1}{c}}$. A similar calculation is used to find n'_s. The value of c is still .73 for the n'_s calculation, but the value b' is .498.

condition, the value for n'_s must be multiplied by the 160 sec of study per replication, and this calculation results in a total study time of 357 sec.[23] Clearly, it is taking more study time to achieve the same level of performance when the presentation time per item is longer, but the number of list replications is fewer. The difference between the two learning conditions is substantial (i.e., the learning is taking 72 percent more study time when the presentation time is longer and the number of training trials is fewer).

The above analysis establishes that the total-time hypothesis is inconsistent with the implications of (1) the Murdock (1960) trade-off in equation (11.1) and (2) the Weibull learning model in equation (11.5). There is no reason to question either of these two regularities about learning, so the total-time hypothesis is most likely wrong. So what is the reason for the benefit of learning when there is a shorter presentation time but more repetitions of the list?[24] The most likely explanation has to do with the advantages of a greater degree of testing. People with more repetition trials with shorter presentation times are also tested more often because after each training trial, there is a testing phase. There now is a substantial experimental literature dealing with the beneficial effects of testing itself. Cooper and Pantle did not take into account the powerful factor of testing on learning. The neglect of the testing effect is probably the reason why the total-time hypothesis is inaccurate. The key research findings about the testing effect are provided in the next section.

11.5   Testing Effects

There are many teachers who are convinced that students learn from testing. For example, it is common in mathematics courses to hear from students that they thought

---

23. Some readers might be concerned in the above calculation that the training trials were non-integer numbers. If n_s and n'_s are rounded to an integer number of training trials, then the participants in the 1-sec condition would have a total study time of 200 sec; whereas the participants in the 4-sec condition would have a total study time of 320 sec. But the levels of learning would be different between the two conditions. For the 1-sec condition, after five trials, the proportion correct is expected to be $P_5 = 1 - \exp[-.239 \cdot (5)^{.87}] = .62$. For the 4-sec condition, after two trials, the proportion correct is expected to be $P_2 = 1 - \exp[-.498 \cdot (2)^{.87}] = .60$. Thus, the 4-sec condition results in a 60 percent greater total study time to achieve a slightly lower level of mastery of the list. Consequently, the use of integer values for the training trials does not alter the conclusion arrived at for this example.

24. Empirically, Roberts (1972) did a test of the total-time hypothesis and found the hypothesis was incorrect because there was faster learning with shorter presentation times and more trials. Thus, the theoretical predictions from the present analysis are supported by the Roberts (1972) study.

they completely understood the chapter, but then they discovered upon trying to solve homework questions that their understanding was incomplete. Certainly learning concepts in an academic course involves more than simply memorizing facts, but even conceptual knowledge requires a development of new memories. Exams and homework assignments are tools for learning.

The benefits of testing is not a recent discovery. For example, John McGeoch noted in his classic book on memory that the feedback provided by a test is valuable because

this information (a) acts as an incentive condition, (b) brings the law of effect directly to bear, (c) favors early elimination of wrong responses, and (d) by informing the subject which items have been learned, promotes a more effective distribution of effort over the material. (McGeoch, 1942, p. 510)

McGeoch's points (a) and (b) deal with the motivational aspect of learning. Earlier in this chapter, in the section about the DEF model, it was stated that learning was suddenly acquired for a memory unit that was within memory span, provided that the person was intending to focus on the item and was given sufficient encoding time. Having a test in an academic course is necessary for student assessment, but it is also important for learning because the students are motivated to get a good grade and thus engage in an active study process.[25] McGeoch's points (c) and (d) deal with a test promoting better learning on subsequent study trials. Knowledge about errors is valuable in correcting wrong answers and directing attention to the harder material that is not yet mastered.

25. The law of effect mentioned in point (b) is simply the principle that responses that are followed by a reward are likely to increase in their frequency of occurrence, whereas responses that result in an unpleasant reaction are likely to decrease in their frequency (Thorndike, 1911). Thorndike, who formulated the law of effect, believed that learning was a connection between the stimulus (S) and the response (R), and he believed that reward altered the strength magnitude of the S-R connection. But McGeoch was well aware that reward might either operate directly on the bond strength between the stimulus and the response or operate as a motivation for future behavior without directly causing the bond in the first place. In fact, earlier in his book, McGeoch suggested that the law of effect phrase can be used to refer to the bond-strength position, whereas the phrase "principle of performance" could be used for the second possible mechanism. Since point (b) mentions specifically the law of effect rather than the principle of performance, McGeoch was essentially claiming that reward is important because it presumedly increases memory strength. Yet I believe that it is more reasonable to consider reward as important for what gets studied and for the amount of time devoted to studying. It is not necessary to assume that reward per se increases a presumed memory strength.

Table 11.3

Proportion recalled[a] as a function of block number m

Block Structure	$m=1$	$m=2$	$m=3$	$m=4$	$m=5$
STST	.41	.65	.75	.80	.85
STTT	.30	.50	.64	.77	.80

[a]Data estimated from a figure reported in Roediger and Karpicke (2006b).

Although McGeoch's observations are sensible, an even older observation stressed the idea that testing might be even more effective than continued studying. For example, Sir Francis Bacon in the seventeenth century stated,

> If you read a piece of text through twenty times you will not learn it by heart so easily as if you read it ten times while attempting to recite from time to time and consulting the text when your memory fails. (Bacon, 1620/2000, p. 143)

The thought experiment suggested by Bacon is a flawed research design because it does not control for the appropriate factors to enable solid conclusions. But experimental psychologists have done careful studies on the effects of testing. For example, Roediger and Karpicke (2006b) reviewed a number of experimental studies that explored the benefits of testing. In a typical paired-associate list-learning experiment, there is a study phase (denoted as S) followed by a test phase (denoted as T). Karpicke and Roediger (2006) studied learning as a function of blocks of four phases. The control block is denoted as STST (i.e., the two trials of a paired-associate list where each trial has a study phase and a test phase). In their experiment, the time for the test phase was equal to the time for the study phase. They compared blocks of STST training versus blocks of STTT training. The correct recall rate is shown in table 11.3 as a function of the number of blocks and the block type. If memory retention were simply a function of the number of times of studying the list, then the performance on one block of STST training (.41) should be comparable to two blocks of STTT training (.50). Moreover, if only the number of study phases mattered, then the performance after two blocks of STST training (.65) should be the same as the performance after four blocks of STTT training (.77). But instead of being equal, there is a benefit from when there is more testing. Thus, some learning is actually occurring from the testing per se.

Testing has also been shown to slow long-term forgetting (Roediger & Karpicke, 2006a). Participants in that experiment had to learn a prose passage that contained 30 idea units. There were two phases devoted to the passage. In one treatment condition, the participants studied the passage in two separate 7-min phases. In the other treatment condition, the participants first studied the passage for 7 min, and later they were given 7 min to recall the passage. For one-third of the participants, there was a final test after a short 5-min delay. Recall performance was better in the study-study condition (.80) than in the study-test condition (.75) for this group of participants. Thus, there was no testing advantage for study-test condition relative to the study-study condition for a short retention interval. The remaining two-thirds of the participants did not receive their final test of the passage until after a delay of either 2 days or 7 days. For the participants in both of these long-delay retention intervals, there was better final recall performance in the study-test condition relative to the study-study condition. It is important to stress that there was a retention effect; hence, the recall was best in the 5-min retention interval, and it decreased with each of the two longer retention intervals. But the decline in the recall rate was slower for the study-test condition than for the study-study condition. For the study-study condition, the recall decreased from .80 to .40 between the 5-min and 7-day retention intervals. For the study-test condition, the recall rate decreased from .75 to .55 over the same two retention intervals. Thus, testing slowed the rate of long-term forgetting.

The higher recall rate for the study-test condition (relative to the study-study condition) is not likely caused by a higher storage rate; although no testing-effect experiment has yet been conducted with model-based measures of storage and retrieval like the ones discussed in part II. Given that there is no test advantage after a retention interval of 5 min, then why should there be a marked storage advantage after a 7-day delay? It is not plausible to suggest that there is a storage gain between the 5-min delay and the 7-day delay. Thus, it is more reasonable to speculate that the advantage for the study-test condition after a long delay is because of a retrieval advantage.[26] But why should 7 min of study, followed by 7 min of recall testing, result in better long-term retrieval compared to the condition where there is a 14-min study period? Presumedly, the items produced in the test phase are encoded again in a testing context. Unlike the original study context, the passage is no longer physically displayed; all that is shown is a blank piece of paper along with any of the previously recalled information written by the participant during the test phase. Thus, the sensory context at test is different, and it is likely that the recalled target ideas are encoded again in this

26. It is reasonable to speculate that a model-based measurement study would show that at a 5-min retention interval, there is a higher storage rate for the study-study condition because the participants would have more study time. If this result were found, then it would also be likely that the retrieval advantage after a 7-day delay for the study-test condition would be due to a larger difference in the retrieval measure than the size of the testing effect found by Roediger and Karpicke (2006a). Thus, the magnitude of the testing effect found for the recall measure might in fact underestimate the magnitude of the testing effect for a pure, model-based retrieval measure.

new context. The reencoding is also at a different temporal context from the original encoding. If the original encoding has a distributed representation like the one shown in figure 11.6, then the reencoded information in memory is an elaborated encoding context. Consequently, the idea units stored in the study-test condition have retrieval cues in addition to those in the study-study condition, and this richer set of cues leads to the improved success rate on the difficult final test. A number of other researchers have similarly argued that testing enhances the relationship between target items and cues (McDaniel & Fisher, 1991; McDaniel, Kowitz, & Dunay, 1989). Bjork (1988) has also suggested that testing increases the number of retrieval routes to the target. Moreover, Karpicke, Lehman, and Aue (2014) have similarly argued that testing provides a new context encoding for the target information, and this additional material aids in retrieval. Consequently, a number of researchers have stressed an enhanced retrieval account for the memory benefit from testing.[27]

Although we have primarily focused on the memory advantage from testing, Roediger, Putnam, and Smith (2011) also delineated many other benefits from testing. Included among these factors were the benefits in (a) transferring knowledge to a new domain, (b) identifying gaps in knowledge, (c) improving subsequent studying, (d) reducing the interference with other learned material, (e) motivating studying in the first place, and (f) providing important feedback to instructors about the current state of student learning. Readers interested in these other effects of testing will find valuable source material in the Roediger et al. (2011) paper.

But not all the effects of testing are positive for the development of an accurate memory representation. For example, Bartlett (1932) studied the repeated testing of prose passages when the participants did not have an opportunity to restudy the original source information. In his famous "War of the Ghosts" experiment, Bartlett had participants repeatedly write their exact recollection of the story. Bartlett found that upon repeated reporting of the story, there were systematic distortions that crept into the participants' reports. Many details about the story were lost with increasing retention interval, and in some cases, the missing details were filled in with reconstructions. Hence, repeated testing without the rechecking with the original source material can result in a false memory of details that are not actually in the story. Roediger et al. (2011) also identified a number of other problems caused by testing in both an educational

27. Karpicke et al. (2014) discussed other alternative accounts of the testing effect before settling upon a retrieval hypothesis that is similar to one described above. I agree with Karpicke et al. (2014) that these alternative hypotheses are not compelling accounts. Among the alternative hypotheses are (a) the bifurcation theory (Kornell, Bjork, & Garcia, 2011), (b) transfer appropriate processing (Morris, Bransford, & Frank, 1977), and (c) the theory of disuse (Bjork & Bjork, 1992).

context and in a memory research context. One negative effect of testing is called retrieval-induced forgetting (Anderson, Bjork, & Bjork, 1994). Finally, Roediger et al. (2011) pointed out that the negative effects of testing can be circumvented by providing feedback and an opportunity for further study.

11.6   Effect of Spacing

Ebbinghaus (1885) extensively reported on his experience when he had to learn new information. He sparked the interest of succeeding generations of researchers in the topic of massed versus spaced practice when he remarked that

with any considerable number of repetitions a suitable distribution of them over a space of time is decidedly more advantageous than the massing of them at a single time. (Ebbinghaus, 1885, p. 89)

Ebbinghaus's observation makes practical sense. Surely if a student needs to study for 10 h to master some new information for an exam, then the massed practice of the material in a single session would be less effective than studying for 10 h but with the study broken up into more sessions of shorter duration. Massed practice can easily lead to fatigue and inattention that would impair effective learning. Also in a practical educational context, if a student has to read four chapters within 2 weeks before a test, then the distributed reading during the 2 weeks would better enable the student to understand the additional lecture information that occurs over this same time period. Despite the reasonableness of these observations, experimental research on the effect of spaced practice versus mass practice has found it difficult to establish a clear set of memory findings that have a straightforward explanation.[28]

Underwood (1961) described his 10-year research program on massed versus distributed practice. He did find effects, but these effects were only under specific, limited conditions. Moreover, he characterized the magnitude of the effects as small.

---

28. Ebbinghaus did not provide formal evidence for his observation, but Jost (1897) analyzed Ebbinghaus's learning data to build a case for superior learning with distributed practice. For example, Jost noted that Ebbinghaus required sixty-eight repetitions to learn a list on a single day, and he then required seven repetitions to relearn the same list on the next day. On another occasion, Ebbinghaus learned a list with thirty-eight repetitions total, but these were distributed over 3 days. The next day, Ebbinghaus required only six repetitions to relearn the list. To Jost, this was evidence for the superiority of distributed practice, but this analysis is unconvincing on many grounds. Ebbinghaus is only a single experimental learner, and the two lists could be quite different in their difficulty. Also, Ebbinghaus's alertness level might have been very different for the two cases examined by Jost. Thus, the idea of the superiority of distributed practice is a hypothesis apparently believed by some researchers without substantial empirical validation.

Underwood defined massed practice as the case where the time between stimulus presentations is less than 8 sec, whereas he considered distributed practice as having a temporal separation between presentations of 15 sec or longer. Among the cases of distributed practice, the amount of the spacing had little effect on the learning performance. The key finding was that mass practice was slightly less efficient than distributed practice.

The research procedure for Underwood's research was the multiple-item list. But that methodology has difficulty in carefully exploring a wide range of spacing conditions. Another approach was to use a frequency-judgment measure, like employed in the Hintzman and Rogers (1973) experiment. These investigators presented colored pictures one, two, or three times over a lengthy series of study trials. For the pictures that occurred more than once, the spacing between the repetitions was filled with zero, one, or five intervening items. The test sequence included novel pictures as well as stimuli from the original learning list. For pictures that were presented twice, the mean frequency rating was 1.2 in the massed-spacing condition and 1.7 when there was a spacing of either one or five intervening pictures. If a participant had perfect memory of the separate events, then the frequency estimate should have been 2. The decrease in the frequency estimate for the zero-spacing condition might reflect a failure to recall a specific occurrence. Alternatively, the memory might be the same for all the spacing conditions, but people might not remember if the picture occurred one or two times. For pictures that were presented three times, the respective mean frequencies for the zero-, one-, and five-spacing conditions were 1.5, 2.2, and 2.5; for this case, perfect performance would result in a frequency judgment of 3.[29] The orderly improvement of the frequency estimates with the degree of spacing might be indicating a corresponding improvement in memory. Yet it is also not clear what a frequency judgment means. Memory for frequency is an interesting research topic in its own right, but one should not confuse a frequency judgment with the probability of picture storage. It might be possible, regardless of the spacing of the target pictures, that there is the same level for the probability of sufficient storage, and the spacing effect for frequency might be solely due to the discriminability of the frequency count. For example, the person might know that the picture is an old item but cannot discriminate whether it occurred two times or three times. Spacing could be having an effect on just the count information as opposed to the item information. Consequently, I do not find the frequency metric a suitable tool for assessing the storage and retrieval success rates for learning as a function of the spacing of the repetition. Yet despite the ambiguity with interpreting

29. The report values here are estimated from a figure in the Hintzman and Rogers (1973) study.

the frequency-of-occurrence measure, the frequency data are nonetheless supportive of the general theoretical approach of the multiple-copy view of memory as exemplified by the representation shown in figure 11.3. If repeated practice of an item were instead represented by a strength model of memory, then it is not clear how people could produce frequency judgments that approximately reflect the actual frequency of occurrence. But the frequency estimates are consistent with the representation shown in figure 11.6 where time-tag information can be used for the judgments of frequency. Thus, frequency judgments are yet another problematic issue for the strength theory of memory.

Glenberg (1976) developed a methodological improvement for the study of spacing effects by using the *continuous pair-associate* task. For this task, each stimulus is presented along with a question mark. The participant has to anticipate the correct response item for that stimulus. After a 3-sec presentation period, the correct response is displayed. Clearly, the person cannot know the right response for the test of a stimulus on the first presentation. But with the feedback, the person can know the correct response for later repeated presentations of the stimulus. Over the session of trials, any given stimulus might be repeated two more times. Thus, each trial is both a study event and a test event. The spacing variable is measured by the number of other intervening stimuli between the first and second presentations. The temporal spacing lag is denoted as t_1, and it is in units of the number of intervening stimuli. Spacing lags (t_1) of 0, 1, 4, 8, 20, and 40 events were examined by Glenberg. For each of the spacing lag values, Glenberg also manipulated the retention interval between the second and third presentations; this lag is denoted as t_2, and it is measured in units of the number of intervening trials. In experiment 1, Glenberg (1976) studied retention intervals (t_2) filled with 2, 8, 32, or 64 intervening events. Overall, there was better recall when there was a spacing value of $t_1 = 40$ than when there was a spacing of either 0 or 1. But the change in recall was only about .06 between these extreme spacing arrangements, and the magnitude of this change was approximately the same across the various retention intervals.[30] The Glenberg study did find a statistically significant (but small effect) between the highly massed training conditions of either 0 or 1 and the widely spaced condition of $t_1 = 40$. Importantly, the overall magnitude of the spacing effect among the larger t_1 conditions was not systematically different.

From the Glenberg study, it is not clear what the causes are for the small spacing effect. To examine if spacing alters the storage probability, Sloboda (2008) did several

30. The magnitude of the differences reported here was estimated from a graph in the Glenberg (1976) paper.

experiments with a modified continuous-recognition procedure. Instead of a recall test for each stimulus, Sloboda required participants to give a YES or NO recognition response to the stimulus and to follow that recognition judgment with a 3-point confidence rating. This experimental procedure enables the estimation of target storage via model 7B from Chechile (2004). Model 7B was described in detail in chapter 7. But without a recall test for the stimulus, it is not possible to estimate the probability of retrieval. Thus, the test protocol used by Sloboda (2008) enabled a correction for guessing for the target storage probability, but it did not provide for a measure of the probability for target retrieval. Furthermore, the testing method focused on the memory for the stimulus, rather than on the memory for an associated response to the stimulus as was done in the Glenberg study. Like the Glenberg study, the experimental design used in the Sloboda study examined the effect of spacing as well as the effect of the retention interval. The storage probability values from her experiment 4 are displayed in figure 11.8. In that experiment, the t_1 spacing lag values were 0, 4, 16, or 20 between the first and second presentations of the memory target. The stimuli were deliberately designed to be difficult to avoid a ceiling effect. The increased difficulty was achieved by using nonsense letter triads like BXT for the memory stimuli. Also, each stimulus was displayed for just 1.5 sec and was followed by the recording of the participants' yes/no response and their 3-point confidence rating. The colored lines in figure 11.8 have a constant spacing t_1 value. The target storage probability is plotted as a function of the retention interval t_2. Clearly, for all spacings, there is a substantial decrease in the storage probability as a function of the retention interval. There is some choppiness in the curves for the various spacing values. To better assess if there is a consistent spacing effect, the data were pooled over the three longest retention intervals. With this grouping, the only statistically reliable storage spacing effect is between the massed spacing condition of $t_1 = 0$ (.134) and the average of the other three spacing conditions (.189).[31] Importantly, there was not a statistically reliable

31. From a Bayesian analysis, it can be said that there is .964 chance that storage is different between these conditions. A difference is considered statistically reliable when the posterior probability of the difference exceeds .95 (Chechile, 1998, 2004). Also, it should be noted that these results are recomputed here because at the time of the Sloboda (2008) study, it was not known that higher levels of statistical accuracy were possible for fitting multinomial models when the data are pooled prior to the model fit as opposed to fitting the model for each participant and then averaging the estimates (Chechile, 2009). Sloboda did not find a reliable difference between the spacing conditions, but that outcome was largely due to greater statistical noise due to the method of model fitting. Readers interested in the issue of pooling versus averaging can find this topic described in detail in the Chechile (2009) paper. It is noteworthy that the pooling method was able to detect a small shift in the storage parameter from .134 to .189 as statistically reliable.

difference among the three distributed spacing conditions (i.e., the storage rates for the t_1 lags of 4, 16, and 20 did not reliably differ). Thus, the only effect on storage is a slight decrease for the massed condition of $t_1 = 0$ relative to any of the other distributed conditions.

So why is there a slightly impaired storage retention for the 0-sec lag condition? The most plausible account for this effect is that participants are sometimes surprised by the re-presentation of the stimulus after they were just tested on the same stimulus. They might be thinking that there is a problem with the computer in recording their last response. In essence, they might be thinking, "Why is the same stimulus being tested again?" If some individuals on some trials were surprised by the immediate retesting, then the benefit for re-presenting the stimulus is missing. Evidence to support this interpretation can be found in the storage probability on the second presentation. In the zero-lag spacing condition, a target occurs on three trials in a row without interruption. For the test on the first presentation, the person should say NO, but on the test for the second presentation, the correct answer should be YES. Nonetheless, the storage probability on the second test is only .771. The decrease of that value from 1.0 indicates that some confusion occurred in this condition. Yet this effect of surprise during target encoding should not occur for any of the other spacings. In fact, there is not a storage-spacing effect beyond that of the $t_1 = 0$ condition. Thus, the failure to find a storage-spacing effect for nonzero spacings implies that the spacing of repetitions does not affect target memory. But it is still possible on a recall task (like the one used in the Glenberg experiment) that there is an effect of spacing due to slightly improved retrieval with distributed practice.

Apart from the common sense of studying course material in a distributed fashion for attentional reasons, there is very little support for a benefit of distributed practice on the storage-learning rate in carefully controlled experiments. But there is a small effect of spacing for improving the retrieval of stored information. This outcome is reasonable because with spaced practice, the time tags in the DEF model of learning are more discriminably different. There are likely to be more temporal retrieval routes relative to the case when all the repetitions occur in the same temporal context. Finally, because the effect of spacing is largely a retrieval phenomenon, spacing has a trivially small or nonexistent effect on recognition memory. All in all, despite the considerable effort to explore the issue of spacing, the magnitude of the observed spacing effect hardly justifies the scope of that research effort.

Consequently, the failure to detect a storage difference between the other distributed spacing conditions is informative because the statistical design and the method of analysis has a high level of discriminability for detecting effects.

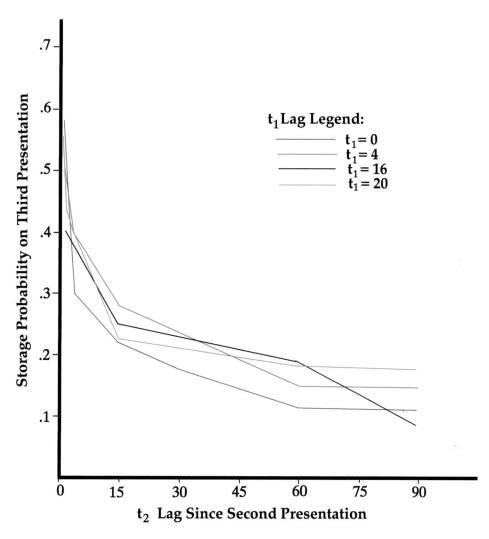

Figure 11.8
Graph of the storage probability on the third stimulus presentation as a function of the spacing (t_1) between the first and second presentations and the retention interval (t_2) between the second and third presentations.

11.7   Summary of Repetition Effects

The study of the acquisition and retention of lists of similar items is a general way for studying the learning process. There is no question that the repetition of the list increases the chance that the items on the list will be stored and retrieved at a later time. Yet we have seen that the theoretical mechanisms underlying this effect have been a challenge for memory researchers to fully understand. A case is made in this chapter against the classic incremental-strength model. But there are also problems with the all-or-none learning model. Instead, the Dynamic Encoding-Forgetting (DEF) model has been advanced as a way to think about the learning of a list of similar items. If people are attending to a target event, which is within their memory span, and if they are given enough encoding time, then it is assumed that two memory traces are produced for the target. One trace is a slower developing neocortical representation called trace **T**. The second trace is a more rapidly acquired hippocampal encoding called trace **T′**. Thus, there is some redundancy in the encoding of the stimulus event. Moreover, the initial encoding is typically a one-trial phenomenon with humans.[32] But both memory traces undergo substantial disruption of their encoding components after their initial imprint. Consequently, the components of the encoding within the two traces undergo dynamic revision. Information is lost, but information is also added and elaborated with repeated practice.

The memory traces have a number of constituent parts. For the **T** trace, there are stimulus-feature elements, contextual elements, propositions, event-binding nodes, associations, time tags, and elements linked to the reaction to the stimulus event. This complex structure is illustrated in figure 11.6. The **T′** trace also includes stimulus-feature elements, contextual components, event-binding nodes, time tags, and components for the reaction to the stimulus event. With repeated training, the informational content becomes elaborated. The encoded information also enables the person to provide frequency-of-occurrence judgments based on the available time tags.

The key idea in the DEF model of rapid encoding followed by massive memory loss is also consistent with a content-addressable view of memory. The same two traces are

---

32. Again, the one-trial encoding claim is for targets that are within the person's memory span and attentional focus. We have seen earlier in chapter 4 that there is a vast amount of information available that is not encoded. Recall from chapter 4 that an information theory analysis of a typical scene resulted in the statement that "our information cup is only less than 0.014% full." Nonetheless, items on experimental lists are easily detected and encoded. But the one-trial encoding does not necessarily hold for nonhuman animals because of differences in their memory span, motivation to learn the target information, and their understanding of the instructions.

generated and accessed because the sensory features of the stimulus are the basis for addressing the traces. But with content-addressable memory (CAM), subsequent similar stimuli can access and overwrite some of the features of a previous memory trace. Thus, CAM can explain the rapid recognition memory speed as well as the dynamic changes in memory storage after an encoding.[33]

The DEF model advocated in this chapter is not a memory strength model, but instead it stresses the elaboration of the informational content with repeated practice. Strength theory has many problems with accounting for the repetition of training. For example, strength theory does not have a ready mechanism for accounting for the ability of people to judge the frequency of occurrence. Also, it is clear that highly practiced learning is qualitatively different from the learning with only a few repetitions. The highly practiced item has more informational content stored (i.e., it is not just a simple association with a greater strength value). Moreover, the finding of a powerful learning effect associated with testing is consistent with the DEF model because the testing of an item is an encoding occasion on its own. Testing provides an encoding opportunity in a different context (i.e., the testing context). With testing, there are additional context cues and time tags, and this elaborated information makes for an enhanced environment for retrieving information from memory. Testing is also the most likely reason why the total-time hypothesis has been disconfirmed.

The study of list learning leads to a simple mathematical function for describing learning as a function of the number of times (n) for studying the list. In terms of the probability of item storage (θ_{se}), the learning equation is

$$\theta_{se} = 1 - \exp(-bn^c),$$

where the b and c parameters are positive constants. The b parameter is directly related to the learning retention rate, and the c parameter is related to what Sloboda and Chechile (2008) call the "learning hazard." The learning hazard is the probability that an item that is not retained before the nth trial will be remembered after the nth training trial. If $c = 1$, then there is a constant learning hazard. This condition is mathematically proven to be impossible without the 100 percent uniformity of all the items on a list in terms of their ease of learning—a condition that cannot be absolutely guaranteed. Sloboda and Chechile (2008) found that generally, the c parameter is less than 1. This result indicates that the memory retention rate after n trials slows as the value for n increases. This outcome is as if people show rapid learning of some items on the list but require many more training trials to learn the rest of the list. It is also possible that each item on the list does individually have a constant learning

---

33. See chapter 5 for a more extensive discussion of content-addressable memory.

hazard, but the items have different learning rates. In that case, the c value for the whole list would be less than 1 (Chechile, 2003). Regardless, the fact that c is usually less than 1 could be due to the fact that the items on a list are unrelated in the standard experimental implementation. But if the informational content of stimulus S_1 had a predictive relationship to stimulus S_2, then it would be reasonable to expect that prior learning would accelerate the learning of the subsequent items and c would be greater than 1.

12 Encoding and Context Factors

12.1 Chapter Overview

In the previous chapter, the focus was on the role of repeated study and repeated test trials on the learning of novel information, whereas in this chapter, the spotlight is on some encoding factors that influence learning without the necessity of repeated study or test. The chapter also addresses the role of changes in context between the time of encoding and the time of test.

12.2 Meaningfulness and Ease of Encoding

Ebbinghaus (1885) began a longstanding tradition of designing experiments with nonsense items to focus on the learning of novel information. For example, consider the following two consonant-vowel-consonant (CVC) triads: XUH and VIK. It is likely that neither of these stimuli has an existing representation in memory. But triads like TOP and MOP are meaningful words within the language and have a rich semantic content, whereas XUH and VIK do not have an existing memory representation. Glaze (1928) was the first investigator to develop a meaningfulness scale for CVC triads. Glaze found that indeed 0 percent of his participants could find anything meaningful about XUH, but surprisingly, he did find that 100 percent of his participants could find something meaningful about VIK. Very likely, the participants associated VIK with the word and semantic concept of VIKING. Several other meaningfulness norms for CVC nonsense syllables were also published after the Glaze report (Hull, 1933; Krueger, 1934; Mandler, 1955; Noble, Stockwell, & Pryer, 1957). But these investigators did not use the same method for scaling meaningfulness. Glaze (1928) and Krueger (1934) measured meaningfulness by the percentage of participants who could provide an association to the CVC triad. But Hull (1933) scaled meaningfulness by the performance of people on a serial-learning task. Noble et al. (1957) scaled meaningfulness by having people provide a 5-point rating judgment about the perceived degree of easiness they would have finding associations to the CVC. But Mandler (1955) scaled meaningfulness by the number of associations that the participant could actually produce to the CVC. This last basis for measuring meaningfulness has come to be called the *production* method. Witmer (1935) used the production method to develop a meaningfulness scale of consonant triads (CCC).

There is variability among the meaningfulness norms. For example, the Hull metric has a low correlation (in the .55 to .72 range) with the other meaningfulness measures for CVC triads (Noble et al., 1957). Yet Noble et al. (1957) reported that the other methods do have better correlations with each other (i.e., in the .81 to .90 range). But variability in the scaled values and correlations less than 1 are reasonable outcomes for a number of reasons. First, many of the studies used a small sample size of participants; consequently, the scaled values have substantial statistical error. Second, the Hull method has been criticized as doubtful because his task was overly taxing on his participants (Underwood & Schulz, 1960). Third, the scales are actually estimating different properties that all happen to be called meaningfulness, so these quantities need not be equal or perfectly correlated.

If the meaningfulness measures are valid, then we would expect the ratings to have predictive utility. That is, do the ratings predict the ease of new learning? Underwood and Schultz (1960) reviewed the paired-associate (PA) learning literature to address this question. They offered two general observations. First, the grouping on scaled meaningfulness values is positively related to the ease of association learning for both the stimulus and response members of the association. Second, the role of meaningfulness is stronger for the response member. An experiment that illustrates both of these effects was conducted by Cieutat, Stockwell, and Noble (1958). After five trials, the percentage of correct responses for the H-H, L-H, H-L, and L-L conditions is nonoverlapping with the respective values of 82, 67, 58, and 19, where H and L refer to high and low meaningfulness. The average difference of high versus low meaningfulness for the response is 36 percent, whereas the average difference of high versus low meaningfulness for the stimulus is 27 percent.

The meaningfulness norms are fine for grouping stimuli into categories that are likely to be more homogeneous. But the rating value is not a perfect predictor for each individual. For example, a stimulus like QJH had a meaningfulness rating of 0 in the Witmer (1935) norm. Yet if I were to see that stimulus on a memory list, then I would immediately think of the encoding "Queen Jack of Hearts." What has to be learned in the memory task is that the stimulus QJH occurred recently, and I am required to know that fact later. The encoding of "(Q)ueen (J)ack of (H)earts" is an excellent way for representing QJH in memory because it can be imaged, and because it is a good way to store both the letter information and the order of the letters. Evidently, none of the twenty-four Yale students studied by Witmer thought of this encoding within the 4 sec that she allowed per item. However, if I were required to learn the word triad GRAPE-BOOK-TIN, then that target would actually be more difficult for me to encode than QJH, despite the fact that it consists of three meaningful words. The problem is that the semantic content of the words does not readily enable a mnemonic for linking the words together. I might try to imagine a grape next to a book, which in turn is next to a tin can. But that internal organization is arbitrary, so later I might recover the three items but not recover the order of the objects. Meaningfulness ratings, at best, are only proxies for predicting the ease of new learning.

Currently, there are a vast number of published stimulus norms. Brown (1976) provided a catalog of 172 studies that presented information for scales for verbal information. Proctor and Vu (1999) listed 142 norms published just within the Psychonomic Society journals. These indexes still underestimate the total number of published stimulus sets from online sources. Norms are a good way of grouping stimuli. There are norms for properties such as word frequency, word association, word meaning, imagery, concreteness, familiarity, and typicality. Norms also exist for a wide variety of stimulus material (e.g., words, anagrams, graphemes, phonemes, letters, letter strings, homographs, homophones, categories, pictures, symbols, faces, and names). Researchers typically want to use a set of stimuli that are nearly equal in terms of their ease of new encoding. Nonetheless, with any stimulus set, it is not surprising to find substantial individual item differences among units that were rated to be similar. This variation is due in part to differences in the effectiveness of the encoding strategy. In other cases, the stimulus material may have a greater ease for encoding because of the participant's unique experiences. For example, one of my participants worked for a company where many of the nonsense letter combinations were similar to the inventory codes used by that company.

12.3   Levels of Processing

The level-of-processing (LOP) approach was introduced by Craik and Lockhart (1972) as a reaction to the Atkinson and Shiffrin (1968) multiple-store model, which was discussed in chapter 3; see figure 3.2. An important distinction made in the multiple-store model was between *structural* properties of memory and *control processes*. Atkinson and Shiffrin argued that the separate memory stores were part of the hardware structure of memory, but the person's encoding strategy is a process for controlling the flow of information between the memory stores. Rehearsal of information in the short-term store is an example of a control process. In chapter 3, the multiple-store model was called the modal model of memory because most psychologists believed it to be correct. Yet there are critics of that view of memory.[1] Craik and Lockhart (1972) argued that the encoding of information varied over levels, and these levels affected the retention rate. The LOP approach is often characterized as an alternative to the multiple-store model, but this position is neither fair nor accurate. Atkinson and Shiffrin did stress the importance of control processes. The LOP approach could easily be accommodated within the multiple-store model by distinguishing between low-level rote or maintenance rehearsal and a higher-level semantic rehearsal where an item in the short-term store is integrated with long-term information and transferred back to the long-term store as a new encoding. In fact, the Atkinson-Shiffrin model was different from earlier mathematical models of memory because of the distinction between structural properties and control processes or strategies. Earlier Markov chain models of memory, like the Atkinson and Crothers (1964) model, did not have provisions for encoding strategy differences. With the distinction that Atkinson and Shiffrin (1968) made between structural properties and control processes, strategies were integrated within an overall mathematical framework. In essence, the distinction between structural features and control processes roughly parallels the computer science distinction between hardware and software. Actually, Craik and Lockhart still adhered to (1) the idea of primary memory, which they likened to the items currently in consciousness, and to (2) the idea of secondary memory, which they regarded as the rest of memory. Their conceptualization of primary and secondary memory was more similar to the James (1890a) distinction rather than the memory stores described in the Atkinson-Shiffrin model. Craik and Lockhart also criticized the idea that information passes from one store to another store. Consequently, the LOP approach is not as much of a challenge to the multiple-store model than it is a different language for discussing memory.[2] The language for discussing memory in the Atkinson-Shiffrin model is borrowed from computer

1. I have been among those critics. Instead of the multiple-store model, I have suggested the two-trace hazard model where both traces are susceptible to storage loss (Chechile, 2006). This model was discussed in chapter 3. Also, a form of the two-trace hazard model is the Dynamic Encoding-Forgetting (DEF) model, which was discussed in chapter 11.

2. Other models like the aforementioned DEF model in chapter 11 and the Chechile (2006) two-trace hazard theory are more appropriate alternatives to the Atkinson-Shiffrin model. Also, the working-memory model elaborates further the structural features of the multiple-store model by its addition of other processing buffers within short-term memory; see figure 3.3 (Baddeley, 2000; Baddeley & Hitch, 1974).

science, where there are information buffers and memory registers. Many experimental psychologists were uncomfortable with the computer analogy and preferred natural language descriptions within the tradition of experimental functionalism.[3]

Craik and Lockhart in their level-of-processing (LOP) approach stressed the interaction between the encoding level and memory retention. For a typical LOP experiment, the participants are induced to encode the target stimulus in different ways so as to vary the depth of processing. At the shallow level, the participants might be asked to judge if a target word is typed in an uppercase or in a lowercase font, whereas to induce a deep level of processing, the participants might be asked if the word fits into the context of a sentence. Between these two extremes, the participants might be asked if the target word rhymes with another word. A number of studies provided support of the general interaction between encoding "depth" and memory retention (Bellezza, Cheesman, & Reddy, 1977; Bower & Karlin, 1974; Craik & Tulving, 1975). For example, Craik and Tulving (1975) did a number of experiments that used different instructions to induce three levels for encoding targets (typographic case, rhyme, or sentence meaning). They found memory retention was generally correlated with the level of encoding.

Although there are studies that supported the prediction that a deeper encoding (i.e., semantic) is remembered better than a more shallow encoding (i.e., physical property), a number of other studies also take issue with the levels-of-processing framework (Baddeley, 1978; Eysenck, 1978; Nelson, 1977; Nelson, Walling, & McEvoy, 1979). For example, Nelson (1977) criticized the prediction of the LOP framework that learning is minimal for a low-level phonological encoding. He found repeated training under a phonemic-encoding rule improved performance to the same degree or better than the corresponding improvement from repetition when a semantic-encoding rule is used. Moreover, Baddeley (1978) and Eysenck (1978) raised questions about the rigidity of the hierarchy of processing levels and the predictive utility of the depth construct.

---

3. Experimental functionalism stresses the examination of effects found for dependent variables as a consequence of the manipulation of an independent variable. Also with this approach, comparisons to a control condition are an essential feature. The functionalist approach also attempts to explore the boundaries of current knowledge by introducing other dependent variables or by refining control conditions. The process-purity approach that was critiqued in chapter 6 is a typical experimental-functionalist approach, whereas the model-based measurement method discussed in chapter 7 is derived from a different tradition that stresses the modeling of the dependent variables in terms of their underlying cognitive processes. Yet with the model-based measurement approach, control conditions and experimental studies that involve the manipulation of independent variables are still important research design features. However, historically many researchers within the experimental-functionalist tradition have been slow to embrace the use of mathematical models of memory.

Also, when it comes to accounting for why there are differences in the retention rate, the LOP framework has less specificity than either the working-memory model or the Atkinson-Shiffrin model.

An experiment by Atwood (1971) nicely illustrates the limitations of the LOP framework for explaining the interaction of memory retention rate with the type of encoding. Participants in that experiment were given thirty-five *auditory* phrases to learn for approximately 5 sec per phrase. For half of the participants, the phrases depicted an imaginary scene, such as "pistol hanging from a chain." Participants were instructed to visualize the phrases. The participants in this encoding condition were divided into three subgroups. Group 1 had a *visually* presented digit (either a 1 or a 2) displayed after each phrase was encoded; the participant simply had to repeat aloud the single digit. Group 2 had an *auditory* digit presented after each phrase; again, the participants had to simply repeat the digit aloud. Group 3 was a control condition where the participants did not have a digit after each phrase. Following the presentation of all thirty-five phrases, the participants were tested with a cued-recall test where the first word from the various phrases was presented as a cue. Atwood reported that the control participants were at an 82 percent accuracy rate. But the recall rate for the visual-digit condition was at a 58 percent level, whereas in the auditory-digit condition, the recall rate was 76 percent. The visual presentation of either the digits of 1 or 2 after the phrase is significantly more disruptive than the auditory presentation of the number. This result is consistent with the dual-code hypothesis (pictorial and propositional) discussed in chapter 4. The presentation of an auditory number was less disruptive to the pictorial image, which participants were required to produce, than was the presentation of a visual number. Atwood also examined the memory retention for thirty-five abstract (nonvisual) phrases such as "the *theory* of Freud is *nonsense*." These participants were instructed to contemplate the meaning of each phrase as a whole. The participants with this encoding rule were divided into three subgroups in the same way as the imagination-encoding participants were divided. That is, the participants were in one of the following conditions: (1) a single visual digit after each sentence, (2) a single auditory digit, or (3) a control condition with no digit between the sentences. Following the presentation of the thirty-five phrases, the participants were given the set of the first nouns from the various phrases as the cues (e.g., the word *theory* from the previous example as a cue for recalling the word *nonsense*). The overall retention rate for the control condition without an interpolated digit was 70 percent, which is significantly reduced from the imagination control mentioned above (82 percent). Consequently, the instruction to focus on the *meaning* of the phrase resulted in poorer performance than the instruction to focus on *imaging* the phrase. More important, the visual presentation of the interpolated digit was less interfering (60 percent) than the auditory presentation of the digit (44 percent). The encoding instruction to focus on the meaning of the phrase presumedly results in a subvocal proposition that can be interfered with, to a greater degree, by an auditory digit than by a visual digit. The Atwood experiment illustrates that the interaction of the retention rate with the "encoding level" is in part due to the type of information that follows the encoded material. The results from the Atwood experiment are difficult to explain from the level-of-processing framework. Yet according to the two-trace hazard model, there is a storage interference process that varies in vigor to the degree of the similarity between the properties of the encoded target and the properties of the items in the retention interval. The Atwood findings are consistent with this framework.

12.4 Contextual Factors

There is a sizable experimental literature on the contextual influences on memory retention. But unfortunately, this area also has experienced a number of misleading hypotheses. In some cases, these hypotheses have garnered unnecessary attention, so it is important to explain both the ideas and my rationale for rejecting these notions. Yet this research area also includes many interesting findings and clever experiments that will be highlighted.

12.4.1 Transfer-Appropriate Processing

Morris, Bransford, and Franks (1977) introduced the idea of transfer-appropriate processing as part of a critique of the levels-of-processing approach. Their objection dealt with the perceived failure to take into account the mismatch of test conditions with the type of encoding. Participants in their experiment heard a series of thirty-two short sentences with the critical target noun missing and replaced by the word BLANK in each sentence. After a sentence was articulated, there was a 2-sec pause, and then a target word was spoken. The participants simply had to respond positively or negatively as to the correctness of each sentence at the time of encoding. In table 12.1,

Table 12.1
Encoding/test for a correct EAGLE answer (Morris et al., 1977)

Encoding Task	Example Sentences	Standard Test	Rhyme Test
Semantic	"BLANK is a large bird."	.844	.333
Rhyme	"BLANK rhymes with legal."	.633	.489

there are two examples of sentences where the response is the word EAGLE; however, in the actual experiment, there were thirty-two different target nouns. Unlike the two examples in the table, there were also sentences where the target word did not match the sentence. For example, for the target EAGLE, the sentence, "BLANK rhymes with book" is a sentence that focuses the encoding on the phonology of the target word and leads to a NO response from the participant. There were also sentences like "BLANK is a mammal" that focuses the encoding on the semantics of the target and results in a NO response. Thus, half of the sentences required a semantic analysis, and half of the sentences required an examination of the phonology. After hearing all the sentences, a recognition test was administered. For half of the participants, the thirty-two targets along with thirty-two novel foils were presented for a standard yes/no recognition test. The other half of the participants received a novel rhyme-recognition test consisting of thirty-two words that rhymed with the target words and thirty-two words that did not rhyme with any of the target words. On the rhyme-recognition task, the participants were instructed to answer YES if there was a target noun on the list that rhymed with the recognition probe; for example, the test item REGAL rhymes with EAGLE. For the standard recognition test, the semantic encoding rule resulted in superior performance relative to the sentences that were encoded with a rhyme focus (.844 versus .633 as shown in table 12.1). But for the new rhyme-recognition test, the sentences encoded with the rhyme focus were better recognized relative to the sentences encoded with the semantic focus (.489 versus .333 as shown in table 12.1).

Morris et al. (1977) argued that the typical finding of better memory retention with semantic encoding is a result that might just be caused by the mismatch between the encoding rule and the manner of test. The authors assume that people normally process a yes/no recognition test probe on the basis of word meaning, but such tests do not optimally match the encodings that were created with a rhyme focus. Thus, they argued that people given a rhyme-encoding task should be tested with a rhyme test and not with a semantic test. Implicit in this argument is the idea that an encoding based on rhyme might be an excellent encoding that has just been incorrectly tested. Yet there are two problems with this interpretation even within the context of the Morris et al. (1977) study. First, although the rhyme-encoding focus was better than the semantic task on the rhyme-recognition test, the participants were nonetheless doing poorly on this difficult recognition test (i.e., the .489 and .333 rates from table 12.1).[4]

4. Morris et al. (1977) stated that these proportions included a correction for guessing (i.e., they subtracted the false-alarm rate from the hit rate). But they did not report the false-alarm rates. Nonetheless, the corrected rates found for this rhyme-recognition test are still very low.

It is reasonable to hypothesize that a person provided with the rhyme-test probe like the word REGAL might retrieve words such as *seagull*, *beagle*, and *eagle*. But the recognition cue REGAL is itself a poor stimulus, within the framework of a content-addressable memory, for eliciting the word EAGLE directly. Second, the rhyme test did not show better performance when the participant encoded the target word as a rhyme rejection. For example, "BLANK rhymes with book"... EAGLE. In this case, a probe like REGAL only resulted in a correct response rate of .184. But if the person had a negative semantic sentence, like "BLANK is a mammal"... EAGLE, then the rhyme probe of REGAL resulted in a better correct response rate of .325 than the corresponding rate for the rhyme test for rhyme encoding. Thus, the improved performance with the rhyme-recognition test for rhyme encoding only occurred for the case of positive sentences. If the Morris et al. (1977) mismatch hypothesis were correct, then the rhyme-recognition test should be also improved even for the case of negative sentences. Consequently, the transfer-appropriate processing idea is a limited hypothesis that sometimes results in inaccurate predictions.

12.4.2 Stimulus Generalization and Encoding Specificity

Despite the criticism of the transfer-appropriate processing hypothesis, there is nonetheless evidence that there is a benefit for encoding information in a close match to the ultimate test condition. Recall from chapter 2 the phenomenon of stimulus generalization decrement with Pavlovian conditioning (i.e., if an animal is trained to a specific stimulus and later tested with a similar but different stimulus, then performance is typically not as good). When the test stimulus is changed from the study conditions, then it is expected that a content-addressable memory will have more difficulty recovering the target memory. The demonstration of stimulus generalization does not mean that the training memory has been altered. In essence, stimulus generalization is a desirable property because the animal or the person is able to respond appropriately even if the stimulus environment has been somewhat changed. It is nearly impossible or even desirable to learn a response in an excessively controlled stimulus context because that response might also be the appropriate action to other similar stimuli. Yet the same response is not always appropriate, so it is again desirable to have a reduced tendency for the conditioned response for altered stimulus contexts. As discussed in chapter 2, the stimulus generalization decrement is likely to involve an active inhibition of the conditioned response because the test stimulus is different from the training stimulus (Bouton, 2004).

Although stimulus generalization is an established phenomenon in animal conditioning, there is a comparable effect for human verbal memory. The effect is referred to as *encoding specificity* (Light & Carter-Sobell, 1970; Tulving & Thomson, 1973). For example, Light and Carter-Sobell (1970) presented people with sentences that contained homographs and tested for the memory of these items. Before discussing the Light and Carter-Sobell study, let us define homographs and briefly review what psycholinguists have learned about them.

Homographs are words such as JAM and BANK that have multiple meanings that do not share any common semantic overlap, for example (strawberry jam, door jam, or traffic jam) and (river bank or savings bank). These homographs are called unsystematic homographs because there is no systematic relationship between the various meanings of the word (Onifer & Swinney, 1981). There are also systematic homographs such as the word FIGHTER, which could be a type of aircraft or a boxer. Systematic homographs would also include noun-verb homographs such as KISS or FARM. It is the unsystematic homographs that are particularly interesting stimuli to study to see if the separate meanings are activated together or separately. Onifer and Swinney (1981) conducted a beautiful study that demonstrated, upon hearing a homograph, both meanings of the word are momentarily activated. The activation was revealed by a semantic-priming effect for both meanings. They used a cross-modal priming procedure where the participants had two concurrent tasks. One task was to listen to sentences. For the other task, the participants had to make a lexical decision for a visually presented letter string. Consider the example sentence: "The broker watched the stock."[5] The word STOCK is a homograph (i.e., it refers either to a traded share of a corporation or to a ranch animal). In a 0-sec delay condition, a visual letter string appears at the same time as the homograph (STOCK) is spoken. Half the time, the letter string is a nonword, and half the time, it is a meaningful word. The participant must decide if the letter string is a word or not as quickly as possible. This is not a difficult task, and people can execute the task easily. Nonetheless, the speed by which the participants make these decisions can reveal interesting properties about the access to lexical memory. Among the stimuli studied in the 0-sec condition were words that were associated with either meaning. For example, COWS is an associate of the ranching sense of the word STOCK, whereas BONDS is an associate of the financial sense of the homograph. There were also control words that were not associated with either meaning (e.g., TRAIN or RUBBER). In a standard lexical task, there is a well-known semantic priming effect, that is, faster lexical decisions when the second word that follows another word is semantically associated with the first word (Meyer & Schvaneveldt, 1971). For example, if the word DOCTOR

5. This example is not one of the sentences actually used in the Onifer and Swinney (1981) study, but it is used simply to illustrate their experiment 2.

follows the word NURSE, then the lexical decision about DOCTOR is faster relative to the case when the word DOCTOR follows the word LAMP. There is a consensus that the associative-priming effect is due to a spread of activation from the presented word to the associates of the word. Consequently, after hearing the word NURSE, there is already some activation started for the associates of NURSE such as the word DOCTOR (Meyer & Schvaneveldt, 1971). Onifer and Swinney found a semantic priming effect in their cross-modal priming task for a word like COWS, which is an associate of the meaning of the homograph that is *not* consistent with the context of the sentence. They also found a priming effect for a word like BONDS, which is associated with the meaning of the homograph that is consistent with the context of the sentence. Furthermore, Onifer and Swinney also studied lexical priming when the visual letter string was delayed by 1.5 sec after the articulation of the homograph. Now the only priming effect found was for the meaning consistent with the sentence context (i.e., the word BONDS in the above example). Thus, this experiment demonstrated (1) immediately upon hearing an unsystematic homograph, there is a spreading activation to all of the meanings of the word, and (2) eventually as time elapses, the initial activation that is inconsistent with the sentence context becomes inhibited, and only a priming effect remains for meaning that is consistent with the sentence context. This remarkable set of findings has been supported by other researchers (Seidenberg, Tanenhaus, Leiman, & Bienkowski, 1982; Simpson, 1984). Notice also the automatic activation of associates of all the meanings is consistent with the general content-addressable architecture discussed earlier in this book.

Our subjective experience when we hear a homograph is to understand the meaning of the word that is consistent with the sentence context. We do not seem troubled by the potential semantic ambiguity of the word. Yet the Onifer and Swinney study has established that unconsciously, both of the meanings of the word are briefly activated. The activation process is rapid, and the context information lags behind the initial activation. Eventually, the context resolves the semantic ambiguity, and the secondary meaning is inhibited. This inhibition happens so fast that the secondary meaning never emerges to a conscious level. Additional studies have placed a limit on the secondary priming effect to be less than 200 msec (Simpson, 1984). Consequently, both the activation and the inhibition of the secondary meaning all occurs in less than 200 msec. Figure 12.1 illustrates the theoretical activation levels for the two meanings of a homograph. The activation of the primary meaning that is consistent with the sentence context is expected to build steadily as a function of the processing time until it is consciously comprehended, but the activation of the secondary meaning initially builds before the contextual constraints disambiguate the conflict and inhibits

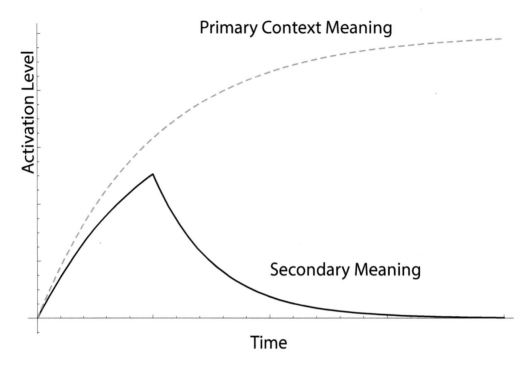

Figure 12.1
A theoretical plot of the activation levels for the two meanings of a homograph.

the secondary meaning. The resolution of ambiguity is a common problem that the brain must solve. Sensory processes are frequently muddled and ambiguous, so the brain needs to interpret the sensory data into a coherent interpretation that is consistent with other contextual information. Thus, it is reasonable to expect that the model illustrated in figure 12.1 is a general roadmap for the resolution of sensory ambiguity.

Now let us return to the Light and Carter-Sobell (1970) study that used homographs. In their experiments, participants studied sentences for semantic understanding and for a later memory test. Each sentence contained a key adjective that described a central noun that was an unsystematic homograph. Consider the following two examples: (1) "Susan opened the <u>PLASTIC JAR</u>," and "David received a <u>BAD GRADE</u>." The participants studied each sentence for 4 sec, which is a time long enough to ensure that only the meaning of the homograph that remains activated is the one consistent with the sentential context. Later the participants were told to respond YES or NO to the test nouns and to ignore the adjectives that were also provided at the time of test. The test nouns included the homographs studied along with some new or foil nouns that were paired with suitable adjectives (e.g., RUSTY CHAIN). There were three types of adjectives provided for old-recognition test trials. In one condition, the same adjective was provided as the one used in the original encoding (i.e., PLASTIC JAR). This condition resulted in a corrected-recognition rate of .80.[6] In another condition, the participants saw a different adjective linked with the same meaning of the homograph (e.g., BROKEN JAR). In this condition, the corrected-recognition measure drops to .63. Finally, in the last condition, the test adjective is connected to the other meaning of the target homograph (e.g., SUDDEN JAR). For this condition, the corrected-recognition rate drops further to the value of .44. Although the participants were told to simply make their recognition response based on the noun and to ignore the adjective, the adjective at test nonetheless is different from the adjective at the time of encoding and results in a generalization decrement. A broken jar is a different mental image from a plastic jar, and sudden jar generates an even more radically different image from the encoded stimulus. Encoding specificity is the idea that the item in memory includes the specific context produced during encoding. Thus, changes in the context at test lead to a lower performance level. But notice this change is not a loss of the target from memory storage, but it is a failure to recover the target due to a change in the context at the time of test. It is more suitably a retrieval loss.

The term *encoding specificity* was coined by Tulving and Thomson (1973), but that paper did not use homographs as stimuli. Instead, their experiments involved the use of strong and weak associates of target words. In my opinion, the Tulving and Thomson (1973) paper is based on problematic experiments that were framed to assess the idea of the generate-decision model of item recall. The generate-decision model is based on the hypothesis that item recall involves first the generation of a candidate item and then deciding if the candidate occurred earlier. Tulving and Thomson also attempted to build a case that there were recognition failures of words that could be recalled.

---

6. Actually Light and Carter-Sobell (1970) used a questionable metric that was designed to correct for guessing, that is, $p_{hit} - p_{fa}$ where p_{hit} is the hit rate on old recognition, and p_{fa} is the proportion of false alarms on foil test trials. This correction method was recommended by Woodworth and Schlosberg (1954), but a better measure can be defined instead. The alternative guessing measure is based on a tree model like the ones described in chapter 7. The alternative metric is equal to $\frac{p_{hit} - p_{fa}}{1 - p_{fa}}$. To see why this measure is an improved correction for guessing, consider the two metrics for the case of a person who detects the target 30 percent of the time and responds YES in those situations but simply guesses at a 50/50 rate when the target is not detected. This person would have a hit rate of .65 and a false-alarm rate of .5. The older correction measure would be .15, whereas the improved guessing correction would have the value of .30, which is the same value as the proportion for target detections. Thus, the Light and Carter-Sobell data were reanalyzed to obtain the value of .80 for the same adjective test condition.

But their purported recall test was one where the participants could have responded at a high level without having studied the targets at all.[7] Also as noted in chapter 7, Tulving and Thomson used a recognition test that had participants pick words from a long list of responses that were intermixed with a large set of novel words, and this testing was done in the absence of the cue words. Consequently, the context was changed between study and test for recognition, but it was not changed for recall. Thus, the Tulving and Thomson study employed a multiply flawed experimental design that did not legitimately contradict either the generate-decision model of item recall or show recognition failures of recallable words. Moreover, it did not expand the findings of the generalization decrement for word stimuli beyond that shown already by the Light and Carter-Sobell study. But Zeelenberg (2005) did implement the right type of experimental controls to properly examine the same type of encoding conditions used by Tulving and Thomson. Participants in the Zeelenberg experiments studied weakly associated pairs such as PULL ROPE. Later at the time of test, the participants had two types of cued-recall tests. For one type of test, the participants are provided with the original cue (e.g., PULL) and are asked to recall the target, whereas for the other type of test, the cue is a strong associate of the target (e.g., JUMP). The reinstated cue (PULL) resulted in a correct response 83 percent of the time, but a different cue (JUMP) resulted in only 48 percent correct recall. Perhaps when encoding the word ROPE, in the context of the verb PULL, the participant imagines a person pulling on a rope. The cue PULL would thus be highly effective for eliciting the recall of the target. But the verb cue JUMP is not consistent with the specific memory encoding of a person pulling on a rope, and it would be less effective as a recall cue. Thus, this simple experiment shows a generalization decrement of a word stimulus when a different cue is presented at the time of test. The decrement is again not a storage problem but rather demonstrates the decrease in performance when the test context omits critical aspects of the encoded item.

12.4.3   The Effects of the Incidental Context

The focus so far has been on the effect of testing memory under conditions where the cues are different from those that were an integral part of the original encoding. It is reasonable to expect that such changes will alter performance, and there is in fact a decrement that can occur for those changes. But is there an effect on memory retention

7. In that study, participants were presented weakly related pairs to learn such as WHISKEY WATER. Later in a supposed recall task, the participants were asked to free associate to the cue LAKE. It was considered to be a correct recall if WATER was the first response. Yet it is likely that WATER would be the first free associate even if the person never had a prior list of associates to study in the first place.

for changing an incidental environmental factor such as the experimental room itself? A number of psychologists have examined the change of a homogeneous incidental feature of the study and test environment. For example suppose people learn information printed on red-colored cards, but later the participants are tested where the stimuli are printed on gray cards. Will such a change of a homogeneous incidental feature of the stimulus environment have an effect on memory retention? A study by Dulsky (1935) did this exact experiment and failed to find a statistically significant change in the memory retention rate. Yet some researchers did find that other environmental or contextual factors could affect the recall rate of the target information (Eich, 1985; Godden & Baddeley, 1975; Smith, Glenberg, & Bjork, 1978). For example, Godden and Baddeley (1975) studied members of the university scuba-diving club to evaluate the context of a dry environment near the edge of the water versus the environment of being 20 feet underwater in their scuba-diving outfit. They found memory recall was better when the environment for testing (either dry or underwater) matched the environment in which the learning took place.

In another paper, Godden and Baddeley (1980) did a follow-up experiment with the university scuba divers, but in this experiment, they found no contextual effects when memory was assessed with a recognition memory test. This pattern of finding an effect for context with a recall measure, but not with a recognition measure, was also reported in several other papers that examined the role of the rooms used for learning and testing (Eich, 1985; Fernandez & Glenberg, 1985; Smith et al., 1978). But Light and Carter-Sobell (1970) found a context effect for recognition memory. However, in that study, the context shift was for cues that were an integral biasing component of the encoding of the target, for example, (at input) PLASTIC JAR but (at test) SUDDEN JAR. In contrast to this type of contextual shift, the room-change studies deal with a homogeneous incidental feature of the learning context and results in a consistent failure to find a contextual effect on a recognition memory test. Yet major shifts in the incidental learning environment between the time of study and the time of test have often been shown to influence performance on a recall test. Remember in chapter 7 that the model for correct recall was a multiplicative product of the storage and retrieval probabilities, whereas the model for old recognition was described by a probability tree involving storage and guessing processes, but importantly, it *did not* involve retrieval or search. If people encode the target information in two different physical environments in the same way, then it would not be expected to see any change in the storage probability for the two learning environments. Yet it might be possible that a change in the homogeneous, incidental encoding context results in a change in retrieval effectiveness. For example, if people form some general additional associations between the features of the environment and the target information, then the absence of those environmental cues at the time of test would impair retrieval. It thus follows that the change in context at the time of test should not affect recognition, but it could affect the recall rate.

Given the above general explanation for the pattern of results from experiments on incidental contextual changes, a question arises about why some experiments such as the Dulsky (1935) study failed to find contextual effects for a recall measure, whereas other studies did find a context effect. The Eich (1985) paper points to a possible explanation. Eich studied the effect of a change (or not) of the experimental room under two different types of imagery instructions. One instruction was called *isolated*, and the other was called *integrated*. For the *isolation* condition, the participants were told to imagine any given target word in its own environment and *not* as a part of the current environment. For example, for the target word KITE, the person must visualize a red kite rising high in the blue sky. For the *integrated* condition, the participants were told to visualize the target within the context of the current room for the study. For example, they might visualize a kite lying on top of the table in the room. Later in the assessment of a change (or not) of the room, Eich did not find a context effect for the *isolated* encoding condition for a recall measure, but he did find a significant context effect for the recall measure for the *integrated* condition. Eich also failed to find any context effect for either type of instructions when testing with a recognition measure. Thus, the experimental studies on incidental features of the context show that there is a retrieval advantage on a later recall test if the person uses aspects of the physical context in the original encoding. Presumedly, the participants in the Dulsky study learned the information on the colored cards and ignored the color of the cards, whereas in the other studies, the incidental environment was a part of the target encoding on enough occasions to produce a retrieval advantage when testing was done in the same environment. But incidental features, such as the physical environment, have not resulted in an advantage on a recognition test. This lack of an effect is a consequence of the fact that recognition does not require a memory search as discussed previously in chapter 7.

12.4.4 Music-Dependent Memory

Background music is another incidental aspect of the learning environment that can influence later memory retrieval. For example, Smith (1985) studied the effect of the sound background on later memory performance. In his first experiment, participants were provided words to learn either with a quiet sound background or with background music (either Mozart or jazz). In his second experiment, the participants learned either with a quiet sound background or with one of two auditory environments (jazz or white noise). On an immediate recall test, the performance was best when the auditory background was the same as the sound background during the study phase. Even variation from the white noise background resulted in less effective recall. Balch, Bowman, and Mohler (1992) did a similar experiment on the effect of musical context on memory. These investigators were interested if the context effect required the same musical selection or would the effect also occur if the musical genre was the same between study and test. They found the context effect was limited to the same musical selection. For example, if jazz was played during the study phase and a different jazz selection was played during the test phase, then there was not an improvement in the recall rate. They also found that the context benefit did not survive after a 48-h delay. Balch and Lewis (1996) also showed that the tempo of the music was particularly important for the music-dependent effect. That is, the context benefit was lost even if the same musical selection was played but the tempo of music was altered. All of these studies can be easily explained in terms of either an encoding specificity or a stimulus generalization account. Variation from the incidental background sound environment can result in a less effective recall of the target items.

12.4.5 Drug State-Dependent Context

Recall from chapter 7 the effect of state-dependent memory where retention is influenced by a change in drug level (Girden & Culler, 1937). For example, if animals learn a task either with or without sodium pentobarbital in their system, then memory retention is best when the animals are tested in the same drug state as the one at the time of the original learning (Overton, 1964). State-dependent memory has also been demonstrated for alcohol with human participants, but the effect was sensitive to the type of test procedure (Goodwin, Powell, Bremer, Hoine, & Stern, 1969). Goodwin et al. (1969) did find more errors for a recall performance measure for rote learning and word-association task when the learning state differed from the test state. But these researchers also evaluated their participants on a picture recognition test. Goodwin et al. (1969) did not find evidence for state-dependent memory for the picture recognition test. Wickelgren (1975) also failed to find evidence for state-dependent memory for alcohol when a recognition memory test was employed. This pattern has also been reported for other drugs such as for the glaucoma medicines physostigmine (Weingartner, 1978) and marijuana (Darley, Tinklenberg, Roth, & Atkinson, 1974). State-dependent memory illustrates the powerful stimulus properties that some drugs produce. The finding of a strong state-dependent memory effect for recall tests but not for recognition tests indicates that retrieval is impaired when there is a change in the drug state from the state at encoding.

It is important to stress that drugs can also impede memory retention beyond the retrieval impairment of testing people in a different drug state from the one that was in place at the time of the original encoding. In other words, a drug state can be more than an *incidental contextual* feature of the learning environment that influences memory retrieval. For example, Gerrein and Chechile (1977) tested people in the same drug state as the drug state of the original learning. This experiment is also described in detail in chapter 7. That study demonstrated an effect of alcohol on both storage and retrieval that was beyond any decrement due to state-dependent learning. Thus, a drug like alcohol can cause a sizable memory impairment on its own.

12.5 Chapter Summary

Some information is easier to learn than other information even when the general physical properties of the stimuli are roughly the same. Initially, psychologists focused on the ease in which an item could be linked to other existing memories as a method for accounting for the variation of item learning. If an item could be interpreted in terms of previous ideas or concepts, then the item was *meaningful*. The norms are snapshots of the success rate of a group of participants either to find an association or to produce a number of associations. The norms are mostly useful as a basis for researchers to select a relatively homogeneous set of items for their experiments. But the stimulus norms for meaningfulness do not capture the process of finding meaning itself. For either the Atkinson-Shiffrin model or the levels-of-processing alternative language, the type of encoding strategy was seen to have an enormous effect on memory retention. For many cases, the memory retention is superior when the participant tries to find a meaningful association for the target items on a list compared to a strategy that only deals with surface characteristics of the target item. Yet the experiment by Atwood (1971) strongly showed that this view is missing an important factor. The Atwood experiment shows that the amount of forgetting is a function of the type of encoding and the properties of the information presented over the retention interval. If the encoding strategy emphasizes particular features that are different from the subsequent information experienced by the person, then the memory loss is reduced. But if the encoding of the memory targets overlaps with the way in which subsequent information is processed, then there is generally more forgetting. The Atwood findings are consistent with the two-trace hazard model discussed in chapter 3.

Meaningful words are perhaps the most widely studied class of memory stimuli. But the meaning of the words is already known and a part of semantic memory. What is novel about a word list is the fact that those particular words were on the current target list (i.e., the words are associated with the current temporal context). Yet from psycholinguistic research, such as the Onifer and Swinney (1981) experiment, it is clear that the presentation of a word automatically activates the meaning of the word in semantic memory. The presentation of the word also unconsciously primes the associates of the word. All the associates, even the ones that are *not consistent* with the current context of the sentence or encoding context, are initially primed. The inconsistent information has an initial spike in activation, but it is followed by an inhibition of that activation. See figure 12.1 for an illustration of the rapidly changing level of activation for the contextually inconsistent information. For homographs, the processing spike for contextually inconsistent meaning and the subsequent inhibition of that meaning occurs within 200 msec and is not consciously perceived. This remarkable finding is in line with the content-addressable view of memory that has been advanced throughout the book. The representation of a stimulus results in a reactivation of the surviving stored information about the stimulus. In fact, only a fragment of the initial stimulus or an associate has a chance to activate the target information. The best reactivation occurs for the intact target stimulus, but some weaker degree of activation does occur for a similar but different stimulus. This mechanism is general and is the basis for the phenomenon of *stimulus generalization.*

A number of one-off hypotheses have been promoted in the memory literature to deal with the effect of context (e.g., *transfer-appropriate processing* and *encoding specificity*). But those alternative frameworks for describing the effect of a changed context between study and test were critiqued in this chapter. Instead, a view was advanced that focused on the degree of similarity between the test stimulus and the stimulus at the time of the initial encoding. This framework successfully accounts for the verbal-learning studies, but it also successfully accounts for the experiments that explore changes in the incidental physical or nonverbal aspects of the initial encoding context. Incidental background information that is present at the time of encoding but altered at the time of test can result in weaker performance on a recall test. Yet reliable evidence of an incidental-context effect for recognition memory was not found. This finding is reasonable because the recognition test has sufficient information available to reactivate the stored target information (if it is available), whereas the recall task is devoid of target-specific features.

13   Generation and Elaboration

13.1   Chapter Overview

In the early years of studying human memory, the research participant was largely treated as a passive receptacle for information that flowed in from the environment. The participant's task typically was to simply learn target items that were provided by the experimenter. But in this chapter, a number of factors are examined that increase the demands on the learner. These greater demands lead to a more active learning process and result in improved memory retention. A common framework for explaining these effects is provided, and there is a discussion of how practical strategies can be implemented to improve memory in educational and social contexts.

13.2   The Effect of Generative Encoding

13.2.1   The Contours of a Phenomenon

There has been a common informal belief that information that is actively created is more memorable than information passively experienced. As Aristotle observed,

The craftsman has to learn how to make things, but he learns in the process of making them. So men become builders by building, harp players by playing the harp. (Aristotle, circa 329, *Nicomachean Ethics*, Book 2, Chapter 1)

But prior to 1978, the idea of self-generated learning was not explored with well-controlled psychological experiments. In 1978, two quite different studies were published that demonstrated the importance of generative encoding (Roenker, Wenger, Thompson, & Watkins, 1978; Slamecka & Graf, 1978).[1] Slamecka and Graf coined the

1. The two papers were published in separate journals, and the authors came from different universities. Moreover, neither group of authors seemed to be aware of the findings from the other group. In fact, in 1978, there was yet another paper that compared *read* versus *constructed* solutions

term *generation effect*, and their paper was the more influential study of the two papers. But both studies compared the memory of self-generated response items versus a control condition consisting of experimenter-provided items. Slamecka and Graf asked participants to read cards and state aloud the verbal response to a stimulus-response pair. They explored five different types of encoding rules relating the stimulus with the response (association, category, opposite, synonym, and rhyme). For the *read* condition, both the stimulus and response words were printed on the card, but for the *generation* condition, the stimulus word was provided along with the first letter of the response word. For example, the word pair might be RAPID - FAST for the control condition and RAPID - F—for the generation condition. The participants were paced to go through the set of cards at the same rate. They found for each of the five encoding relationships that there was superior recognition accuracy for the generation condition relative to the control condition. The subsequent experiments in their paper also showed that the generation effect occurred for free and cued recall as well as for cued and uncued recognition. They argued that strength theory and the total-time hypothesis could not account for their results because those theories did not have a mechanism in which an impoverished fragment viewed for the same amount of time resulted in better memory retention than an intact response. They further argued that a level-of-processing account does not have a ready explanation of the effect because the generation benefit occurred for all five encoding relationships between the stimulus and the response. For example, the generation effect also resulted when a rhyme-encoding relationship linked the stimulus and response words. They further ruled out the differential-attention hypothesis because the generation effect was limited to the response item (i.e., the memory for the stimulus item of the pair was not statistically different). Had participants paid greater attention in the generation condition, then it would seem that the memory for the stimulus items would be also improved, but it was not. Moreover, they argued that a differential-rehearsal hypothesis could not account for the generation effect because there was actually a greater opportunity to rehearse in the control condition. Slamecka and Graf thus demonstrated a novel encoding effect that was not uniquely predicted from existing accounts within memory theory.

---

to cue-word puzzles (Jacoby, 1978). However, it is easy to overlook this paper in the generation effect literature because Jacoby did not focus on the comparison between these two conditions; the paper was about remembering versus solving problems in the context of repetition learning. Again, Jacoby came from a different university and did not seem to know of the work of the aforementioned authors. Consequently, it seems that 1978 was a year when the role of self-generated information was a topic of converging interest to cognitive psychologists. These papers also sparked even more interest in the topic of generative encoding.

Roenker et al. (1978) used an unconstrained generation procedure and found an additional puzzling effect. These investigators studied congruent and incongruent answers to three types of questions (structural, rhyme, and categorical). In the *read-control* condition, a question like "Does the word contain a P and a K?" is a congruent-structural probe about a word like PANCAKE that is shown on the card. The participant's task for this type of test was to circle either Y or N on the card. For an incongruent-structural probe, the same question might be shown for the word WRINKLE. For the self-generation counterpart to these read-structural instructions, Roenker et al. asked participants either to produce a word containing a P and a K (the congruent condition) or to generate a word that did not contain a P and a K (the incongruent condition). As an example of a rhyme-control condition, the participants might get a question like "Does the word rhyme with STRIFE?" with either a congruous answer (WIFE) or an incongruous answer (CAREER). Again, there were also instructions to generate a word that either rhymes (the congruous case) or does not rhyme with STRIFE (the incongruous case). Later, the participants had to free recall the responses that were either provided on the card or self-generated. They found a generation effect for each question type (i.e., structural, rhyme, and categorical). More important, they found a larger generation effect for the words produced in the incongruous condition. Schulman (1974) previously studied the memory for encoding either a congruous question such as "Is a SPINACH green?" or an incongruous question like "Is SPINACH sudden?" Later, the participants had to recall the uppercase words. Schulman found better recall for the words encoded to the congruous questions. But Schulman was not asking for self-generated answers as in the Roenker et al. study. In contrast to the Schulman findings, Roenker et al. found that if participants have to generate a solution to an instruction like "—is not a tree," and if they come up with the answer CLOWN, then later the response CLOWN is highly memorable.[2] Thus, the larger generation effect for incongruous encoding is another puzzling finding that will be discussed later.

In addition to words as memory items, Graf (1980, 1982) also demonstrated the generation effect for meaningful sentences. For these studies, the participants either read sentences or constructed sentences from a scrambled word list. Participants were

2. There is a glaring difference between the control and self-generated conditions because the responses were not constrained to be the same as in the Slamecka and Graf (1978) study. Consequently, the generation effect found by Roenker et al. is potentially an artifact of a difference in the words between the generation and the control conditions. But other investigators solved this issue of an appropriate control condition, and these investigators found results that replicated the findings from Roenker et al. (Ayers, 1991; Chechile & Soraci, 1999; Soraci et al., 1994).

told that the sentences were of the form of an article (THE), an adjective, a noun, a verb, an article (THE), and a final noun (e.g., THE TINY MOUSE FRIGHTENED THE COOK). In the read-control condition, the sentence was displayed, whereas in the generation condition, the four nonarticle words were arranged in a vertical stack in a scrambled order. Besides these meaningful sentences, Graf also studied meaningless or anomalous sentences (e.g., THE CHEERFUL CARPET EXCHANGED THE MOUSE). Because of the restriction on the structure of the four key words, anomalous sentences could also be solved by the participants in the generation condition. In both papers, Graf found a generation effect for meaningful sentences but was unable to show a generation effect for the anomalous sentences. The Graf studies are similar to an anagram, so it is not surprising that the generation effect has also been found for anagrams (Foley, Foley, Wilder, & Rusch, 1989). But the generation effect for anagrams only occurs for easy anagrams, so this is yet another puzzle that requires explanation.

Although the vast majority of the experiments on the generation effect have used verbal material, a number of studies also have demonstrated the generation effect for nonverbal material. For example, McNamara and Healy (1995) showed that the generation effect occurred for the memory of numerical answers to simple arithmetic problems. During encoding for the generation condition, the participants had to solve a simple problem such as $3 \times 7 = ?$ and say the correct response, whereas in the control condition, the participants read aloud the answer to the problem $3 \times 7 = 21$. Later, the participants had to recall the numerical answers for the arithmetic problems. These authors found better recall for the generation condition.[3]

The generation effect has also been shown for pictures (Peynircioğlu, 1989; Wills, Soraci, Chechile, & Taylor, 2000). For both of these studies, the participants had to generate pictures in a fashion similar to the simple game that children play in connecting numbered dots. In contrast to this generation condition, the participants in the control condition were shown the same dot pattern but with the lines already drawn. In the control condition, the participants still have to trace the lines, so the same motor activity is required in both conditions. Peynircioğlu (1989) found a generation effect for both meaningful figures as well as for "nonsense" figures. But her description of the nonsense figures demonstrated that they were completely captured by a meaningful propositional encoding of familiar shapes, so the characterization of those

---

3. Although there was a reliable finding of a generation effect for multiplication, there was not a consistent finding of the generation effect for addition problems such as $11 + 21 = ?$. For addition, they found that the generation effect would occur only when the participants were provided with an instructional hint just before recall testing. The hint consisted of the suggestion of trying to remember the original problems rather than just trying to remember the answers. With that hint, the participants were able to remember the solutions at a higher rate in the generation condition.

figures as nonsemantic is not convincing. Wills et al. (2000) also found a generation effect for meaningful figures. Moreover, Wills et al. found that the magnitude of the generation effect was larger when the participants did not know the picture that would emerge after the dots were connected. They tested the generation effect either when the participants were given or not given a preview of the final picture before the connect-the-dots problem started. The generation effect is larger when the problem solution is not known in advance. This finding will be discussed later in the overall explanation of the generation effect.

Overall, there is clear evidence of the positive benefit of generative encoding for memory retention as demonstrated by a comprehensive meta-analytic review of eighty-seven papers published in the research literature (Bertsch, Pesta, Wiscott, & McDaniel, 2007). Yet the literature also shows that a positive improvement in memory retention does not always result from generative encoding. For example, a number of investigators failed to find a generation effect for nonwords (Gardiner, Gregg, & Hampton, 1988; McElroy & Slamecka, 1982; Nairne, Pusen, & Widner, 1985; Payne, Neely, & Burns, 1986). One might wonder how a person was able to generate a nonword in the first place. In the McElroy and Slamecka experiments, the person is told to reverse the letters of the stimulus member of a pair and to attach that letter string to a provided initial letter; the resulting construction was a nonword. All the studies showed that the generated nonword was not remembered better than an experimenter-provided nonword.

A number of researchers have claimed that the generation effect can be substantially reduced or eliminated altogether when the conditions are separated into different groups or different lists (Begg & Snider, 1987; Hirshman & Bjork, 1988; Slamecka & Katsaiti, 1987). But this claim is not universally accepted. As usual, claims of a null effect are difficult to establish without a sophisticated statistical analysis such as the computation of a Bayes factor, but that type of analysis was not employed in those studies.[4] Most important, other researchers did find a statistically significant generation effect when different groups were used for the *read* and *generation* conditions (Chechile & Soraci, 1999; Soraci et al., 1994, 1999; Wills et al., 2000). Consequently, the reports of a weak-to-no generation effect for a between-groups methodology is not itself a reliable effect. The failure of those studies to produce a generation effect for between-list or between-group comparisons is likely a shortcoming of either the statistical sensitivity of the study or the type of generative encoding task used.

---

4. It usually is the case that the manipulation of a variable in general is more statistically sensitive for detecting an effect when it is done on the same individual. Moreover, if within a list there are words better encoded than other words, then the recall of the more poorly encoded items has an additional disadvantage because the recall resources are devoted more on the better developed items. But this does not mean that the generation effect only occurs for within-list experiments.

A number of researchers reported a *negative generation effect* (Schmidt & Cherry, 1989; Steffens & Erdfelder, 1998). But there are complications with many of these experiments that compromise a straightforward interpretation of a negative generation effect. For example, in the Steffens and Erdfelder study, the participants either completed the missing letters of the response word of a pair or read an intact word pair. These researchers used either a rhyme or category basis for relating the stimulus and response words. For rhyme encoding, the people studied pairs such as WIFE-KN—in the *generation* condition, whereas people saw pairs like WIFE-KNIFE in the *read* condition. The participants in the *read* condition were instructed to repeat aloud the stimulus and response words for 7 sec, whereas in the *generation* condition, the participants were instructed to generate the response word and then repeat the pair of words aloud for the remainder of the 7-sec interval. Steffens and Erdfelder reported that the *read* condition resulted in significantly better free recall. Yet this procedure is unusual in that the people in the control condition had to repeat a simple word pair aloud for 7 sec. Later in this chapter, another encoding effect will be discussed (i.e., the *production effect*). Briefly, the production effect is the better memory retention for prose material when people read it aloud rather than reading the material silently (MacLeod, Gopie, Hourihan, Neary, & Ozubko, 2010). Moreover, 7 sec is an unusually long time for encoding a pair of words, and this method of study makes it possible for more repetitions of the word pair in the *read* condition. Mulligan and Duke (2002) also examined the same type of stimulus-response pair as used in the Steffens and Erdfelder study, except the participants in the control condition read the pair silently and wrote the response word on a sheet of paper, whereas people in the generation condition had to generate the solution and write the response word. Mulligan and Duke did find a lower initial recall rate for the generation condition relative to the control condition, but the participants in their study also had to recall the studied list again four more times *without further study* of the information. The people in the generation improved over the five test trials. This improvement in the memory with repeated testing without restudying is called *hypermnesia*. With repeated recall tests, the difference between the read and generation conditions disappeared. Importantly, in a related experiment, Mulligan and Duke tested memory retention by means of a recognition test. Recognition memory was actually better for the generation condition (i.e., a positive generation effect). Consequently, the report of the negative generation effect from the Steffens and Erdfelder study is unconvincing.

The study by Schmidt and Cherry (1989) found in five experiments a positive generation effect for the response member of the stimulus-response pair, but these investigators found that the participants in the *read* condition did better on recalling an intact pair of words. They referred to the poorer recall of the stimulus-response pair as a negative generation effect. Yet Burns (1992) argued that generation had a *positive* effect on both item memory and the association between the cue and the response, but he reported that the responses recalled were less clustered on a free-recall test than the items in the control condition. This reduction in output clustering for the items generated is what Burns calls a negative generation effect. But Soraci et al. (1994) also measured free recall and several measures of response clustering, and in that study, generation increased both the item recall and the output clustering.[5] Thus, it would seem that Schmidt and Cherry (1989) as well as Burns (1992) were eager to label some property of generation as a negative generation effect, but in both cases, the attribute that showed a higher rate for the control condition was not replicated. Moreover, it is not the goal of the participants in a free-recall test to output their responses into nice clusters; rather, their goal is to remember the response items.

Other researchers have also reported a deficit for generative encoding for some other attribute of memory beyond that of item memory, and they have labeled that difference as a negative generation effect. For example, Jurica and Shimamura (1999) found for generative encoding an improved memory for item information, but they found a poorer memory for the identification of the source of the information. Yet Geghman and Multhaup (2004) found an improved memory with generative encoding for both item information and for the information source. Other investigators argued for a negative generation for memory context (Mulligan, 2004; Mulligan, Lozito, & Rosner, 2006), but Marsh, Edelman, and Bower (2001) found evidence for improved memory for the context for generative encoding. Also, some researchers assert that generative encoding results in impaired memory for the order of the words (Mulligan, 2002; Nairne, Riegler, & Serra, 1991), but Marsh (2006) found evidence that generative encoding improved memory for list location. Consequently, despite a concerted effort to find problems with generation, it is difficult to consistently demonstrate that generative encoding leads to impaired memory retention for item information. The evidence instead strongly favors the conclusion that active encoding improves memory. In the next section, the reasons for this improvement are examined.

---

5. There are a number of metrics developed to study the response order of free-recall data in order to infer the organization of the information in memory (Frankel & Cole, 1971; Roenker, Thompson, & Brown, 1971). Burns (1992) and Soraci et al. (1994) used different metrics, but it is unlikely the difference in the metrics was the reason for the opposite findings between those experiments. It is more likely that the instructions for generation were the reason for the differences in the output clustering. The Soraci et al. study employed an unconstrained generation method, whereas the Burns study used a constrained generation procedure.

13.2.2 Explaining the Generation Effect

A number of researchers have proposed multiple-factor accounts of the generation effect (Hirshman & Bjork, 1988; McDaniel, Waddill, & Einstein, 1988). In the Hirshman and Bjork version, generative encoding strengthens the response items, and it also strengthens the relationship between the stimulus and response. McDaniel et al. (1988) add a third factor of enhanced whole-list processing. Although these multiple-factor accounts are widely cited, I believe that they do not account for many key empirical results. More important, the hypothesized factors buys into a strength theory of memory, which has so far in this book not fared well as a theoretical concept when examined closely. An alternative theory of the consequences of generative encoding was developed in part from a paper by Chechile and Soraci (1999).

Chechile and Soraci studied unconstrained generative encoding in terms of storage-retrieval measures. The participants in the *generation condition* had to generate either a related or an unrelated response to a stem word and then type their answer on the computer keyboard. In the related condition, the participants were instructed to produce an "extremely related word to the stem," whereas in the unrelated condition, the participants were instructed to produce an "extremely unrelated word to the stem." The stem word and the response remained visible for 2 sec after the response was inputted. For both types of generation, the participants were instructed to select a response that did not repeat a previous response and did not repeat a previously presented stem. There were sixty stem words on a list, and the whole list was blocked to be either related or unrelated encoding. Half of the participants in the generation condition did the related encoding before doing the unrelated encoding, whereas the other participants did the two conditions in the opposite order. Because the responses differed for each participant in the generation condition, there was a yoked control-condition person for each generation participant. For example, if one participant in the unrelated generation condition came up with the response CLOWN to the stem TREE, then there was a yoked-control participant who would be provided with the intact word-pair TREE CLOWN. The control participants also had to type in the response word, and the pair of words remained on the screen for 2 sec after the response word was typed. Following the encoding phase for the sixty-item list, the participants were allowed 270 sec to free recall any of the response words by typing their answers. The computer determined which words were correctly recalled, and these words were not tested further. But the remaining words that were not recalled were tested one at a time with a contingent forced-choice task. The computer had a file of words from the Kucera and Francis (1967) norm. A total of 120 words were randomly selected to be stem words; the same 120 words were used for all participants. The remaining words minus any of the response words entered during the encoding phase were the pool of words that could be used as foils on the contingent forced-choice lineup test. The initial lineup test had one response word that the participant generated or saw during encoding and three randomly selected foil words from the computer pool of available words. For the four-alternative forced-choice test (4-AFC), the participant had to pick the word that he or she previously typed. If the participant was correct, then feedback was provided and testing shifted to another of the words not recalled on the free-recall test. But if the participant selected an incorrect alternative, then (1) feedback was provided, (2) the incorrect choice was removed from the lineup list, and (3) the participant was instructed to pick again (i.e., it became a 3-AFC test). If the participant was correct on the three-alternative test, then testing shifted to another word that was not recalled. If the participant was again incorrect, then he or she was given feedback, the incorrect choice word was removed, and the participant was given a 2-AFC test.

The elaborate set of test protocols used in the Chechile and Soraci experiment enabled a storage-retrieval analysis for each encoding condition. The process-tree model is shown in figure 13.1. The measurement model has parameters for the sufficient storage probability (θ_S), the probability for successful retrieval (θ_r), the fractional storage probability (θ_F), the probability of nonstorage (θ_N), and the conditional probability when there is fractional storage that the partial information is sufficient for eliminating two of the three foils on the 4-AFC task (θ_{f2}). The probability of $1 - \theta_{f2}$ is the conditional probability that fractional storage enables only the elimination of one of the three foils. The case where all three foils can be eliminated based on stored information occurs when there is sufficient storage but unsuccessful retrieval, that is, $\theta_S \cdot (1 - \theta_r)$. The case when knowledge is insufficient to reject any of the three foils is a pure-chance guessing state; the proportion of items in this state is the nonstorage parameter θ_N.

Figure 13.1 also shows the five possible outcomes for any item. Two of the five outcomes are unique in that they occur by means of only one pathway. One unique outcome is correct recall (cell 1), which occurs when the item is sufficiently stored *and* successfully retrieved, and the probability for this outcome is $\theta_S \cdot \theta_r$. The other unique outcome is a wrong response on the 2-AFC task (cell 5), which has a probability of $\frac{\theta_N}{4}$. With either sufficient storage or fractional storage, the person can use that knowledge to avoid the case of always being incorrect on the contingent forced-choice tests. Cell 5 can only occur when there is no storage and the person is guessing at a pure-chance rate. If there is sufficient storage but unsuccessful retrieval, then that will result in a cell 2 outcome. But an outcome in cell 2 can also occur by guessing when there is either partial storage or no storage. That is the reason why a process-tree model is

A.

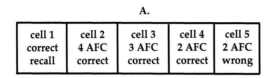

cell 1	cell 2	cell 3	cell 4	cell 5
correct	4 AFC	3 AFC	2 AFC	2 AFC
recall	correct	correct	correct	wrong

B.

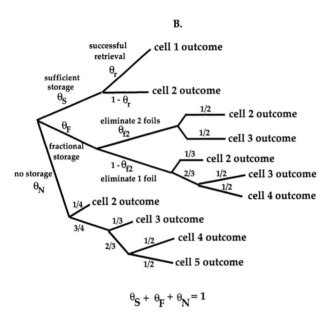

$$\theta_S + \theta_F + \theta_N = 1$$

Figure 13.1
(A) The outcome data structure of the Chechile and Soraci (1999) experiment along with (B) the corresponding process-tree model in terms of the probabilities for storage θ_S, retrieval θ_r, fractional storage θ_F, and nonstorage θ_N.

required to correct for the alternative ways a particular outcome can occur. It is outside the scope of this section about the generation effect to discuss the tree shown in figure 13.1 in detail. In essence, the process-tree model automatically corrects for guessing and enables the estimation of the latent processes of sufficient storage (θ_S), retrieval (θ_r), fractional storage (θ_F), and nonstorage (θ_N). The statistical estimation of the model parameters is discussed in detail in the Chechile and Soraci paper.

Chechile and Soraci reported that the correct recall rate is significantly higher in the *related generation* condition compared to the *related read* condition (.265 versus .181), and the recall rate is also higher in the *unrelated generation* condition compared to the *unrelated read* condition (.364 versus .201). Moreover, there is a significantly higher recall rate for the *unrelated or incongruous generation* condition than the *related*

Table 13.1
Process-tree results from Chechile and Soraci (1999)

Parameter	Generative Encoding		Yoked Read Control	
	Unrelated	Related	Unrelated	Related
θ_S	.879	.880	.612	.603
θ_r	.413	.301	.328	.301
θ_F	.103	.086	.186	.165
θ_N	.018	.034	.201	.232
θ_{f2}	.690	.666	.492	.731

or congruous generation condition (.364 versus .265).[6] Chechile and Soraci thus replicated the findings from the Roenker et al. (1978) study. But the Chechile and Soraci study also allows for an examination of the underlying process probabilities for these effects. The values for the underlying memory probability parameters are shown in table 13.1 for each condition. The storage component of the recall rate is substantially different for the two generation conditions relative to the two control conditions. The sufficient storage measure (θ_S) was about .88 in the two generation conditions, and it was about .61 in the two control conditions.[7] Thus, generative encoding resulted in improved storage to the same degree for both the *related* and the *unrelated* conditions. But the correct recall rate is higher in the unrelated generation condition than it is in the related generation condition. This effect is entirely due to the improved retrieval for the *unrelated generation* condition.[8] Thus to account for the effects of generative encoding, it is necessary to explain why there are changes in both storage and retrieval processes.

The measurement model also detected other condition differences. All the response items in the *read* condition were successfully typed and stayed visible for an extra 2 sec, but nonetheless, many of these items were no longer stored in memory at the time of test. In fact, 21.7 percent of the responses in the *read* condition degraded entirely to the state of nonstorage (θ_N), whereas with the *generation* condition, only 2.6 percent of items degraded to the level of nonstorage. In terms of fractional storage, there is not a statistically reliable difference between the two generation conditions, and there is not

6. All three of these differences in the correct recall rate are statistically significant with an α level less than .00005.

7. A Bayesian analysis of the storage measures detected a highly reliable difference between the *generation* and the *read* conditions, that is, $P[\theta_S(generate) > \theta_S(read)] > .999$.

8. A Bayesian analysis of the retrieval parameter resulted in the finding of $P[\theta_r(unrelated\ gen.) - \theta_r(related\ gen.)] > .98$.

a reliable difference between the two control conditions, but there is a reliably higher fractional storage probability in the *read* conditions.[9] Thus, the active cognitive process of creating a response to a stem word produces a memory trace that is less susceptible to storage loss compared to the *read* control items. Finally, the measurement model has a conditional parameter about the type of fractional storage (θ_{f2}). But given the low rate of fractional storage in the generation condition, this parameter is not very informative, and it has wide variability.[10]

Chechile and Soraci (1999) detected strong changes in both storage and retrieval, but there is still a need to explain these effects. For example, why does generation result in less information being lost from storage? Moreover, why does incongruous generation result in more retrievable memory traces than congruous generation?

Let us focus first on the issue of less storage loss for response items that arise from generative encoding. To understand this effect, consider the cognitive steps required of the participants in the *read* control condition.

Read Trial Processing Steps for the Chechile and Soraci (1999) Study:

- View the stem-response pair.
- Type the response word.
- View word pair for 2 sec.

In contradistinction to these three simple steps, consider the cognitive operations on a generative encoding trial.

Generative Processing Steps for the Chechile and Soraci (1999) Study:

- View the stem word.
- Consider a candidate solution word.
- Test if the candidate and stem words have the instructed relationship.
- If the above test fails, then a new candidate is tried and tested.
- Type in the candidate word, and recall word spelling from lexical memory.
- View word pair for 2 sec.

Clearly, the cognitive operations required to generate, test, and spell a candidate word in the *generation* condition requires the utilization of semantic memory and lexical memory. It also requires testing the semantic relationship between the stem and the response. The memory trace contains information from these extra cognitive steps, so

---

9. A Bayesian analysis of the fractional storage parameter resulted in the conclusion that $P[\theta_F(read) - \theta_F(generate)] > .975$.

10. This parameter is only a nuisance probability with no memory interest. The parameter deals with the number of foils that can be rejected on the 4-AFC task when there is partial target storage.

it has a rich, elaborative structure. But in the *read* condition, the cognitive steps do not necessitate semantic processing, stem-response testing, or lexical/spelling memory access. In many previous studies discussed in this book, we have seen that newly stored information in memory is susceptible to storage loss. It is reasonable to expect that an elaborated memory encoding will be more different from other encoded items on the list, and thus the amount of intralist storage interference would be less than the level of storage interference in the *read*-control condition. In this account, the generated responses are not stronger, but they have a more elaborated structure and hence are less susceptible to storage loss. For example, if a person in the *unrelated generation* condition produces the response word CLOWN to the stem word of TREE, then it is not because the word CLOWN has a stronger bond strength to the word TREE or a stronger bond to the general experimental context. Rather, it is argued that the person develops a chain of associated thoughts until a generated candidate word is sufficiently unrelated to the stem word to satisfy the requirement of being unrelated. It is argued that all the extra cognitive operations are part of the memory representation of the events triggered by the presentation of the stem word (i.e., the memory trace structure is more elaborated rather than just stronger).

Alternatively, the storage generation effect can be thought of as an encoding deficit for the *read* condition. The participants in the *read* condition do not spontaneously process the stem and response words as much as they could. Support for this interpretation can be found in the Donaldson and Bass (1980) study. These investigators found a typical generation effect in their first two experiments, but in their third experiment, the effect disappeared when the participants in the *read* condition had to make an additional relatedness judgment. The participants were asked after writing the response word to supply a 3-point numerical rating about the relatedness between the stem and the response word. With this additional task for the *read* control, the recall rate improved to the same level of the *generation* condition. It is not likely that a relatedness judgment would strengthen the response word, but it is likely to initiate a set of cognitive operations to check the relationship between the stem and response. Those extra cognitive processes thus make the *read* condition more similar to the encoding processes acquired in the *generation* condition. The generation instruction thus forces the participants to encode an elaborated memory representation and thus avoid the "lazy" encoding that people often do in the *read* condition.

Let us now consider why incongruous generation results in better memory retention than congruous generation. We know from the Chechile and Soraci study that the storage probability is the same in the incongruous and congruous conditions, but the retrieval probability is greater for the incongruous condition. Previously, Soraci et al. (1994) found support for a multiple-cue account for generation. The multiple-cue hypothesis was born from a pilot study that we conducted with congruous and incongruous generation. In the pilot study, the participants first were provided with a very simple jigsaw puzzle, and they were instructed to continuously talk as they solved the puzzle. Next the participants were told to produce either a very related or a very unrelated word to a provided stem word. Again, it was stressed that they should talk aloud as they did this task. For the related condition, no participants produced an intermediate thought before they gave their response. However, in the unrelated condition, a rich set of intermediate thoughts occurred for all but one of the participants. For example, one participant to the stem word SHIP said,

Ship. All I can think of is water but that's very related… hmm CAT.

Another participant to the stem CHAIR said,

Chair. I see myself in the dorm on my chair. I look outside the window and picture myself outside. I look up… SKY.

Another participant said to the stem HAMMER,

Hammer. I see my workshop. There are many things to fix. Too many. HEADACHE.

Notice that these verbal responses are consistent with the hypothesis of an elaborate structure for generative encodings. Soraci et al. (1994) hypothesized that the content of this intermediate information can serve as additional retrieval cues during free recall. Because there is a greater number of intermediate items for the incongruous condition, it follows that the retrieval parameter should be larger in that condition.

A number of experiments support the multiple-cue explanation of the improved retrieval from generative encoding (Soraci et al., 1994; Soraci et al., 1999). Soraci et al. (1994) studied incongruous generation both with and without constrained encoding instructions. For unconstrained generation, the participant is free to select any suitable item that meets the instructional requirement. Chechile and Soraci (1999) as well as Roenker et al. (1978) used unconstrained generation, but most other researchers employed a constrained generation method by presenting a letter fragment that restricted the answer to a single item. Thus, with constrained generation, the same words are produced in the generation and control conditions. Three types of cue stems were studied with the constrained procedure in the Soraci et al. (1994) study. One type was a *single congruous cue* such as C__P: a policeman. Another type was a *single incongruous cue* such as C__P: not a hat. The third type was a *double incongruous cue* C__P: not a hat, not a vessel for drinking. All the target responses could be solved by three words (e.g., CUP, CAP, COP). The single congruous cue is likely to only elicit the correct solution (e.g., COP). But the single incongruous cue of "not a hat" is likely to elicit the response CAP prior to rejecting that solution and focusing on a correct solution (i.e., either COP or CUP). Moreover, the double incongruous cues of "not a hat" and "not a vessel for drinking" are likely to elicit the two wrong solutions of CUP and CAP before focusing on the correct solution of COP. In summary, the generation cuing conditions are:

- *Single congruous* C__P: a policeman
- *Single incongruous* C__P: not a hat
- *Double incongruous* C__P: not a hat, not a vessel for drinking

The correct recall rates in the *single congruous, single incongruous,* and *double incongrous* generation conditions are, respectively, .28, .35, and .43.[11] The increase in the correct recall rate with the number of potential retrieval cues is statistically significant. Consequently, incongruous generation results in better memory retention for the constrained generation procedure as well as for the previously discussed unconstrained procedure. Moreover, this effect is consistent with the expectation that more retrieval routes should increase the recall rate.

Soraci et al. (1999) also investigated the multiple-cue hypothesis in the context of *constrained-congruous generation*. Again, there were three types of constrained-generation conditions. One condition had one cue with a single semantic referent (1C1R); this is exemplified as B____: a winged mammal. Another condition had two cues but a single semantic referent (2C1R); this is illustrated by the fragment cue of B____: a winged mammal, a nocturnal flyer. The third condition had two cues and two semantic referents (2C2R) as illustrated by the example of B____: a winged mammal, a wooden club. In summary, the three congruous generation cues are:

- *1C1R 1 cue, 1 referent* B____: a winged mammal
- *2C1R 2 cues, 1 referent* B____: a winged mammal, a nocturnal flyer
- *2C2R 2 cues, 2 referents* B____: a winged mammal, a wooden club

Of course, there were also *read*-control conditions as well. All the response words were homographs such as BAT, FORK, and DATE. The pooled correct recall rates for the critical third and fourth experiments for the *1C1R, 2C1R,* and *2C2R* conditions, respectively, are .43, .45, and .57. The recall rates for the *1C1R* and *2C1R* conditions are not significantly different, but the *2C2R* condition did result in a higher recall rate. Consequently, it is the number of semantic referents rather than the number of cues that is important

11. These proportions were obtained from pooling the results of experiments 3 and 4 from Soraci et al. (1994). The results from those two studies were nearly identical.

for improving the recall rate. From the perspective of the multiple-cue hypothesis, it is the number of retrieval routes that should aid free recall, and the number of retrieval routes is likely to be greater in the case of multiple semantic referents. For the example given above, it is likely that the cue "a winged mammal" would trigger the response of a BAT as a flying creature at the time of encoding, and the cue of "a wooden club" would also trigger the thought of a baseball bat during encoding. Later in free recall, if either the category of a flying animal or the category of sports were produced, then that information would have a chance to trigger the response word of BAT. More stored cues about a target item improves the retrieval success rate.

Additional support for the multiple-cue hypothesis comes from two studies using visual rather than verbal stimuli (Carlin, Soraci, Dennis, Chechile, & Loiselle, 2001; Wills et al., 2000). Participants in the Wills et al. experiment produced finished pictures by connecting the dots. The magnitude of the generation effect was larger if participants did not know the nature of the final constructed figure. Participants were either shown in advance the final picture or were not shown the picture until after they connected all the dots. If the participants did not know the final solution, then they could potentially entertain educated guesses about the identity of the solution. These intermediate hypothesized solutions might be incorrect, but those incorrect ideas should also be encoded in memory. Later on a recall task, the additional stored guesses should increase the chance of retrieving the correct target information. If the participants did not know the solution in advance, then the recall rate was higher for the connect-the-dots condition than for the control trace condition. The study by Carlin et al. (2001) also provided evidence of the benefit of entertaining intermediate hypotheses while attempting to identify an ambiguous picture. Participants in this study viewed a dynamically changing picture. In one condition, the picture was out of focus and gradually faded into focus, whereas in another condition, the picture started in focus and gradually faded out of focus. The total viewing times for the *fade-in* and *fade-out* conditions were equal. Consistent with the multiple-cue hypothesis, the later recall rate was higher for the *fade-in* condition. Importantly, Carlin et al. also examined the recognition rate for these two conditions. The multiple-cue hypothesis is an explanation of the increase in the retrieval component of the correct recall rate. But recognition taps storage and guessing processes rather than retrieval. Consequently, it is not expected that a recognition measure would show a difference between the *fade-in* and *fade-out* conditions. There were in fact no significant differences found between those two conditions on the recognition test. Although this study was not really a generation experiment per se, it nonetheless shared some similarity of presenting a learning event in either a puzzling out-of-focus state or an understandable in-focus state. The out-of-focus state presents the person with a problem that is likely to result in the implicit generation of intermediate hypotheses that become part of the encoding of the event. These extra thoughts about the event can thus provide additional retrieval routes that can be helpful on a later recall test.

For much of the history of learning and memory research, there is an implicit mechanistic framework in which well-defined stimuli are presented to the learner who was considered to be a passive biological system that recorded the input event. From this perspective, an ambiguous, fragmentary input signal would not be the best condition for achieving an accurate account of an event. Yet we have seen evidence that the learner is an active processor of information. The learner is generating novel hypotheses and interpretations of the input information. These cognitive acts of developing predictions about ambiguous information are not a rare or unusual activity; rather, it is common for people to play a strong interpretative part in encoding information. The generation task actually places people in a more usual learning environment than does the standard learning task. It is surprising that it took researchers so long to discover the advantages of generative encoding. Generative encoding prompts the learner to elaborate hypotheses about the current puzzling environment. The additional elaboration makes the memory representation less likely to be lost due to subsequent storage interference, and it makes the items that do survive more easily retrieved.

13.3   Other Generation-Like Encoding Effects

The Carlin et al. (2001) task of *fade-in* versus *fade-out* viewing of visual pictures is not really a generation-effect study per se. Nonetheless, it shares some similarity to a generation task, and it shares some theoretical factors in common with the generation-effect studies. Two other phenomena are also generative-like. These phenomena are the *aha effect* and the *self-choice effect*. Like the generation effect, these phenomena require more from the participant, and this extra activity results in improved memory. These phenomena are thus generation-like encoding effects.

13.3.1   Aha Effect

In problem solving, there often is a long period of confusion and perhaps later a sudden insight about the solution. This moment of insight or *aha* experience is often described as a pleasant or happy event. Several neuroscience researchers have begun to identify changes in brain-scan metrics that occur at the moment of insight, although these changes do little to advance a theoretical understanding of the insight process beyond the detection that it involves a changed neural state (Aziz-Zadeh, Kaplan, & Iacoboni,

2009; Kounious & Beeman, 2009; Qiu et al., 2010; Sandkühler & Bhattacharya, 2008). The question arises about the memory of insight or aha events. Auble, Franks, and Soraci (1979) studied encoding and memory retention for puzzling sentences. They were interested in the memory of events when the person moved from a state of non-comprehension to a state of sudden insight. Participants in their experiments read sentences such as

People departed because of the sign.

This statement is ambiguous. Why are the people leaving? What is the sign? Where are the people? Who are the people? Does the word *departed* mean that the people died? How can a sign cause people to depart? Participants in the experiment would see the solution to the ambiguous statement either before or after reading the sentence. The aha effect would be expected only when the solution was withheld until after the sentence was read. For this example, the solution is the phrase **ON STRIKE**. Auble et al. reported that the recall rate was higher when the solution cue was delayed rather than when shown in advance (i.e., 18 percent for the control condition of knowing the solution in advance and 30 percent for the condition of the delayed solution).[12] Currently, there has not been a storage-retrieval analysis of the aha effect, so the underlying understanding of this effect is still unresolved. If retrieval is improved in the delayed-solution case, then the multiple-cue hypothesis for the retrieval component would be a plausible explanation of the aha effect just as it was for the generation effect. For the example above, perhaps the participant might consider other possible candidate ideas prior to learning the correct answer of the ON STRIKE sign. These other potential solutions can aid the later retrieval process because there would be additional retrieval routes in memory. The multiple-cue explanation for the aha effect is thus a similar explanation of the dynamic fade-in versus fade-out effect found by Carlin et al. (2001). Yet there might also be a storage basis for the aha effect as well. Future research on the aha effect in terms of the storage and retrieval components of memory will be needed before a more detailed understanding of the effect is achieved.

13.3.2   Self-Choice Effect

The *self-choice effect* is another phenomenon that results in improved memory retention from a generation-like encoding (Takahashi, 1991; Watanabe, 2001; Watanabe &

---

12. Auble et al. (1979) studied the percentage of the participants who could solve the ambiguous sentences and varied that factor in their experiments. The aha effect was detected for both easy and hard ambiguous sentences. The above example was not one of the sentences used by Auble et al.

Soraci, 2004). Suppose a person is presented with a word triplet such as **HAMMER WRENCH SAW**, and the person is asked to circle one of the words for a later memory test. The counterpart control condition consists of the same words but with one of the words circled already. The participant in the control condition is instructed to trace a line around the circled word and is told that there will be a later memory test. The experimental and control conditions are labeled as the respective *choice* and *forced* conditions. Takahashi (1991) consistently found improved memory retention for the *choice* condition. Watanabe (2001) did a number of experiments to study cuing in the context of this effect. For his first experiment, Watanabe used the same type of stimuli studied in the constrained-generation experiments by Soraci et al. (1994) (i.e., CUP CAP COP), but there were no word fragments for any of the conditions. In the *choice* condition, the participants saw three words and read two cues to guide their word choice. The participants had to write their chosen word in a blank space. The corresponding *forced* encoding trial was identical except that one of the words was underlined. The task in the *forced* condition was to read the cues and write the underlined word in the blank space. For one of the conditions, the two cues reduced the choice to two of the three words. In another condition, the two cues reduced the selection to a unique item. In the last cuing condition, the two cues did not cause any of the words to be eliminated from consideration. For each type of cuing, the *choice* condition resulted in a higher recall rate than the *forced* condition. Yet in the double-rejection cuing condition, the participants really did not have a choice. Nonetheless, they did not know this fact in advance. They had to read and process the cues before deciding on the solution. Watanabe's second experiment showed that the recognition rate also resulted in a choice effect for each type of cue. In experiment 3, Watanabe found that the choice effect resulted in improved memory retention for both the chosen item and the nonchosen items. Thus, the self-choice effect is a robust phenomenon that does not depend on the type of cuing.

The fact that the choice effect even occurs for the double-rejection type of cuing means that the effect is not due to an inherently stronger item being selected in the choice condition because both the *choice* and the *forced* conditions are constrained ahead of time to the same word. Consequently, a strength hypothesis is unlikely. Let us see how the task requirements for the *choice* and *forced* conditions can result in a different amount of propositional elaboration. Suppose for this hypothetical example, the word triplet is **HAMMER WRENCH SAW**. First let us examine the case when there is double-rejection cuing (i.e., suppose the cues are: Not a tool for cutting. Not a tool for turning bolts). The person after reading these cues would choose the only available option (i.e., **HAMMER**).

The reasoning might be like the following hypothesized thoughts:

Choice Condition *Double-Rejection* Hypothesized Thinking:

HAMMER, WRENCH, SAW. They are all tools. "Not a tool for cutting." A saw is a cutting tool, so it cannot be SAW. "Not a tool for turning bolts." A wrench is used for turning bolts, so it cannot be WRENCH. That leaves HAMMER as my choice. I need to write HAMMER in the space.

Forced Condition *Double-Rejection* Hypothesized Thinking:

HAMMER, WRENCH, SAW. HAMMER is circled. I need to write HAMMER in the space. "Not a tool for cutting." "Not a tool for turning bolts."

Note in the *choice* condition, the hypothetical person is checking the word list after each cue and processing the implications of the cue. But in the *forced* condition, there is no need to process the implications of the cues or to check back with the word list. Thus, the *choice* condition requires more processing than the *forced* condition. Alternatively, the *forced* condition can result in a lack of elaborative reasoning (i.e., it can result in relatively lazy processing).

The above discussion of the choice effect for the Watanabe initial cuing experiment can be expanded to include other types of cuing that have also been used in choice-effect studies. In each case, the propositional encoding for the *choice* condition is more elaborate than the corresponding control condition. But it is important to remember that the choice effect can also occur when there are no cues (Takahashi, 1991). Let us consider how propositional encoding might differ for the *choice* and the *forced* conditions when there are no cues.

Choice Condition Hypothesized Thinking without Cues:

HAMMER, WRENCH, SAW. They are all tools. I use a hammer the most often, and it is alphabetically before WRENCH and SAW, so it is easier to remember later. I need to write HAMMER in the space.

Forced Condition Hypothesized Thinking without Cues:

HAMMER, WRENCH, SAW. HAMMER is circled. I need to write HAMMER in the space.

Without knowing in advance the solution, each word on the list is attended to more frequently in the *choice* condition, and there is greater processing of common features shared by the words. There might also be additional unconscious activation of other tools when the person notices that each of the words is a tool. For example, perhaps the word SCREWDRIVER was implicitly activated in the *choice* condition. This greater degree of elaboration in the *choice* condition should make the encoding of the chosen word more distinctive in memory from other words on the list of memory words. But it should also augment the encoding of all the words, even the nonchosen words. This greater degree of elaboration should result in less storage interference compared to the relatively "lazy" level of processing for the selected word in the *forced* condition.

If there is a greater degree of processing of other nonpresented items, such as the word SCREWDRIVER, then it is possible that people will falsely think that SCREWDRIVER was on the list. Roediger and McDermott (1995) found a false memory effect in a traditional memory learning procedure when a number of words on the list came from the same category, but the list did not contain the common associate itself. For example, suppose on a standard learning experiment the person hears the words BED, REST, AWAKE, TIRED, DREAM, and SNOOZE. Roediger and McDermott found that people tend to think incorrectly that the word SLEEP was also on the list. When a word is perceived, the other associates of the word are also implicitly activated. Remember the previous discussion of unconscious priming of associates that occurs for a lexical decision task (Meyer & Schvaneveldt, 1971; Onifer & Swinney, 1981). For example, if a word like STOCK is heard, then all the associates of the word show a priming effect even if the associate is not connected to the current contextual meaning. For example, Onifer and Swinney showed a faster lexical decision to a word like COWS even for sentences in which the word STOCK meant a share of a company. In memory studies, Roediger and McDermott found that if enough implicit activations occurred for the nonpresented critical lure by the presentation of associates to the word, then there is a strong tendency for the participant to think falsely that the critical lure was also on the list. This false memory effect occurs for both recall and recognition tasks.[13]

Watanabe and Soraci (2004) wanted to see if false memories would occur with the choice-effect procedure. They presented word triplets with a single congruous cue to constrain ahead of time the choice to a unique word. For example, if the triplet were **HAMMER WRENCH SAW**, then the cue "A tool for hitting nails." would uniquely identify HAMMER as the solution. But the participants in the *choice* condition would not know in advance this fact, whereas in the *forced* condition, the word HAMMER was circled. Later, the participants were presented with a list of words, and they were asked to circle any items previously studied. On the list were old nonchosen words from the study phase such as WRENCH and SAW. The list of words also included lures

13. This false memory is a robust effect that has been extensively researched. But it is only mentioned here because of the prediction that it should also occur to a greater degree in the *choice* encoding condition. The procedure of presenting similar items and later testing for the nonpresented critical lure that was associated with list items is a method that has come to be called the *DRM* procedure after the initials of the three researchers who first discussed the effect. The researchers are Deese (1959) and Roediger and McDermott (1995). This false memory effect is also discussed in the glossary.

unrelated to any of the words previously studied. The hit rate for detecting list items was 71 percent in the *choice* condition, but the hit rate was significantly lower in the *forced* condition (i.e., 37 percent). Thus, there was a strong choice effect for detecting nonchosen words that were previously shown among the studied word triplets. Importantly, among the words on the list were related nonpresented words like the word SCREW-DRIVER. Watanabe and Soraci reported that there was a significantly higher false-alarm rate for these related words in the *choice* condition (21 percent) compared to the corresponding false-alarm rate in the *forced* condition (14 percent). But the 14 percent rate for the *forced* condition was a significant false memory effect because the false-alarm rate for unrelated words like RUBY was only 2 percent. Thus, a related word, which was not presented on the study list, elicited a false memory effect in both the *choice* condition and in the *forced* condition, although the effect was larger for the *choice* condition. The increase in the false-alarm rate for the *choice* condition is support for the trace-elaboration account for the encodings produced with the choice procedure. Thus, the *choice* encoding condition results in a greater memory retention, but it also results in a greater false-alarm rate. This pattern has also been detected with some other studies that used the DRM procedure (Toglia, Neuschatz, & Goodwin, 1999).[14]

It is reasonable to speculate that there is a higher storage probability for the items encoded in the *choice* condition compared to the items encoded in the *forced* condition because the encoding process results in a more elaborated memory representation. If there is a more elaborated encoding, then the storage inference effect that occurs for recently studied items should be reduced because the encoding events are more dissimilar. Nonetheless, there has not been an experiment to date that has examined the choice effect with an experimental design that enables the estimation of storage and retrieval probabilities. Consequently, more research on the choice effect will be needed to see which fundamental processes are different between the *choice* and the *forced* conditions.

13.4   Generative Learning in Education

One way to look at the generation effect is as an enhancement of memory retention that results from the learner's problem-solving engagement. But another way to view this research literature is to focus on the low level of encoding that results from passive reading. Normal reading and studying is usually highly ineffective for building a

---

14. Although there is a greater false memory rate for the choice-effect procedure, the standard generation-effect procedure did not result in an increase in the DRM false memory rate (Soraci, Carlin, Toglia, Chechile, & Neuschatz, 2003).

lasting representation. Studies of the memory for prose passages have demonstrated that much of the idea units in the text are not remembered on an immediate memory test (Kramer & Kahlbaugh, 1994; Meyer & Rice, 1981). For example, in the study by Meyer and Rice, the recall rate for idea units within a prose passage varied from 28 percent to 48 percent depending on the level of the idea. A high-level fact corresponds to a gist-level concept, whereas a low-level fact is a specific piece of information from the text. In the Kramer and Kahlbaugh study, the recall rate for idea units varied from 5 percent to 25 percent, depending on the age of the participant and the formal structure of the prose passage. They also found that the average percent correct on true/false questions was only 78 percent; this performance level is particularly troubling because the guessing rate for these questions is 50 percent. Given the low rate of memory retention for generic prose, it is reasonable to worry about the retention rate of college textbook information, which is likely to be much more technically challenging. In some courses, the professor requires homework to force students to put the information to use, but there are many cases where the professor wants the student to read a chapter or an entire book. The memory representation of this information is likely to be too low if the material is just read. The educational problem is not limited to inefficient reading of textbooks because there is also a problem with passive listening to a lecture. Passive listening is like passive reading (i.e., both result in a substantial loss of information).

There is no reason why a reader cannot be more generative. By adopting an improved reading strategy, it should be possible to make the encoding from reading similar to the elaborative encoding established from a generative task. The experiments on the generation effect have controlled the total processing time available for encoding, but the people in the *read* condition have not spontaneously taken the opportunity to elaborate the encoding in a fashion similar to the encodings created in the *generation* condition. Such elaboration should improve memory and student learning. A study by deWinstanley and Bjork (2004) demonstrated that students can be quickly taught to be more generative in reading. These investigators required students to read prose passages consisting of phrases that were presented one at a time. Each phrase contained a critical word that was printed in red. All the other words in the phrase were in black. For half of the phrases, the critical word was intact, whereas for the other half of the critical words, there were missing letters. For each case, the student had to write the critical word in a booklet. The critical words for consecutive phrases alternated between the *generate* and the *read* conditions. The students were given a recall test for the critical words after a 2-min retention interval. The test showed a significantly higher recall rate for the words from the *generation* condition. Because of the critical word test, many of the students became aware of their better performance in the *generative* condition. Next, deWinstanley and Bjork asked the students to learn a second passage about a different topic. Again, the critical words alternated between the *read* and the *generation* conditions. The later memory test for the second set of phrases showed that there was no difference between the *read* and *generation* conditions. The students had learned to encode the critical words in the *read* condition in a more elaborative fashion. The deWinstanley and Bjork study thus illustrates that students can learn to avoid passive reading with some training. Donaldson and Bass (1980) also found that requiring people to make a relatedness judgment between key words caused the control *read* encoding to improve to the level of the *generation* condition. Thus, reading does not have to be passive.[15]

13.5   The Enactment Effect

Historically, there has been a strong inclination in memory research to focus on the learning of perceptual events. Investigators typically required people to learn a list of words or pictures. Yet in everyday life, the memorization of unrelated stimuli is not common, even for students. Rather than memorizing a set of grocery items, we more likely write down what we need on a list before we go shopping. The list itself is our stored representation. Also, our daily tasks often require actions, and we need to remember how to execute those actions. In the early 1980s, three independent research laboratories began to study the memory for actions (Cohen, 1981; Engelkamp & Krumnacker, 1980; Saltz & Donnenwerth-Nolan, 1981). These experimenters

15. I cannot resist making the comment that some current university practices have made lectures more *read-like* rather than generative. Smooth PowerPoint presentations can be too polished for effective student learning, especially when the slides are packed with information. A masterful lecture from an erudite professor is an impressive performance, but do the students learn and remember the detail points, and do they develop demonstrable skills? The essence of the Socratic method is to engage the audience in a debate around problems. It is the opposite of a lecture where the professor does all the work while the students listen and watch. Toward that end, I found it beneficial even when teaching statistics to withhold answers or critical content and instead query the class for their solution. A student proposal might be in error, but a discussion as to why it is mistaken is particularly beneficial for student reconceptualization. In essence, education can be improved by forcing the students to be more generative and more engaged with the topic. Flexible lecturing and debate with built-in mini tests evokes both generative encoding and the benefit of testing. The class debate also helps develop student reasoning skills. Empirical support that this generative style improved student comprehension and memory for the content of a statistics course was documented in a number of papers (Cohen, 1998; Cohen & Chechile, 1997; Cohen, Chechile, & Smith, 1996; Cohen, Chechile, Smith, Tsai, & Burns, 1994; Cohen, Smith, Chechile, Burns, & Tsai, 1996; Cohen, Tsai, & Chechile, 1995; Smith et al., 1997).

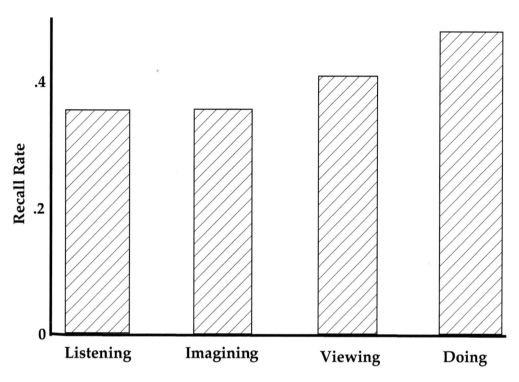

Figure 13.2
These results are adapted from a paper by Engelkamp and Zimmer (1985). The means were computed across six different experiments that were separately discussed. The final correct recall rate is shown for each encoding instruction (i.e., listening, imagining, viewing, or doing).

instructed people to remember events that consisted of propositions such as "open the drawer" or "pour the coffee." The participant might be asked to do any of the following tasks: (1) listen to a statement, (2) imagine doing a statement, (3) watch someone else doing a task, or (4) enact the stated instruction. The consistent finding is that the best memory retention occurs when the person acts out the instruction. See figure 13.2 for the overall average recall rate for these encoding instructions. The enactment effect refers to the increase in memory retention from the acting of the stated task. This effect occurs for both recall and recognition memory (Mohr, Engelkamp, & Zimmer, 1989; Zimmer, 1991). There are a number of reviews of the research in this area (Chatley, 2013; Engelkamp & Zimmer, 1989; Nyberg, 1993; Peterson, 2009; Zimmer et al., 2001).

Some enactment-effect researchers have claimed that they were studying a different system of episodic memory because of the involvement of motor activity with the encoding (Cohen, 1981). The argument is based on the failure for finding some effects that were previously found with standard verbal memory. In essence, the argument is another example of the experimental dissociation approach. But as shown previously in this book, experimental dissociations are meaningless. No conclusion can be based on finding an experimental dissociation. Model-based measurement is required to understand the underlying cognitive processes because the available dependent variables tap multiple cognitive processes and because independent variables can affect more than one latent cognitive component of memory. The history of experimental psychology is replete with examples where researchers initially argued that some new experimental procedure was tapping a different memory system, but later evidence was advanced that undermined that claim. For example, we have seen previously that the multiple-store model of memory was incorrect in assuming that the storage susceptibility is dichotomized with no loss for items in the long-term store and 100 percent loss for the items in the short-term store (Chechile & Ehrensbeck, 1983). With model-based memory, a case was made that all memories are susceptible to storage loss, but some memories nonetheless survive for a very long time. We have also seen that implicit memory was not a separate and independent storage system from explicit memory because information that once was explicit can degrade to become an implicit memory (Chechile, Sloboda, & Chamberland, 2012). Moreover, we have seen that information from episodic, event memory can generalize to become a semantic memory representation that is no longer tied to specific temporal events. Also, information that was once in semantic memory can be lost from memory storage. Consequently, it is not surprising that the claim of a separate system for enacted memories was not universally supported by other researchers in this area. For example, Nyberg (1993) has highlighted the similarities rather than the differences between enactment-based memories and verbal memories. Thus, the experimental dissociations in the enactment area have not been replicated. Consequently, there is no strong basis to claim that the encodings established with gesturing are in a separate memory system.

It is important to stress that it is not surprising that people can remember an extensive sequence of actions. There are ample examples from the performing arts of impressive feats of motor memory. Since antiquity, actors in theatrical plays have learned how to recite many pages of dialogue along with the accompanying action. Ballet dancers and concert soloists implement an intricate and lengthy sequence of motor responses. But this type of stunning action memory is not what the researchers in the enactment area are studying. Their focus instead is on how simple gestures can improve the memory for verbal statements that could be remembered without the gestures. In essence, enactment research simply uses a procedure of comparing the memory of commands that are either enacted or not enacted. Thus, it is not the study of motor memory per se. In fact, other experimental procedures for studying memory also involve a motor component. For example, researchers in the area of the generation effect require participants to write or type the target word. This step is done to be certain that the person has solved the puzzle or has properly read the target item. For both the generation and the enactment procedures, the participant must initiate and guide the motor response. But the motor response of typing the target word is not the same as enacting a command such as "point to the door." The typing response is the spelling of a single item, whereas the acting response is a gestural representation of an entire command. The acting response is equivalent to a verbal phrase or even a few sentences. For the enactment case, the action chunks the whole propositional relationship of the episodic event into a single coherent action, whereas the typing response is not interpretable as a concept or a command. Thus, a key difference between the generation and the enactment procedures is the nature of the *content* of the motor component. For the enactment procedure, the action is a redundant alternative memory representation to the verbal command (i.e., another representation of the episodic event but in a different format). In contradistinction to the enactment case, the motor component of a typing response is similar to the response for any other typed message. Moreover, it is unlikely that people actually have a pure motor memory representation of the typing movements that survives past the span of a second or less.

Previously, it was argued that the generation effect occurs because the memory elaboration that results from solving a problem enables the memory to have potentially both storage and retrieval advantages. The storage advantage occurs because the elaborated encodings for the different generated targets are more dissimilar to one another, so there was less storage interference. The retrieval advantage occurs when the learner considers a possible but incorrect solution before coming up with a suitable solution. The incorrect intermediate candidates are nonetheless helpful because they provide alternative retrieval routes on a free-recall test. As discussed previously, Chechile and Soraci (1999) showed that generative encoding in fact results in positive changes for both storage and retrieval. Because enacted memory encodings are elaborated representations relative to the memory for verbal commands, it is thus reasonable to also expect that both storage and retrieval are involved with the enactment effect. The two representations of enacted memories should benefit storage for several reasons. One reason for enhanced survival is because there is a memory backup. Consider the case when the main electrical power fails for a building that has a backup generator. Whenever the main power fails, the backup generator comes to the rescue. Systems are more reliable when there are backup systems. An enacted memory representation is a backup representation, so it should similarly be a more robust encoding and less susceptible to storage loss. Another mechanism to reduce the storage loss for the enacted memories is the action itself elaborates the representation; thus, the encodings are more likely to be distinctly different from one another. If the memories are dissimilar, then we expect less storage interference among other memory targets as well as less interference from extra-list items. Either of these mechanisms would augment the chance that the enacted memories survive the storage interference process. There are also several reasons why enacted memories should be more successful in memory retrieval. First of all, the extra representation of the action provides an alternative retrieval route for recovering the verbal message. In general, additional retrieval routes increase the likelihood of a successful memory search. Moreover, actions might have some organizational similarities that could further improve the retrieval probability. For example, among the commands, perhaps there is a subset of actions that involve the participant's hands (e.g., "draw a circle," "cross your fingers," "clap your hands"). These separate commands can be regarded as a hand-movement category. Later in free recall, the recovery of any one of these memories in this category would be a powerful retrieval cue for the recall of other memories in the category. Hence, there are solid reasons to expect that storage and retrieval processes are enhanced with the enactment of a verbal command. But at the current time, no storage-retrieval separation measurement model has been used to study the enactment effect.

13.6   Distinctiveness and the Production Effect

The production effect occurs as the result of a particular type of motor response (i.e., the reading aloud of some select items on a list compared to the silent reading of the other list items) (Gathercole & Conway, 1988; Hopkins & Edwards, 1972; MacDonald & MacLeod, 1998; MacLeod, Gopie, Hourihan, Neary, & Ozubko, 2010). The initial studies in this area found that the improved memory was restricted to the case where there were mixed lists (i.e., a list with some items that were read aloud, whereas the other items were read silently). If there were pure lists (either all items read aloud or all items read silently), then the production effect was not detected (MacLeod et al., 2010). Researchers argued that the items read aloud were more distinctive and hence more memorable. But the concept of distinctiveness is a slippery idea; at first it seems clear, but as soon as one begins to study it carefully, there are more questions raised than answered. So let us first scrutinize the concept of distinctiveness to understand its problems as an explanation for memory retention effects.

13.6.1   Distinctiveness Defined as a Contrast

In general, the term *distinctiveness* refers to the property that makes an entity stand out from all of the other entities. A simple example would be a single object such as a sailboat seen against the homogeneous visual background of the ocean. Attention is immediately drawn to the sailboat; it perceptually pops out. Remember that sensory neurons are designed to detect changes. Homogeneous stimulation of sensory neurons leads to sensory adaptation and the lack of a neural response. The sailboat is different from the rest of the visual field, so it is a visual change, and it is perceptually distinctive.

Although the concept of perceptual distinctiveness is clear, the application of the concept to the learning of items in a list is a more subtle matter. For example, consider the following two lists of items:

List 1:   74   23   18   DUQ   65   97   32
List 2:   ZAM   RAJ   HEG   DUQ   TIV   YOX   SOU

Undoubtedly, we detect DUQ on list 1 as the different item, but that same item is not distinctively different from the other items on list 2. Yet the letter triad DUQ is unique on both lists. Why is DUQ distinctive on list 1 but not on list 2? To be distinctive requires two attributes. First, the other stimuli need to be easily grouped, and second, the target stimulus needs to be encoded very differently from what is common among the other background items. The distinctive stimulus can also be a number or a word from a different encoding category, as can be seen from the following two examples:

List 3:   ZAM   RAJ   HEG   59   TIV   YOX   SOU
List 4:   GRAPE   CHERRY   PLUM   LAMP   APPLE   PEACH   PEAR

For both lists 3 and 4, there is an isolated item that satisfies the two criteria for distinctiveness. In general, there is better memory retention for a single isolated item in a list; the phenomenon is called the *von Restorff effect* in recognition of the classic paper on the topic (von Restorff, 1933). Interestingly, there were earlier papers that found a similar effect, but these investigators attributed the effect to the *vividness* of the isolated stimulus (Calkins, 1894, 1896; Jersild, 1929; Van Buskirk, 1932). But von Restorff challenged the idea of perceptual vividness and introduced the more cognitive and configuration-based idea of distinctiveness. Distinctiveness is not a property of the stimulus per se as was the implication of the concept of vividness, but instead distinctiveness was seen as a property of the relationship between the stimulus and all the other stimuli. Notice also that distinctiveness in memory does not require all the stimuli to be presented together at one time. For memory studies, the items are usually presented in a serial order, but still the isolated item stands out, and it is better remembered.

13.6.2 Problems with Distinctiveness

Distinctiveness has come to be used as an explanatory factor by some memory researchers; nonetheless, few have thought systematically about what the idea really means. But the Hunt (1995) paper is an exception to this general characterization. Hunt notes that few researchers have studied the von Restorff paper carefully, probably because the paper was never translated from its original German. Hunt also believes that this omission in scholarship has led to a tendency for circular reasoning (i.e., explaining retention changes as a consequence of distinctiveness while at the same time using memory effects as a sign of distinctiveness differences). Hunt does believe that distinctiveness is a theoretically useful idea, but he warns of problems in the use of this subtle construct. Yet in my opinion, there are more troubling aspects of distinctiveness than the points raised by Hunt. These issues will be explored in the subsequent analysis.

Although the above examples appear to make the concept of distinctiveness clear, a closer analysis reveals some of the difficulties with this idea. To begin, there are problems with the initial step of von Restorff using distinctiveness in a memory sense and her deliberate rejection of the idea of vividness. One stimulus can be vivid, but one item cannot be distinctive because distinctiveness requires a background set of items. Yet one memory item can be presented so that it is more detectable by varying the physical intensity or by varying the duration of the presentation. An item flashed in 20 msec is less vivid than the same item displayed for 1,000 msec. An auditory memory item would be more vivid if articulated at a 60-decibel level rather than at a 20-decibel level. A visual item can be displayed with a degree of focus (i.e., from being sharply in focus to being blurred out of focus). Thus, vividness can vary for a single item in either a memory sense or a perceptual sense, but distinctiveness cannot be applied for the memory of a single item. Hence, vividness has an advantage over distinctiveness in the case of single memory targets. Distinctiveness can only be possible for memory when there is a list of three or more items. If there were only two items on a list and the two items were different, then distinctiveness would still be ambiguous because either of the two items might be regarded as distinctive.

To avoid circularity in the use of distinctiveness, it would be important to have a measure of the distinctiveness level of an item prior to its use in an experimental study. Heretofore, we saw that *meaningfulness* values were available ahead of time because of the preexisting meaningfulness norms.[16] The question arises, can we similarly measure distinctiveness in advance and thus avoid circular reasoning? The fact that there are many norms for meaningfulness but currently no norm for distinctiveness is a tipoff of the likely answer to this question. The problem is that distinctiveness cannot be defined in the absence of at least two other items; individual items do not have distinctiveness apart from the context of all the items. Nonetheless, some might argue that it is unnecessary to measure the distinctiveness in order to construct lists of stimuli in which there is a distinctive item. After all, we have already done that in the construction of lists 1, 3, and 4. But the inability to measure distinctiveness means that a host of questions cannot be addressed. For example, the following questions are pertinent issues that are unanswerable due to the failure of measuring distinctiveness ahead of time:

• Is distinctiveness all-or-none or does it have a continuous value?
• How many items can be distinctive on a list?
• What is the proportion of distinctive items possible?
• Can a homogeneous list of items from different word categories have a distinctive item due to its position within the list?
• If encoding is a dynamic process, then when does the distinctiveness of an item emerge?
• Why did the encoding of some items result in a distinctiveness difference?
• Is distinctiveness stable or does it change as a function of the retention interval and the type of information presented in the retention interval?

If the failure to address the above questions were not troubling enough, then we also need to face the fact that distinctiveness by itself cannot explain a difference in the memory retention rate. Distinctiveness is not a cognitive process. At best, distinctiveness is a property of some items relative to some other items. But how does the difference among the list items translate to different rates of either storage retention or retrieval success? This question must be answered in order to use distinctiveness as a factor for explaining a memory effect. It is insufficient to simply state that some items were more distinctive, and thus those items were better remembered. How did their distinctiveness result in improvements for either the storage or retrieval rate of information? There are several theoretical reasons why distinctive items could have a

16. Actually, there were a number of meaningfulness metrics based on different operational definitions, and these metrics were not perfectly correlated with the ease of learning or with each other. Yet any particular norm provided a method for constructing, in advance, a stimulus list of items that reduced the variability among the items on the list. Also, the norms provided a method for constructing lists of differing meaningfulness levels.

higher rate for the probability of storage, and there are also several different theoretical processes that could result in a higher rate of information retrieval for the distinctive items. Hence, the finding of superior memory retention for distinctive items is not a surprising effect. Without identifying the underlying cognitive processes involved, the memory effect is not explained. This point can be better understood by elaborating some of the storage and retrieval mechanisms that could be altered by having some distinctive items embedded among a larger set of other items.

13.6.3   Possible Storage Factor for Distinctive Items

At least two possible mechanisms could result in a higher storage probability for distinctive or isolated items. The first mechanism is related to the phenomenon of *proactive interference* (PI) (Butler & Chechile, 1976; Chechile & Butler, 1975; Wickens, 1972, 1973). This point can be illustrated by list 4. For that list, the first three items belong to the fruit category, but there is a category shift for the fourth item (i.e., LAMP). When items are encoded along a common basis, it has been shown to result in progressively poorer memory. *Proactive interference* is the increased difficulty for encoding and remembering the current item because of the interference effect from previously learned similar items. Proactive interference is said to be *released* if there is a shift to a different encoding type, for example, shifting from the fruit category to the word LAMP. Proactive interference has been shown to occur for both storage and retrieval processes (Butler & Chechile, 1976; Chechile & Butler, 1975). Figure 13.3 illustrates the buildup and release of proactive interference. The data displayed in the figure come from Chechile and Butler (1975). That experiment used the Chechile-Meyer test protocols to obtain separate storage and retrieval probabilities for the overall correct recall rate. The encoding order refers to the successive trials for encoding target items presented and tested by means of the Brown-Peterson task. The four encoding order positions are thus four successive, separate Brown-Peterson trials where each trial had its unique memory target items for learning. For the first three trials, the target stimuli were selected from the same word category, whereas for the fourth trial, the target stimuli were from a different word category.

Following an idea introduced by Wickens (1972), an approximate estimate of the magnitude of the release of proactive interference can be obtained from the formula $\frac{c_4(new) - c_4(old)}{c_1(old) - c_4(old)}$, where $c_1(old)$ and $c_4(old)$ are the correct recall rates on the old encoding basis for the respective first and the fourth event positions and where $c_4(new)$ is the correct recall rate in the fourth event point when there is a shift to a new encoding basis. See table 13.2 to see estimates of the magnitude of the proactive interference release; these values were gleaned from a variety of research reports. It is important

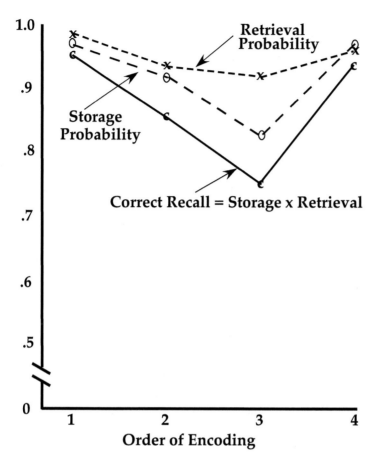

Figure 13.3
The buildup and release of proactive interference as a function of serial order. These data are replotted from the Chechile and Butler (1975) study that obtained separate storage and retrieval components of the overall correct recall rate by using the Chechile-Meyer experimental task.

Table 13.2
Estimated proactive interference (PI) release

Encoding Dimensions	PI Release Estimate
Words-number shift[a]	.98
Auditory-visual shift[b]	.97
English-Spanish shift[c]	.60
Word polarity shift[d]	.56
Gender shift[e]	.50
Shift in # of syllables[e]	.19

Note: The input data for the PI release calculations were estimated from information in papers by (a) Reutener (1972), (b) Wittlinger (1967), (c) Goggin and Wickens (1971), (d) Wickens and Clark (1968), and (e) Wickens (1972).

to observe that a shift in a number of stimulus properties results in a reduction of the level of proactive interference. Consequently, isolated items within a single memory list might be better remembered due to the release of proactive interference for the isolated items. Because proactive interference buildup and release effects occur for both storage and retrieval, it follows that one possible mechanism for improved storage for the distinctive items is as a consequence of a reduction in the level of proactive interference.[17]

Another potential storage mechanism that might underlie the distinctiveness advantage is the effect of a greater degree of trace elaboration. For example, suppose that a subset of the items on a list are elaborated by reading those items aloud. That motor

17. It is outside the scope of the current discussion to fully explore what is going on with the "release" of proactive interference. Wickens developed and used the term *release of proactive interference*, but this terminology implies that proactive interference has been removed. But has the interfering effect from similar previous learning been eliminated by a single release event? That is, suppose after the "release" event that the encoding returned to the original basis, then would the proactive interference be like the first time with that encoding (as would be implied by a release concept) or would the proactive interference be like the event just before the release event? Surprisingly, this question has not yet been addressed in the research literature. If proactive interference returns to the same level as before the "release" event, then it is clear that the proactive interference effect of earlier learning has not been removed. At the time of writing this book, the above question is in the process of being explored in my laboratory. Although the data have yet to be submitted for publication, the evidence is clear that returning to the original encoding category results in a continuation of PI. Thus, the phrase "release of PI" is an inaccurate description; it would be better to use instead the term *interruption of PI*. One trial of studying information from a different encoding category is an insufficient basis for removing the proactive interference effect.

activity, along with its auditory output, makes those items distinctively different from the other list items that are silently read. But it also makes those traces elaborated to a greater degree than the memory representations of the other list items. It is reasonable to expect that this elaborated subset has an improved chance to survive storage interference for several reasons. First, the elaboration can be a backup representation. With a backup representation, the hazard risk of storage loss due to storage interference is reduced. Second, elaborated items are also likely to be more dissimilar from one another, so there should be less storage interference. The vigor of the storage interference process is likely to be a function of the similarity between the target memory representation and the representation of other subsequent events. If items within a list are more dissimilar, then the storage interference is assumed to be reduced. Thus, elaborated memory traces are less likely to be similar to one another, so they are more likely to survive the storage interference process.

A key reason for delineating the above two storage mechanisms is to illustrate that a claim that some items are distinctive could result in a possible storage basis for the distinctiveness effect. Both of the two storage mechanisms discussed above cannot be ruled out without some compelling evidence. In fact, other storage mechanisms might also be possible. There are many ways in which a subset of items can be distinctive, and in each case, the theoretical account of the memory difference could be due to either storage or retrieval processes. Without using a storage-retrieval measurement method, the finding of a distinctiveness effect cannot be identified. But even in the case where it is known that distinctiveness alters trace storage, further research is still needed to identify the actual storage mechanism. For example, clever experiments would be needed to test the storage proactive-interference hypothesis versus other rival storage accounts.

13.6.4 Possible Retrieval Factor for Distinctive Items

When there is a distinctive subset of items stored in memory, it is possible that there is a benefit for retrieving those items. But as with the storage case, multiple possible mechanisms could underlie a retrieval-based distinctiveness advantage. One possible mechanism relates to the benefit of restricting the size of the search set (Shiffrin, 1970). The intuition underlying this idea is straightforward. Suppose that you are looking in your house for your missing car keys. The search is likely to be more successful if the search can be narrowed to a specific room. It is easier to find a specific item when you are able to take advantage of some knowledge to immediately rule out many locations. The recall task is similarly a search process driven by hypotheses that can guide the search. If the focus of the search is simply the recent temporal context, then there could be many activated features and items. Information experienced prior to the experiment, information from the prior experimental trials, information from the instructions, and other extraneous thoughts along with the desired memory targets are all part of the recent temporal context. The desired stored memory targets are embedded within a larger pool of incorrect candidates. But suppose that the person knows that a subset of the words come from a particular word category, then that knowledge can narrow the search of candidates and thereby improve the chance of correct recall recovery.

Another mechanism for a retrieval-basis distinctiveness effect is related to the case where the memory traces are distinctive because of a greater degree of trace elaboration.[18] An elaborated trace has more components that can improve its chance of being sampled in a random memory search. Once retrieved, the trace might be updated (i.e., further elaborated). Moreover, in a random search, it is also possible that a later recovery attempt resamples this same trace again. The vivid distinctive targets tend to be selected over the nonelaborated and less distinctive items. This relative contrast effect and competition would be bypassed by a recognition task, but on a recall test, the elaborated/vivid traces absorb much of the search effort. Hence, there is a possible retrieval advantage for the elaborated traces and an interference effect for the nonelaborated traces.

13.6.5 How Should Distinctiveness Be Used?

The idea of distinctiveness is not sufficient as an explanation of some memory effects because it does not adequately account for the reasons for the effect. Nonetheless, the idea of distinctive features within a memory trace is an important theoretical concept. As we have seen previously, the memory trace has many components and aspects that can vary dynamically after the initial encoding. Some features can be lost or altered while others are intact. Also, the encodings themselves vary in terms of their adequacy. Poststudy events can further alter the content of memory. In some cases, there might be something stored about a specific event, but that information might only include some general features that are insufficient to identify the details about the original event. But if the features of the encoding are sufficient to identify the target event from all other events stored in memory, then the event is said to be sufficiently stored. The features of the trace that differentiate that encoding from all others can also be said to be a set

---

18. This argument is similar to the idea that retrieval competition favors the "strong" memory traces over the weak traces. But in this book, I have avoided using the concept of memory strength because it is not needed when there is evidence that there is trace elaboration.

of the distinctive features or components.[19] The distinctive features are the subset of all the stored components that encode the explicit information of the experience or event.

In the implicit-explicit separation (IES) model of Chechile et al. (2012), there are four nonoverlapping states for the representation of any target event. These states are (1) explicit storage, (2) implicit storage, (3) fractional storage, and (4) nonstorage. The memory representation that supports explicit storage is a set of distinctive features. For the explicitly stored state, there can also be redundant information and unimportant incidental information stored as well as the essential distinctive features. A loss of any of the unnecessary components would not alter the explicit storage of the event provided that the distinctive components are still available. But the loss of some or all of the distinctive features results in the representation being insufficiently stored. Nonetheless, there may still be enough other stored components to support either implicit or fractional memory. When the information is either implicit or fractional, it cannot be correctly recalled, and it would not even be confidently recognized. Consequently, the idea of distinctive features is a valuable theoretical construct rather than using the notion of distinctiveness for a subset of memory traces within a larger set of items on a list.[20]

13.6.6 The Production Effect

As mentioned earlier, the production effect is the retention benefit that comes from reading list items aloud as opposed to silent reading. The study by Hopkins and Edwards (1972) was an initial paper in this research area. These investigators found the production effect for item recognition when there was a mixed list of words with some items vocalized and some silently read. But they reported not finding the production effect when there were pure lists of items that were either all vocalized or all silently read. Hopkins and Edwards did not use the term *production effect* as a label for this phenomenon. Rather, the label was first coined 38 years later by MacLeod, Gopie, Hourihan, Neary, and Ozubko (2010). MacLeod et al. essentially found the same results as Hopkins and Edwards. The production effect has also been found for recall tests (Icht, Mama, & Algom, 2014; Jones & Pyc, 2014). Moreover, several investigators have

19. Previously, I argued for a general representation of a target event as a pair of traces T and T' that consists of a set of stimulus features, contextual features, propositions, associations, time tags, and event-binding features. See the illustration in figure 11.6.
20. The idea of distinctive features versus gist knowledge is also an essential notion in the *fuzzy-trace theory* (Brainerd & Reyna, 1990). However, in the fuzzy-trace theory, one of the two traces contains the verbatim information, and the other trace contains gist information, whereas in the two-trace hazard model, both of the traces associated with a target event contain distinctive and incidental features.

expanded the domain of the production effect. For example, Forrin, MacLeod, and Ozubko (2012) found evidence that typing, writing, spelling, mouthing, and whispering all improved memory retention relative to silent reading. Jamieson and Spear (2014) found that typing and imagining typing resulted in improved memory retention compared to silent reading.

Because a number of researchers failed to find a production effect for pure lists where all the items are either vocalized or silently read, they argued that this effect is due to the fact that the vocalized items on mixed lists are more distinctive (Dodson & Schacter, 2001; Hopkins & Edwards, 1972; MacLeod et al., 2010; Ozubko & MacLeod, 2010). But this argument is questionable on a number of grounds. As mentioned earlier, distinctiveness is not an information-processing mechanism, so it cannot be used as a sufficient explanation of the production effect. Another problem is that the claim of a failure to find a production effect for pure lists requires a sophisticated statistical analysis such as a Bayes factor; otherwise, classical statistics cannot compile evidence for a null finding. But none of the studies reporting a null effect for pure lists used an appropriate statistical analysis to justify this conclusion. But even more important, other researchers *did* find a production effect for pure lists (Bodner, Jamieson, Cormack, McDonald, & Bernstein, 2016; Fawcett, 2013; Gathercole & Conway, 1988).[21] Fawcett (2013) did a meta-analysis of the pure-list tests of the production effect. A meta-analysis is a statistical analysis of the combined results across a set of different experiments. There were twelve studies available at the time of the meta-analysis. The net effect for

21. One might wonder how it is possible for researchers to find an effect, whereas others do not. There are a number of reasons for this phenomenon. First, empirical research is subject to statistical sampling variation and statistical noise. That is why probability models are used in the analysis of the data. Given probabilistic variation, it is possible (in some cases expected) to find that a weak effect does not show up for some samples but is present in other samples or experiments. But in addition to these statistical reasons, the various laboratories may be using experimental treatments of varying degrees of effectiveness. One might speculate about the differences among the various research laboratories, but it usually is not known what might be the cause of the differences. For example, the MacLeod et al. (2010) laboratory found that the pure-list vocalizing condition resulted in a hit rate of .697 with a false-alarm rate .260, whereas Gathercole and Conway (1988) found a pure-list hit rate for vocalization of .91 with a false-alarm rate of .15. There are differences in the universities between these two studies, but there may also be differences in the motivation and attention to the task for the participants in the two studies. Gathercole and Conway informed their participants that they were part of a control condition examining stimuli that may be used for brain-damaged patients. This bit of deception is not necessary and is not commonplace, but it might have induced a higher level of attention to the task. This hypothesis is only speculation. Typically, we never know why some laboratories get results that other laboratories have difficulty achieving.

the *difference* of the hit rate between the two encoding conditions was .37, and the 95 percent confidence interval was (.16, .57). Because the confidence interval is entirely within the range of positive values, it follows that it is likely that there is a production effect for pure lists. The idea of distinctiveness does not apply for pure lists because the concept is based on a contrast among items within a person's memory. But there were separate groups of people in the two encoding conditions for the pure-list studies, so distinctiveness cannot be evoked. Consequently, the distinctiveness hypothesis must be dismissed as the sole reason for the production effect. Yet distinctiveness could still be a contributing factor when there is a mixed-list design where some items are vocalized and others are silently read. But even in the case of mixed lists, it is important to identify the production effect in terms of storage and retrieval processes rather than the simple claim that vocalized items are more distinctive.

Regrettably, the excessive attention on testing the distinctiveness hypothesis has impeded the uncovering of the underlying information-processing mechanisms involved with the production effect. Nonetheless, clues about some of the processes seem to be altered by vocalizing the target stimuli. Because of the consistent finding of a higher recognition rate for words read aloud, it is likely that this form of encoding increases the probability of item storage. As argued previously, if an item is sufficiently stored, then the presentation of the target item on a recognition memory test will automatically reelicit the stored representation. This idea is the central supposition of a content-addressable memory system, and it has led to successful storage-retrieval measurement models. But without employing a measurement tool, like those associated with the Chechile-Meyer task, there currently is no experiment that has examined if there is also an improvement in item retrieval with the vocalization of the target stimuli. Like in the Chechile and Soraci (1999) experiment on the generation effect, it is possible that both storage and retrieval are improved with the reading of items aloud.

So why is there an increase in the storage probability for vocalized items? The most promising hypothesis is trace elaboration.[22,23] The motor response along with

22. The previously discussed proactive interference hypothesis is not credible for mixed lists when the ratio of spoken words to nonspoken words is 1-to-1 (Icht et al., 2014). Proactive interference is plausible when there only are a few isolated items, such as in the case of the von Restorff effect. Moreover, the production effect has been found for pure lists, but the proactive interference hypothesis cannot be made when there are different groups of participants for the two types of encoding.

23. Instead of the elaboration hypothesis, some researchers have simply evoked memory strength theory as the basis for the production effect (Bodner et al., 2016). From this theoretical position, the vocalization of items is assumed to strengthen the memory trace. But we have seen a number of problems in this book with the concept of memory strength; hence, I do not favor that approach

its auditory output becomes a part of the encoding of the spoken items. This added information should increase the resistance to storage loss for two reasons. First, the additional features added by vocalization are a backup representation. Whenever there is a redundancy of stored information, it is more difficult to lose both representations than to lose only one representation. Second, an elaborated representation is likely to be more dissimilar to the other representations. For example, if a stimulus were encoded as a "red flat object," then there are many other similar stimuli. However, if the encoding were instead "a red stop sign on the corner of the street in front of a statue of a large horse," then the encoded stimulus is less likely to be similar to other encoded events. Elaborations should better define the encoded event. Moreover, it is expected that storage interference is a function of the similarity among the items. A recent similar event to an older target item might distort or even contribute to the loss of the older memory. But if the items are elaborated, then it is likely that the items are made more dissimilar; thus, the items are less susceptible to storage interference. It is also possible on a recall task that the elaboration of the vocalized items can make those items more easily retrieved, especially in a mixed-list design. Furthermore, the magnitude of the retrieval advantage should be inversely related to the time allowed for recall. With only a limited search time, it is more likely that the elaborated traces are recovered. But with a longer search time, the retrieval advantage would tend to be reduced. Currently, no study has examined this hypothesis.

13.7   Self-Reference and Survival Effects

The self-reference and survival effects are discussed together in this section because both involve the direct personal relevance of the target information to the individual. The self-reference effect was reported about 30 years before the survival effect. Given the similarity of the experimental manipulation in the two research areas, it is reasonable to explore if the survival effect is the same as or different from the self-reference effect (Nairne, Thompson, & Pandeirada, 2007). But before addressing the relationship between the effects, let us first discuss them separately in the chronological order of their discovery.

---

for the production effect. Also, strength theory does not have an explanation why vocalization is better than silently mouthing the words, which in turn is better than the silent reading of the words (Forrin et al., 2012). Yet from an elaboration framework, the degree of elaboration is directly related to the order of those three conditions.

13.7.1 Self-Reference Effect

The self-reference effect emerged from research about personality (Rogers, Kuiper, & Kirker, 1977). Participants in the Rogers et al. study answered questions about adjectives. For any given adjective, the questions could be structural, phonemic, semantic, or self-referencing. A structural question dealt with the size of the font. The phonemic question asked if the adjective rhymed with another word. The semantic question asked if the adjective was a synonym of another word. The self-referencing question asked if the adjective described the participant personally. After the questions about the adjectives, the participants were asked to free recall all the adjectives. The mean recall rate for adjectives in the structural, phonemic, and semantic conditions was 20 percent, whereas the percentage correct for the adjectives in the self-referencing condition was 32 percent. This effect has been replicated under a variety of conditions (Ganellen & Carver, 1985; Halpin, Puff, Mason, & Marston, 1984; Kendzierski, 1980; Klein & Kihlstrom, 1986; Klein, Loftus, & Burton, 1989). This improvement in memory when people relate the adjective to themselves has been found for recognition measures as well as for recall (Glisky & Marquine, 2009; Keenan & Baillet, 1980; Rogers, 1977).

The magnitude of the self-reference effect has been shown to be dependent on the comparison condition (Bower & Gilligan, 1979; Symons & Johnson, 1997). For example, in the Bower and Gilligan study, adjectives were rated with different orienting questions. Two self-reference groups were asked either (1) if they could remember a personal experience that exemplified the trait adjective or (2) if the adjective describes them. There also were two other comparison groups. For one condition, the participants were asked if the adjective described Walter Cronkite, who was a well-known news anchorman at the time of the research. For the other comparison condition, the participants were asked if they could remember (or were told about) an experience in which their mother exemplified the trait adjective. The mean on the later recall test for the two self-reference conditions was .315, whereas the recall mean for the Walter Cronkite condition was .220. Thus, the means in the self-reference condition and the Walter Cronkite condition were remarkably close to the findings in the Rogers et al. (1977) study. But the recall rate for the mother-reference control condition was .290, which is not very different from the recall rate in the self-reference condition (.315). Based on the *mother-control* condition, there is not an improved memory for self-referencing, but based on the *Walter Cronkite–control* condition, there is an effect for self-referencing.

The findings of the Bower and Gilligan study bring up a number of questions about the underlying causes of the self-reference effect. We are all experts about ourselves. We have a rich episodic and semantic storehouse of facts and experiences that apply directly to us. Consequently, when attempting to rate if an adjective applies to us, it is reasonable to recover some appropriate autobiographical experience in which the trait adjective would apply. So the adjective is encoded automatically in a rich elaborative context. Thus, without trying to encode the target adjective, the information is nonetheless likely to be encoded with more than the usual level of elaboration. But it is unlikely that one can answer a question about Walter Cronkite's personality by recalling a specific memory. He was known mostly as someone reading the news on television. Unlike an actor, he did not play a complex role with many remembered scenes and episodes that defined his character, whereas people are experts on their mother. So is the memory benefit due to the involvement of the self as the frame of reference, or is the effect caused by the retrieval of specific information from memory? For personality researchers, the concept of the self is an important psychological construct (Roger, 1997). But memory researchers are less inclined to believe that the self-concept drives the memory effect. For example, the results from experiment 2 of the Bower and Gilligan paper indicate that it is expertise or familiarity about the frame of reference that is the critical reason for the memory advantage. Moreover, given the hypothesis that memory retrieval implicitly underlies the decisions about trait adjectives, it would seem that positive cases affirming the characteristic should be different from cases where the trait is rejected as being suitable. For example, suppose the answer to a question is NO because of the lack of support from autobiographical memory, then the later recall of the adjective should be reduced because the answer did not have the same level of trace elaboration formed at the time of encoding. In fact, in experiment 1 from the Bower and Gilligan study, the recall rate is substantially lower for the recall memory of the NO-rated adjectives. This difference between the YES and NO judgments about the adjectives has been replicated for the recall rate (Ganellen & Carver, 1985), as well as for the recognition hit rate (Glisky & Marquine, 2009).

Finally, it is reasonable to hypothesize that the self-referencing effect is a special case of a more general unconstrained generation effect. Previously, we have discussed the unconstrained generation task where the participants are provided with a stem word, and they are required to type in a response word to the stem that fits a particular encoding guideline (Chechile & Soraci, 1999; Roenker et al., 1978). Each trial for the unconstrained generation task is thus a mini problem-solving question that induces the participant to produce intermediate ideas prior to a decision. For example, recall the previously discussed example of a participant who responds to a cue of CHAIR with the unrelated solution of SKY because the participant stated, "Chair. I see myself in the dorm on my chair. I look outside the window and picture myself outside. I look up… SKY." Similar to the generation experiments, the typical self-reference task also presents a question for the participant to answer. Instead of the task being an explicit learning task as in the generation-effect research, the self-reference task conceals the true purpose of the study. The problem for the participant in the self-reference task is one of deciding if an adjective applies to him or her, whereas in the case of a semantic task, the participant simply has to decide if the word means the same thing as another provided word. Despite some procedural differences, the self-referencing task shares similarity to the unconstrained generation task. It is likely that participants answer questions about themselves by retrieving information from their autobiographical storehouse of experiences. For both generative encoding and self-referencing encoding, a set of propositions is generated before reaching their final answer. The underlying information processing is better understood for the unconstrained generation task because a model-based measurement experiment linked the unconstrained generation effect to changes in both storage and retrieval (Chechile & Soraci, 1999). Using the Chechile and Soraci study as a template, a set of predictions follows about the self-referencing effect. First, it is likely that the self-reference effect also results in improvements in both storage and retrieval processes. The elaboration of the trace by the recollection of a related autobiographical event to guide the decision about the self-reference question should make the memory trace less susceptible to a storage loss. Second, the extra memory content generated to form a judgment about an adjective also provides additional retrieval cues.

13.7.2 Survival Effect

In the biological sciences, there is perhaps no more important theory than that of evolution (Darwin, 1859). Complex structures in biology are not accidents, but they come about from a long evolutionary process guided by random genetic variation and the mechanism of natural selection. Those genetic traits and capabilities that increase the fitness of the individual and improve the chance of sexual reproduction have a chance of becoming trait features and capabilities of the next generation. Clearly, memory is a powerful capability, and it greatly increases the chance of survival both for the individual and for the species. Human memory requires a complex biological substructure that exists as a result of a long evolutionary history of change from more simple structures (Cosmides & Tooby, 2003).

The above evolutionary perspective is well established and unquestionably true, but some memory researchers go further to hypothesize that certain ideas and words are evolutionarily important, so these items are easily remembered. For example, if a list of words includes stimuli like WATER or FOOD, then those items would be better remembered because of an assumed priority for remembering information important for our survival. This position is controversial. But before directly discussing the survival effect, it is valuable to first deal with some issues from philosophy, animal ethology, neurobiology, and classical conditioning.

13.7.3 Philosophical Background

Although evolution motivates the survival effect in memory research, the notion of innate ideas first arose in the philosophy of Plato (Samet, 1999). For Plato, geometric concepts and the existence of an immortal soul were nativistic knowledge that only required reason to discover (Samet, 1999). The concept of innate ideas was also advanced by René Descartes and Gottfried Leibniz. The belief in innate ideas is a philosophical position that has come to be called rationalism (Schwartz, 1999).[24] But the empiricist tradition in philosophy, led by John Locke and David Hume, strongly disputed the claim of innate ideas and argued that all ideas are derived from experience and learning (Schwartz, 1999). The rationalist versus empiricist debate in philosophy, as with other philosophical issues, is a seemingly unresolvable dispute on the basis of purely philosophical argumentation. But the Irish scholar William Molyneux, in a letter to Locke, proposed a test of the conflicting predictions between Descartes and Locke about innate ideas. The test is now called the Molyneux problem.

Suppose a man born blind, and now adult, and taught by his touch to distinguish between a cube and a sphere of the same metal, and nighly of the same bigness, so as to tell, when he felt one and the other, which is the cube, which is the sphere. Suppose then the cube and the sphere placed on a table, and the blind man be made to see: query, Whether by his sight, before he touched them, could he now distinguish and tell which is the globe, which the cube? (Locke, 1694, p. 67)

Locke predicted that a blind man, who previously only understood objects by his touch, would have to learn about the visual objects afresh if he were suddenly given sight, whereas Descartes predicted that the newly sighted man would immediately understand the meaning of the new visual sensation. Molyneux's problem has been experimentally tested (Held et al., 2011). In the Held et al. study, five newly sighted people were immediately tested to see if they could distinguish items based on their sight of

24. It is interesting that reason is required to discover innate ideas within the rationalistic tradition. Why should reason be required if the knowledge were hardwired in our biology? Our DNA genetic heritage is a code for building all the cells of our body. Reason is not required for any of this information. So why should Plato's belief in an immortal soul require reason for its discovery? If it were truly an innate idea, then it would be in the DNA code, and reason would not be required. Mathematics is based on some initial assumptions, and then with reasoning, new results and conclusions are derived. But the axioms of mathematics are abstractions of other structures that had its origin in experience. Thus, reason can enable many new inferences, but it is constructed within a system that has its foundations rooted in experience.

the objects. Held et al. reported that the newly sighted people were at chance levels on a discrimination test for object identification. A number of other less formally conducted experiments have also found that the newly sighted individuals are confused by their new visual sensation. For example, in the May 10, 1993, issue of *The New Yorker* magazine, neurologist Oliver Sacks describes a case study of Virgil, a man who was made to see after 45 years of blindness. Sacks provided a lengthy description of the slow, frustrating process that Virgil experienced in trying to understand his new confusing visual sensation. Thus, Locke's prediction has been supported. Visual perception differs from visual sensation because with perception, we bring previously acquired knowledge concerning that sensation to bear to process new information. Consequently, when it comes to sensation, the philosophical notion of innate ideas is not accurate.

13.7.4 Animal Ethology

Although human perception requires experience, there is still evidence that animals have hardwired programs for complex behavior that is activated by a triggering stimulus. Ethologists, who study the behavior of animals in their natural habitat, have demonstrated that some species exhibit nonrandom behavior soon after birth and without learning (Lorenz, 1965, 1966). For example, after hatching, a newborn gosling is "imprinted" to the large animal seen nearby, and the gosling follows that animal. Usually, that animal is its mother, but ethologist Konrad Lorenz demonstrated that goslings can be imprinted to different animals. In his classic experiment, Lorenz divided the eggs of a goose into two groups. One group of eggs was hatched normally with the mother. The other batch of eggs was placed in an incubator. When these goslings were hatched, they became imprinted to Lorenz. Even when the two groups of goslings were recombined later, the first group continued to follow its mother, whereas the second group followed Lorenz. This example illustrates *species-specific behavior*, and such behaviors became the focus of ethological studies. A species-specific behavior is a hardwired behavior that depends on the evolutionary history of the species. These responses are instincts like the human sucking response. Instincts are not regarded as acquired memories, but instead they are like the unconditioned stimulus and the unconditioned response of classical conditioning. Associative learning and memory are exhibited when neutral stimuli become learned.

13.7.5 Neurobiological Hardware for Language and Memory

Human language is a powerful expressive system that is often used for the encoding of information. From the neurosciences, it is clear that regions of the brain specialize in language processing. Although the acquisition of our natural language requires years of experience, we are nonetheless born with structures such as the Broca and Wernicke regions of the brain that are respectively critical for enabling the production and comprehension of natural language (Benson, 1985). Damage to these brain regions or damage to the neural pathways between these regions results in *aphasia*—a term for impaired language processing (Benson, 1985). Without these specialized brain regions, language is hopelessly impaired. Similarly, we have already extensively discussed the necessity of both the hippocampus and the neocortical association areas for the formation of a memory representation. The patient Henry Molaison, who we discussed previously, illustrates the severe learning deficit that results when one of those critical regions (the hippocampus) is removed. Thus, there clearly is specialized neural hardware for processing language and building new memories. These complex structures are the product of a long evolutionary history.

13.7.6 Garcia Effect

The Garcia effect is an interesting example of a possible evolutionary influence on the learning of a neutral stimulus. The context for this effect is classical conditioning, a topic that we have discussed extensively. From a behavioristic position, the neutral conditioned stimulus **CS** can be any stimulus that does not elicit the conditioned response **CR** ahead of time. It could be a light, a sound, or a taste. In classical learning theory, it was thought that any stimulus can be conditioned, provided that the animal can detect the stimulus and the stimulus is presented in an effective temporal relationship to the unconditioned stimulus **US**. But John Garcia and his collaborators showed that there is a strong bias on the conditioned stimulus for any given context (Garcia, Ervin, & Koelling, 1972; Garcia, Kimeldorf, & Hunt, 1961; Garcia & Koelling, 1972). In the Garcia and Koelling study, rats were placed in a chamber in which the animals had access to both a flavored water and an audiovisual stimulus. Next, the animals were X-rayed to induce illness. The animals developed a conditioned avoidance response to the flavored water but not to the audiovisual stimulus. In another experiment reported by Garcia and Koelling, the same two stimuli were available in a chamber, and later the animals received a foot shock. In this case, the animals learned to avoid the audiovisual stimulus but continued to drink the flavored water. Thus, there is a strong bias to associate taste and food with illness even when these stimuli are not responsible for the illness. Sickness can produce an unpleasant internal sensation that functions as the unconditioned stimulus. In this context, it seems as if the animals are biased to associate the unpleasant sensation with a recent taste or food rather than an external audiovisual stimulus. But shock is an external stimulus, so in that context, an external audiovisual stimulus is the object of the conditioned avoidance response. It is reasonable to hypothesize that the animal evolved with some priorities for the type of stimuli that are more likely to be conditioned in any given context. But perhaps the bias is cognitive rather than a genetic predisposition. Previously, it was argued that classical conditioning in rats reflected a signaling relationship rather than a mechanical altering of the strength of association. So perhaps the Garcia effect is mediated by biases in the linking of causes with effects in the animal's cognitive model. At the present time, it is simply unknown if the Garcia effect is in the genes of the rats or the effect is due to a cognitive model based on their experience.

13.7.7   Survival Effect in List Learning

Let us now return to the question about the priority of processing survival-related words and concepts. Do humans have a built-in tendency to process some concepts, such as words associated with water and food, so that these conceptual categories become particularly easy to encode in new situations? Memory is a model of our experience, and it is a model of the environment. But do certain ideas have a high priority in the memory model building? Some memory researchers believe that they can demonstrate the advantage of information pertaining to the topic of human survival. The survival effect in the context of the learning of verbal material began with the Nairne et al. (2007) paper. In that study, participants were asked to rate words such as BOOK FINGER APARTMENT CATFISH COTTAGE SCREWDRIVER SWORD ⋯ on a 5-point scale from 1 indicating completely irrelevant to 5 indicating extremely relevant. The participants rated the words in the context of two different scenarios (i.e., survival and moving). For the *survival scenario*, the participants were instructed to imagine that they are stranded on a grassland of a foreign land and without any of the provisions needed for survival such as food, water, and protection from predators. In the *moving scenario*, the participants were told to imagine that they are moving to a new home in a foreign land and that they need to obtain a new home and to move their belongings. There was also a third group (*pleasantness condition*) in which the participants simply were asked to rate the words for pleasantness; there was no imagined context for this condition. The people in all the conditions were not told to remember the words; their task was to simply rate the words. Following the rating of the words, the participants had to do a digit-recall test for nine-digit sequences; the digit-recall task took about 2 min. Finally, the participants were given a surprise free-recall test for the words that were rated in the first phase of the experiment. In experiment 1, the recall rate for the *moving* and the *pleasantness* conditions was about 53 percent, whereas the recall rate for the *survival* condition was higher (about 59 percent). In experiment 2, the significant difference between the *survival* and *moving* conditions was replicated with an experiment where the participants had experience with both of the scenarios. In experiment 3, a recognition test was used to assess the *survival effect*; again, a significant difference was found.[25] The above survival effect findings has been replicated in other papers (Nairne & Pandeirada, 2008; Nairne, Pandeirada, & Thompson, 2008; Weinstein, Bugg, & Roediger, 2008). Furthermore, Fellner, Bäuml, and Hanslmayr (2013) found a behavioral as well as a neuroimaging difference between survival and semantic processing.

Somehow the procedure of rating words in terms of a grassland-survival scenario results in an improved chance of memory retention. But does the human memory system retain some hardwired preference for dealing with survival issues like those encountered by our genetic ancestors in our distant evolutionary past, or can any engaging imagined survival context elicit an improved memory encoding? A study by Soderstrom and McCabe (2011) provides evidence that modern imagined survival scenarios can lead to even better memory retention. These investigators created three new scenarios to compare with the standard *grassland predators* scenario. To obtain a *grassland zombies* scenario, they merely replaced the word *predators* with the word *zombies*. They also produced *city zombies* and *city attackers* scenarios by replacing either one or two words from the standard *grassland predators* narrative. The *city attackers* and *grassland predators* scenarios resulted in a nearly identical recall rate. But the *grassland zombies* and *city zombies* conditions resulted in even higher recall rates. Thus, it seems that any engaging survival scenario results in enhanced memory retention; neither grasslands nor predators are necessary to get an enhanced level of memory encoding.

Are the above studies on the survival effect sufficient for concluding that the general survival category has a genetically favored priority, or is there a more prosaic explanation for the effect in terms of ordinary memory processes? A study by Kroneisen and Erdfelder (2011) makes a convincing case that the survival effect is due to an elaboration or richness of encoding. These investigators examined two survival scenarios. The setting for one narrative (called *original survival*) is one of being abandoned on a foreign grassland and having to protect oneself from predators for months and locate food and water. The setting for the second narrative (called *alternative survival*) is one of being washed ashore on a foreign grassland with other objects but without clean water. Kroneisen and Erdfelder also examined a third scenario as a control condition; this was the standard narrative about moving to a foreign country. In three experiments, they found a consistent difference between the *original survival* scenario and the *moving* scenario. But the difference between the *alternative survival* scenario and the *moving*

25. Nairne et al. (2007) also examined the difference between *survival effect* and the *self-reference effect*. The recall rate was higher in the *survival* condition.

scenario was weak to nonexistent. If the survival effect were a robust effect, then the trivial alternative phrasing of the narrative should not disrupt the effect. Moreover, in the first two experiments, Kroneisen and Erdfelder report the final recall rates as a function of the initial relevancy ratings. The authors did not discuss the implications of the magnitude of the survival effect in terms of the relevancy ratings, but their data actually pose a problem for a genetic-fitness interpretation of the effect. The difference in the recall rate between the *original survival* condition and the *moving* condition is largest for the lowest rated words for relevance. In other words, the strong survival effect found is for the words judged to be totally *irrelevant* for survival. For example, for experiment 1, there is an overall difference of .13 between the *original survival* and *moving* scenarios, but the differences between these conditions as a function of the relevancy ratings from 1 to 5 are, respectively, .25, .10, .07, .06, and .04.[26] Although there is a statistically significant difference for the overall recall rate, the difference between the conditions is not significant for the words that received a relevancy rating of 5. Thus, the sizable difference between the conditions is mostly due to the words that are not relevant for either a survival or a moving context. This outcome is not consistent with the genetic-fitness hypothesis for the survival effect.

Kroneisen and Erdfelder also found that the lower memory retention for the *moving* condition was related to the time to make the relevancy rating. The rating in the *moving* condition was 331 msec faster on average than a rating in the *original survival* condition. Given that the participants are in an unconstrained state of freely attempting to assess if the provided word is relevant, then it follows that there would be less information generated in a shorter time period. Thus, compared to the *survival* scenario, the *moving* scenario would be less likely to develop as many ideas for the possible use for the word in the imagined context. Alternatively, the extra time for the *original survival* condition is likely to produce a more elaborated encoding. Consequently, like the *self-reference effect*, the most likely account for the memory difference for the *survival effect* is the enhanced amount of elaborated information generated and incorporated as part of the encoding of the word. Thus, the survival effect as defined by the difference between two imagined scenarios is not evidence for a special evolutionary augmentation for processing certain concepts. Like the innate ideas of Plato, real ideas are built from experience and augmented by reasoning. However, later in the discussion of emotion and memory, we will return to the question of a special enhancement of memory that does have a biological underpinning.

26. These numeric values are visually estimated from a bar chart provided in the Kroneisen and Erdfelder (2011) paper.

13.8 Organizational Factors

There is a consensus among memory researchers that organization is important for learning and memory retention. In general, event memory is linked to other information in the current temporal setting along with generated propositions and associations. The intrusion of extraneous information from the current or past is distracting. Consequently, it is an excellent encoding strategy to ignore nonessential facts and to concentrate on the pertinent information and to link it to other information that will be also important to know at a later time. This commonsense encoding principle is essentially stressing the importance of organization as an effective learning strategy. Imagine a law office manager who organized data in a haphazard fashion where the facts about one client are scrambled along with the facts about all the other clients. It would not be easy to search for information and to deal with the current legal cases because the relevant information for each client would be randomly scattered rather than grouped together in one place. Organization is essential for any information storage system—including the human memory system.

Initially, it was thought that the capacity of human memory span placed a limit on the number of items that can be grouped together. Memory span is the longest list length where the person is at a 50 percent level for repeating back the entire list in the right order. In his classic paper on memory organization and chunking, George Miller (1956) observed what he called the limit of the magic number 7 ± 2. Miller noted in a humorous article that humans have an affinity to the number 7, that is, number of days in the week, the number of digits in a telephone number (without the area and country code), the number of wonders of the world, the number of notes on the musical scale, and so on. He suggested that these examples might reflect a processing capacity limit. But subsequent research has indicated that these limits can be exceeded, and the number of 7 itself is not the best estimate (Cowan, 2015; Mandler, 1967). For example, the memory span depends on both the individual and the type of stimuli. The span for letters is about six, but the span for three-letter words is less than four (Crannell & Parrish, 1957). Warren (2015) showed that the memory span for familiar shapes was four, but the span for unfamiliar pictograms was only one. But with chunking, people can easily remember much longer sequences. For example, suppose one needed to learn the following sequence of letters:

FBINASAIRSCIA

This sequence of thirteen letters can be easily chunked as FBI NASA IRS CIA; these chunks are four abbreviations for familiar agencies of the U.S. government.

Processing information at the rate in which events unfold is limited by the size of the memory span. As discussed previously in this book, only a small fraction of the information in the environment can be encoded. From a previous discussion of the information content in a 1-sec time slice of a real-world event in a French town, it was estimated that the event had more than 14,000 bits of information, but the memory-span capacity limited the immediate memory to about 19 bits of information per second.[27] Miller (1956) suggested that this processing limitation might place a limitation on the number of items within a category and the number of available categories. Yet it is puzzling why a real-time processing rate limit is relevant either to the number of conceptual categories available or to the number of exemplars. Miller did not produce a cognitive model to support this hypothesis, and in my opinion, this magic-number-7 hypothesis is clearly wrong. A good counterexample is the size and organization of semantic memory. With enough practice, facts can be learned to such a high degree that the information becomes part of world knowledge or semantic memory. In most cases, the facts have become generalized, so it is not about a specific moment in time or about a specific event. Nonetheless, the information has a vast number of categories and exemplars.

As another example, let me describe some of my experiences studying in my earlier student days. In any course in which performance was based on examinations, I developed a highly organized structure of all of the course material. The reading assignments were studied, and some sections were read again. Mnemonic notes for the textbook readings were written, as well as notes about selective homework assignments were added. These notes focused on the vital factual and conceptual information. Classroom lecture notes were similarly rewritten to focus on the essential content. Thus, notes were generated for the entire corpus of the material that I needed to know and understand completely. There were many pages of this first level of summarizing information. But this information had a structure and could be arranged into a structured hierarchy. Consequently, I reexpressed the first-pass summarizing notes into a second-pass higher-level summary that was restricted to be only about two to three pages in length. This reorganization and distillation of information sometimes required

---

27. The relationship between the number of possible items N and the bits of information b can be expressed either as $N = 2^b$ or $b = \frac{\log N}{\log 2}$. For a sequence of four letters, there are $N = 26^4 = 456,976$ possible items, so the information content $b = \frac{\log 456,976}{\log 2} = 18.89$; thus, a four-letter sequence has 19 bits of information. Note that the number 456,976 in base 10 is equal to 1,101,111,100,100,010,000 in binary arithmetic. The bits of information are the number of places in base-2 arithmetic for expressing the value of N.

considerable effort and rewriting. The final set of highly distilled notes were studied and practiced until I could recall the entire three pages from memory. I tested myself to ensure that the high-level notes could be recalled. I had memory access without my notes to all the key higher-level ideas, which in turn triggered the other levels of facts, previous homework problems, and pertinent equations. Hence, with enough practice, a large amount of information could be stored in an organized fashion. At any given second of processing time, there is a severe limit on the processing capacity, but with repeated practice and written records, the amount of information learned and organized can be vast.

13.8.1   Evidence of Spontaneous Organization

People do not need to be instructed to discover the advantages of using organization to improve their recall rate. If people are given a list of words that are selected from different categories, but the words are presented in a scrambled order, then the output free recall tends to be remembered in terms of the separate categories (Bousfield, 1953). This phenomenon is called *clustering*. The free-recall task is an ideal procedure for obtaining a clustering measure (Bousfield & Bousfield, 1966; Frankel & Cole, 1971; Roenker, Thompson, & Brown, 1971). The free-recall task simply asks people to output the items that they were previously presented in any order. While the output order is ignored for the scoring of item correctness, the order is nonetheless essential for the measurement of clustering. One metric for organization is the *adjusted ratio clustering* (ARC) score, which was developed by Roenker et al. (1971). To use this metric, each target item on a learning list should belong to a fixed number of identifiable conceptual categories. Typically, there are a number of exemplars from each category. At the time of learning, the list could be presented in a random arrangement for the category exemplars. For example, the learning order for the exemplars from categories A, B, and C might be in the following scrambled order: BCABACBCABCACBACABCABACB. The formula for the Roenker et al. (1971) ARC clustering metric is given as

$$ARC = \frac{R - E(R)}{R_{max} - E(R)}, \tag{13.1}$$

where R is the number of category repetitions, R_{max} is the maximum number of category repetitions, and $E(R)$ is the average number of repetitions when there is a random output order. Bousfield and Bousfield (1966) stated that the value for the mean number of repetitions given only randomness is

$$E(R) = \frac{\sum_{i=1}^{k} n_i^2}{N_{tot}} - 1, \tag{13.2}$$

where n_i is the number of words recalled from category i, and N_{tot} is the total number of words recalled. Note that the value for N_{tot} is strictly the number of items recalled and is not the total number of items on the original study list. The maximum number of repetitions R_{max} is equal to $N_{tot} - k$, where k is the number of categories recalled. The ARC metric is equal to 1.0 for the case of perfect clustering, such as when the output order is BBBBBAAAAACCC. Note for this example of perfect clustering that the value for N_{tot} is 13, and the respective values for n_1, n_2, and n_3 are respectively 5, 5, and 3. Also, $k = 3$, so $R_{max} = 13 - 3 = 10$. From equation (13.2), the value for the expected number of repetitions given randomness $E(R)$ is

$$E(R) = \frac{5^2 + 5^2 + 3^2}{13} - 1.$$
$$= \frac{59}{13} - 1 = 3.5385.$$

Finally, the value for the number of repetitions in the above perfect sequence is $R = 10$. Based on these values for R, $E(R)$, and R_{max}, the value for the ARC score from equation (13.1) is

$$ARC = \frac{10 - 3.5385}{10 - 3.5385} = 1.$$

But suppose the output sequence were instead BAACCABCABBAB. Now the number of repetitions would be $R = 3$ and the ARC score from equation (13.1) would be $\frac{3-3.5385}{10-3.5385} = -.083$. Clearly, this sequence does not provide evidence for clustering.

In current practice, researchers typically report positive values for a clustering measure (such as the ARC score) as evidence for clustering. But this practice has a statistical problem. To illustrate the difficulty, let us consider a hypothetical case where the participant outputs the following sequence of categorical responses:

AABBCCAACAACBBCCBB.

On the surface, this sequence looks like evidence for clustering. There are $R = 8$ repetitions with $R_{max} = 18 - 3 = 15$ and $E(R) = \frac{6^2 + 6^2 + 6^2}{18} - 1 = 5$. The clustering metric is $ARC = \frac{8-5}{15-5} = .3$. But it can be demonstrated that obtaining an ARC score of .3 is not a rare event.

With Monte Carlo sampling, it is possible to assess what type of clustering scores can occur when $k = 3$, $N_{tot} = 18$, and $n_i = 6$ for $i = 1, ..., 3$. This sampling is implemented via a computer simulation of a random card shuffling of a deck of eighteen cards with six As, six Bs, and six Cs. In table 13.3 are the results from 40,000 simulated "deals" after random shuffling of the deck of cards. In order for there to be significant

Table 13.3
Monte Carlo statistics for clustering example for $n_i = 6$, $i = 1, \ldots, 3$

Repetitions R	ARC	$P(R \geq R_{observed})$
5	0	.5944
6	.1	.3851
7	.2	.2120
8	.3	.0943
9	.4	.0362
10	.5	.0110
11	.6	.0026
12	.7	.0005
13	.8	$<10^{-4}$
14	.9	$<10^{-4}$
15	1	$<10^{-4}$

clustering, the probability for the number of repetitions equal to or greater than the observed repetition number $R_{observed}$ should be less than .05, which is the traditional significance level for classical statistical tests. For the observed repetition number of 8, there is a probability of .0943, which is not rare enough to be a statistically significant amount of clustering. That is, of the 40,000 simulated random shuffles of the deck of eighteen cards, there were a total of 3,772 outcomes where the repetitions were eight or more; thus, the probability of eight or more repetitions has an estimated value of $\frac{3772}{40000} = .0943$. It is surprising that heretofore, researchers ignored the probabilistic assessment of the possibility that the degree of clustering could be simply due to random output as opposed to deliberate clustering. To reject the null hypothesis that output is simply random, we would have to have nine or more repetitions if there were six exemplars recalled for each of the three categories.[28]

13.8.2 Subjective Organization

The context for the topic of subjective organization deals with verbal lists that do not have experimenter-provided categories. Consider, for example, the following list of eight words:

CHILDREN FOREST KING WINDOW CHAIR BIRD SMALL ELEVEN.

If the participants are allowed to free recall their memory of the list, then it is possible that they might connect the words in some idiosyncratic fashion. Suppose,

28. It is also possible to do a Bayesian analysis of the degree of clustering. But that analysis is not yet published and is more technical than suitable for the current discussion in this book.

for example, that a participant encodes these words into the following mnemonic sentence:

(The) KING (sits on his) CHAIR (and looks out the) WINDOW (, and he sees a) FOREST (and a) SMALL BIRD (near) ELEVEN CHILDREN.

This natural language sentence can be an effective way to chunk the unrelated words into a meaningful and highly memorable unit. If the participant engaged in this type of idiosyncratic organization, then it would be expected that the eight words would be recalled together in an order that differs from the serial order of the actual word presentation. If the participant were asked to recall the words again after another learning trial, then it is likely that the participant would output the same words in the same order as he or she did before (i.e., KING CHAIR WINDOW FOREST SMALL BIRD ELEVEN CHILDREN). For this example, the participant is exhibiting a *subjective organization*.

Tulving (1962) introduced a metric for subjective organization that can be obtained from an analysis of the free-recall order data over a series of consecutive test trials. Gorfein and Blair (1971) added another metric for subjective organization. Let us illustrate these measures by means of a simple example. Suppose that a participant outputs the following words over four testing trials.

Trial 1	Trial 2	Trial 3	Trial 4
LAMP	GRAPE	POEM	POEM
GRAPE	TREE	TREE	TREE
TREE	POEM	GRAPE	LAMP
POEM	LAMP	LAMP	GRAPE

Tulving represented the data for the word adjacencies in the form of a matrix. See the matrix T_{ij} in equation (13.3) for the above example.

$$
T_{ij} = \begin{bmatrix}
   & E & W_1 & W_2 & W_3 & W_4 \\
B & 0 & 1 & 1 & 0 & 2 \\
W_1 & 2 & 0 & 2 & 0 & 0 \\
W_2 & 1 & 1 & 0 & 2 & 0 \\
W_3 & 0 & 1 & 1 & 0 & 2 \\
W_4 & 1 & 1 & 0 & 2 & 0 \\
n_j & 4 & 4 & 4 & 4 & 4
\end{bmatrix},
\tag{13.3}
$$

where $W_1 = $ LAMP, $W_2 = $ GRAPE, $W_3 = $ TREE, and $W_4 = $ POEM. The entries in the column with the E label represent the end word in each of the four output orders. Because the word LAMP (W_1) is the last word for trials 2 and 3, there is a count value of 2 for the entry in row W_1 in the E column. Because the words POEM and GRAPE are the last words for trials 1 and 4, respectively, the row entries for rows W_2 and W_4 are each 1. The row labeled B represents the data for the beginning word for each trial. The other entries in the rows labeled with words are the count values for the number of trials where a particular word is followed by another word. For example, the entry in row W_1 and column W_2 is 2; this value is because the word GRAPE (W_2) follows the word LAMP (W_1) for two trials (i.e., trials 1 and 4). Tulving denoted the count values for word-order adjacency as n_{ij}. Tulving also denoted row B and column E with a 0 subscript. Finally, the row labeled n_j is the sum of the count values for column j, that is, $n_j = \sum_{i=0}^{4} n_{ij}$, for $j = 0, \ldots, 4$. Based on an information theory, Tulving defined a metric for subjective organization as

$$SO = \frac{\sum_{i=0}^{L} \sum_{j=0}^{L} \log n_{ij}^{n_{ij}}}{\sum_{j=0}^{L} \log n_j^{n_j}}, \tag{13.4}$$

where L is the number of words recalled. Because $0^0 = 1$ and $\log 1 = 0$, there is no contribution from any n_{ij} entries that are 0. Because $\log 1 = 0$, there also is no contribution from any of the cells where $n_{ij} = 1$. Consequently, for our simple example,

$$SO = \frac{6 \log 2^2}{5 \log 4^4} = .3.$$

If the participant had the same order for recalling the four words on each trial, then the SO metric would be 1. But in the extreme case where each word is recalled in a different serial position across the four trials, then the matrix would be

$$\mathbf{T}_{ij} = \begin{matrix} & E & W_1 & W_2 & W_3 & W_4 \\ B & 0 & 1 & 1 & 1 & 1 \\ W_1 & 1 & 0 & 1 & 1 & 1 \\ W_2 & 1 & 1 & 0 & 1 & 1 \\ W_3 & 1 & 1 & 1 & 0 & 1 \\ W_4 & 1 & 1 & 1 & 1 & 0 \\ n_j & 4 & 4 & 4 & 4 & 4 \end{matrix} . \tag{13.5}$$

For this extreme case, the SO metric is 0. Thus, the subjective organization measure is a value on the [0, 1] interval.

Gorfein and Blair (1971) suggested that a measure for subjective organization should examine bidirectional adjacency. Sternberg and Tulving (1977) provided a formula that accomplished this goal. Their formula was also expressed in terms of a matrix representation of the forward adjacencies, that is, a matrix like $\mathbf{T}_{ij}$ from equation (13.3).

This measure is called SO2, and it is defined as

$$SO2 = \frac{\sum_{i=0}^{L} \sum_{j=0}^{L} \log(n_{ij} + n_{ji})^{n_{ij}+n_{ji}}}{2 \sum_{j=0}^{L} \log n_j^{n_j}}.$$
(13.6)

For the matrix in equation (13.3) the SO2 measure is .7254. Like the SO metric, if the output is the same for each trial, then the SO2 metric is 1. But if each word were recalled in a different serial position across the four trials (i.e., the maximally disordered arrangement like in equation (13.5)), then the SO2 measure would be .5 rather than 0. Hence, the SO2 measure is *not* on the [0, 1] interval. This feature is a major problem.

There are a number of theoretical and statistical problems with the measurement of subjective organization. Unlike the ARC clustering metric for categorized lists, subjective organization for unrelated lists cannot be measured with the SO metric on the basis of just a single trial. In fact, the SO measure cannot be defined when there are only two trials because the numbers in the adjacency matrix would be either 0 or 1, and thus both the numerator and denominator of the SO computation in equation (13.4) would be 0. Hence, the number of output trials must be equal to or greater than three. The problem is memory changes over three learning trials. Even if the researcher had the participants study the list once and test repeatedly, then the literature on testing effects has shown us that there will be improvements in memory retention. Consequently, the fundamental method of measuring subjective organization by examining the order of information over repeated test trials is contaminated by the fact that memory is dynamically changing. Only when the list is learned to mastery does it make sense to examine if there is a consistency in the order of the output. Yet even in this case of mastery, the repeated testing of the list can have an effect (i.e., different output chunks might develop as a more effective way to retrieve all the list items). In my opinion, this problem is a serious shortcoming of the SO metric.

Yet another problem with the SO metric is the lack of statistical assessment. Like with the ARC clustering measure, it is essential that we know what type of SO values are possible when there is only randomness for the output order. To understand better how randomness affects the SO metric, a Monte Carlo simulation was developed for the very simple case of four words and four trials. Based on a random arrangement across the four trials, it is possible to find the critical 95th percentile point for SO. This number would be the critical SO value to assess if the observed SO value was statistically significant because only 5 percent of the random arrangements would have that value or higher. Based on the Monte Carlo simulations, the critical SO value was .51. Thus, as with the ARC metric, researchers have reported positive SO values as evidence for subjective organization without assessing if that conclusion is justified. This practice fails to build evidence for subjective organization.

In light of this simulation result and the theoretical problem with defining subjective organization with the SO metric, it is doubtful that many of the claims about subjective organization that were based on the SO measure are rigorously valid. Yet I do not doubt that participants can group unrelated words together, but the empirical study of that process requires other measurement tools and methods beyond the SO measure. Perhaps a suitable measure can be produced by asking participants to output any subjective chunks that they developed, although that approach also has the undesirable effect of implanting a suggestion that they ought to form a subjective group.[29] But perhaps a careful examination of the output versus input serial order might uncover the presence of subjective organization without placing a single number on the whole list. In fact, a metric called *lag conditional response probability* (lag-CRP) has successfully revealed a consistent structure in the output order of single-trial free-recall data for unrelated words (Kahana, 1996; Kahana, Howard, & Polyn, 2008; Sederberg, Miller, Howard, & Kahana, 2010). For example, suppose that the input list includes the word subsequence of PICKEL VOLCANO STUDENT TRAIN SOLID, and the participant responds in consecutive output positions with the words STUDENT TRAIN, then that would be a *lag* $+1$ response. But if the participant responded STUDENT PICKEL, then that would be a *lag* -2 response. Kahana and associates have demonstrated that independent of list length, presentation time, word frequency, and mode of stimulus presentation, there is a typical lag-CRP function. Approximate lag-CRP values for lags of $-5, -4, \cdots, -1$, respectively, are .05, .05, .06, .07, and .13, whereas for the lags of $1, 2, \cdots, 5$, the respective lag-CRP values are .23, .10, .07, .06, and .05.[30] Importantly, Kahana and associates were able to place a confidence interval about each lag-CRP mean, so the issue of potential statistical artifacts is properly assessed with this metric. If the words were recalled in a random order, then the lag-CRP function should be flat. But clearly, that is not what happens. Thus, the lag-CRP function demonstrates that participants are chunking or grouping the words during encoding, and this organization is based on the major organizational structure provided (i.e., the input position itself).

29. Whenever the measurement system makes the participants cognizant of what the researcher expects or how they should behave, that system is said to have a *demand characteristic* problem. The problem is this awareness might alter the participants' behavior. The demand-characteristic problem could be avoided by introducing a surprise probe question after a particular free-recall trial. By probing different participants on different trials, the researchers might see the frequency of subjective word groups (i.e., groupings that do not correspond to the serial order during learning).

30. These values were estimated from figure 1 of the Sederberg et al. (2010) paper.

Even unrelated word lists still have a temporal or ordinal structure underlying the learning episode. Although the organization can be backward (i.e., a negative lag), there is a substantially larger forward organization. The forward bias is reasonable if the person is trying to integrate subsequent words in terms of the words that were previously presented.

13.9   Mnemonic Techniques and Imagery

Up to now, the focus has been on the factors that professional memory researchers have examined because those factors alter the rate of memory retention. The context for those studies has been that of controlled experiments where participants are instructed to learn information. Yet we also have real-world challenges of learning information for either our job or our class. But for these tasks, some might wonder if we even need to have a good memory when we have electronic storage devices within easy access. Admittedly, we have useful external storage systems for our appointment schedule, recipes, groceries, tax, and financial records. As another example of external storage, a pill box loaded with the daily medications provides a simple means for taking the right medicine at the appropriate day and time. But despite all these examples, there are nonetheless many other cases where there is no alternative to having the information stored in our memory system. A good memory is usually important for effectiveness on the job, and in some cases, a quick retrieval of the pertinent information is essential for our survival. For example, consider the pilot who has to deal with an unexpected problem that requires a rapid response. As another example, consider a doctor or a nurse who might need to implement a medical intervention in an emergency. In these cases, there is no time to learn what the problem is and to learn what steps are needed to solve the problem. A good human memory along with supplemental electronic memory, when time allows, is a far more potent information-processing combination than a confused human attempting to solve a complex problem that requires considerable knowledge. Also, there are cases where external memory storage is not allowed. For example, consider a student required to take an exam in a college course. As another example, consider the athlete who is required to learn a thick book of plays that might be called at a moment's notice during a game. The learning in all these examples is over a vast domain of facts and concepts, many of which are novel. Thus, remembering information is a common problem in everyday life, and it is a problem that can be made easier by using effective encoding strategies called mnemonics.

A number of practical books are available for coaching people to develop improved skills in memory (e.g., Cermak, 1975; Higbee, 1977; Lorayne & Lucas, 1974). The methods advanced in these books make sense in light of the memory research literature. The initial point stressed in all the methods is that a good memory requires an active awareness to encode the pertinent information. It is not a passive witnessing of incoming events, but rather it requires an effortful attention to features of the target stimuli and a skill to transform the stimuli as perceived into some other more memorable form. Yet this statement is easier said than done, so practice at encoding is essential for improving the encoding efficiency.

13.9.1   The Method of Loci

The oldest known memory improvement technique is the method of loci. It is a mnemonic technique that is attributable to Simonides of Ceos, a fifth-century BC Greek poet (Bower, 1970).[31] There are four components to the method of loci: a *memory palace* with *imaginary objects* that are *linked* together and *placed along a route* through the memory palace. The imaginary memory palace need not be a building, but it should be a distinctive spatial arrangement. Perhaps it is the route from your house to some other known location. It could also be an imaginary sequence of created images of pairs of items. It is further recommended that the image and the route be distinctive. To be distinctive, the images can be exaggerated and out of proportion. Bizarre images are considered more memorable than commonplace images that are highly similar to other familiar images. In addition, it is helpful if the objects and characters in the series of mental images are linked with action. The action should sequence the images throughout the mental palace. For example, suppose again that we want to learn the following words in the exact order as presented:

CHILDREN FOREST KING WINDOW CHAIR BIRD SMALL ELEVEN.

Previously, we used a mnemonic sentence to illustrate a subjective organization of these words in an order that was different from the presented form. Now let us use the method of loci to encode the words in the exact order as presented. To accomplish this task, we might imagine **children** walking in an imaginary **forest** and encountering a huge **king** who is looking through a very large **window** into a house. You further imagine that the children walk into the house and see a large **chair** with a huge **bird** sitting on the back of the chair. The bird is pecking marks into the wall. The marks are very **small** but look like two straight lines and a space with the pattern repeating all over the wall, that is, it looks like **11 11 11**···.

---

31. It is not surprising that a Greek poet originated a major mnemonic method. Poets in ancient Greece were thought to recite from memory long heroic tales to the audience.

There is little doubt that this rich elaborated encoding that was created in one trial will be retained later. Using modern psychological assessment, Bower (1970) found that indeed the method of loci is an effective means for learning the target information. The problem is that it is difficult to do for most people unless they really concentrate on the encoding task. But if the person practices using this mnemonic technique, then the skill to build a memorable encoding will improve. Nonetheless, there will be wide variability among individuals as to their ability to implement this mnemonic strategy. For example, there are some individuals who suffered brain damage that resulted in an impairment in their ability to create mental images (Farah, 1984; Zeman et al., 2010). Also, there are other individuals who never had the ability to construct images in the first place, a condition that has been medically described as *congenital aphantasia* (Zeman, Dewar, & Della Sala, 2015). Because imagery is a key part of the method of loci, these imagery-impaired individuals will not be able to use this mnemonic method.

13.9.2 The Substitution Mnemonic

Although the method of loci can work for common words that have a well-understood meaning, the method does not work for unknown words. This problem might arise either with rare words in your native language or with unknown words in a foreign language. As an example of the substitution method, Lorayne and Lucas (1974) describe one of the memory feats achieved by Harry Nelson Pillsbury (1872–1906), a famous world-class chess player, who also could amaze audiences with his remarkable memory. Chernev and Reinfeld (1949) reported that Pillsbury was once asked to remember the following list of words: *Antiphlogistine, periosteum, takadiastase, plasmon, ambrosia, Threlkeld, streptococcus, staphylococcus, micrococcus, plasmodium, Mississippi, Freiheit, Philadelphia, Cincinnati, athletics, no war, Etchenberg, American, Russian, philosophy, Piet Potgelter's Rost, Salamagundi, Oomisillecootsi, Bangmamvate, Schlechter's Nek, Manzinyama, theosophy, catechism,* and *Madjesoomalops.* Pillsbury was able to accomplish this challenge. It is not known what method Pillsbury used to execute this memory challenge, but Lorayne and Lucas suggest that the substitution method can be used for the rare words on the list. This method creates alternative word phrases that sound like the unfamiliar words. For example, Lorayne and Lucas suggest that the word *antiphlogistine* be substituted for the phrase **Auntie flog a stein**. They also suggested that one can imagine your favorite Aunt flogging a stein of beer.[32] Each word in

---

32. Many of the words on his list do not require the substitution method because Pillsbury surely knew the words. It is not known if Pillsbury knew the meaning of *antiphlogistine*. At the time of his memory test, the old phlogiston theory of combustion in chemistry was long out of date, but

the list that requires substitution can then be linked with an image and placed into a sequence via the method of loci.

Harry Lorayne and Jerry Lucas are themselves highly skillful mnemonists. They describe an elaborate system in which number sequences can be encoded by means of chunking and substitution. In their system, each digit is substituted with a letter or a combination of letters. These associations need to be learned first. In the Lorayne-Lucas system, the letters of either b or p are substitutes for the digit 9. The letter m is a substitute for the digit 3. Vowels in their system are "wild cards" in that they can be used for encoding but are ignored in decoding. A string of digits is chunked and then transformed to words. For example, suppose one is challenged to remember an arbitrary twelve-digit sequence in only one study trial, and the sequence is 939627184758. This task is well beyond the human memory span, so few would be able to execute this challenge without an advance mnemonic strategy. In the Lorayne-Lucas system, the twelve-digit sequence can be chunked into four numbers, that is, (939), (627), (184), and (758). Each of these numbers can then be recoded into words (e.g., BOMB CHUNK DIVER GOLF). Finally, the four words can be linked into an image; perhaps one might imagine a bomb breaking into pieces that fall into water and then an underwater diver picks up the pieces and magically turns them into a golf club. This imagined sequence is easy to remember, and it is a means for recovering the key words; those words in turn are decoded back to the original digits. This elaborate system is clearly challenging to implement, and it is unlikely to be executed by most people without extensive practice. Yet this demonstration does demystify how some mnemonists implement their impressive memory feats. It is unknown what proportion of ordinary people can learn to use this mnemonic method when given sufficient training and practice.

13.9.3 Rhyme and Peg-Word Mnemonics

A rhyme mnemonic is a sentence or a poem that provides the desired information. For example, to learn the treble clef notes of E G B D and F, music students are taught the rhyme "Every Good Boy Does Fine." Also, for remembering the number of days in each month, children are often taught the following familiar Mother Goose poem:

Thirty days hath September,

April, June, and November.

Pillsbury still might have known of it. If he did, then antiphlogistine is obviously one who is against the phlogiston theory.

All the rest have thirty-one,

Except for February alone,

Which hath but twenty-eight days clear,

And twenty-nine in each leap year.

Using the rhyme mnemonic, it is easy to associate the digits 1 to 10 with concrete objects (i.e., *One Bun, Two Shoe, Three Tree, Four Door, Five Hive, Six Sticks, Seven Heaven, Eight Gate, Nine Line, Ten Hen*). Once these digit-object associations are well learned, they can then be used for the peg-word mnemonic. Suppose, for example, that a person is asked to learn the following words in order: POOL BRICK SPOON BOOK STOVE PENCIL. With the peg-word mnemonic, the word POOL would be associated with BUN. One might imagine cartoon-like buns in sunglasses sitting around a swimming pool. Perhaps for BRICK, one might imagine two brick shoes with shoelaces on the feet of an imaginary man. Each list word is similarly associated with the peg word for that ordinal position. Later if you were asked what the second word was, you could remember that Two is Shoe, which is associated with our imaginary man with brick shoes. In essence, the peg words play a similar role as the imagined locations along the route in the method of loci.

13.9.4 Tips for Face-Person Associations

Many people wish that they were better at remembering the names as well as other facts about the people that they met. Although the recognition memory for faces is usually reasonably accurate, it is more difficult to remember the context in which you met the person as well as other information about the person such as his or her name. It is socially awkward and perhaps professionally a liability not to remember a person's name. Conversely, people are generally pleased to be remembered. So how can one improve the learning of names?

In many cases, names are arbitrary, like words of a language that you do not understand, so to learn a name will require a high level of attention and concentration. The substitution method can be helpful for establishing a meaningful phonetic representation of a name. My name is a good example of an unusual name because it is pronounced as *Cakely* despite the fact that it is spelled *Chechile*.[33] Someone using the substitution method when first being introduced to me might imagine a birthday cake with the letters LY on the cake. My first name can be added to the cake via another

---

33. The reason for this unusual pronunciation is due to the original Italian roots of the family name. In Italian, *ch* is pronounced like *k* when it precedes either an *e* or an *i*.

meaningful symbol (i.e., I usually introduce myself as Rich Chechile, so a $ sign can also be put on the cake). The resulting image should be an effective encoding of the name. In other cases, the name is less unusual (e.g., *Ruth Smith*). For *Ruth*, one might imagine *Babe Ruth*, and for *Smith*, one might imagine a *blacksmith*. Admittedly, it is not always easy to form a vivid encoding of a person's name because people can speak too rapidly or too quietly. But it is okay to ask them to repeat their name or even engage in a discussion about the name itself. These conversations are normal, so relax and make sure that the name is adequately encoded at the beginning.

Knowing the name is half of the battle of learning the face-name association. Next one needs to concentrate on studying the face in an effort to identify the person's distinctive features. Unusual features are features that make the face stand out from other faces. For example, if the person has a high forehead and curly red hair, then those features can be used to link the face with the name. Cartoonists had long known that it only takes a few distinctive features to make a caricature or a sketch with just a few exaggerated lines. In fact, Rhodes (1996) showed that participants identified a caricature more rapidly than a detailed realistic drawing of the same face. Perhaps one can imagine a caricature of the person and link that image with the image constructed for the name. Later when you are alone, you might make some notes about the face, name, and other key facts that you learned about the person that you met earlier. The taking of notes is a rehearsal trial for the new information.

As with other memory tasks, one can improve in associating the name and face with practice. Newspapers are an excellent source of novel faces and names. You can hone your skills in face-name encoding by practicing with these novel stimuli without the pressure of a real social interaction. Moreover, you can study how well you can retain this information after a delay.

13.9.5   Mnemonic Methods in Context

The encoding of information varies widely in quality and effectiveness. It can also vary across individuals and with the approach used. With mnemonic techniques, such as the method of loci, substitution, and peg words, a remarkably high performance level can be achieved. What is different between these representations and ordinary, less effective encodings? To remember an event, it is necessary to store information about the event and to be able to recover that information later. Mnemonic techniques provide a means for storing the event information in such a way that the features are elaborated and made highly distinctive. The mnemonic method further stresses the storage of links for the later recovery of the information. For example, with the peg-word mnemonic, if I want to recall the first word on a list, then the problem is reduced to remembering the chain of *1 → Bun* and then remembering what was associated with *Bun* (i.e., *Pool*). But the peg-word mnemonic would be less effective if the first word in the sequence were not elaborated in its properties and if it were not linked to the peg word. Imagery and exaggeration are generative processes that produce a distinctive elaboration of event. It is amusing and easy to remember the cartoon buns in sunglasses lounging around a swimming pool. That encoding is dissimilar to other memories, so there is likely to be less storage interference.

Yet ordinary experience is quite different from a memory experiment. Events unfold at an uneven pace, and most of the vast sea of information that we encounter is not important enough to encode into a lasting memory. So there is little need to be in a hypervigilant state of attention. Although we are not routinely concentrating on forming a durable encoding, there can be a rapid change in the memory demands. For example, we might meet a new person and want to remember his or her name along with other information that we learned from the encounter. This occasion might trigger an uptick in our attentiveness. The occasion is a time when elaborative encoding and a mnemonic strategy would be helpful.

13.10   Chapter Summary

In this chapter, attention has been directed toward very effective methods for encoding information. We have seen enhanced memory retention associated with the effects of generation, aha, self-choice, enactment, production, self-reference, distinctive imagery, subjective organization, and mnemonic strategies. A common feature among these methods is a change from a passive encoding of the memory targets to a more elaborated memory representation. These procedures did not alter postencoding processing and did not alter the strategy of memory search. Rather, what differs from a more passive encoding is the content of what was initially stored in memory. The methods create a more elaborate encoding that is linked with other memories. The enhanced information representation is more distinctively different from the subsequent other events that occur over the retention interval, so there is a reduced storage interference effect. Also, the elaborated and more distinctive encodings have more retrieval routes to aid subsequent remembering. This common account thus avoids employing separate theories to explain each of these effects of elaborative encoding. This account also does not employ the notion of memory strength as an explanation. Active encoding produces an elaborated and more distinctive encoding rather than just a stronger memory representation.

14   Mood and Emotional Factors

14.1   Chapter Overview

In the real world, events are not always emotionally neutral and routine. Some events occur in a highly emotional context. How are these events stored and retrieved? The spotlight in this chapter is on the impact of emotion on memory encoding and retention. We will see that emotional events elicit additional physiological processes that influence memory. Moreover, the analysis of emotion and memory has also uncovered problems with some widely held beliefs about the relationship between emotional arousal and memory retention.

14.2   The Question of Mood-Dependent Memory

Can a homogeneous mood state, such as being angry, happy, or sad, play a role in influencing memory recall? Given that there is evidence that incidental features of the learning context can influence memory retrieval, then it would seem that the person's mood should also exhibit cuing properties. Yet the experimental literature on mood-dependent context effects has been fraught with problems and confounded factors, so it has been a challenge to establish the phenomenon of mood-dependent memory.

One complication for studies on mood and memory is the distinction between *mood-congruent* memory and *mood-dependent* memory. Mood congruency refers to the matching of the affective mood at the time of recall with the *mood state implied by the informational content of the target* (Blaney, 1986). For example, a participant might be asked to recall an autobiographical event that was a joyful experience. In terms of this example, mood congruency is the idea that the memory of this encoding of a happy event being generated in the laboratory would be easier to recall later if the person is also in a happy mood rather than in a sad mood. In contradistinction to mood congruency, the informational content of the to-be-remembered material is affectively neutral for studies on mood-dependent memory. For a mood-dependency experiment, the participants themselves are induced to be in a particular mood state before studying some affective neutral target information. For example, in a mood-dependent study, a person might be induced to be in a negative mood and then asked to learn a set of neutral words. Perhaps a day later, the person is asked to remember the previously encoded neutral words when in either the same mood as the original encoding or in a different mood state. Mood-dependent memory is the idea that the memory for affectively neutral material studied in a particular mood-state condition is better remembered at a later time if the person is again in the same mood state as for the original encoding (Blaney, 1986). For mood-dependent memory, the mood is an incidental feature because there is no connection between the mood and the target information that is encoded. Thus, mood-congruent memory is not an example of state dependency like that of the drug state; rather, it is simply a cuing effect. Given $Mood_1$, there was a mood-relevant $Target_1$ encoded, and given $Mood_2$, there was a different mood-relevant $Target_2$ encoded. If the person is in the $Mood_1$ state, then it would be expected that $Target_1$ is more likely recalled than would be $Target_2$ because $Target_1$ is directly related to the content of the $Mood_1$ state.

The attempts to demonstrate mood-dependent memory in the laboratory have been problematic. There were reports of limited success that were followed by failures to replicate the successful cases. For example, Bower, Monteiro, and Gilligan (1978) performed three experiments to find mood-dependent memory. In their first two experiments, the participants studied a single list once they were induced via hypnosis procedure to be in a particular mood state (either happy or sad). Later, the participants were tested in either the same mood state or in the opposite state. There was no evidence for mood-dependent memory for either experiment.[1] Other investigators also failed to find evidence for mood dependency for single lists (Isen, Clark, Shalker, & Karp, 1978; Schare, Lisman, & Spear, 1984). In the third experiment in the Bower et al. (1978) paper, the participants studied two lists of words and had experience with both mood states. This procedure has come to be called the interference method for studying mood dependency. Bower et al. (1978) did report finding mood dependency in their third experiment, but Bower and Mayer (1985) failed to replicate that experiment.[2] This

1. The word lists were questionable because not all the words were affectively neutral. The twenty words on the lists consisted of eight happy words, eight sad words, and four neutral words. If these investigators found that a mood shift between the encoding phase and the test phase altered the recall rate, then that difference might have been caused by the degree of mood-congruency cuing.
2. The Bower et al. (1978) third experiment was further compromised by the fact that each of the two lists had affectively laden words.

null effect is particularly troublesome because it is a replication failure from the same research laboratory. Moreover, this failure to find a mood-dependency effect for the interference method has been confirmed by other researchers (Løhre, 2011; Wetzler, 1985).

Several reviewers of the research literature about mood and memory have concluded that there is good support for mood congruency but weak to no support for mood-dependent memory (Blaney, 1986; Forgas & Eich, 2012). Yet I believe that the study by Eich, Macaulay, and Ryan (1994) successfully demonstrated a mood-dependency effect by using a novel experimental procedure. These investigators used music to induce either a happy or a somber mood. The music played throughout the entire study phase. The researchers also instructed the participants to generate important personal memories from their past that could be associated with a set of emotionally neutral words. Later during memory testing, a different musical selection was played, but the music was linked to either the same mood state as in the study phase or a different mood state. In both of their experiments, Eich et al. (1994) found a mood-dependency effect. The effect was demonstrated after a 2-day retention interval but not after a 7-day interval. Critics might wonder if the effect found was due to the use of music to induce the mood. Yet it is important to note that the music-dependent effect described earlier does not occur when there is a change from the specific musical selection (Balch, Bowman, & Mohler, 1992; Balch & Lewis, 1996). Because the Eich et al. study used different music at the time of test from the music played during the encoding period, their evidence for mood dependency cannot be dismissed as a music-dependency effect. But there is another novel feature of the Eich et al. experimental protocol in its use of self-produced, autobiographical associations to the neutral words presented. This procedure is a generative encoding method. The term *generative* is used for this type of encoding because the participants supply their solution to a cue or a puzzle. It is not always obvious which personal memory to link with an experimenter-provided neutral word, so it is likely that several alternative candidates might be considered before one memory is selected. Soraci et al. (1994) found that this type of generative encoding produces additional intermediate candidates that can later be helpful as retrieval cues. Generative encoding can thus provide a retrieval-rich encoding environment. Moreover, the collection of this network of cues is produced when the person is in a particular mood state. A shift in the mood state at the time of test can thus be particularly disruptive.

So what can we conclude about the issue of mood-dependent memory? One would think that mood-dependent memory is a natural extension of drug-dependent memory, but after eliminating the confounding of mood congruency, it has been difficult to show that the learning of neutral information is altered by a shift in the mood state. Perhaps moods induced in the laboratory are not sufficiently extreme to produce an effect. But I am not claiming that mood is unimportant for learning, and there is a consensus that mood congruency is a reliable effect. Yet the moods induced within controlled experiments have not been an effective retrieval cue for targets that are emotionally neutral. Perhaps there is little to differentiate laboratory-induced moods from countless other occasions when the person experienced a similar mood. Yet it only requires one good study to establish the possibility of mood-dependent memory. I believe that the Eich et al. (1994) study did show us how to produce a mood-dependency effect with their use of music and generative encoding. I suspect that the generative encoding is the critical feature of their experimental protocol for producing a retrieval-rich set of meaningful information that was linked to the mood state. If there are lots of retrieval cues produced in a particular mood state, then a shift in the mood state at the time of test is more likely to result in a retrieval deficit. It remains an open question if mood-dependent memory can be produced with other experimental protocols that do not rely on the powerful effect of generative encoding.

14.3   Effects of Arousal and Emotion on Encoding

It is traditional to initiate the discussion of emotion and encoding with the Yerkes-Dodson law. But there are good reasons for questioning the correctness of this law. Hence, before addressing the memory of emotionally arousing information, let us first critically examine the Yerkes-Dodson law.

14.3.1   The Questionable Veracity of the Yerkes-Dodson Law

It is widely accepted that there is an inverted U-shaped function between performance and the level of arousal. Performance is believed to be at a low level for low arousal levels and is low again for very high arousal levels, but between the two extremes in arousal, the performance is considered optimal. The attribution for this claim is based on the findings of the Yerkes and Dodson (1908) paper, and the effect is referred to as the Yerkes-Dodson law. The inverted U-shaped function between performance and a general level of arousal is common in Introductory Psychology textbooks (Winton, 1987). There is a second part of the Yerkes-Dodson "law" that states that the value of the optimal arousal level decreases as the learning task becomes more difficult. After more than a century since the original paper was published, the Yerkes-Dodson law is still widely accepted. However, it is doubtful if succeeding generations of psychologists have actually carefully read the original Yerkes and Dodson paper because it does not support this conclusion. Rather, it seems like psychologists have repeated their understanding of what other psychologists have taught or written about this paper, like a version of the Bartlett (1932) experiment where the succeeding retelling of the the War of the Ghosts story across different participants resulted in an increasingly distorted version of the story. So it is important to carefully consider the Yerkes-Dodson study to see what was actually done.

For any trial in the Yerkes-Dodson learning study, an animal was placed in a chamber, and it was allowed to return to a nest-goal chamber by either one of two paths. One path was white, and the other was black. If the animal picked the black pathway, then it experienced an electrical shock. It was assumed by the authors that the shock was aversive at all intensity levels, but there was not a separate assessment of that assumption. They considered the correct choice was the selection of the white pathway because it was free of shock. The independent variable manipulated was the stimulus intensity level of the electric shock. The dependent variable was the number of incorrect choices in ten consecutive test trials. They were interested in the number of training trials required before the animal reached the criterion of three consecutive days in which the animal always selected the correct path. They reported finding that the best learning was for an intermediate shock level when the apparatus was either dimly lit or when it had an average room brightness. But if the apparatus was brightly illuminated, then the best learning occurred for the condition with the most intense shock. Thus, it was only in the case of either dim or average brightness did they find that the best learning occurred for an intermediate shock level. Thus, the independent variable studied was shock intensity and not arousal or motivation.

A number of peculiar aspects of the Yerkes-Dodson study immediately should raise a skeptical reaction. First of all, the species of animal used in their experiment was a mutant strain of hyperactive deaf mice that are called *dancing mice*; these animals engage in an obsessive, repetitive circular running that could interfere with their learning of a discrimination task based on locomotion (Hancock & Ganey, 2003). In fact, subsequent studies with other species did not find support for the inverted U-shaped function. For example, following closely the original research protocols, Cole (1911) did not find an inverted U-shaped function for chicks; instead, the relationship was an increasing function of shock intensity. Dodson (1915) also failed to find evidence of an inverted U-shaped function in another replication study with kittens. Thus, the function is not clearly an inverted U. Moreover, the amount of data as well as the vagaries of one experimental condition further compromises the value of the evidence in the Yerkes-Dodson study. More specifically, in their first experiment, the strength of the stimulus condition was unknown. That problem was resolved in their second and third studies. But in those experiments, only a few animals were tested and no statistical analyses were provided. Nonetheless, the data are available for anyone to do a statistical analysis. I did a classical nonparametric analysis of their data and did not find a statistically significant result for either experiment.[3] Hence, there is not a strong case that the Yerkes-Dodson data support any curved function across the various levels of the independent variable. Bäumler and Lienert (1993) did a different nonparametric analysis of the Yerkes-Dodson data and also rejected the hypothesis of a curvilinear function.

At a theoretical level, the Yerkes-Dodson study was neither escape training nor appetitive training. Shock was administered only when the animal took the black pathway. But for escape training, the animal is immediately placed in an aversive condition and must learn to be free of the aversive stimulus. But in appetitive conditioning, the animal is seeking to obtain the goal object. If one presumes that the animals were seeking to return to their home nest, then it can be considered an appetitive-like study. From this perspective, the shock stimulus was a feature associated with one of the pathways. But to assess the relationship between learning and the intensity of the motivating stimulus, it is better to employ either pure escape training or pure appetitive training. For the case of pure escape training, a later well-controlled experiment found that the magnitude of the shock has no influence on the rate of learning (Theios, Lynch, & Lowe, 1966). In escape training, an error trial is considered any trial in which the animal took more than 5 sec to escape shock (Theios & Dunaway, 1964). Thus, we can consider the Theios et al. (1966) experiment as a more sophisticated study of the learning rate versus the level of shock intensity, and that study found no statistical difference with shock intensity. In fact, the animals learned to achieve a perfect level of learning in only 7.9 trials on average, and it took 1.1 sec to escape from shock, regardless of the shock intensity level. Now let us consider the case of pure appetitive conditioning. The effect of reward depends on the amount of food deprivation, the type of food, and the magnitude of the food reward. Beier (1958) investigated the speed of maze running as a function of the magnitude of the food reward for food-deprived rats. The

---

3. I used a nonparametric analysis of the rank-order information. This analysis does not require making assumptions about the statistical error structure because the data are not available to assess the distributional properties of the statistical error. The dependent variable was the number of trials before a run of ten correct trials, and the statistical test was the Kruskal-Wallis one-way analysis of variance (see Siegel, 1956). The test statistic was $H = 5.57$ in experiment 2, which is not significant, $p > .23$. For experiment 3, the test statistic was $H = 6.167$, which is also not significant, $p > .10$. For classical tests, the null hypothesis is rejected when it is rare (i.e., $p < .05$). See the glossary for more details about classical statistics. Since these results are not rare, there is not a significant effect.

unit of food was in terms of the number of small food pills. He found that the running speed increased with the number of food pellets over the range tested (i.e., 1, 7, or 13 pills). The Beier experiment is thus an appetitive version of the learning rate versus the magnitude of the motivating stimulus. Unlike the escape training, Beier found that as the intensity of the reward increased, there was an increase in the performance level. Importantly, for neither the Theios et al. (1966) study nor the Beier (1958) study was there an optimal level of performance for an intermediate level of the independent variable. Neither study was consistent with the Yerkes-Dodson inverted U-shaped function.[4] Thus, the Yerkes-Dodson law is a psychological myth.[5]

Why then has the Yerkes-Dodson law been so influential and so mistakenly accepted by many psychologists even today? It is difficult to know entirely, but in part it seemed to fit within a general theory of arousal that was popular in the 1950s. For example, Hebb (1955) advanced a drive theory as an inverted U-shaped function. He hypothesized deep sleep as the low point on the x-axis for arousal and emotionally disturbed anxiety as the other extreme on the arousal axis. He considered learning as poor at these extremes, but he felt that a high level of learning occurred for people who were alert and positively interested in the task. Implicit in this simplistic folk psychology hypothesis is the notion that there is a continuous arousal axis and a corresponding continuous learning function. Hebb offered no experimental support for this framework. But Lacey (1967) has carefully critiqued the hypothesis of a unidimensional continuum for arousal in light of physiological evidence. In Lacey's analysis, the concept of a unidimensional arousal function is not consistent with the physiological data. Admittedly, a person who is asleep, depressive, disinterested, or totally inattentive to the stimuli in an experiment will have inefficient encoding. Moreover, a person who is suffering from either test anxiety or ADHD (attention-deficit hyperactivity disorder) is also not likely to focus and concentrate on the target stimuli in a memory experiment. But it is a mistake to believe that there is a smooth continuum between these extremes. It is also a mistake to believe that the experimenter can dial in a level of arousal and

4. Neiss (1988) argued that the Yerkes-Dodson law is too flexible because arousal cannot be accurately measured in most studies. In essence, he argued that the Yerkes-Dodson law was flawed because it was immune from experimental falsification. I share Neiss's critical view of the Yerkes-Dodson law but disagree with the premise that it is not falsifiable. I consider the studies by Theios et al. (1966) and Beier (1958) as suitable tests of the Yerkes-Dodson law, and the law was disconfirmed.

5. I am not alone in criticizing the Yerkes-Dodson law (Christianson, 1992; Hancock & Ganey, 2003; Teigen, 1994). For example, Christianson concluded that the Yerkes-Dodson law is no longer useful for the psychology of emotion.

continuously vary the arousal level by changing the intensity of a stimulus. It is better to abandon the Hebbian belief in an inverted U-shaped function and to focus instead on studying discrete emotional states.

Although there is not a smooth continuum for arousal, there is evidence that stress hormones can nonetheless have a graduated influence on learning. For example, Gold and Van Buskirk (1975) studied the effect of epinephrine dosage level on the retention of a one-trial foot-shock avoidance response. Epinephrine or adrenaline is normally produced by the adrenal gland as part of a stress response, but in the Gold-Van Buskirk study, the dosage level of epinephrine was experimentally varied. The dosages were .001, .01, .05, .10, and .50 milligrams per kilogram of body weight. There was also a saline control condition. All animals were administered the drug or saline via a subcutaneous injection after the initial training. After a 24-h interval, the animals were given a retention test. The dosage levels of .001 and .50 mg/kg did not differ from the saline control, whereas the animals performed at a higher level for the other three dosage levels. The best performance was for .01 mg/kg. Thus, despite the fact that the studies on the magnitude of foot-shock intensity by Theios et al. (1966) did not support an inverted-U performance curve, the variation of epinephrine at a fixed shock level does support the fact that an intermediate dosage level yields a better memory retention 24 h later. We will return to discuss other effects of stress on memory in a later section; however, it is important here to differentiate the Yerkes-Dodson law from the variation in memory retention caused by epinephrine. Many drugs have an optimum dosage level. Epinephrine can facilitate later memory retention, but the magnitude of this effect is dosage level dependent.

14.4 The Easterbrook Hypothesis

We have already discussed the effect of context for emotional moods (i.e., the relationship between the mood at the time of the encoding and the mood at the time of test). The *mood-congruent effect* is the improved recall for an event when the participant is in the same mood state at the time of test as the mood state associated with the content of the encoding (Blaney, 1986; Forgas & Eich, 2012). But information that is emotionally evocative might also have important effects on attention, which might alter the encoding (Easterbrook, 1959). Easterbrook hypothesized that there is a relationship between the arousal level and the range of cues utilized in memory encoding. More specifically, he speculated that an increase in the arousal level results in a corresponding shrinkage of the attentional focus (i.e., at higher arousal levels, there is more attention distributed over fewer features). Interestingly, Easterbrook introduced this hypothesis as a way to explain the Yerkes-Dodson law. To Easterbrook, the narrowing of the focus from low to moderate arousal levels improves encoding because there is a focus on the relevant features and less attention to the irrelevant features. But at even higher arousal levels, the further narrowing of the attentional focus was thought to impair the encoding effectiveness. In light of the above critique of the Yerkes-Dodson law, there is a problem with the Easterbrook hypothesis of a continuous narrowing process for cue utilization as a function of arousal level. It is more likely that arousal has discrete states rather than a continuous gradation. Also in light of the different functions for appetitive versus aversive conditioning, it would seem that the Easterbrook hypothesis would also have a major problem accounting for positive emotions.

14.5   Attentional Spotlight, Cognition, and Emotion

The *spotlight* and *zoom lens* model of attention (Eriksen & St. James, 1986) provides a mechanism for a dynamic focusing effect during the processing of information. These investigators were inspired by the observations of William James, who discussed his introspective experience of visual attention. James describes a gradient of visual clarity that can dynamically vary with conscious control.

It has been said, however, that we may attend to a object on the periphery of the visual field and yet not accommodate the eye for it.[6] Teachers thus notice and the acts of children in the school-room at whom they appear not to be looking.... Usually, as is well known, no object lying in the marginal portions of the field of vision can catch our attention without at the same time 'catching our eye'—that is, fatally provoking such movements of rotation and accommodation as will focus its image on the fovea or point of greatest sensibility. Practice, however, enables us, *with effort*, to attend to a marginal object whilst keeping the eyes immovable. The object under these circumstances never becomes perfectly distinct—the place of its image on the retina makes distinctness impossible—but (as anyone can satisfy himself by trying) we become more vividly conscious of it than we were before the effort was made. (James, 1890a, p. 437)

Similar to the above description, the spotlight/zoom lens model also stresses the idea of zones of perceptual clarity that can dynamically vary (Eriksen & St. James, 1986). Initially, the spotlight model emphasized the idea that there was a sharp attentional focus that was surrounded by a fringe area with lower resolution. Surrounding this fringe area is the marginal region where the information does not reach conscious attention (Eriksen & Hoffman, 1972). But later, the size of the spotlight could also vary (Eriksen & St.

6. The accommodation that James is referring to deals with the ability to consciously control the visual focus point in space (either far or near) by changing the shape of the ocular lens; this phenomenon is called *visual accommodation* (Duane, 1922).

James, 1986). As the spotlight shifts its location, the person processes more of the visual scene. The goals of the viewer can influence what is processed next after scanning the current spotlight region (i.e., there is a temporal sequence of scans). Thus, this framework is consistent with the temporal encoding process that was discussed previously with the syntactic pattern-recognition model. But in addition to cognitively guided scanning, emotional content is particularly salient; in fact, stimuli that have a high emotional arousal have a processing priority (Mather, 2007). In fact, highly threatening stimuli are more rapidly detected (Mather & Knight, 2006; Öhman & Mineka, 2001).

14.6   Neural Processing of Emotional Stimuli

There are neurological reasons why fear and arousing stimuli have an attentional and processing priority (Mather, 2007). The amygdala are brain regions that play an important role in detecting and representing emotionally important information (Blair, Schafe, Bauer, Rodrigues, & LeDoux, 2001; Cahill et al., 1996; Kensinger & Schacter, 2007; Maren, 1999).[7] For example, Cahill et al. (1996) used positron emission tomography (PET) to examine which brain regions were engaged at a higher rate when encoding emotionally arousing information. The PET scan is an neuroimaging method that can detect changes in glucose metabolism, which is indicative of local neural activity. Cahill et al. reported that the amygdala were more engaged when participants were viewing an emotional film rather than a neutral film. Similarly, Kensinger and Schacter (2007) used functional magnetic resonance imaging (fMRI) to examine the brain regions that were more engaged with emotional rather than neutral information. The fMRI signal is based on a change in cerebral blood flow via a method that is sensitive to the blood oxygen level. Kensinger and Schacter also reported that the amygdala were more active for emotional stimuli. The studies by Blair et al. (2001) and Maren (1999) examined fear conditioning in animals. More specifically, they investigated which brain regions were involved when the animals processed a previously neutral stimulus that was associated with an aversive unconditional stimulus. These studies reported evidence for long-term potentiation (LTP) in the amygdala. Since LTP is considered important for the construction of a memory trace, the development of LTP in the amygdala supports the hypothesis that a previously neutral stimulus, which was associated with a negative emotional event, has a representation in the amygdala. Thus, for negative

---

7. The amygdala are almond-shaped brain regions near the hippocampus (Haines, 1991). In mammals, each hemisphere has an amygdalae.

emotional information, the amygdala play an important role for both the detection and the representation of learned fear stimuli. However, it should be stressed here that these studies continue to find that the hippocampus is engaged in new learning as has been discussed previously in this book, but negative emotional events also have the additional involvement of the amygdala. In the Teyler and DiScenna (1986) indexing model, there normally is a rapidly developing hippocampal trace and a slower developing neocortical trace. Yet negative emotional events have an advantage of having possibly a third representation in the amygdala. Yet the LTP evidence in the amygdala for fear conditioning might be just an encoding of the emotional content of the event rather than a representation of the whole event. For example, McGaugh, Cahill, and Roozendaal (1996) argue that the amygdala play a facilitatory role to the memory representation of the event in the hippocampus and other brain regions. The mechanism of this facilitation is via adrenal stress hormones triggered by the amygdala (McGaugh & Roozendaal, 2002). Thus, it is not clear if the amygdala have a representation of the whole event or just facilitate a vivid event representation elsewhere in the brain.

There is also evidence that the amygdala are involved with positive emotions (Nishijo, Ono, & Nishijo, 1988; Schoenbaum, Chiba, & Gallagher, 1998; Schoenbaum, Setlow, & Ramus, 2003). But these studies indicate the neural activation pattern for appetitive conditioning is different from the pattern for fear conditioning because it involves the orbitofrontal cortex, the amygdala, and the hippocampus. The neural pattern is also different because there is a different subregion within the amygdala associated with appetitive conditioning from the subregion linked with fear conditioning. Nonetheless, the amygdala play a contributing role for both positive and negative emotional events.

In light of the spotlight model and the neurological boost for emotional events, there should be some different features associated with the encoding of arousing information. Emotionally significant stimuli should be detected quickly. But what happens next might be very different for positive affect than for negative affect. So let us examine these two cases separately.

14.7   Negative Affect: The Weapon Focus

The key idea of the Easterbrook hypothesis is the narrowing of the attentional focus when detecting an arousing stimulus. This idea would seem to make sense for the case of negative or threatening stimuli. For example, many legal experts felt that eyewitnesses of a violent crime were inattentive to other features of the crime scene because the witness was focused instead on the threat object (Loftus, 1979). A weapon is likely to trigger a response from the amygdala and cause the person to center his or her attentional spotlight on the threat object. Moreover, after detecting the weapon, the person is unlikely to shift his or her attentional focus; consequently, he or she might miss other aspects of the information environment.

Despite the reasonableness of an attentional focus effect for threat objects, a careful experimental demonstration of the *weapon focus* phenomenon was not established prior to the study by Loftus, Loftus, and Messo (1987). In that study, participants saw a customer at a fast-food restaurant point either a gun or a check at the cashier. They tracked the eye movements of the participants and later tested the memory of the scene. The information in the two conditions was contained in a series of eighteen slides that were identical except for four slides showing either a gun or a check. Loftus et al. (1989) reported that the eye fixations on the weapon was longer than on the check. The participants were asked to answer a series of twenty questions about the slides. For each question, there were four alternatives. The questions dealt with features such as the color of the customer's coat, the type of soft drink, the name of the restaurant, and the identity of the customer. Loftus et al. reported that there is a better memory for the *check* condition. The extra fixation on the gun resulted in a poor retention of the other aspects of the scene. There are numerous other demonstrations of the weapon-focus effect under many different conditions (e.g., Maass & Köhnken, 1989; Mitchell, Livosky, & Mather, 1998; Pickel, 2009).

14.8 Memory of Traumatic Experiences

It is fitting here to observe that in general, the memory of horrific event is usually very well remembered, as would be expected in light of the special encoding priority of highly emotional experiences (Berntsen, 2001; McNally, 2003; Peace, Porter, & ten Brinke, 2008; Schelach & Nachson, 2001; Wagenaar & Groeneweg, 1990). The memories of abuse are vivid and frequently not compartmentalized experiences from the past because the victims of abuse can relive those horrific events with flashbacks (Berntsen, 2001). In essence, the traumatic memories can be reexperienced. In some cases, the person might be clinically suffering from posttraumatic stress disorder (PTSD). But as we have discussed previously, the retrieval of information is a new encoding opportunity that could either reinforce the existing memory representation or result in a reconstructed and distorted memory of the past (Bartlett, 1932). Consequently, traumatic memories can be vivid and accurate, but they also can be altered to be inaccurate. For example, the studies of concentration camp survivors report that the victims of abuse have accurate memories of some features of their experience and a poor memory of

other features (Schelach & Nachson, 2001; Wagenaar & Groeneweg, 1990). For example, Wagenaar and Groeneweg examined the testimony of seventy-eight witnesses in
a trial against a Nazi concentration camp official. Fifteen witnesses were interviewed
in the 1940s and interviewed again in the 1980s. Overall, the witnesses had a detailed
memory of Camp Erika in The Netherlands. However, Wagenaar and Groeneweg also
reported a number of dramatic discrepancies between the two interviews. For example,
one witness reported in the 1940s that he was beaten so badly that he was unable to
work for a year, but in the 1980s, he was unable to remember the name of his persecutor.
Another witness reported that two guards, who he remembered by name, drowned a
prisoner in a water trough, but 40 years later, he could not remember that crime, and he
denied reporting it earlier. It is not surprising that there is some forgetting even of traumatic events. In general, there is a good memory of the traumatic experience, contrary
to the Freudian hypothesis of a repressed memory, but distortions and forgetting can
still occur for these events. A similar analysis was made by Schelach and Nachson (2001)
about the memories of the survivors of the Auschwitz extermination camp during the
Second World War. Schelach and Nachson agree with the overall assessment from the
Wagenaar and Groeneweg study. They further recommend that traumatic memories
should be admitted as evidence in courts of law, but this eyewitness memory should
also be examined by the same standards as nontraumatic memories.

14.9 Positive Affect: Attentional Broadening?

It is reasonable to expect that positive affect could have very different consequences for
memory retention. In joyful or playful occasions, the person is not likely to have the
amygdala trigger the adrenal stress hormones, so the hippocampus would not receive
the facilitatory benefit of the stress hormones. Moreover, there can be a different attentional response for the case of positive emotions. In fact, a number of psychologists
have hypothesized that positive affect can lead to an attentional broadening (Derryberry & Tucker, 1994; Fredrickson, 1998; Gasper & Clore, 2002). In essence, these
researchers are suggesting that in the case of positive emotions, there is the opposite
effect from that suggested by the Easterbrook hypothesis. Instead of narrowing the
focus of attention, they are suggesting a broadening of the attentional focus. However, there is not yet a convincing demonstration of the broadening effect on memory
encoding. The reason that I characterize the research on attentional broadening as
unconvincing is because there has not been a study (to my knowledge) like the Loftus
et al. (1987) experiment where eye movements are recorded as the participant views
a scene while being in a positive affective state. Researchers have instead focused on

higher-level reasoning and perceptual tasks. Perhaps the best case for broadening is for the perceptual-flanker task (Rowe, Hirsh, & Anderson, 2007). For this task, the participants are asked to focus on a central figure and ignore the flanking stimuli. The failure to ignore the flankers is taken as evidence of a broadening of attentional focus. Yet this demonstration does not tell us how much time is devoted to processing flankers. Without eye-movement tracking of the processing while in a positive affective state, it is difficult to know that the encoding is widely sampling components of the visual display.

14.10   Postencoding Stress and Memory

Up to now, we have focused on the encoding of stimuli that have heightened emotionality. But there is also an important research literature that has examined the role of stress after the learning trial is completed. The effects of postencoding stress are complex and can have potentially important theoretical implications.

There are several hormones released by the adrenal gland as part of the fight-or-flight response to acute stress (Cannon, 1953). These hormones help prepare the organism to deal with a threat by increasing heart rate, releasing glucose as an energy source, and inhibiting processes associated with growth and reproduction (Kemeny, 2003). We will focus here on the memory effects associated with two of these hormones—epinephrine and cortisol. The time for the release of these two hormones is different. Epinephrine is released seconds after the stressor event and quickly reaches its peak level, whereas cortisol is released slowly and takes about 30 min to reach its peak level (Kemeny, 2003).

Stress is typically induced experimentally in people by either the Trier Social Stress Task (TSST) or the Socially-Evalutated Cold-Pressor Task (SECPT) (Dickerson & Kemeny, 2004). The TSST method requires people to give a speech and to perform difficult arithmetic problems before a panel of critical judges, whereas the SECPT method requires people to keep their hand in an ice bath while undergoing a social evaluation. Both methods reliably produce a stress reaction (Dickerson & Kemeny, 2004). For human memory stress studies, the retention of items that are either emotionally neutral or emotionally evocative can be studied as a function of stress in the postencoding time period.

Compared to the human stress-inducing methods, animal stress studies use different experimental protocols. The animal experiments, unlike the human research, usually involve emotionally significant stimuli. For example, learning to avoid foot shock or learning to find a hidden platform in a water maze are escape-training procedures

(Gold & Van Buskirk, 1975). These learning tasks are stressful. After the training trial is completed, the retention interval is typically a time of relative calm. Researchers inject the animals at this time with various substances to assess the influence of postencoding stress hormones.

14.10.1 Epinephrine and Memory

Gold and Korol (2012) concluded in their review of the research literature that adrenaline (i.e., epinephrine) has an inverted U-shape positive effect on recent memories. Yet adrenaline injected into the bloodstream does not cross the blood-brain barrier (Axelrod, Weil-Malherbe, & Tomchick, 1959), so the mechanism of the memory effect is itself an interesting scientific question. Gold and Korol hypothesized that the influence of epinephrine on memory occurs because it triggers an increase in the blood glucose, which does cross the blood-brain barrier. In support of this idea, Gold (1986) found that posttraining glucose injections also had an inverted U-shaped function on memory retention. Perhaps peripheral adrenaline affects memory indirectly by causing a spike in blood glucose. Regardless of the mechanism, a peripheral injection of epinephrine results in improved performance relative to a saline injection. But higher levels of both epinephrine and glucose result in a weaker effect relative to an intermediate dosage level (Gold & Korol, 2012).

The inverted U-shape dose-rate function of epinephrine is not a surprising outcome. Many substances have an inverted U-shaped dose response (Calabrese & Baldwin, 2001). Such a response function simply means that the substance has an optimal range of effectiveness. When the epinephrine dosage is either too high or too low, the effect on memory is not an impairment; rather, it is less beneficial. Consequently, the more important question is why the influence of postencoding adrenaline is positive in the first place. This effect of peripheral epinephrine has been interpreted by some researchers as support for the consolidation theory of memory (Gold, 1986). In the consolidation theory, a recent memory is like a wet painting that requires time to dry into a stable painting. This hypothesis was critically discussed earlier in chapter 3. A consolidation theory account for postencoding stress hypothesizes that the weak, unstable memory of the target event is somehow strengthened by the combination of epinephrine and glucose. Consolidation theorists argue that support for this interpretation comes from the finding that chemically blocking the effect of epinephrine results in an impairment of memory (Clayton & Williams, 2000; Quirarte, Roozendaal, & McGaugh, 1997). However, there is another theory that can account for the beneficial effects of epinephrine in the postencoding stage.

The consolidation theory account of the epinephrine benefit is unconvincing and problematic. The central assumption of consolidation theory is the idea that the encoding is initially poor. Yet in well-controlled experiments using the Brown-Peterson procedure, the initial storage probability after encoding is generally excellent, as shown earlier in chapter 7. See, for example, figure 7.8. But all memories must undergo a postencoding storage interference process. From the two-trace hazard theory perspective discussed in chapter 3, this process is characterized by a hazard function. Memory hazard is the risk that a stored memory trace becomes insufficient as a result of another experience in the retention interval.[8] Chechile (2006) provided evidence that the hazard function is peak shaped. As the number of postencoding events increases, there is a corresponding increase in memory hazard. Eventually, this increase in hazard reaches a peak, and it decreases as a function of the number of postencoding events. From this theoretical position, the passage of time does not strengthen an initially weak trace. Rather than the target event changing in strength, the interference process is seen as dynamically changing. Storage interference is a probabilistic process that is more vigorous toward recent memories. Hence, after an initial rise in the memory-hazard rate, the risk of a memory loss from intervening activity is reduced in vigor. Thus, unlike consolidation theory, the memory representation in the two-trace hazard model is initially a good representation that undergoes attempts to dislodge the features of the target event. The interference process selectively attacks existing memories, with the most recent memories being at a greater risk. The trace can either withstand the storage interference risk or it can degrade in quality. Other factors can also play a role during the postencoding retention interval. For example, rehearsal during the retention interval can result in a memory benefit by elaborating the target trace rather than interfering with the target encoding.

So how can the two-trace hazard theory account for the epinephrine benefit in the postencoding timeframe? To answer this question, it is important to remember that for animal studies, the training is usually emotionally evocative, whereas when the trial is finished, it is a time of relative calm. Consequently, the experience of having a supplemental boost in adrenaline might reestablish the initial context of the training trial. The adrenaline rush is a stimulus itself, so it is reasonable to hypothesize that this internal stimulation can retrigger the memory of the recent arousing activity that the animal experienced earlier in training. In essence, the retriggering of the memory is a rehearsal event. Rehearsal between the time of training and the time of test

8. Memory hazard is the conditional probability that a stored item becomes insufficiently stored at a particular point in the retention interval given that it has survived up to that time.

should improve memory. Thus, the two-trace hazard model provides an alternative explanation to the account advanced by consolidation theorists.

The findings from several studies raise problems for the consolidation account of the effect of postencoding stress hormones (e.g., Cahill & Alkire, 2003; Okuda, Roozendaal, & McGaugh, 2004). Consolidation theory is based on the idea that stress hormones in the postencoding period strengthen an initially weak representation. This benefit should occur regardless if the original training involved heightened emotionality or if training was affectively neutral. The finding that corticosterone does not aid memory for a nonemotional event is thus a problem for the consolidation position (Okuda et al., 2004). In that study, animals were either allowed to become habituated to the test chamber or not. It is a stressful experience for rats to be put into a new environment. Animals in the study by Okuda et al. were eventually provided an opportunity to observe objects in the experimental environment. The animals, which were allow to first habituate to the experimental chamber, were presumedly exploring the objects without feeling stress, but the animals, which were not allowed to habituate first to the environment, were likely exploring the objects while under stress. The posttraining administration of a stress hormone helped later memory retention only for the animals that explored the objects while being stressed. It is interesting to observe here that the two-trace hazard account for the postencoding storage effect is consistent with the findings of this study because the cuing effect in the postencoding period should only aid a retriggering of the training experience if the training in fact involved stress.

Another study by Cahill and Alkire (2003) also found that epinephrine in the postencoding time period only aided the items that elicited higher arousal. This experiment was conducted with humans who were asked to learn a list of words. The investigators also monitored the participants' electrodermal skin response, which is an ongoing measure of emotionality. The researchers found that during learning, only the beginning of the word list elicited increased stress, as measured by the skin monitor. Furthermore, the memory benefit was limited to the beginning part of the word list. Once again, consolidation theory cannot account for why only the emotional items are improved by stress hormones in the postencoding timeframe. If the hormones trigger a biological strengthening of an initially fragile encoding, then both affectively neutral items and items of greater emotionality should be strengthened. The failure to find improved memory for neutral items is thus a problem for the consolidation account for the postencoding stress benefit. But the two-trace hazard model can explain the memory benefit if the hormones reactivate the memory of the training. This reactivation is only expected if the original learning also involved stress.

14.10.2 Delayed Cortisol and Memory Recall

A number of investigators studied the effect of inducing stress long after the original encoding. These experiments typically generate stress 24 h after the original encoding (Schwabe, Joëls, Roozendaal, Wolf, & Oitzl, 2012). Schwabe et al. report that recall is impaired when cortisol is at its peak level. Presumedly at this time period, the effect of stress should be mainly on memory retrieval processes. The educational implications of this effect are important. Often students study information days prior to an examination in an unstressed context. But later during an examination, the students need to recall the studied information while feeling a stress response. Performance is not expected to be optimal under these conditions (Vogel & Schwabe, 2016). But a study by Smith, Floerke, and Thomas (2016) provides evidence that this delayed-stress impairment of cortisol can be eliminated by an improved encoding procedure.

If memory retrieval is impaired when a person is under the influence of cortisol, then perhaps practice at retrieving the information would be beneficial. Recall in chapter 11 that the testing effect demonstrated that a testing trial results in a substantial improvement in learning. Consequently, in the Smith, Floeke, and Thomas study, participants studied a list of words or images under two different encoding conditions. One group was called the Study-Practice group, and these participants had to study a list of target items on four consecutive trials. The other group was called the Retrieval-Practice group, and these participants studied the items only once and then they were given three consecutive recall-test trials. No feedback or corrections were provided on the test trials. Hence, the Study-Practice group received four study trials and no test trials, whereas the Retrieval-Practice group received one study trial and three test trials. After a 24-h delay, all participants returned for further assessment. The Trier Social Stress Task was used to induce a stress reaction for half of the people, whereas the other half of the people engaged in a different task that did not induce stress. Remember that the Trier procedure produces a quick epinephrine response, but the cortisol reaction requires another 25 min to develop. To study the effect of stress under both hormone conditions, half of the stress participants were tested immediately, whereas the other half was tested 25 min later when the cortisol reaction was in effect. For the immediate test, there was neither an impairment nor a benefit caused by stress. The Retrieval-Practice group recalled more items than the Study-Practice group, but stress did not significantly alter the recall rate as a function of the type of study. A different outcome occurred for the delayed test when cortisol was in effect. When cortisol was present, people in the Study-Practice group exhibited a significant recall decrement. Importantly, the Retrieval-Practice group was not impaired in their recall even when there was cortisol in their system. Thus, Smith et al. showed that (1) cortisol 24 h after learning can impair memory recall only if the participant studied the items repetitively without earlier testing, and (2) delayed epinephrine neither helped nor impaired recall. Cortisol makes memory retrieval more difficult, so it is reasonable that retrieval practice helps offset the negative effect of stress. The implications of this experiment for student study habits are obvious. Students should engage in self-testing earlier while they are learning information in order to practice retrieving the information.

14.11 Summary of Affect and Memory

There are a number of memory effects related to emotion. In general, emotion improves the chances that some information about the event will be remembered later. It would not be adaptive if important stimuli were ignored or treated identically as unimportant features of the environment. Emotion serves as a valuable adaptive function by augmenting the learning and memory of affectively significant stimuli.

The study of information with positive and negative emotional content is complex and depends upon what the person attends. The attentional-broadening hypothesis for positive emotions is a reasonable idea, but currently it is not a compelling phenomenon. A person viewing a photograph of a loved one might focus on many more features of the photograph than if it were instead a scene of strangers who did not elicit a positive emotional response. If the picture of the loved one is about a social event, then there is no reason why the face of the loved one is the sole center of the scanning of the picture. The viewer might instead be scanning the whole picture carefully. So unlike the weapon-focus phenomenon, the positive-emotional context might lead to a broadening of attention as was suggested by the aforementioned researchers. Yet it might also be the case that neutral components of the photograph might be missed in the attentional-spotlight samples. For example, the type of furniture in the room or the type of drapes on the window might be missed because the attention is captured instead by the affective components of the scene. Perhaps these neutral contextual features would be viewed more completely if the entire photograph consisted of emotionally neutral items. So there might be an attentional narrowing for this case with positive emotion that is similar to the weapon-focus effect. Yet if there were a neutral stimulus that was a part of the social interaction, then perhaps the viewers would sample the stimulus in their scanning. That is, they might be curious why their loved one is using this object. Even in the negative-affective context, it might be the case that people look away from a horridly vivid negative stimulus. In that case, the negative stimulus might lead to an attentional broadening effect rather than a narrowing effect like for the case of a weapon focus. In fact, Harmon-Jones, Gable, and Price (2013) suggested that there is an interaction between motivation level and attentional scope (i.e., under conditions of low motivational intensity, there is a broadening effect, whereas under high motivational intensity, there is a narrowing effect). This hypothesis might be valid, but it has not been demonstrated with eye-movement tracking.

Hopefully, future studies on the encoding of complex scenes in the context of positive and negative emotions will clarify the interaction of affect and memory encoding. Currently, there are more questions than demonstrated facts. I suspect that emotion and motivational intent guide the viewing of complex scenes. What gets encoded is likely to be the information that dominates the time within the attentional spotlight scans.

Furthermore, postencoding stress immediately after learning can also improve memory retention. The theoretical interpretation of this effect is still an ongoing issue. In this chapter, both a consolidation theory account and an explanation in terms of the Chechile (2006) two-trace hazard theory are provided. But a case is made in this chapter against the consolidation position.

Finally, while there are many positive memory effects for learning information with higher emotional content, there is also a problem caused by a delayed cortisol reaction caused by stress. Yet this effect appears to be limited to impairing retrieval, and it can be mitigated by retrieval-practice encoding.

14.12   Summary of Part III

Learning and memory are interconnected. Learning is the acquisition of new information and the integration of that information with other existing knowledge. Memory is the stored record of the learning process. In part III, the focus has been on the learning process. *Encoding* as a verb is the action of transforming the input information into a personalized memory representation. *Encoding* as a noun is the result of that learning process. We have seen in part III that the process of encoding varies widely in its effectiveness. In general, the memory representation was roughly considered to be a distributed representation like the theoretical illustration that was first introduced in chapter 4 and reintroduced in chapter 11; see figure 11.3. Given sufficient attention and presentation time, some distributed set of micro-features, elaborated propositions, temporal tags, associated facts, contextual-event connections, and motor responses can all be rapidly activated in the brain, even on a single-trial basis. The evidence from physiological psychology points toward a redundancy in the stored memory as illustrated in figure 11.5 because there is both a hippocampal and a neocortical representation. An alternative system-level description of this dual-trace model was developed in part III.

The system-level description is in terms of a linked pair of vectors; see figure 11.6. The vector elements correspond to the components of a distributed representation like the one shown in figure 11.3. The elements of the vectors can be destroyed by other subsequent events, so the target encoding undergoes dynamic variation. Thus, a storage interference process can degrade the encoding to the point where it is no longer sufficiently stored, although it might still be implicitly stored or there might be a surviving fragment of the target event. Insufficient storage is a common fate for encodings that are not particularly well developed. But as the person engages in enhanced encoding activity, the representation becomes augmented with additional component features. Lost components might be restored and new features added. If the memory representation is sufficiently elaborated and connected to other memory traces, then the encoding becomes distinctively different from other encodings. If a trace is less similar to other subsequent traces, then there should be less storage interference. An elaborated trace that survives is also likely to have more retrieval routes that can improve the success rate on a later recall test. Thus, it was argued in part III that the improvement in encoding is a result of a more elaborated representation, not because the trace has greater strength.

15 Epilogue

In this book, an effort was made to look at several big-picture questions about memory. How is it that animals can learn about their environment, store information about that learning, and later access that information? How can we measure the underlying cognitive processes linked with the storage and retrieval of information? What do the measurements of the latent memory processes tell us about the development of memory and the loss of information? What learning and encoding methods are effective, and why are these methods superior? These are tough questions that require a host of technical skills. For some students and even for some researchers, the technical aspects of the work in neurobiology and mathematical modeling are impediments to achieving a deeper understanding of the ideas explored in this book. Yet hopefully, the context of the verbal discussion helps to clarify these issues. All too often in education, difficult concepts are edited out of the textbooks and courses in an effort to reach a broader audience. This volume has attempted to reach a wide audience without either trivializing the treatment or ignoring certain topics altogether. Memory is too complex and fascinating a research topic to be understood by half measures. As mentioned in the Preface, previous books on memory have not attempted to cover the full range of relevant topics. Researchers typically deal with a narrow set of problems and do not concern themselves about problems from different subdisciplines. Specialization helps promote progress on isolated issues, but the thesis advanced in this book is that progress is also gained from the effort to bridge the separate research areas and to seek a synthesis by the process of critical analysis. Although considerable ground has been covered in the book, there is still a sizable body of research that remains to be addressed. So the effort toward memory synthesis via the process of critical analysis goes on.

Glossary

acetylcholine.   Acetylcholine is a chemical neurotransmitter that occurs throughout the nervous system.

achromatopsia.   Achromatopsia is a loss of an understanding of color that results from brain damage.

aha effect.   The aha effect is the improvement in memory when there is an initial puzzle that is eventually resolved. For example, suppose that a research participant reads the sentence "People departed because of the sign." This sentence is ambiguous because it is not clear why the people departed. But after a short delay, the participant reads the implied reason why the sign caused the people to depart (i.e., ON STRIKE). For the control condition for the aha effect, the solution is provided before the ambiguous sentence is read.

allocentric spatial encoding.   This type of memory is a representation where objects and locations in space are understood in terms of their actual physical relationship.

all-or-none learning.   The all-or-none model was developed in reaction to the continuity theory of learning. The all-or-none hypothesis assumes that learning occurs abruptly despite the fact that the learning curve is a gradually improving function. Theorists who advanced this hypothesis argued that the continuous learning curve is the result of the statistical averaging over people and items. In its simplest form, the all-or-none model assumes that the basic performance measure is a mixture of either 0 for the nonlearned state or 1 for the learned state. As the number of trials increases, the mixture of 1s increases and accounts for the apparently gradual improvement in learning. This approach to learning is linked to Markov chain models of learning. The DEF model and the Sloboda-Chechile model are compromise approaches to learning that reside between the continuity theory and the all-or-none theory. See **Markov chain**, **DEF model**, and the **Sloboda-Chechile learning equation**.

amygdala.   The singular form of *amygdala* is *amygdalae*. The amygdala are almond-shaped brain regions in the temporal lobe and near the **hippocampus**. The amygdala play a critical role in the detection and representation of emotionally important stimuli.

anisomycin.   A drug that inhibits protein synthesis. It has been used in animal studies of memory consolidation.

anterograde amnesia. Anterograde amnesia is the difficulty in learning new information that occurs after a serious head injury.

aphasia. A problem with the production, comprehension, or repetition of language that is caused by brain damage.

Aplysia. *Aplysia* is a species of marine snails that have been extensively studied because they possess large neural cells that are convenient for neurobiologists to study. *Aplysia* have commonly been used for research on habituation and sensitization.

apperceptive agnosia. Although visual perception of a stimulus feels like a unitary event, the neurological evidence is nonetheless quite clear that the brain is examining the retinal image along a parallel set of independent, elemental pathways. Brain damage can result in a syndrome called apperceptive agnosia where the components of the visual event are detected, but the linkage of these components is not integrated into a perceptual whole.

ARC score. The *adjusted ratio clustering* or ARC score is a measure of the degree of clustering of free recall output. See **free recall** and **clustering** for additional information.

associative agnosia. Associative agnosia is an impairment in understanding the meaning of a stimulus because of brain damage that affects the access to the memory. For example, it might be a visual agnosia where the concept of the stimulus is still available but not elicited by visual perception. For example, a patient with visual agnosia might not know what car keys are when the keys are displayed to the patient. But the patient would immediately recognize the object as keys if the neurologist jingles the keys, thus allowing access to the concept via an undamaged auditory pathway.

Atkinson-Shiffrin model. See the **multiple-store model.**

autoassociative network. An autoassociative network is a special case of a heteroassociative network that maps a N-dimensional vector to itself. See **Hopfield networks** and **heteroassociative networks.**

axon. The threadlike part of a neural cell that passes an electrical impulse to terminal endings.

Bayes factor. The Bayes factor is a decision-making metric that emerged from a Bayesian perspective for doing statistical analyses. Given two specific rival hypotheses for the data H_0 and H_1, the Bayes factor is the ratio $\frac{P(D|H_1)}{P(D|H_0)}$ where $P(D|H_1)$ and $P(D|H_0)$ are the respective likelihoods for the data given model H_1 and model H_0. Each likelihood requires the integration over the possible values for the parameters of the model. When models differ in the number of free parameters, the Bayes factor automatically takes into account the intrinsic complexity differences between the models for describing the data. A Bayes factor greater than 1 favors model H_1, whereas a Bayes factor less than 1 favors model H_0. However, the magnitude of the Bayes factor carries differing levels of decisiveness regarding the two models. Kass and Raftery (1995) suggest the following: (1) a Bayes factor between 1 and 3 is not worthy mentioning, (2) a Bayes factor between 3 and 20 is positive, (3) a Bayes factor between 20 and 150 is strong, and (4) a Bayes factor greater than 150 is very strong. The Bayes factor can be used to build a case for either model depending on the data.

Bayesian statistics.   Bayesian statistics is a systematic method of using Bayes theorem for making a statistical inference about the population parameters based on data from an empirical study. The key idea underlying the Bayesian approach is the assumption that there are probability distributions for the parameters of a statistical model. Consequently, before the study is conducted, there is already a probability distribution over the possible values for the population parameters. This distribution is called the prior distribution because it is the distribution before the data are collected. The other key probability used in Bayesian statistics is the likelihood function. The likelihood function is the same function used in both Bayesian statistics and in frequency-theory statistics (i.e., classical statistics). After the assumption of the prior probability, the rest of Bayesian theory is rigorously grounded on proven mathematical results. Bayes theorem enables the calculation of a posterior probability given the data and given the prior probability. Although the resulting conclusions in a Bayesian analysis are dependent on the prior probability distribution, Bayesian practice tends to use standardized "ignorance" prior distributions that result in the posterior distributions being mostly determined by the sample results. Bayesian statistics test hypotheses about the model parameter with probability statements about the posterior distribution. Bayesian statistics provides interval estimates of the model parameters with an interval that has the highest posterior probability density over the marginal distribution for the population parameter. Thus, a 95 percent highest density interval is a unique interval that has a probability of .95 that the population parameter is within the interval limits; the interval is unique because it is the smallest interval that contains 95 percent of the posterior distribution. Bayesian statistics is also a method for doing meta-analyses. In such cases, the posterior probability distribution from one study is usually taken as the prior distribution for analyzing the data from the next study.

Bayesian statistics grew gradually in popularity in the field of statistics over the latter half of the twentieth century and was in stark contrast to classical or frequency-theory statistics. The theory of probability as a measure of information gave the Bayesian approach a strong intellectual grounding. One of the key differences between Bayesian and classical statistics has to do with the question of what the knowns and unknowns are for the problem of statistical inference. To Bayesian statisticians, the population parameters are unknown, so they are represented with prior and posterior probability distributions. What is known are the sample data from an empirical study. But classical statistics reverses the status of these two components of the inference process. In classical statistics, the parameters are regarded as fixed constants, so they cannot be represented with a probability distribution. Instead, classical statistics is based on the likelihood function. See **likelihood function**. For the likelihood function, there is an assumed value for the population parameters, and those values are used to compute the probability of all the possible empirical outcomes. In classical statistics, the inference procedures evaluate the likelihood of the observed data plus the more extreme outcomes that were not observed. To Bayesian statisticians, the only likelihood that is computed is the likelihood of the observed data. Hence, Bayesian statisticians treat the population parameters as having a probability distribution but regard the empirical outcome of a study as a fixed result, whereas classical statisticians treat the population parameters as fixed, but they treat the outcome of a study as having a probability distribution. See **classical statistics.**

Another difference between Bayesian and classical statistics is the procedure for assessing scientific hypotheses about the population parameter. In classical statistics, a null hypothesis is *assumed* to compute the likelihood of the observed data plus more extreme data outcomes. If that summed likelihood is low, then the null hypothesis is rejected. Because the null hypothesis is assumed, evidence cannot be built in favor of the null hypothesis. In contrast to classical statistics, Bayesian statistics can test directly the alternative hypothesis as well as the null hypothesis. Given two rival hypotheses H_0 and H_1, the Bayesian statistician can compute the Bayes factor to assess both hypotheses. For further details, see **Bayes factor**.

Becker (1980) model.   This model is an example of a **short-list model** of lexical access. Unlike the pure **direct-access model** of word recognition, the Becker model assumes a partial direct-access process that narrows the word candidates in lexical memory to a small set of words. Next the candidate words are serially compared (one at a time) to the input stimulus to see if there is a match. If there is a match, then the semantic content in memory is elicited, and the stimulus is recognized. If there is not a match, then another candidate word on the short list is compared in a similar fashion. If the entire set of words on the short list are tested and no match is found, then the person produces a nonword response.

between-subjects design.   The between-subjects design is a research plan where different, independent groups of participants or subjects are used in each experimental condition. In statistics, this research plan is called a completely randomized design.

bits.   See **information coding**.

Blackwell model.   Blackwell (1963) developed a threshold model for **signal detection**. The person is assumed to have two discrete states for detection; one state is below threshold, and the other state is above threshold. There is a probability (δ) that a target stimulus can trigger the above-threshold state, but it is assumed that the noise distribution has no chance for triggering the above-threshold state. This theory predicts an **ROC** curve that is a straight line between the points $(0, \delta)$ and $(1, 1)$. Evidence for curvature in empirical ROC plots was widely interpreted as evidence against this model.

blocking effect.   This effect occurs when there is first a conditioned stimulus that is paired with the unconditioned stimulus. Later, the same stimulus is combined with a different stimulus, and the two stimuli are still paired with the unconditioned stimulus. Finally, only the second stimulus is tested. From an information theory perspective, the second stimulus has no new information value. Hence, it should not elicit a classically conditioned response. However, from classic association theory, the second stimulus should have developed an association to the conditioned response. Thus, behavioral association theory predicted a conditioned response for the second stimulus. Kamin (1969) found no conditioned response as would be expected from an information-theory perspective.

bottom-up.   The term *bottom-up* refers to a process that is data driven by the current sensory input. For example, the **RBC model** of pattern recognition is a bottom-up model where the pattern is recognized because of the automatic activation of its features. The bottom-up approach is the opposite of the **top-down** approach where a process is cognitively or conceptually driven. For example, if the person generates a hypothesis or an expectancy based on the current time and context, then that cognitive knowledge can affect the processing of new sensory input.

Broca's region. Broca's region is a left-hemisphere brain region that is important for the production of speech.

Brown-Peterson task. The experimental procedure where a single subspan target item is presented and followed by a rehearsal preventing task, such as counting backward or saying aloud a string of digits. After the retention interval, the participants are tested for their knowledge of the current target item.

Bush-Mosteller linear operator model. This model describes learning with a mathematical function of the form $p_{n+1} = p_n + \alpha(p_\infty - p_n)$ where p_n is the accuracy on the nth trial, α is a learning rate, and p_∞ is the asymptotic level possible with the training program. The model is based on the idea that the change in memory is a fixed proportion α of the remainder that needs to be learned.

CAM. See **content-addressable memory**.

cAMP. Cyclic AMP or cAMP is an important chemical involved with **receptor trafficking** in neural cells.

central executive. The central executive is a component of working memory. See **working memory** for more details.

CHARM. The *Composite Holographic Associative Recall Model* or CHARM is based on the Metcalfe-Eich (1982) paper. Memory in the CHARM framework is inspired by a holograph—not an actual visual holograph but by the mathematical operations that apply to holographs. Memory in CHARM is based on a distributed representation of the stimulus in terms of a vector with many feature components. Items are associated by means of the convolution operation. The rule for the convolution operation is derived from probability theory, and it relates to the addition of two random variables. Testing memory in CHARM is related to the correlation operation. The rule for the correlation operation is based on the difference of two random variables in probability theory.

Chechile-Meyer task. The experimental procedure of randomly intermixing recall, old-recognition, and new-recognition test trials in a fashion that the participants cannot predict how they will be tested on any given trial. Moreover, for the old- and new-recognition tests, the participants are required to provide a 3-point confidence with the guidelines that 3 is a certain response, 2 is an educated guess, and 1 is a blind guess.

chunking. Chunking refers to the grouping of a sequence of items into meaningful units. For example, if one were required to learn the sequence FBINASAIRSCIA, then an effective chunking strategy is to encode the letter string as FBI, NASA, IRS, and CIA, which are well-known abbreviations for some U.S. agencies.

classical conditioning. Classical conditioning is the same as **Pavlovian conditioning**. With this experimental procedure, the animal is first presented with a conditioned stimulus (CS) and later presented with an unconditioned stimulus (US). The unconditioned stimulus elicits a consistent response. For example, a puff of air on the eyelid elicits in humans a blink reflex, which is called the unconditioned response (UR). In contrast to the unconditioned stimulus, the conditioned stimulus (CS) is a neutral stimulus that does not elicit a response like the unconditioned response (UR). For example, a CS of a brief tone does not elicit in humans an eye-blink response. However, the repeated pairing of the CS (tone) with the later presentation of the US (air puff to the eye) results in the person learning to anticipate the onset of the US. In preparation of the US, the person learns to blink to the tone (i.e., the CR occurs before the onset of the US). This learned response is called the conditioned response or CR.

classical statistical theory. Classical statistics or frequency theory is a system of statistical inference based on (1) the rejection of the idea in Bayesian statistics that parameters can be described by probability distributions and (2) that probabilities are reserved only for the likelihood of the data given hypothesized values for the population parameters. Hypotheses are assessed by assuming the null hypothesis about the population parameter(s). The null hypothesis is rejected if the likelihood of the observed data and more extreme unobserved outcomes are unlikely under the assumption of the null hypothesis. If the null hypothesis is rejected for a specific sample, then the outcome is declared statistically significant. Interval estimates in classical statistics are based on the theory of the confidence interval (or confidence region in the multivariate case). The confidence interval is not a probability for the population parameter because the parameters cannot have probability distributions in frequency theory. Instead, the $1 - \alpha$ confidence interval is constructed via a rule-based procedure that should, when repeatedly applied over separate random samples from the population, result in containing the population parameter on at least $1 - \alpha$ percentage of the intervals. Another key concept in frequency theory is that of statistical power. Power, like the confidence interval, is defined not as a probability but as the proportion of significant outcomes over separate random samples when a specific alternative to the null hypothesis is true. For a comparison with Bayesian statistics, see **Bayesian statistics**.

clustering. For the free-recall task, the participant is free to output his or her memory of list items in any order. If the items on the learning list have exemplars from different categories, then it is possible that the participants can use those categories to organize their output. If participants employ such a strategy, then the resulting output is said to be clustered. There are a number of measures for quantifying clustering. One common measure is called the *adjusted ratio clustering* (ARC) score. The formula for the ARC score is

$$ARC = \frac{R - E(R)}{R_{max} - E(R)},$$

where R is the number of category repetitions, R_{max} is the maximum number of category repetitions, and $E(R)$ is the average number of repetitions when there is a random output order.

complementary storage model. See **Teyler-DiScenne indexing theory**.

compound cue theory. This model is an extension of the **SAM model** (McKoon & Ratcliff, 1992; Ratcliff & McKoon, 1988). The theory is designed to account for **semantic priming**.

conditional probability. A conditional probability is a probability given some stated constraint. The probability of a correct response given a high confidence rating is an example of a conditional probability. A conditional probability of event A given event B is denoted as $P(A \mid B)$.

conditioned inhibitor (CS⁻). Suppose that a stimulus (like a tone) is always followed by an **unconditioned stimulus**, but whenever the tone occurs with another neutral stimulus (like a light), the unconditioned stimulus does not occur. Eventually, the animal learns a conditioned response to the tone but does not respond to the combination of the tone and light. Pavlov observed whenever the light is added to any conditioned stimulus, then the animal does not exhibit a conditioned response. The light has become a conditioned inhibitor. The conditioned inhibitor is denoted as CS^-. Like the conditioned stimulus, the conditioned inhibitor is denoted as a vector because the stimulus consists of many elemental sensory features. See **classical conditioning**.

conditioned response (CR). A conditioned response is a learned association to the **conditioned stimulus (CS)** that comes from the temporal relationship between the conditioned stimulus and the **unconditioned stimulus (US)**. The classic example from Pavlov of a conditioned response is that of a dog salivating to the sound of a dog biscuit rattling in a bag. The conditioned response is denoted as **CR** in order to stress that it consists of a vector of responses. See **classical conditioning**.

confidence interval. See **classical statistical theory**.

congenital aphantasia. This term refers to the medical condition where the individual never develops a mental image.

connectionist models. The connectionist approach uses a parallel distributed artificial neural network for representing learning and memory. There are many arrangements for this class of models. Connectionist models have also been used as unsupervised machine-learning systems. Learning and memory is represented in the network by a change in the activation strengths or weights between nodes. Connectionist models are bottom-up or data-driven cognitive architectures rather than a system that builds a propositional representation as part of the cognitive architecture. Connectionist models can be a form of a feature-space model that identifies a concept with the activation of a set of elemental features.

consolidation theory. Consolidation theory is the hypothesis that characterizes the state of a recent memory representation as initially fragile and easily disrupted, similar to the susceptibility of wet paint to distortion. Eventually, if the encoding survives long enough, then the memory becomes hardened like dried paint, and the encoding becomes resistant to loss. However, some researchers found that the additional practice of a presumedly hardened trace makes the trace susceptible again to memory loss. These researchers argue that the restudy of a hardened trace transforms the trace back to a fragile state and hence requires a reconsolidation process to harden the trace again.

constrained generation task. The participants in this task are typically provided with a stem word or an instructional cue, and they are required to generate the response word, but the response word is constrained by providing some of the letters of the response. A POLICEMAN: C__P is an example of a congruous constrained generation trial, whereas NOT A HAT, NOT A VESSEL FOR DRINKING: C__P is an example of a two-cue, incongruous generation trial.

content-addressable memory. Content-addressable memory (CAM) is a system where the information content itself is used to automatically access the information in memory along with other associated data. This type of memory avoids the need to search locations one at a time in order to find the desired match as would be the case for random-access (RAM) memory. Both CAM and RAM systems can be implemented on machines. To store some information in a RAM system, the next available computer memory address is used. To find this information later requires a search process over locations (i.e., memory addresses). In contrast to RAM computers, a CAM arrangement uses the properties of the present information to localize a storage address. Of course, this means that new information can interact with older data previously located at that address. There are both advantages and disadvantages with the CAM arrangement. The advantage of the CAM system is its speed of access. The disadvantage of a CAM system is that there might be some distortion caused by the storage of new data into an address where other information was previously stored.

contiguity.   Contiguity is the learning principle based on the idea that mental associations can be constructed based on what occurred together either in time or in space.

CPEB.   CPEB is a protein that activates messenger RNA.

CR.   CR is the **conditioned response**. It is an associative learned response. See **classical conditioning** for more details.

CS.   CS is the conditioned stimulus, and it is sometimes denoted as CS^+. The conditioned stimulus is a neutral stimulus that does not initially elicit a response that is like the unconditioned response; however, in a classical conditioning training experiment, the animal learns to give a conditioned response when presented with the **CS**. The conditioned stimulus is denoted as **CS** in order to stress that it consists of a vector of sensory components. See **classical conditioning** for more details.

CS⁻.   See **conditioned inhibitor**.

cumulative probability.   For a continuous probability distribution on dimension x, the cumulative probability is the integral of the probability density from the lowest possible value for the variable up to the value x_0. The cumulative probability $F(x_0)$ is the area under the left tail of the probability density display for the region $x < x_0$. See **probability density**. Cumulative probability is also defined for a discrete random variable on an ordered dimension.

d′.   In signal-detection theory, when the signal and noise distributions are assumed to have equal variance on a latent strength axis, d' is defined as the separation between the signal and the noise latent strength distributions divided by the standard deviation of the distributions. In memory research, d' has been applied to recognition memory studies under the assumption that memory targets have a strength distribution that is different from an assumed strength distribution for the non-studied foil items.

declarative memory.   Declarative memory is an explicit memory of either an episodic event or a semantic fact. Usually, declarative memory can be verbally recalled.

DEF model.   The Dynamic Encoding-Forgetting (DEF) model is designed to account for the learning of paired associates. Given sufficient encoding time, the intention to learn, and a subspan event to learn, it is assumed that the encoding of the event is 100 percent initially, but when time and/or other events occur, the storage of a target pair can undergo a loss. The functional form for explicit storage of a target pair is given as $\theta_{se} = 1 - exp(-bn^c) + exp(-bn^c - dx^2)$ where b and c are parameters associated with the learning rate, d is a parameter associated with the speed of forgetting, n is the number of training trials, and x is a proportional value on the (0, .99) interval that varies as a function of the time between repetitions of the target pair. The DEF model is a compromise between the continuity theory of learning and the all-or-none theory. In the DEF model, learning is 100 percent successful after the presentation of the target pair. In fact, the DEF model assumes that there are two memory traces established for each target event. However, after the encoding, there is also a dynamically varying storage interference process. Thus, explicit storage for any given pair of items can degrade to another state of a reduced storage quality. The reduced states include implicit storage, fractional storage, or nonstorage. Yet when the target pair is presented again, the explicit storage of the pair returns to the 100 percent level.

delayed conditioning. Delayed conditioning is a Pavlovian conditioning arrangement between the **conditioned stimulus (CS)** and the **unconditioned stimulus (US)** where the **CS** occurs prior to the **US** but remains on before and during the presentation of the **US**. See **classical conditioning**.

demand characteristic. Whenever a psychological experiment contains a subtle cue that makes the participant aware of what the researcher's expectations are, there is a danger that the participant is influenced by this cue. This problem is referred to as a demand characteristic.

dendrite. The dendrite is the input section of a neuron. Dendrites have many small protrusions that are called **dendritic spines**. Neurons do not physically touch one another because there is a gap between the neurons that is called the **synapse**. However, chemical neurotransmitters can cross the synaptic gap and penetrate the next neuron's dendrite. See **neuron**.

dendritic spines. Dendritic spines are small protrusions of the dendrite wall. The spines increase the surface area on the dendrite wall. The spines also contain receptor gates that provide a channel for neurotransmitter chemicals to pass from the **synapse** to inside the neuron. Additional floating receptor gates that are in the fluid of the dendrite can attach to the dendritic spines to increase the number of receptor gates. But receptor gates already attached to the dendritic spine can also detach and go back into the fluid within the dendrite. This process is called **receptor trafficking**. See **dendrite**.

dentate gyrus. This structure is part of the **hippocampus**, and it is thought to be involved in the formation of event memories. The dentate gyrus of mammals is also a brain region where **neurogenesis** occurs.

dependent variable. A dependent variable is a measured response. It is called a dependent variable because the values are influenced by other variables or factors. In experimental design, the dependent variable is contrasted to an **independent variable**, which can be controlled by the experimenter. For example, the experimenter can control the meaningfulness of a stimulus (an independent variable) and record the learning rate of the participant (the dependent variable).

dichoptic masking. The dichoptic masking procedure was developed by Moutoussis and Zeki (2002). With dichoptic masking, the participant is shown in one eye a stimulus such as a black cross within a red square background while the other eye is shown a black cross in a green square background. The presentation of the images is simultaneous and brief (i.e., 50 msec). After a 150-msec blank interval, the pair of stimuli are presented again for 50 msec, but the background colors are reversed to the eyes. If the participants cannot consciously detect the cross, then they perceive instead a solid yellow square. Because of the use of opponent colors, there is binocular color fusion between the red and green backgrounds (i.e., red and green fuse to become yellow). If the participant can detect the cross, then it appears as a black cross within a yellow background.

direct-access model.   A direct-access model is a **bottom-up** activation model like the **interaction-activation model** of word recognition. It is also an example of a **content-addressable memory** system. For a direct-access model, a search process is not needed because the perceptual detection of the stimulus elicits the process of activating the stored information in memory.

dishabituation.   After habituation training to a stimulus, the response rate of a reflex is diminished, but if at that time another strong, potentially threatening different stimulus is presented, then the animal returns to responding to the original stimulus. This return to responding is called dishabituation.

distinctiveness.   Distinctiveness in memory research refers to the property that a memory encoding stands out as different from other memories. An encoding can be distinctive because many additional features of the target item are integrated with the basic information about the target. Because an encoding is different from others, there is a reduced tendency for storage interference. Highly distinctive items can be also easier to find in a memory search.

distributed practice.   See **massed versus distributed practice.**

dopamine.   Dopamine is a neurotransmitter and plays an important role in reward behaviors.

dorsal occipitoparietal stream.   The dorsal occipitoparietal stream is a broad neural pathway in humans from the occipital lobe to the parietal lobe. This pathway is associated with the visual processing of the location of the object in space.

double experimental dissociation.   If there is an experiment where levels of a factor only affect task Y or dependent variable Y and there is another experiment where levels of a factor only affect task X or dependent variable X, then this condition is called a double experimental dissociation. The **process-purity** hypothesis assumes that tasks or measures X and Y are purely tapping different processes. However, the process-purity hypothesis is not a valid conclusion from finding a double experimental dissociation because many counterexamples can be generated.

drug state-dependent memory.   Some drugs have strong stimulus properties. For such drugs, if there is some original learning that occurs in a particular drug state, then there can be a deficiency in recall later when the drug is no longer present. The finding of a strong state-dependent memory effect for a recall test but not for recognition tests indicates that retrieval is impaired when there is a change in the drug state from the state at encoding.

dual-process model.   Although there are a number of dual-process models of recognition memory, the term *dual-process model* has come to be associated with the theory advanced by Yonelinas

(1994, 1997, 2004). The two processes discussed in those papers are familiarity and recollection. For the representation of foil items, a normal distribution is assumed like in the standard form of **signal-detection theory**. But the corresponding model for old items is a mixture of two processes. It is assumed that on R_0 proportion of the old-recognition test trials, the person is able to recall the target information and produce a YES response. But for all of the other old-recognition test trials, the person has a familiarity process like what would be expected for the equal-variance form of signal-detection theory.

Dynamic Encoding-Forgetting model.   See **DEF model**.

Easterbrook hypothesis.   The Easterbrook hypothesis is the idea that the focus of attention narrows as the arousal level is increased. Easterbrook (1959) developed this hypothesis as a way to account for the **Yerkes-Dodson law**, but that law was shown to be incorrect. Nonetheless, the Easterbrook hypothesis is consistent with the phenomenon of **weapon focus**. However, the degree of narrowing of the available cues is not a continuous function. But the Easterbrook hypothesis is not consistent with the broadening attention to cues in the case of high positive emotion.

echoic trace.   The echoic trace is a brief auditory persistence that decays within 300 msec. The echoic trace is treated as a pure sensory memory before the information is perceptually classified. The echoic trace is the auditory counterpart to visual **iconic memory**.

ecphory.   Ecphory is a concept introduced by Semon (1921). Semon hypothesized that each presentation of information forms separate memory traces or engrams. The later activation of the memory involves the elicitation of the system of interdependent engrams; this process he described by the term *ecphory*. To Semon, ecphory refers to the simultaneous activation of the memory representations that are produced by the triggering event. Ecphory is analogous to the vocal response of an entire chorus of singers rather than the sound of just a solo singer.

EEG.   EEG denotes an electroencephalogram. As neurons in the brain produce ionic currents, there is a change in the magnetic field. Small flat electrodes placed on the scalp can detect this changing field, and the detector produces a small electrical voltage response. Since the magnetic fields travel at the speed of light in the medium, the EEG has outstanding temporal sensitivity. The EEG signal from any one detector is sensitive to neural current flows from all parts of the brain; thus, it is not known where the spatial origin is for the neural ionic current based on the response of this single detector.

egocentric-spatial encoding.   This type of encoding positions objects and spatial locations relative to the current position of the person or animal.

eidetic imagery.   Eidetic imagery is a particularly vivid visual image that is remembered. In a very rare case, an individual with an unusually strong eidetic imagery ability can recall an old image and can fuse that image with a current stimulus to generate a new perception.

empiricism.   Empiricism is a philosophical tradition associated with John Locke, and it is based on the concept that knowledge is acquired from sensory information. This approach was a reaction to the rationalism approach, which believed that innate ideas could be discerned by means of reason.

enactment effect. For enactment-effect experiments, the participants are instructed to enact simple imaginary tasks such as "pour the coffee." There is a consistent finding that the enacted condition results in better memory retention than (1) listening to the statement, (2) imagining performing the statement, or (3) watching someone else doing the statement.

encoding specificity. Encoding specificity is the idea that a stimulus learned in a particular cue context will be best remembered later if the same cue is present. For example, if one were to learn a homograph on a memory list like the word BRIDGE with the cue context of a game of cards, then at a later time, the cue of a span across a river would be less effective as a retrieval cue.

entanglement problem. The entanglement problem is the methodological challenge caused by the fact that available dependent variables reflect multiple underlying cognitive processes. The recommended solution to the entanglement problem is to employ **model-based measurement** in order to model how the underlying processes are functioning in the various experimental tasks.

episodic buffer. The episodic buffer is a component of working memory. See **working memory**.

episodic memory. Episodic memory is the memory of an event.

ERP. The event-related potential (ERP) is an average EEG signal. Typically, there is a common event that occurs over repeated trials for the same participant or over different participants. The EEG responses for these separate episodes are averaged. The ERP signal is the result of this signal averaging. See **EEG** for more information about the EEG signal.

exemplar model. The exemplar model deals with the processes of categorization and recognition. For this model, the categorization decision is based on the aggregated value of the similarity between the new item's representation in a feature space and all the other stored exemplars of the conceptual class, whereas the recognition decision is a function of the summed similarity between the probe item and all the items in memory. The memory representation assumed for the exemplar model is that of a vector of component features. The structure among the features is not computed, but rather the dissimilarity between vectors is simply treated as a distance in a **feature space**.

extinction. Extinction for appetitive **instrumental conditioning** refers to the learning that reward will no longer result from executing the previously learned response. Similarly, in **classical conditioning**, extinction is the learning that the unconditioned stimulus will no longer follow the conditioned stimulus.

extended process-dissociation model. The extended process-dissociation model was developed by Buchner, Erdfelder, and Vaterrodt-Plünnecke (1995) in order to obtain estimates of explicit and implicit memory. The authors identified explicit memory with *conscious* memory and implicit memory with *unconscious* memory. The model is a measurement method based on the **extended process-dissociation task**. The model is an improvement of an earlier Jacoby (1991) process-dissociation model that did not have a correction based on foil test trials.

extended process-dissociation task. In a typical experiment, the participants are first presented with target information from two sources (e.g., list 1 and list 2). At the time of test, some participants are given the *Exclusion Instructional Rule* that directs them to only respond YES for items from list 2. Other participants are given the *Inclusion Instructional Rule* that directs them to respond YES for any of the old items regardless of the source. At the time of test, the participants are tested with items from both lists as well as nonstudied foil items. Thus, in the *Exclusion* condition, the participants should respond NO both to foils as well as to target items from list 1.

feature space.   Feature spaces are models for a memory representation that can be expressed as a point in a high-dimension space where the dimensions are defined by values associated with elemental perceptual features. A stimulus is assumed to automatically elicit the activation of the point in feature space because the stimulus is regarded as simply the collection of the features. Thus, a feature-space point is activated by a bottom-up data-driven process. Connectionist models are examples of a feature space, but models where there are language-like propositional encodings are not feature-space models. The distance between points in the feature space is assumed to be a function of the dissimilarity between the memory representations. Exemplars from any particular category are also assumed to be contained within a localized region of the feature space, which is not the same as the bound region for a different category or concept.

fMRI scan.   Functional magnetic resonance imaging (fMRI) is a neural scan procedure that is based on the detection of the cerebral blood oxygenation level.

free recall.   Free recall is a test procedure in which the participants are free to recall the information in any order of their choosing.

fuzzy-trace theory.   The fuzzy-trace theory is a memory framework developed by Brainerd and Renya (1990). In this theory, it is hypothesized that the memory encoding of an event consists of two memory representations where one is a gist-based trace and the other one is a verbatim trace.

GABA.   GABA is an amino acid that acts as a neurotransmitter in the central nervous system. It plays an important role for inhibition.

Garcia effect.   The Garcia effect is the bias for developing a taste or food aversion when the unconditioned stimulus is an internal sense of illness. This association is formed even if the illness was not caused by the food and even if there is another potential conditioned stimulus available such as a light or a sound. But if the aversive unconditional stimulus were a foot shock, then there is a bias to avoid a light or a sound rather than avoiding the food. The Garcia effect demonstrates that only some neutral stimuli develop a conditioned response, and the nature of the unconditioned stimulus determines the bias as to which stimulus develops the conditioned avoidance response.

Gaussian distribution.   The Gaussian distribution is an important probability distribution because from the central limit theorem, it has been established that the sample mean obtained from all other probability distributions converges (as the sample size increases) to a Gaussian distribution. The Gaussian distribution is also commonly called the **normal distribution**. The distribution has two population parameters—its mean μ and its standard deviation σ. The probability density for the Gaussian is $f(x) = \frac{1}{(2\pi\sigma)^{1/2}} exp\left(-\frac{(x-\mu)^2}{2\sigma^2}\right)$. The shape of the distribution is a symmetric bell-shaped distribution centered about the mean. It is known that the Gaussian distribution has a monotonically increasing **hazard function**.

generation effect.   The generation effect is the improved memory with generative encoding. See **generative encoding.**

generative encoding.   Generative encoding requires the participant to solve a problem in order to determine the identity of the target item. This type of encoding can be done either in a constrained fashion, where there is a strong cue to determine the unique target solution, or in an unconstrained fashion, where there are many possible solutions. See the **constrained generation task** and the **unconstrained generation task.** For both types of generative condition, there is a corresponding control condition in order to assess the effectiveness of generative encoding. Typically, there is better memory retention with generative encoding.

geons.   In the recognition-by-components model or **RBC model**, it is assumed that there are visual features that are like the phonemes for language. The presumed visual features are called geons.

glutamate.   Glutamate is a neurotransmitter that typically is contained within neural cells and signals by means of the chemical reaction of other compounds called **glutamate receptors.**

Groves-Thompson model.   Groves and Thompson (1970) model is a dual-process model for the interaction of habituation and sensitization. The essential assumption of the model is any stimulus results in two independent processes; one causes a decrement in the motor response and the other causes an excitatory response. The decremental component can be thought of as pure habituation, and the excitatory component is a pure sensitization component. The resultant behavior of a motor reflex is a mixture of these two latent components.

habituation.   Habituation is a nonassociative memory that is exemplified by the reduction of a response to a harmless stimulus that occurs due to experience. For example, many animals have a startle reflex when they hear a loud, surprising sound, but this innate response weakens or disappears if that sound occurs more regularly. Developmental psychologists have also used the word *habituation* to described the attention span of infants. This process is considered different from the reduced vigor of a reflex. Thus, to avoid confusion, it is recommended that for the developmental context, the suitable term should be called **novelty habituation.** See **novelty habituation** for the reasons for this distinction.

habituation/sensitization strength.   See **system model of habituation/sensitization.**

habituation/sensitization strength threshold.   The system model treats the concept of strength as the theoretical counterpart to the increasing of the ionization rate of neurons. It thus follows that there should also be a threshold value for strength that is sufficient to trigger a response. If the threshold is denoted as ς_c, then the unconditioned response is expected when $\varsigma_k \geq \varsigma_c$ where ς_k is the strength for the kth training trial. The magnitude of response is a function of the difference between the strength and the threshold value. Also, when $\varsigma_k < \varsigma_c$, a motor response does not occur. It is also possible, when there is extensive habituation training, that ς_k is substantially less than ς_c. In this case, the habituation can be decreased to a point that Humphrey (1930) called "beyond the zero point." See **system model of habituation/sensitization.**

hash functions.   A hash function is used in computer science to map generic information to a location in a hash table or hash array. Based on the content of the information, a function typically maps the information to an integer between 0 to M where the number is the address in the hash table. A hash function thus simulates a **content-addressable memory** system.

hazard function. For an ordered dimension x, such as time or strength and for a property that can undergo a state change, the hazard function is the conditional probability that the state change occurs at x_0 given that the change did not occur for $x < x_0$. For a continuous probability function with probability density $f(x)$ and cumulative probability $F(x)$, the hazard $h(x) = \frac{f(x)}{1 - F(x)}$. The term $1 - F(x)$ is the right-tail area of the probability distribution, and it is called the survivor function. For the dimension of time after encoding, a hazard function can be defined for memory; the function would be the risk of a state change from sufficient storage to insufficient storage at time t_0 given that memory was intact up to the time of t_0. Hazard can also be defined for discrete probability distributions on ordered counts.

Henry Molaison (H.M.). Henry Molaison (1926–2008) was an individual who had his left and right hippocampus removed. The result of this surgery left him in a state in which he did not form new explicit memories. He continued to have a good memory of events learned before his surgery.

heteroassociative networks. A heteroassociative network is a type of a distributed artificial neural network where a N-dimensional vector is mapped to a K-dimensional vector. The input stimuli are represented by a set of N features, and the output response is a K-dimensional vector. There are connection strengths between the nodes in the network that are altered as a function of the learning rule. See **connectionist models**.

highest density interval (HDI). The HDI is a Bayesian probability interval for a model parameter. See **Bayesian statistics**. See **probability density**.

hippocampus. The hippocampus is a brain region in the temporal lobe. Each hemisphere has a hippocampus. The hippocampus plays an important role in building a rapid encoding of an event.

homographs. See **lexical ambiguity**.

homophones. See **lexical ambiguity**.

Hopfield network. A Hopfield network is an artificial neural network that connects a N-dimensional vector to itself. It is an autoassociative network. (See **autoassociative networks**.) The nodes in the network are connected to all other nodes, and the connection between node i and node j has a strength or connectivity weight w_{ij}. The weights between the nodes are symmetric (i.e., $w_{ij} = w_{ji}$). For the network to converge, it is necessary to assume that the weight between a node and itself is 0. A Hopfield network is an example of a content-addressable memory. See **content-addressable memory**.

iconic memory. Iconic memory is a brief visual persistence that rapidly decays after about 200 msec. The iconic trace is treated as a pure sensory memory.

IES model. The implicit-explicit separation model was developed by Chechile, Sloboda, and Chamberland (2012). It is a multinomial processing-tree model for a specific experimental procedure, and it results in separate probability estimates for explicit storage, implicit storage, fractional storage, and nonstorage.

IES task.   This task provides for the data needed for the IES model. The test procedure begins with a yes-no recognition decision based on a lineup of four items. The lineup consists of either a single old item along with three novel foils or a lineup of four novel foils. Three-point confidence judgments are also collected for the yes-no recognition decision. If the lineup in fact contains an old memory target item, then there is also a forced-choice task where the person must select one of the four items. This is called a four-alternative forced-choice (4-AFC) task. If the person is incorrect, then additional forced-choice tests are done (i.e., 3-AFC or 2-AFC).

incidental context effect.   Incidental context refers to some feature of the encoding other than the central stimulus. For example, if the participant is required to learn a list of words, then the words would be the central target stimuli, but the room in which the experiment was conducted is a part of the incidental context. An incidental context effect is the better memory retention for the central targets when tested in the original incidental context of the encoding.

independent variable.   An independent variable is a factor or a measure that can be controlled by the experimenter. For example, the experimenter can control the meaningfulness of a stimulus (an independent variable) and record the learning rate of the participant (the **dependent variable**).

information coding.   In the formal theory of information, the amount of information is measured in bits where bits are the number of binary digits needed to represent a stimulus. For example, a single character of text can be any one of twenty-six letters plus eight other symbols for either spacing or punctuation. The number 34 in binary is 100010, so a single text character can be expressed by a binary string of length six, so there are six bits of information per text character. In general, the relationship between the number of possible outcomes N and the bits of information b can be expressed either as $N = 2^b$ or $b = \frac{\log N}{\log 2}$.

innate ideas.   The belief in an innate idea is the philosophical position that some knowledge can be arrived by reason, and it is not based on experience. See **rationalism**.

instrumental conditioning.   Instrumental conditioning is the same as **operant conditioning**. With this procedure, the animal is first placed in a motivational state in order that an object can serve as a reward. For example, animals might be placed on a restricted diet so that they are motivated to obtain a small amount of food for completing the task. To achieve this training goal, the overall task is broken into more simple components, which are initially rewarded. This process is called **successive approximation**. For example, to train a rat to press a lever in a test box, the researcher might reward the animal for just looking in the direction of the lever. This operation will typically result in the animal looking toward the lever most of the time. Later the requirement is increased, so reward might be given only if the rat is directly in front of the lever. Again that operation will result in the rat typically spending most of its time near the lever. Finally, the researcher might increase the requirement for obtaining the reward so that it is supplied only if the animal presses the lever. The animal will thus be trained to press the lever to obtain the goal.

interaction-activation model.   This model was developed by McClelland and Rumelhart (1981) and Rumelhart and McClelland (1982). It is an artificial neural network model of a class that they called a parallel distributed processing system. The model was designed as a word recognition system. A key feature of the model is a distributed representation over three different levels (i.e., visual features, letters, and words). Elemental visual features were separate nodes at the lowest level in the model. These features activated the letter nodes at the next level. The letter nodes activated the word nodes at the top level in the model. The interaction-activation model is a bottom-up, direct-access model for word recognition.

interneurons. An interneuron receives input from another neurons and its electrical discharge influences other neurons.

ISI. ISI means interstimulus interval (i.e., the delay between the first and second stimulus).

Jost's law. Jost's law is the hypothesis that if there are two memories of equal strength but with different ages, then the more recent memory is more likely to be lost.

Korsakoff syndrome. Korsakoff syndrome is a severe memory disorder for learning new information. This syndrome occurs as a result of brain damage from a vitamin B-1 deficiency.

lag conditional response probability (lag-CRP). The lag-CRP is a metric used with free-recall lists to see the influence of subjective organization. For example, suppose that the input list includes the word subsequence of PICKEL VOLCANO STUDENT TRAIN SOLID, and the participant responds in consecutive output positions with the words STUDENT TRAIN, then that would be a *lag* + 1 response. But if the participant responded STUDENT PICKEL, then that would be a *lag* − 2 response.

law of effect. Any behavior that is followed by a reward is likely to increase its frequency of occurrence, and any behavior followed by an aversive outcome is likely to have a reduced frequency of reoccurrence.

learning/performance distinction. Behavior might not reflect what an animal has learned. The animal might acquire some knowledge from experience but not demonstrate that learning until the animal has a motivation to use that information.

levels-of-processing (LOP) framework. This memory framework was developed by Craik and Lockhart (1972) as an alternative to the multiple-store model of memory. The LOP approach stressed the interaction between the encoding level and memory retention. For a typical LOP experiment, the participants are induced to encode the target stimulus in different ways so as to vary the "depth" of processing. At the shallow level, the participants might be asked to judge if a target word is typed in uppercase or lowercase font, whereas to induce a deep level of processing, the participants might be asked if the word fits into the context of a sentence. Between these two extremes, the participants might be asked if the target word rhymes with another word.

lexical ambiguity. Lexical ambiguity is any word where there are different meanings for either the same printed form or for the same spoken form. A homograph is an example of lexical ambiguity for printed words (e.g., *bank*, which could be a monetary institution or a side of a river). An example of a homophone is *knew* and *new*.

lexical decision task. The lexical decision task requires the participant to decide as rapidly as possible if a presented letter string is a word or not.

likelihood function.   The likelihood function is the conditional probability for some possible outcome (or data) given a specific value for the population parameter(s). The likelihood function is the basic model for the data for both classical statistics and Bayesian statistics. The maximum likelihood estimate is the value for the model parameter(s) that has the highest likelihood value for the observed data.

long-term depression (LTD).   Long-term depression (LTD) is a decrease in the synaptic potential to fire that develops as a function of experience. LTD is regarded as a basic mechanism of learned inhibition. The opposite of LTD is **long-term potentiation (LTP).**

long-term potentiation.   See LTP.

long-term store.   See the **multiple-store model.**

LOP.   See **levels-of-processing (LOP) framework.**

LTP.   Long-term potentiation (LTP) is an increase in the synaptic potential to fire that develops as a function of prior cellular activation. LTP is thus a synaptic strengthening due to experience. LTP is regarded as a basic mechanism of learning. The opposite of LTP is **long-term depression (LTD).**

magnitude estimation.   Magnitude estimation is a method for scaling a continuous intensity dimension such as the loudness of a sound. With this method, the participant is given a standard stimulus, and the participant is told to give that stimulus a certain value (e.g., the standard is arbitrarily instructed to be a 10). For a stimulus that is judged to be twice as loud as the standard, the participant is instructed to rate it as a 20, whereas for a stimulus that was judged to be half as loud, the participant is instructed to give a rating of 5. Thus, all other stimuli are rated with numeric values that reflect the relative perceived intensity relative to the standard stimulus.

Markov chain.   A Markov chain is a Markov model with constant values for the transition probabilities. See **Markov model.**

Markov model.   A Markov model is a probability model that has a finite number of states. The probability that the system (on trial $n + 1$) is in any particular state is a function of the probability distribution for the various states on trial n, along with a matrix of transition probabilities. If the transition probabilities are independent of n, then the Markov model is said be a Markov chain. The condition of constant transition probabilities is called stationarity.

masking.   Masking is a procedure in which a visual (masking) stimulus follows immediately after a target stimulus. This type of masking is called backward masking. The masking stimulus erases the iconic image of the target. If the target is very briefly presented (less than 50 msec), then the masking stimulus can prevent (on most trials) the conscious detection of the target. Forward masking is the procedure of presenting the masking stimulus before the target.

massed practice.   See **massed versus distributed practice.**

massed versus distributed practice.   Massed versus distributed practice deals with the time schedule of practice for some new information that is being learned. For massed practice, the new information is studied without interruption. For distributed practice, the learning time is broken up, so there are other activities between learning sessions. Massed practice is less efficient than distributed practice for a number of reasons. Related to the distinction between massed practice versus distributed practice is the idea of the spacing effect. Consider the memory accuracy of a new association after a second presentation of the information where the spacing between the first and second presentations is denoted as t_1 and the delay after the second presentation is t_2. The spacing effect is the change in the retention rate for a fixed t_2 retention interval as a function of the spacing interval t_1. Massed practice is the special case when $t_1 = 0$. Although the massed case is less effective than the case when $t_1 > 0$, the magnitude for spacing values for nonzero spacings is a weak and inconsistent effect that depends on the value of the retention interval t_2.

matching-to-sample. Matching-to-sample is a training procedure where the animal sees an initial stimulus (i.e., the sample). Later at the time of test, the animal is provided with a set of response choices, of which one is the original sample.

maximum likelihood estimate. See **likelihood function**.

MCMC. Markov chain Monte Carlo (MCMC) is a class of algorithms for the approximate sampling from a desired distribution. The method is based on a Markov chain, and the quality of the estimation improves with the number of steps in the chain. MCMC algorithms have enabled the sampling from probability distributions that were heretofore difficult to implement directly.

memory span. Memory span is the longest list length where the person is at a 50 percent level for repeating back the entire list in the right order. The size of memory span depends on the information content and the individual. Crannell and Parrish (1957) showed that the memory span for familiar three-letter words was less than four, but the span for letters is about six. Warren (2015) showed that the memory span for familiar shapes was four, but if the shapes were unfamiliar pictograms that could be drawn easily in one or two fluid line strokes, then the memory span was only one.

memory trace. The term *memory trace* refers to the stored encoding of some target information.

messenger RNA. Messenger RNA carries the chemical blueprint for construction of a specific protein.

meta-analysis. Meta-analysis is a statistical procedure for combining information over several selected studies to develop a single statistical conclusion.

metathetic. Stevens (1961) called a sensory continuum a metathetic dimension if changes on the dimension reflect changes in the activation of either the detector cells or the combination of detector cells. For example, hue is a metathetic continuum because color changes are a function of different detector cells being activated. If a sensory continuum reflected changes in only the intensity of the activation of the detectors, and it can be organized on a unidimensional axis, then the continuum is called **prothetic**.

method of loci. The method of loci is a mnemonic technique. There are four components to the method: a *memory palace* with *imagined objects* that are *linked* together and *placed along a route* through the memory palace.

Minerva 2 model. Minerva 2 is a computer model of memory developed by Hintzman (1984, 1988). In this model, each memory event is treated as a vector of features; the values for these features are 1, 0, or -1. Importantly, the same target is assumed to have a separate feature vector for each presentation of the item. For this model, there is an echo response to a probe, and this response has a strength or intensity value. The echo intensity is equal to the sum of the activation strength values from all the vectors in the entire memory system. The individual activation values for a memory trace are a function of the similarity between a test probe and the memory trace. The model does not include relational and elaborative propositions as a property of the memory encoding. It is a feature-space model, and it uses a strength theory of memory.

mirror effect. This effect often occurs for recognition memory experiments where there are two conditions that have different degrees of memory retention. Let us called one condition the "strong" condition and the other condition the "weak" condition. The **signal-detection theory** analysis of the data typically finds that the mean of the "strong" condition is shifted to the right on the strength axis relative to the mean of the "weak" condition. However, the corresponding display of the foil means has the mean for the "strong" condition shifted to the left on the strength axis relative to the mean for the "weak" condition. Thus, the means for the foil distributions demonstrate a mirror reversal of the order for the target means. This pattern is thus called the mirror effect.

mixed design. A mixed design is a research plan for some multifactor experiments. Multifactor experiments have two or more independent variables. A mixed design has at least one between-subjects variable and at least one within-subjects variable. See **between-subjects design** and **within-subjects design**. In statistics, the mixed design is also called a split-plot research plan.

mixed versus pure lists. Pure lists are homogeneous lists where the type of items and the encoding instructions are the same for all the items. Mixed lists have a mixture of either item types or encoding rules.

MLE. See **maximum likelihood estimate** and **likelihood function**.

modal model of memory. The modal model is called by this name because it is believed by most memory researchers. See the **multiple-store model**.

model 6P. This model is for obtaining storage and retrieval parameters for the **Chechile-Meyer task**. The model has six parameters. This model is a special case of **model 7B**. The special case is when the probability for using the high-confidence rating of "3" is the same regardless of a YES or NO recognition response.

model 7B. This model is the general model for obtaining storage and retrieval probability measures for the **Chechile-Meyer task**. The model has seven parameters.

model-based measurement. Model-based measurement is the general approach to measurement that employs a model of how the various latent variables interact on any given experiment. Typically, there are different experimental conditions that are separately modeled, and from the combination of these conditions, the separate latent measures can be statistically estimated provided that the full model is statistically identifiable. Model-based measurement is the recommended solution to the **entanglement problem** that occurs in cognitive psychology.

model identifiability. There are several types of identifiability of scientific models. The most rigorous standard for identification is when no two configurations of model parameters have the same **likelihood function**. Chechile (1998) developed a method for assessing model identifiability. The method is based on making the assumption that two configurations of model parameters exist that have the same likelihood function. These points are described in terms of vectors $(\theta_1, \theta_2, \ldots, \theta_m)$ and $(\theta_1 + \varepsilon_1, \theta_2 + \varepsilon_2, \ldots, \theta_m + \varepsilon_m)$. If it can be proved that $\varepsilon_i = 0$ for $i = 1, \ldots, m$, then the two points are the same. Thus, no two configurations of model parameters exist that have the same likelihood function (i.e., the model is identifiable). It is impossible for any scientific model to be identifiable in terms of this standard if the number of parameters exceeds the number of the degrees of freedom in the data. However, just having more degrees of freedom in the data than the number of model parameters does not guarantee that the model is identifiable. The mathematical structure of the model can make some models unidentifiable despite having fewer parameters than statistical degrees of freedom. Furthermore, the local maximization of the likelihood function also does not guarantee that a model is identifiable.

Molyneux problem. The Molyneux problem was proposed in 1690 as a way to test if there are innate ideas or concepts. The proposed experiment deals with a person who is born blind and learns the meaning of solid shapes by touch, and later sight is given to the individual. Molyneux suggested that the person either would immediately understand the new visual sensations or would be substantially confused and have to learn the meaning of the sensations. The Molyneux problem is thus a critical test of the concept of innate ideas. See **innate ideas**.

mood-congruent memory. Mood congruency refers to the matching of the affective mood at the time of recall with the mood state associated with the target item. For example, if a participant is asked to recall a specific autobiographical event that was a joyful experience, then the recall is easier when the person is in a joyful mood.

mood-dependent memory. Mood-dependent memory is the hypothesis that a change in the mood state of the *person* at the time of test from the mood state at the time of learning will lead to a recall deficiency of emotionally neutral material. Although mood-dependent memory ought to be a special case of state-dependent memory and music-dependent memory, it has been difficult to find support for the effect. Mood-dependent memory is different from **mood-congruent memory** because the content material for learning is neutral, and the mood state of the participant is altered between learning and testing.

Morgan's canon. Morgan's canon is a general principle that psychologists should use as simple an account of animal behavior as possible. For example, before assuming a human-like attribute to the animal, the psychologist should try to explain the animal's behavior with a more simple account if possible.

motor neuron. The discharge of a motor neuron directly affects a muscle or a gland; thus, the output of the motor neuron is not another neuron. See **neuron**.

MPT models. Multinomial processing-tree models are designed for specific tasks to measure underling cognitive processes in terms of probability trees. For example, psychological constructs, such as storage and retrieval, are treated as probability values that can be estimated from the data.

multinomial processing-tree models.   See **MPT models.**

multiple-cue model of generation.   The context for the multiple-cue model is a generation task where there are different semantic cues linked to the target. This model also applies in an unconstrained generation task. See **unconstrained generation task.** The multiple-cue model is the hypothesis that the additional cues that were formed during encoding improved subsequent memory retrieval because there are more retrieval routes.

Multiple-store model.   The model hypothesized by Atkinson and Shiffrin (1968) in which there are separate stores for memory. The memory stores are (1) sensory memory store (which they called the **sensory register**), (2) the **short-term store**, and (3) the **long-term store.** Long-term memory was permanent, whereas information is 100 percent susceptible to loss if it is only in the short-term store or in the sensory register. Information cannot go directly from the sensory register to the long-term store. Information from the environment that is eventually stored in the long-term memory goes first to the sensory register, then to the short-term store, and finally to the long-term store. The short-term store is limited in the number of items that can be simultaneously kept there, whereas the long-term memory is unlimited in time and capacity. The hardware of the memory system can be used with a variety of strategies, which are called control processes.

Murdock learning equation.   A special case of the Chechile-Sloboda learning equation where there is a constant learning rate. See **Chechile-Sloboda learning equation.**

Murdock trial 1 trade-off equation.   Based on empirical fits of paired-associate learning studies, Murdock (1960) found that on average, the number of items correctly recalled after *one* trial of learning is R_1 and it is equal to $.06 P_T L + 6.1$, where P_T is the presentation time per word and L is the number of pairs on the list.

music-dependent memory.   Music-dependent memory is a specific type of incidental context effect. If there is background music at the time of learning, then a variation in the musical background at the time of test can result in a less effective recall of the target items.

N400.   The N400 is negative-going **ERP** response. This peak or component of the ERP has been associated with a number of cognitive processes. A common interpretation of the N400 is that it is triggered by an event that requires a greater degree of semantic processing.

negative reinforcer.   If an animal is in an aversive context that it is trying to escape, then the escape or removal of the aversive stimulus is a negative reinforcer.

neural adaptation.   See **sensory adaptation.**

neurogenesis.   Neurogenesis is the process of new neural cell growth in the adult brain of mammals. The new cells are typically **interneurons.** Neurogenesis has been found in the **olfactory bulb** as well as in the **dentate gyrus** of the **hippocampus.** This process has been linked with the learning of a new event.

neuron.   A neuron is a cell that has a nucleus, a dendrite, an axon, and axon terminals. Neurons do not physically touch other neurons because there is a tiny gap between the neurons called the **synapse.** The electrical potential in the dendrite varies as a function of neurotransmitter chemicals that it received from the synaptic inputs to the **dendrite** of the cell. When the electrical potential is sufficiently elevated, there is an electrical spike discharged down the axon that causes the release of neurotransmitter chemicals from the axon terminals, thus influencing the likelihood of activation of the adjoining neuron. The influence of an electrical discharge can be either excitatory or inhibitory. Generally, there are three types of neurons: **sensory neurons**, **motor neurons**, and **interneurons**.

nondeclarative memory.   Nondeclarative memories are the memories that are difficult to recall in terms of verbal statements. An example of nondeclarative memories would include motor skills and habits such as tying one's shoelaces or typing. Nondeclarative memory is often described as an implicit memory.

norepinephrine.   Norepinephrine is a hormone and a neurotransmitter. It plays a role in the adrenal stress response.

normal distribution.   See **Gaussian distribution**.

novelty habituation.   To avoid confusion over the use of the term *habituation*, it is recommended for research in developmental psychology that the clarifying term *novelty habituation* should be used instead. Infants are engaged in exploring their environment. For example, if the child is shown a plastic blue triangular object, then the infant will orient and visually track the new object. This type of response is called an **orienting response**. Eventually, the object becomes a familiar stimulus, and the infant explores other features of the environment. The blue triangular object, however, does not produce an innate response such as in the other cases of habituation (e.g., the gill withdrawal reflex triggered by gently touching the syphon of the *Aplysia* or marine sea snail). Consequently, a different phrase (*novelty habituation*) is used for the reduction of the novelty response.

oddity concept.   For an oddity concept-learning task, the correct response is always the stimulus that is different. Consequently, a given stimulus can be either correct or incorrect depending on its relationship to the other stimuli.

olfactory bulb.   This structure is the neural region in the forebrain of vertebrates that is involved with the detection of the sense of smell. The olfactory bulb is also a region where **neurogenesis** has been observed.

operant conditioning.   See **instrumental conditioning**.

orbitofrontal cortex.   The orbitofrontal cortex is a region in the frontal lobes of humans. This region has been linked to the cognitive processes associated with decision making as well as with appetitive conditioning (Kringelbach, 2005).

orienting response.   Many animals (including humans) notice a novel stimulus. The response of attending to the novel stimulus is called the orienting response.

paired-associate task.   For this task, the person must learn a list of pairs. Each pair has a stimulus item and a response item. Usually, the whole set is presented in a random order and followed by a test phase. The list is repeated until the person is trained to the desired level of proficiency.

Pavlovian conditioning.   See classical conditioning.

P(coh).   See PPM.

peg-word mnemonic.   This encoding strategy is based on concrete objects that are associated with the digits. These concrete objects are the peg words for later learning. Suppose later that an ordered list must be learned. With the peg-word mnemonic, the list items are associated with the corresponding peg words.

PET scan.   Positron emission tomography (PET) is an imaging procedure that detects gamma rays that are produced from a radioactive substance injected into a person. The gamma rays are emitted along with positrons, which are like electrons except they have a positive charge. The radioactive substance is fludeoxyglucose, which behaves chemically like glucose. Neural cells that have been particularly active need more glucose. Consequently, the PET scan is a three-dimensional map of the brain regions of the highest glucose metabolism level.

phonological loop.   The phonological loop is a component of working memory. See **working memory** for more details.

PI.   PI or Proactive Interference is the increased difficulty in encoding and remembering the current event because of the interference effect from previously learned similar events.

place learning.   Place learning is the idea that an animal, which has learned to run to a goal box in a maze, learned the spatial arrangement of the maze rather than just a set of responses. The mechanical learning of the running responses without regard to place learning is called **response learning**.

posttraumatic stress disorder (PTSD).   PTSD is a clinical psychiatric condition caused by traumatic experiences or abuse. The symptoms include intrusive flashback memories, upsetting dreams, negative moods, and difficulty concentrating (American Psychiatric Association, DSM 5, 2013).

power.   See **classical statistics** for a description of statistical power.

PPM.   The population parameter mapping (PPM) method is a statistical procedure for estimating the parameters of a scientific model by a two-step Bayesian inference. The first step is the sampling of a vector of parameter values from the population of model parameters for the statistical data structure. For example, for categorical data, the statistical parameters for the population proportions are denoted as subscripted ϕ values, and a vector of these proportions has a known prior and posterior distribution. The second step is to functionally map that vector of values to the corresponding vector of parameters for the scientific model (i.e., the subscripted θ parameters). This mapping might violate a model rationality condition, in which case it is rejected. The two-step process is repeated thousands of times. The proportion of successful mappings to the parameters of the scientific model is an estimate of the coherence of the model itself. In the PPM method, the coherence probability is denoted as $P(coh)$. The point and distributional information about the scientific parameters are obtained from the collection of successfully mapped vectors. Thus, unlike a standard Bayesian analysis, the scientific model is not assumed a priori, but instead the statistical data model for the parameters is assumed. With the PPM method, the value of $P(coh)$

can be computed without any data (the prior value for the metric), and it can be computed after the data are obtained from an experiment (the posterior value). If the posterior-to-prior ratio for the $P(coh)$ metric increases, then the experiment has increased the degree of belief in the scientific model.

proactive interference. See **PI**.

probability density. For a continuous probability distribution, the probability density $f(x)$ is the derivative of the cumulative probability $F(x)$. A density display is a plot of the density as a function of x. The familiar bell-shaped normal curve is a probability density display. The area under a probability density display between point x_1 and x_2 where $x_2 > x_1$ is the probability between those points. See **cumulative probability**.

procedural memory. This type of memory is considered a subset of long-term memory, and it concerns the stored information associated with a well-learned motor task such as typing or riding a bicycle. This information is typically not verbal, and therefore it is not regarded as part of **declarative memory**.

process-dissociation model. See **extended process-dissociation model**.

process-purity hypothesis. The process-purity hypothesis is the belief that a **double experimental dissociation** occurs because **dependent variables** Y and X are pure measures of an underlying cognitive process. See **double experimental dissociation**. This hypothesis is an invalid conclusion because there are numerous counterexamples.

production effect. This effect is the memory improvement for words read aloud as opposed to words read silently.

proposition. A proposition is a knowledge representation or a memory encoding that has the structure of a simple declarative sentence (i.e., a subject, an action or relationship, and a direct object). Propositions have a natural directionality (i.e., Fred hired Larry).

prosopagnosia. Prosopagnosia is the inability to recognize the faces of familiar people because of brain damage.

prothetic. Stevens (1961) called a sensory continuum a prothetic dimension if changes are purely intensity-level changes from the same set of activated detector cells. For example, brightness is a prothetic continuum because the hue of the light is the same but the light varies only in its intensity. A prothetic continuum reflects intensity changes on an ordered unidimensional axis, whereas changes in the quality of the perception are a **metathetic** feature.

prototype model. The prototype model deals with concepts and categories. The model assumes that all items in memory are points in a **feature space**. Existing categories have a prototype representation in feature space that is an average of the exemplars of that conceptual category. A new item is categorized based on a function of the distance between the vector for the item and the nearest prototype. As with the **exemplar model**, the structure among the features of the stimulus is not computed for the prototype model.

PTSD. See **posttraumatic stress disorder**.

RAM. Random-access memory (RAM) is described in the section on **content-addressable memory.**

random-access memory. See **content-addressable memory.**

rationalism. Rationalism is a philosophical tradition associated with Descartes, and it is a philosophical tradition that believes in innate ideas that can be discovered by means of reason.

RBC model. The *recognition-by-components* or RBC model by Biederman (1986, 1987, 1990) is an example of a **bottom-up** theory of pattern recognition. It is hypothesized in this model that all visual patterns are made up of elemental features, which are called **geons.** The entire collection of these features is activated in parallel and leads to the recognition of a familiar pattern. The RBC model is also an example of a **feature-space** approach for recognition. The finding of a **syntactic complexity effect** is evidence against this model.

receptor trafficking. Receptor trafficking refers to the flow of receptor gates either from a fluid inside the neural cell to the **dendritic spine** or from the dendritic spine back to the fluid within the **dendrite.** When a receptor gate is attached to the dendritic spine, it enables the flow of chemical neurotransmitters to pass from the synapse to the dendrite. See **neuron.**

reconsolidation theory. See **consolidation theory.**

redintegration. Redintegration is a process where the original encoding is reconstructed from a partial representation.

REM model. The *Retrieving Effectively from Memory* or REM model was developed by Shiffrin and Steyvers (1997). For the REM model, information in the long-term store about a memory image is captured by a vector of features. The values for the components of the vector are the frequency values for the various features. On recognition testing, a participant must decide if a test probe is old or new. For the REM model, the decision is based on the Bayes rule, that is, the ratio

$$\Phi = \frac{P(old|data)}{P(new|data)},$$

$$= \frac{P(old)}{P(new)} \frac{P(data|old)}{P(data|new)},$$

where the second term is the likelihood ratio. For most experiments, the probability of old and new items is equal (i.e., $P(old) = P(new)$), so the Bayes rule reduces to a decision based on the likelihood ratio. The likelihood ratio is considered by Shiffrin and Steyvers (1997) to be the average likelihood ratio for the last L individual memory images. Thus, they set $\Phi = \frac{1}{L} \sum_j \lambda_j$, where λ_j are the individual likelihoods of matching to mismatching features. The individual likelihood is computed as

$$\lambda_j = \left(\frac{\alpha}{\beta}\right)^{m_j} \left(\frac{1-\alpha}{1-\beta}\right)^{q_j},$$

where α is the probability of a match given the storage for a presented trace, and β is the probability of the match given a nonstored item. The numbers m_j and q_j are the respective frequencies for matches and mismatches. Diller, Nobel, and Shiffrin (2001) also showed that the probability for sampling the image I_i is $p_s(I_i)$, and it is given as

$$p_s(I_i) = \frac{\lambda_i^\gamma}{\sum_k \lambda_k^\gamma},$$

where γ is a parameter. The $p_s(I_i)$ strength quantity is then treated in the same fashion as the comparable term in the **SAM model**.

Rescorla-Wagner model.   This model is a modification of the standard association theory in an effort to address the informational aspects of Pavlovian conditioning. The following equations characterize the association strength $V_n(\mathbf{A})$ for the conditioned stimulus $\mathbf{A}$ on the nth pairing with the unconditioned stimulus:

$$V_n(\mathbf{A}) = V_{n-1}(\mathbf{A}) + \Delta V_{n-1}(\mathbf{A}),$$

$$\Delta V_{n-1}(\mathbf{A}) = \alpha_\mathbf{A}\beta[\lambda - V_{n-1}(total)],$$

where $\alpha_\mathbf{A}$ is a parameter for the relative salience of $\mathbf{A}$, β is a learning rate parameter, λ is the asymptotic level for the strength of the association, and $V_{n-1}(total)$ is the current strength expectation level.

response learning.   Response learning is the hypothesis that animals simply learn a set of reinforced responses rather than a cognitive map of the spatial arrangement of their environment. See **place learning** for a contrasting hypothesis about learning.

retroactive interference.   The difficulty in remembering earlier information due to the interference effect of more recent information.

retrograde amnesia.   Retrograde amnesia (RA) is the memory loss that can occur after a head injury or electroconvulsive shock. With RA, the most recent memories before the injury are lost. Retrograde amnesia has also been called **Ribot's law**.

rhyme mnemonic.   This memory aid uses a rhyming sentence or a poem in order to remember the key facts.

Ribot's law.   This law is the same as **retrograde amnesia**.

Rizzuto-Kahana model.   The Rizzuto-Kahana model is a modified Hopfield artificial neural network model for paired associated learning.

ROC.   The receiver operator characteristic (ROC) curve is a plot of the hit rate versus the false-alarm rate. This curve plays a role in **signal-detection theory**.

SAM model.   Raaijmakers and Shiffrin (1981) developed the *Search of Associative Memory* or SAM model that dealt with the retrieval of items from the presumed long-term store of the **multiple-store model**. Guillund and Shiffrin (1984) further developed the SAM model. Forgetting in the SAM model is considered a function of a retrieval failure to generate the target information based on the cues available at the time of test. Event information for this model is represented as a "memory image" consisting of item information, associative information, and contextual information. Besides the item information itself, the content of the stored representation is a set of strength values. There is a strength developed between an item and the context. This strength is proportional to the time that the item is in a particular stimulus *context*. More specifically the strength between the ith item (denoted as I_i) and the context cue Q_c is $S(I_i, Q_c)$ and is equal to $\frac{at_i}{n}$, where the a parameter is a context-learning parameter, t_i is the time that the item and context are together in the short-term buffer, and n is the size of the buffer. Other stimuli that are together in the short-term buffer develop an interitem-associative strength that is proportional to the time in the buffer. The proportionality parameter for the acquisition of the interitem strength is the b parameter, and the equation for this learning is similar to the context-learning formula. There is also a self-association strength for each item. Again, there is a similar equation for the development of this strength component except that there is a c proportionality parameter used for the self-association strength equation. After learning is completed, there are no storage losses. In the SAM model, the above-mentioned strength values do not decay, and there are no losses due to item interference. But the magnitude of the strength can randomly vary due to the addition of a random strength "noise" factor; there is a y parameter for this random component. The combined strength $A(i)$ between the memory image for the ith stimulus and all the cues present at the time of test $(Q_1, \ldots, Q_m)$ is assumed to be

$$A(i) = \prod_{j=1}^{m} S(I_i, Q_j).$$

$$= S(I_i, Q_1) \cdot S(I_i, Q_2) \cdots S(I_i, Q_m).$$

Importantly, there is also a strength $A_r(i)$ value for the ith item *relative* to all the other items activated by the test cues, and this quantity is

$$A_r(i) = \frac{A(i)}{\sum_k A(k)}.$$

The relative strength feature of the SAM model leads to its characterization as a global-matching model because the strength of any item probed generates the activation of the strength values for *all the other items in memory*. However, the product rule for the strength for an item results in an activation value of zero if there is zero strength between the item and any of the cues in the memory probe. In addition to these assumptions about the memory representation, there are additional parameters in the SAM model associated with the search process. The details of the search process differ with the type of test probe. For a simple recall task, the SAM model assumes that there is random sampling of information from the long-term store. The random sampling process has a probability for sampling the I_i image $p_s(I_i)$, and this quantity is

$$p_s(I_i) = 1 - [1 - A_r(i)]^{L_{max}},$$

where L_{max} is the maximum number of attempts for sampling from the long-term store. Finally, the probability of correctly recalling the stimulus associated with the I_i memory image is $P_{recall}(i) = p_s(I_i)(1 - \exp[-\sum_{j=1}^{m} S(I_i, Q_j)])$. For a recognition test probe, Gillund and Shiffrin (1984) assume that there is a familiarity value that is a function of the total activation strength produced by the probe item. Thus, for recognition memory, the SAM model employs a strength-based approach in the tradition of the models of memory that use the **signal-detection theory** framework. The **compound cue theory** is an extension of the SAM model of recognition testing to account for **semantic priming**.

SDT.   SDT denotes **signal-detection theory**, which is described in detail in section 8.3.

self-choice effect.   Suppose a person is presented with a word triplet such as *hammer, wrench,* and *saw,* and the person is asked to circle one of the words for a later memory test. This condition is called the *choice* condition. The counterpart control condition consists of the same words but with one of the words circled already. The person is instructed to circle the previously circled word for a later memory test. This control condition is called the *forced* condition. The self-choice effect is the consistent finding of an improved memory for the *choice* condition over the *forced* condition.

self-reference effect.   In a self-reference experiment, participants usually are asked if some trait attribute is a good description of them personally. In a control condition, the participants might be asked if the trait adjective means the same thing as another word. The self-reference effect is the improved memory retention for the adjectives when the words were assessed as personal trait attributes.

semantic memory.   Semantic memory is the memory of specific facts that are highly overlearned to the point that the facts are no longer linked to a specific episode.

semantic priming.   The presentation of a word automatically elicits associates of that word. If an associate actually does occur later, then the processing of the associate is facilitated. For example, in a lexical decision task, the decision about the letter string NURSE is more rapidly completed if the word DOCTOR occurred earlier because the word NURSE has been primed. The influence of priming is said to spread to the semantic associates.

sensitization.   Sensitization is the opposite effect from **habituation** (i.e., it is an increase in an innate response after an unpleasant or aversive experience). If there is a weak response to a stimulus but the effect of that stimulus is unpleasant or aversive, then the vigor of the response to the stimulus is increased. Like habituation, sensitization is a change in the rate of an innate response as a function of experience.

sensory adaptation.   Sensory adaptation is the same as **neural adaptation**. The phenomenon is the dropping out of a sensory response whenever there is constant sensory input. As an example of sensory adaptation, consider the failure to detect an odor if the chemical basis for the odor is not varied.

sensory neuron.   A sensory neuron converts external stimuli into an electrical discharge; hence, sensory neurons receive input from a sensory organ rather than from another neuron. See **neuron**.

sensory preconditioning.   Sensory preconditioning is demonstrated by a training procedure that begins with the pairing of two neutral stimuli. Next, one of the two stimuli is paired with an **unconditioned stimulus** until there is a learned **conditioned response**. Finally, the other of the two neutral stimuli is presented, and it triggers immediately the conditioned response. The learning of the conditioned response can be said to occur because of the initial pairing of the two neutral stimuli. See **classical conditioning**.

sensory register.   The sensory register is the first storage system in the Atkinson-Shiffrin model of memory. It is equivalent to either the **iconic memory** or **echoic trace**. Information in the sensory register rapidly decays unless it is transferred to the **short-term store**.

sensory threshold.   A sensory threshold is the stimulus intensity level necessary for the stimulus to be detected.

serotonin.   Serotonin is a chemical neurotransmitter.

short-list model.   A short-list model is a lexical-access model that uses a partial direct-access assumption along with a serial-search process. See the **Becker (1980) model** for more details and an example of a short-list model.

short-term store.   See **multiple-store model**.

signal-detection theory.   There is a class of signal-detection models, but the case with Gaussian distributions for the signal and noise conditions is the standard form. See section 8.3 for a detailed explanation of this theory.

Sloboda-Chechile learning equation.   This equation developed by Sloboda and Chechile (2008) is designed to account for the probability of explicit storage (θ_{se}) for the items on a paired-associate list as a function of the number of training trials (n). The equation is $\theta_{se} = 1 - exp(-b\,n^c)$, where b is a parameter related to the ease of learning, and c is a parameter related to the learning "hazard." Learning hazard is defined as the probability for learning to occur on a particular trial given that it was not remembered prior to that trial. When $c < 1$, the hazard is decreasing with increasing n, which means that learning slows. If $c > 1$, then the learning quickens with increasing training trials. If $c = 1$, then the learning rate is a constant. In the special case of $c = 1$, the Sloboda-Chechile equation is equal to the Murdock (1960) learning equation. The Sloboda-Chechile equation is based on the Weibull probability model. The Sloboda-Chechile equation is related to the DEF model, but where intratrial changes in memory retention are ignored. See **Weibull distribution, Murdock learning equation, DEF model,** and **hazard functions**.

SO.   SO is a measure of subjective organization. See **subjective organization**. The metric is based on the consistency of the order information of items on a free-recall list. For the mathematical basis for computing the SO metric, see equation (13.4).

SO2.   SO2 is an alternative measure of subjective organization that is sensitive to both the forward and backward order adjacency of words on a free-recall list. See **subjective organization**. The computational formula for SO2 is given by equation (13.6).

social learning.   Social learning is the learning by example from other members of the same species.

spacing effect.   See **massed versus distributed practice**.

species-specific behavior. Species-specific behavior is an instinct that depends on the evolutionary history of the species.

spontaneous recovery. If the stimulus that was previously habituated is withheld for some length of time, then the reintroduction of the stimulus can again elicit a response.

spotlight/zoom lens model of attention. In this model, there is an attentional spotlight that can vary in size. At the center of the spotlight, there is crisp attention to all the objects. Surrounding this center, there is a fringe area where the attention has low resolution. Outside of the fringe area, there is a marginal area where the information does not reach conscious levels of attention. The location of the spotlight can shift with the processing of a visual scene.

spreading activation. See **semantic priming**.

state-dependent memory. Animal studies have shown a state-dependency effect for a number of drugs. These studies demonstrate that a drug can effectively produce an internal stimulus state that is a part of the original encoding of the learned event. Thus, if the drug state at the time of test is different from the time of learning, then there can be a retrieval impairment. This effect is called state-dependent learning.

stationarity. See **Markov chain** and **Markov models**.

Stentor. The *Stentor* is a single-cell protozoa that is capable of **habituation** and **spontaneous recovery**.

stimulus generalization. If an association is learned to a training stimulus, then other similar stimuli will also produce the response but generally with a reduced intensity. For example, if an animal is classically conditioned to a 1,000-hertz tone, then the animal will also exhibit a conditioned response to tones of other frequencies above and below the training stimulus.

stimulus-stimulus expectancy. Stimulus-stimulus expectancy refers to the learning of what follows a specific stimulus. As an example of this type of learning, consider the sight of one's front door and the expectation of what is on the other side of the door.

storage interference. Storage interference is the idea that a previous encoding becomes altered or lost altogether by the storage of an even more recent item.

subjective organization. In a free-recall task, the participant might chunk or group unrelated items together into a cluster. If that is the case, then there ought to be evidence of that grouping in the order information. Metrics such as **SO**, **SO2**, and **lag conditional response probability** have been used to quantify subjective organization.

subliminal conditioning. Subliminal conditioning refers to the classical conditioning that occurs without conscious awareness.

subliminal perception. Subliminal perception refers to the detection of a stimulus that occurs without conscious awareness.

substitution mnemonic. This encoding technique replaces an unknown word with a phrase that sounds like the unknown word.

successive approximation.   A technique of animal training. See **instrumental conditioning.**

survival effect.   The survival effect is the improved memory for words that were rated initially in terms of their usefulness for a hypothetical survival scenario where one needs to avoid predators and to obtain water and food. For the control condition, the words are rated in terms of their usefulness for a nonsurvival scenario such as moving to a new apartment. Typically, there is a higher recall rate for the survival condition.

synapse.   A gap between two neural cells that allows the passage of neurotransmitter chemicals that can affect the potential of a neural electrical response.

synaptic plasticity.   Synaptic plasticity refers to changes due to learning that occur either to the axon terminals leading to the synapse or to the receptor gates on the dendrite of the adjoining neuron. These changes alter the synaptic efficiency and thus are a part of a memory representation. See **neuron.**

synesthesia.   Synesthesia is a sensation when another atypical sensory modality is generated in addition to the typical sensation. If one were to see colors when hearing sounds, then that would be an example of synesthesia.

syntactic complexity effect.   See **syntactic-pattern recognition.**

syntactic-pattern recognition.   This approach was originally developed in the field of computer science, and it was called *syntactic-pattern recognition* because it borrowed ideas from syntactic theory in linguistics and applied them to the field of machine recognition. This framework for the representation of a visual pattern emphasizes the generation of **propositions** about the relationships between the features of the pattern. These propositions are produced by a **top-down** analysis of the pattern, which takes time to conduct. A pattern is judged to be an old pattern if the analysis of the current stimulus reactivates a prior analysis. Thus, from this perspective, it should take longer to recognize a pattern with more propositional details (i.e., a **syntactic complexity effect**). The syntactic-pattern recognition approach is a contrary position to the idea that the recognition of a pattern is done via a **bottom-up** activation of the features that are processed automatically and in parallel. The **RBC model** by Biederman (1986, 1987, 1990) is an example of a bottom-up feature-detection model of recognition where the recognition is a result of the activation of the features of the pattern in parallel. Chechile, Anderson, Krafczek, and Coley (1996) found experimental support for the syntactic-pattern recognition approach.

system-level model.   A system is any structure or process that can be described by input and output measures or variables. For example, the whole organism can be treated as a system without delineating the neurological substructure.

system model of habituation/sensitization.   The system model of habituation is a mathematical model based on an overall strength value denoted as ς. The value of the strength changes as a function of the number of trials in which the stimulus leads to a harmless outcome. The strength on the current trial is equal to the strength from the previous trial plus an incremental change value. For habituation, the change amount is negative, and it is proportional to a learning rate parameter. The change is equal to the learning rate parameter times the difference between the current strength value and a positive minimum strength value, which itself is equal to a function of the number of training trials. The model for sensitization is similar except the change increments are positive. Also, for sensitization, the change is proportional to a different learning rate parameter. The sensitization increment is equal to the learning rate times the difference between a strength maximum value and the current strength value. The sensitization maximum is also a function of the sensitization training trials.

TAP.   See **transfer-appropriate processing framework.**

testing effect.   The testing effect is the memory benefit that accrues from the attempt to retrieve some recently presented information. This effect is also referred to as either the retrieval practice effect or test-enhanced learning. Although there can also be negative consequences of testing, in general, these are more than offset by appropriate feedback.

Teyler-DiScenna indexing model.   In the Teyler-DiScenna indexing model, the hippocampus is in an ongoing interaction with the neocortical association areas. An episodic event activates a distributed set of neocortical cells, which in turn activate automatically a set of hippocampal cells. The hippocampus also receives input from subcortical areas (i.e., the amygdala, medial septum, thalamus, hypothalamus, raphe nuclei, and locus coeruleus). This theory supports the hypothesis that any event has two complementary memory representations—one is the rapidly developing hippocampal trace and the other is a slower neocortical trace.

θ_f.   This parameter is the probability of fractional storage. It is a component in the **IES model.**

θ_k.   This parameter is the probability of having enough information about the studied targets to be able to reject a foil item with a high-confidence NO response.

θ_r.   This parameter is the probability of a successful retrieval of a sufficiently stored item in memory on a recall task.

θ_s.   The probability of sufficient memory storage is denoted as θ_s. This probability represents the proportion of the studied items that are available in memory at the time of test that have enough information content to support the correct recall of the item, provided that the person can retrieve this information.

θ_{se}.   This parameter is the probability of explicit storage in the **IES model.** It is also equal to the probability of sufficient storage in **model 7B.**

θ_{si}.   This parameter is the probability of implicit storage in the **IES model.**

θ_n.   This parameter is the probability of nonstorage in the **IES model.**

TODAM model.   The *Theory of Distributed Memory* or TODAM model is discussed in the Murdock (1995) paper. The TODAM model is perhaps the ultimate vector representation of memory because all memories are stored into a single common N-dimensional vector **M**. The value for N is assumed to be a large odd integer, that is, $N = 2L + 1$, where L is a positive integer. The indexing of the vector is from $-L$ to L in steps of 1. There is also an indexing for the states of the common memory vector; hence, the state of the common memory vector after $j - 1$ events is denoted as $\mathbf{M}_{j-1}$. Individual event memories also consist of a N-dimension vector. Let the $j-1$th and jth stimulus events be represented as respectively $\mathbf{f}_{j-1}$ and $\mathbf{f}_j$ with components

$$\mathbf{f}_{j-1} = (a_{-L}, a_{-L+1}, \ldots, a_0, \ldots, a_L),$$

$$\mathbf{f}_j = (b_{-L}, b_{-L+1}, \ldots, b_0, \ldots, b_L).$$

The TODAM storage after the jth event is

$$\mathbf{M}_j = \alpha \mathbf{M}_{j-1} + \gamma \mathbf{f}_j + \omega_j \mathbf{f}_j * \mathbf{f}_{j-1},$$

where the $*$ symbol denotes the *convolution* operation between the past two event vectors. The rationale for the convolution operation originates from probability theory, and it is related to the sum of two random variables (Metcalfe-Eich, 1982). The α parameter in the storage equation is a forgetting parameter, and it has a value on the $(0, 1)$ interval. Hence, after an event, each component of the common memory vector is decreased because it is multiplied by α. But there are also two positive increments that arise from the other two terms of the storage equation. The second term is connected to the contribution of the jth event, provided that the person is attentive to the stimulus. The γ parameter is an attention parameter, so if there is full attention, then $\gamma = 1$. Finally, the last term in the storage equation is related to the learning of the association between the last two events. The ω_j parameter is the weight for the associative information relative to the item storage component. Murdock (1995) also describes the response of the TODAM model to various retention test probes. Finally, the TODAM model has purely a vector representation of knowledge, so it cannot encode the learning of a **proposition**.

top-down. The term *top-down* refers to a cognitively or conceptually driven process. For example, if the person generates an expectancy based on the current time and context, then that cognitive knowledge can affect the processing of new information. Top-down is the opposite of a **bottom-up** process, where the incoming sensory information drives the process.

total-time hypothesis. The total-time hypothesis was proposed by Cooper and Pantle (1967). They proposed that the total time to learn a list is only a function of the number of items on the list, the time of study per item, and the number of replications of the list. Moreover, they claimed that if one were to have less time per item and more training trials, then the total time to achieve a given learning level was the same. This hypothesis is inconsistent with the findings of the testing effect.

trace conditioning. Trace conditioning is a Pavlovian conditioning arrangement between the **conditioned stimulus (CS)** and the **unconditioned stimulus (US)**, where the CS turns on prior to the US, but it is also turned off before the onset of the US. See **classical conditioning**.

transfer-appropriate processing framework. The transfer-appropriate processing framework was developed by Morris, Bransford, and Franks (1977) as part of a critique of the levels-of-processing approach. They stressed the importance of matching the method of testing with the method for encoding. For example, they argued that a rhyme test rather than a meaning test should be used if the original encoding was based on a rhyme strategy.

two-high-threshold model (2HTM). The two-high-threshold (2HTM) model by Bröder and Schütz (2009) is a process tree model for recognition memory. It is a strength model that has separate thresholds for (1) when the strength is high enough to trigger a YES response and (2) when the strength is so low that the person gives a NO response. Strength values between the two thresholds result in guessing.

two-trace hazard model. In the two-trace hazard model, all items are initially encoded with two concurrent memory representations—trace 1 and trace 2. These two traces have qualitatively different hazard properties. The hazard function for trace 1 is monotonically increasing, but if there is another study trial, then trace 1 is refreshed or elaborated. The hazard properties for trace 2 are monotonically decreasing as a function of time, but if there is a rehearsal or another study trial, then this trace is also reset to the initial condition. This process is ongoing over the life of the individual, so there is always some risk that an old memory is lost. But the risk of memory loss for an old memory is increasingly small. The overall hazard property of the system of two traces is a peaked-shaped memory hazard function. The probability of memory storage (θ_S) as a function of time since the last presentation (t) is

$$\theta_S = 1 - b(1 - exp[-a\,t^c])(1 - exp[-dt^2]),$$

where $0 < c < 1$, $0 < b < 1$, $a > 0$, and $d > 0$. The combination of the two traces is consistent with Jost's second law, which stresses the increased risk of contemporary learning. See **hazard function** and **Jost's law**.

unconstrained generation task. For the unconstrained generation task, the participants are provided with a stem word, and they are required to generate a response word to the stem that fits a particular encoding rule. Each trial for the unconstrained generation task is thus a mini problem-solving question. For each participant in the generation condition, there is a corresponding yoked-control participant who is provided with the stem and response words produced by his or her counterpart. With a congruous encoding instruction, the participants are directed to generate a related word to the stem word, whereas with an incongruous encoding instruction, the participants are asked to generate an unrelated word.

UR. UR is the unconditioned response in Pavlovian conditioning. The unconditioned response is denoted as **UR** to stress that it consists of a vector of responses. See **classical conditioning** for greater detail.

US. US is the unconditioned stimulus in Pavlovian conditioning. The unconditioned stimulus is denoted as **US** to stress that it consists of a vector of a sensory component. See **classical conditioning** for more details.

ventral occipitotemporal stream. A broad neural pathway in humans from the occipital lobe to the temporal lobe. This stream is associated with the visual processing of information from its physical properties to its memory understanding.

visual accommodation. The lens of the human eye is elastic and can change its shape, and this change alters the focus point in space (either near or far). Normally, the changes are automatic, but it can also be changed voluntarily. Visual accommodation is discussed in the context of the theory of attention by William James, and the idea has been revived and extended in the **spotlight/zoom lens model** of attention.

visuospatial sketchpad.　The visuospatial sketchpad is a component of working memory. See **working memory** for more details.

von Restorff effect.　This effect is the better memory for an item that is isolated or distinctively different from the other items on a list.

weapon focus.　Weapon focus is the phenomenon of a poor memory of other features of a crime scene because the eyewitness has focused instead on the threatening object. This phenomenon is consistent with the **Easterbrook hypothesis**.

Weibull distribution.　The Weibull probability distribution is a continuous function for a variable x, which is usually the positive real axis (i.e., $x \geq 0$). The cumulative probability is $F(x) = 1 - e^{-ax^c}$, where a and c are positive parameters for the distribution. The shape of the distribution is determined by the c parameter, which also has an important connection to the hazard properties of the distribution. If $c = 1$, then the Weibull distribution has constant hazard. If $c > 1$, then the distribution has a monotonically increasing hazard. If $c < 1$, then the distribution has a decreasing hazard. Consequently, the Weibull distribution is a valuable tool for studying hazard systems. Because of the linkage with hazard, the Weibull distribution is also valuable for describing dynamically varying processes. There can also be an application of the Weibull distribution to discrete integer values, such as the number of training trials, n. In this case, the modified Weibull distribution becomes a discrete model that only has probability values at integer values for n. Because learning can be considered a state change from unlearned to learned and because state changes are associated with a hazard function, the Weibull cumulative distribution is thus an ideal model for representing learning, that is, $P_n = 1 - e^{-bn^c}$, where P_n is the proportion of a list that is correct after n training trials.

Wernicke's area.　Wernicke's area is a region in the left hemisphere that is important for the comprehension of spoken language.

within-subjects design.　A within-subjects design is a research plan where the subject is tested in all the levels of the independent variable. In statistics, this research plan is also called a repeated-measures design or a randomized block design.

working memory.　Working memory is an elaboration of the **Atkinson-Shiffrin model**, and it is associated with the papers by Baddeley and Hitch (1974) and Baddeley (2000). In the working-memory model, there is a hypothesized module for a **central executive function** as well as separate stores depending on the stimulus. In the Baddeley-Hitch version of working memory, the other two buffers are a **phonological loop** and a **visuospatial sketchpad**. Baddeley (2000) added another component to working memory; this component he called the **episodic buffer**. The central executive module along with phonological loop, visuospatial sketchpad, and episodic buffer constitute *fluid working memory*, and it is an elaboration of the short-term store of the Atkinson-Shiffrin model. The long-term store of the Atkinson-Shiffrin model is the *crystallized long-term memory*. Crystallized long-term memory consists of *visual semantics*, *episodic long-term memory*, and *language*.

Yerkes-Dodson law. The Yerkes-Dodson law is the idea that learning is an inverted U-shaped function of the amount of arousal. This law is still widely accepted but has been shown to be incorrect on a number of grounds.

zero-contingency effect. This effect refers to the fact that no conditioning occurs when the onset of the conditioned stimulus and the onset of the unconditioned stimulus occur simultaneously. Without a temporal delay between the conditioned stimulus and the unconditioned stimulus, the conditioned stimulus cannot serve as a signaling event for the later onset of the unconditioned response.

Bibliography

Aimone, J. B., Wiles, J., & Gage, F. H. (2006). Potential role for adult neurogenesis in the encoding of time in new memory. *Nature Neuroscience, 9,* 723–727.

Allaby, M. (2004). *A chronology of weather.* New York: Facts on File Books.

Allport, G. W. (1924). Eidetic imagery. *British Journal of Psychology, 15,* 99–120.

Allport, G. W. (1928). The eidetic image and the afterimage. *American Journal of Psychology, 41,* 418–425.

Ally, B. A., Hussey, E. P., & Donahue, M. J. (2013). A case of hyperthymesia: Rethinking the role of the amygdala in autobiographical memory. *Neurocase, 19,* 166–181.

Altman, J., & Das, G. D. (1965). Post-natal origin of microneurones in the rat brain. *Nature, 207,* 953–956.

Alvarez, P., & Squire, L. R. (1994). Memory consolidation and the medial temporal lobe: A simple network model. *Proceeding of the National Academy of Science, 91,* 7041–7045.

American Psychiatric Association. (2013). *Diagnostic and statistical manual of mental disorders* (5th ed.). Arlington, VA: American Psychiatric Publishing.

Anderson, J. R. (1976). *Language, memory, and thought.* Hillsdale, NJ: Erlbaum.

Anderson, J. R. (1980). Concepts, propositions, and schemata: What are the cognitive units? *Nebraska Symposium on Motivation, 28,* 121–162.

Anderson, J. R. (1981). Interference: The relationship between response latency and response accuracy. *Journal of Experimental Psychology: Learning, Memory, & Cognition, 7,* 326–343.

Anderson, J. R. (1983). A spreading activation theory of memory. *Journal of Verbal Learning and Verbal Behavior, 22,* 261–295.

Anderson, J. R., & Bower, G. H. (1973). *Human associative memory.* Washington, DC: Hemisphere Press.

Anderson, M. C., Bjork, R. A., & Bjork, E. L. (1994). Remembering can cause forgetting: Retrieval dynamics in long-term memory. *Journal of Experimental Psychology: Learning, Memory, and Cognition, 20,* 1063–1087.

Anooshian, L. J., & Seibert, P. S. (1996). Conscious and unconscious retrieval in picture recognition: A framework for exploring gender differences. *Journal of Personality and Social Psychology, 70,* 637–645.

Antes, J. R. (1974). The time course of picture viewing. *Journal of Experimental Psychology, 103,* 62–70.

Apostle, H. G. (1980). *Aristotle's categories and propositions (De Interpretatione).* Grinnell, IA: Peripatetic Press.

Aristotle. (350 BCE). On memory and reminiscence. In J. Barnes (Ed.) (1984) *The Complete Works of Aristotle* (Vol.1, pp. 714–720). Princeton, NJ: Princeton University Press.

Aristotle. (circa 329 BCE). *The Niomachean Ethics, Book 2.* In J. A. K. Thomson (Ed.) (1953). *The Ethics of Aristotle* (pp. 55–75). Baltimore, MD: Penguin.

Ashby, F. G. (1992). Multidimensional models of categorization. In F. G. Ashby (Ed.), *Multidimensional models of perception and cognition* (pp. 449–483). Hillsdale, NJ: Erlbaum.

Atkinson, R. C., & Crothers, E. J. (1964). A comparison of paired-associate learning models having different acquisitions and retention axioms. *Journal of Mathematical Psychology, 1,* 285–315.

Atkinson, R. C., & Shiffrin, R. M. (1968). Human memory: A proposed system and its control processes. In K. W. Spence & J. T. Spence (Eds.), *The psychology of learning and motivation: Advances in research and theory* (Vol. 2, pp. 89–195). New York: Academic Press.

Atwood, G. (1971). An experimental study of visual imagination and memory. *Cognitive Psychology, 2,* 290–299.

Auble, P. M., Franks, J. J., & Soraci, S. A. (1979). Effort toward comprehension: Elaboration or "aha!"? *Memory & Cognition, 7,* 426–434.

Austen, J. (1814). *Mansfield Park.* London: T. Egerton.

Axelrod, J., Weil-Malherbe, H., & Tomchick, R. (1959). The physiological disposition of H3-epinephrine and its metabolite metanephrine. *Journal of Pharmacology and Experimental Therapeutics, 127,* 251–256.

Ayers, T. J. (1991). *An incongruity effect in recall of constrained generated targets.* Paper presented at the annual meeting of the Psychonomic Society, San Francisco in November.

Aziz-Zadeh, L., Kaplan, J. T., & Iacoboni, M. (2009). "Aha": The neural correlates of verbal insight solutions. *Human Brain Mapping, 30,* 908–916.

Baayen, R. H. (2008). *Analyzing linguistic data: A practical introduction to statistics using R.* Cambridge, UK: Cambridge University Press.

Bacon, F. (2000). *Novum organum.* Cambridge, UK: Cambridge University Press. (original work published 1620)

Baddeley, A. D. (1978). The trouble with levels: A reexamination of Craik and Lockhart's framework for memory research. *Psychological Review*, **85**, 139–152.

Baddeley, A. (2000). The episodic buffer: A new component of working memory. *Trends in Cognitive Sciences*, **4**, 417–423.

Baddeley, A. D., & Hitch, G. (1974). Working memory. In G. A. Bower (Ed.), *The psychology of learning and motivation* (Vol. 8, pp. 47–89). Hillsdale, NJ: Erlbaum.

Baddeley, A. D., & Warrington, E. K. (1970). Amnesia and the distinction between long- and short-term memory. *Journal of Verbal Learning and Verbal Behavior*, **9**, 176–189.

Bahrick, H. P. (1984). Semantic memory content in premature: Fifty years of memory for Spanish learned in school. *Journal of Experimental Psychology: General*, **113**, 1–29.

Bailey, C. H., & Chen, M. (1983). Morphological basis of long-term habituation and sensitization in *Aplysia*. *Science*, **220**, 91–93.

Bailey, C. H., & Chen, M. (1988). Morphological basis of short-term habituation in *Aplysia*. *The Journal of Neuroscience*, **8**, 2452–2459.

Bailey, C. H., & Chen, M. (1989). Time course of structural changes at identified sensory synapses during long-term sensitization in *Aplysia*. *The Journal of Neuroscience*, **9**, 1774–1780.

Balch, W. R., Bowman, K., & Mohler, L. A. (1992). Music-dependent memory in immediate and delayed word recall. *Memory & Cognition*, **20**, 21–28.

Balch, W. R., & Lewis, B. S. (1996). Music-dependent memory: The roles of tempo change and mood mediation. *Journal of Experimental Psychology: Learning, Memory, and Cognition*, **22**, 1354–1363.

Baldwin, M. (1960). Electrical stimulation of the medial temporal region. In E. R. Rsmsey & D. S. O'Doherty (Eds.), *Electrical studies on the unanesthetized brain* (pp. 159–176). New York: Hoeber.

Balota, D. A., & Neely, J. H. (1980). Test-expectancy and word-frequency effects in recall and recognition. *Journal of Experimental Psychology: Human Learning and Memory*, **6**, 576–587.

Barbizet, J. (1970). *Human memory and its pathology*. San Francisco: Freeman.

Bartlett, F. C. (1932). *Remembering: A study in experimental and social psychology*. Cambridge, UK: Cambridge University Press.

Batchelder, W. H. (1975). Individual differences and the all-or-none vs incremental learning controversy. *Journal of Mathematical Psychology*, **12**, 53–74.

Batchelder, W. H., & Riefer, D. M. (1990). Multinomial processing models of source monitoring. *Psychological Review*, **97**, 548–564.

Batchelder, W. H., & Riefer, D. M. (1999). Theoretical and empirical review of multinomial process tree modeling. *Psychonomic Bulletin & Review*, **6**, 57–86.

Bäumler, G., & Lienert, G. A. (1993). Re-evaluation of the Yerkes-Dodson law by nonparametric tests of trend. *Studia Psychologica*, **35**, 431–436.

Bayen, U. J., Murnane, K., & Erdfelder, E. (1996). Source discrimination, item detection, and multinomial models of source monitoring. *Journal of Experimental Psychology: Learning, Memory, and Cognition*, **22**, 197–215.

Becker, C. A. (1980). Semantic context effects in visual word recognition: An analysis of semantic strategies. *Memory & Cognition*, **8**, 493–512.

Becker, S. (2005). A computational principle for hippocampal learning and neurogenesis. *Hippocampus*, **15**, 722–738.

Begg, I., & Snider, A. (1987). The generation effect: Evidence for generalized inhibition. *Journal of Experimental Psychology: Learning, Memory, and Cognition*, **13**, 553–563.

Beier, E. M. (1958). *Effect of trial-to-trial variation in magnitude of reward upon an instrumental running response.* Unpublished doctoral dissertation, Yale University.

Békésy, G. von (1960). *Experiments in hearing.* New York: McGraw-Hill.

Bellezza, F. S., Cheesman, F. L., & Reddy, B. G. (1977). Organization and semantic elaboration in free recall. *Journal of Experimental Psychology: Human Learning and Memory*, **3**, 539–550.

Belmont, J. M., & Butterfield, E. C. (1969). The relation of short-term memory to development and intelligence. In L. P. Lipsitt & H. W. Reese (Eds.), *Advances in child development and behavior* (Vol. 4, pp. 30–83). New York: Academic Press.

Bender, R. H., Wallsten, T. S., & Ornstein, P. A. (1996). Age differences in encoding and retrieving details of a pediatric examination. *Psychonomic Bulletin & Review*, **3**, 188–198.

Benson, D. F. (1985). Aphasia. In K. M. Heilman & E. Valenstein (Eds.), *Clinical neuropsychology* (pp. 17–47). New York: Oxford University Press.

Bergson, H. (1911). *Matter and memory.* New York: Macmillan.

Bernardo, J. M., & Smith, A. F. M. (1994). *Bayesian theory.* Chichester, UK: Wiley.

Bernbach, H. A. (1969). Replication processes in human memory and learning. In G. H. Bower & J. T. Spence (Eds.), *The psychology of learning and motivation: Advances in research and theory* (Vol. 3, pp. 201–239). New York: Academic Press.

Bernbach, H. A. (1970). A multiple-copy model of post perceptual memory. In D. A. Norman (Ed.), *Models of human memory* (pp. 103–116). New York: Academic Press.

Berntsen, D. (2001). Involuntary memories of emotional events: Do memories of traumas and extremely happy events differ? *Applied Cognitive Psychology*, **15**, S135–S158.

Berry, C. J., Shanks, D. R., Speekenbrink, M., & Henson, R. N. A. (2012). Models of recognition, repetition priming, and fluency: Exploring a new framework. *Psychological Review*, **119**, 40–79.

Berryhill, M. E., Fendrich, R., & Olson, I. R. (2009). Impaired distance perception and size constancy following bilateral occipitoparietal damage. *Experimental Brain Research*, **194**, 381–393.

Bertsch, S., Pesta, B. J., Wiscott, R., & McDaniel, M. A. (2007). The generation effect: A meta-analytic review. *Memory & Cognition*, **35**, 201–210.

Biederman, I. (1986). Recognition by components: A theory of visual pattern recognition. In G. H. Bower (Ed.), *The psychology of learning and motivation: Advances in research and theory* (Vol. 20, pp. 1–54). Orlando, FL: Academic Press.

Biederman, I. (1987). Recognition-by-components: A theory of image understanding. *Psychological Review*, **94**, 115–147.

Biederman, I. (1990). Higher-level vision. In D. E. Osherson, S. M. Kosslyn, & J. M. Hollerbach (Eds.), *An invitation to cognitive science: Vol. 2. Visual cognition and action* (pp. 41–72). Cambridge, MA: MIT Press.

Biró, T. S. (2011). *Is there a temperature? Conceptual challenges at high energy, acceleration and complexity*. New York: Springer.

Bjork, R. A. (1968). All-or-none subprocesses in the learning of complex sequences. *Journal of Mathematical Psychology*, **5**, 182–195.

Bjork, R. A. (1988). Retrieval practice and the maintenance of knowledge. In M. M. Grunberg, P. E. Morris, & R. N. Sykes (Eds.), *Practical aspects of memory: Current research and issues* (Vol. 1, pp. 396–401). New York: Wiley.

Bjork, R. A., & Bjork, E. L. (1992). A new theory of disuse and an old theory of stimulus fluctuation. In A. Healy, S. Kosslyn, & R. Shiffrin (Eds.), *From learning processes to cognitive processes: Essays in honor of William K. Estes: Vol. 2* (pp. 35–67). Hillsdale, NJ: Erlbaum.

Blackwell, H. R. (1953). Psychological thresholds. Experimental studies of methods of measurement. *Bulletin of the Engineering Research Institute of the University of Michigan*, No. 36.

Blackwell, H. R. (1963). Neural theories of simple visual discriminations. *Journal of the Optical Society of America*, **53**, 129–160.

Blair, H. T., Schafe, G. E., Bauer, E. P., Rodrigues, S. M., & LeDoux, J. E. (2001). Synaptic plasticity in the lateral amygdala: A cellular hypothesis of fear conditioning. *Learning & Memory*, **8**, 229–242.

Blake, R., & Sekuler, R. (2006). *Perception*. New York: McGraw-Hill.

Blaney, P. H. (1986). Affect and memory: A review. *Psychological Bulletin*, **99**, 229–246.

Bliss, T. V., & Lømo, T. (1973). Long-lasting potentiation of synaptic transmission in the dentate area of the anesthetized rabbit following stimulation of the perforate path. *Journal of Physiology*, **232**, 331–356.

Blough, D. S. (1977). Photoelectric recording of pigeon-peck response to computer-driven visual displays. *Behavior Research Methods*, **9**, 259–262.

Bodner, G. E., Jamieson, R. K., Cormack, D., McDonald, D-L., & Bernstein, D. M. (2016). The production effect in recognition memory: Weakening strength can strengthen distinctiveness. *Canadian Journal of Experimental Psychology*, **70**, 93–98.

Bottjer, S. W. (1982). Conditioned approach and withdrawal behavior in pigeons: Effects of a novel extraneous stimulus during acquistion and extinction. *Learning and Motivation*, **13**, 44–67.

Botvinick, M. M., & Plaut, D. C. (2006). Short-term memory for serial order: A recurrent neural network model. *Psychological Review*, **113**, 201–233.

Boulis, N. M., & Sahley, C. L. (1988). A behavioral analysis of habituation and sensitization of shortening in the semi-intact leech. *The Journal of Neuroscience*, **8**, 4621–4627.

Bousfield, A. K., & Bousfield, W. A. (1966). Measurement of clustering and sequential constancies in repeated free recall. *Psychological Reports*, **19**, 935–942.

Bousfield, W. A. (1953). The occurrence of clustering in the recall of randomly arranged associates. *Journal of General Psychology*, **49**, 229–240.

Bouton, M. E. (2004). Context and behavior processes in extinction. *Learning & Memory*, **11**, 485–494.

Bower, G. H. (1961). Application of a model to paired-associate learning. *Psychometrika*, **26**, 255–280.

Bower, G. H. (1970). Analysis of a mnemonic device: Modern psychology uncovers the powerful components of an ancient system for improving memory. *American Scientist*, **58**, 496–510.

Bower, G. H., & Gilligan, S. G. (1979). Remembering information related to one's self. *Journal of Research in Personality*, **13**, 420–432.

Bower, G. H., & Karlin, M. B. (1974). Depth of processing pictures of faces and recognition memory. *Journal of Experimental Psychology*, **103**, 751–757.

Bower, G. H., & Mayer, J. D. (1985). Failure to replicate mood-dependent retrieval. *Bulletin of the Psychonomic Society*, **23**, 39–42.

Bower, G. H., Monteiro, K. P., & Gilligan, S. G. (1978). Emotional mood as a context for learning and recall. *Journal of Verbal Learning and Verbal Behavior*, **17**, 573–585.

Bower, G. H., Thompson-Schill, S., & Tulving, E. (1994). Reducing retroactive interference: An interference analysis. *Journal of Experimental Psychology: Learning, Memory, and Cognition*, **20**, 51–66.

Box, G. E. P., & Tiao, G. C. (1973). *Bayesian inference in statistical analysis*. Reading, MA: Addison Wesley.

Brady, T. F., Konkle, T., Alvarez, G. A., & Oliva, A. (2008). Visual long-term memory has a massive storage capacity for object detail. *Proceedings of the National Academy of Science USA*, **105**, 14325–14329.

Brainerd, C. J., & Howe, M. L. (1978). The origins of all-or-none learning. *Child Development*, **49**, 1028–1034.

Brainerd, C. J., & Reyna, V. F. (1990). Gist is the grist: Fuzzy-trace theory and the new intuitionism. *Developmental Review*, **10**, 3–47.

Breuer, J. & Freud, S. (1895). *Studien über Hysterie*. Leipzig: Feanz Deutricke Publisher.

Brion, S. (1969). Korsakoff's syndrome: Clinico-anatomical and physiopathological considerations. In G. Talland & N. C. Waugh (Eds.), *The pathology of memory* (pp. 29–39). New York: Academic Press.

Bristol, A. S., Sutton, N. A., & Carew, T. J. (2004). Neural circuit of tail-elicited siphon withdrawal in *Aplysia*. I. Differential lateralization of sensitization and dishabituation. *Journal of Neurophysiology*, **91**, 666–677.

Brock, D. C. (2014, April). Dudley Buck and the computer that never was. *IEEE Spectrum*, pp. 55–69.

Bröder, A., & Schütz, J. (2009). Recognition ROCs are curvilinear—or are they? On premature arguments against the two-high-threshold model of recognition. *Journal of Experimental Psychology: Learning, Memory, and Cognition*, **35**, 587–606.

Brogden, W. J. (1939). Sensory pre-conditioning. *Journal of Experimental Psychology*, **25**, 323–332.

Brown, A. S. (1976). Catalog of scaled verbal material. *Memory & Cognition*, **4(1B)**, 1S–45S.

Brown, G. D. A., Preece, T., & Hulme, C. (2000). Oscillator-based memory for serial order. *Psychological Review*, **107**, 127–181.

Brown, J. (1958). Some tests of the decay theory of immediate memory. *Quarterly Journal of Experimental Psychology*, **10**, 12–21.

Brown, W. L., Overall, J. E., & Gentry, G. V. (1959). "Absolute" versus "relational" discrimination of intermediate size in the rhesus monkey. *American Journal of Psychology*, **72**, 593–596.

Bruce, V., & Young, A. (1986). Understanding face recognition. *British Journal of Psychology*, **77**, 305–327.

Bryan, T. H. (1974). Learning disabilities: A new stereotype. *Journal of Learning Disabilities*, **7**, 304–309.

Buchanan, T. W., Lutz, K., Mirzazade, S., Specht, K., Shah, N. J., Ziles, K., & Jäncke, L. (2000). Recognition of emotional prosody and verbal components of spoken language: An fMRI study. *Cognitive Brain Research*, **9**, 227–238.

Buchner, A., Erdfelder, E., & Vaterrodt-Plünnecke, B. (1995). Toward unbiased measurement of conscious and unconscious memory processes within the process dissociation framework. *Journal of Experimental Psychology: General*, **124**, 137–160.

Buchner, A., & Wippich, W. (1996). Unconscious gender bias in fame judgments? *Consciousness and Cognition*, **5**, 197–220.

Buffart, H., Leeuwenberg, E. J. L., & Restle, F. (1981). Coding theory of visual pattern completion. *Journal of Experimental Psychology: Human Perception and Performance*, **7**, 241–274.

Buonomano, D. V. (1999). Distinct functional types of associative long-term potentiation in neocortical and hippocampal pyramidal neurons. *Journal of Neuroscience*, **19**, 6746–6754.

Burgess, N., & Hitch, G. J. (1992). Towards a network model of the articulatory loop. *Journal of Memory and Language*, **31**, 429–460.

Burgess, N., & Hitch, G. J. (1999). Memory for serial order: A network of the phonological loop and its timing. *Psychological Review*, **106**, 551–581.

Burns, D. J. (1992). The consequences of generation. *Journal of Memory and Language*, **31**, 615–633.

Bush, R. R., & Mosteller, F. (1955). *Stochastic models for learning*. New York: Wiley.

Butler, K., & Chechile, R. (1976). "Acid bath" effects on storage and retrieval PI. *Bulletin of the Psychonomic Society*, **8**, 349–352.

Byrne, J. H., & Kandel, E. R. (1996). Presynaptic facilitation revisited: State and time dependence. *The Journal of Neuroscience*, **16**, 425–435.

Cahill, L., & Alkire, M. T. (2003). Epinephrine enhancement of human memory consolidation: interaction with arousal at encoding. *Neurobiology of Learning and Memory*, **79**, 194–198.

Cahill, L., Haier, R. J., Fallon, J., Alkire, M. T., Tang, C., Keator, D., Wu, J., & McGaugh, J. L. (1996). Amygdala activity at encoding correlated with long-term free recall of emotional information. *Proceedings of the National Academy of Science, USA*, **93**, 8016–8021.

Cahill, L., & McGaugh, J. L. (1998). Mechanisms of emotional arousal and lasting declarative memory. *Trends in Neurosciences*, **21**, 294–299.

Cai, D. J., Aharoni, D., Shuman, T., Shobe, J., Biane, J., Song, W., Wei, B., Silva, A. J. (2016). A shared neural ensemble links distinct contextual memories encoded close in time. *Nature*, **534**, 115–118.

Calabrese, E. J., & Baldwin, L. A. (2001). U-shaped dose-responses in biology, toxicology, and public health. *Annual Review of Public Health*, **22**, 15–33.

Calkins, M. W. (1894). Association. *Psychological Review*, **1**, 476–483.

Calkins, M. W. (1896). Association: An essay analytic and experimental. *Psychological Review Monograph Supplement*, **2**, i-56.

Campbell, N. R. (1920). *Physics: The elements*. Cambridge, UK: Cambridge University Press.

Campbell, N. R. (1928). *An account of the principles of measurement and calculation*. London: Longmans, Green, & Associates.

Campbell, N. R., & Jeffreys, H. (1938). Symposium: Measurement and its importance for philosophy. *Proceedings of the Aristotelian Society, Supplement,* **17**, 121–151.

Cannon, W. B. (1953). *Bodily changes in pain, hunger, fear, and rage.* Boston, MA: Branford.

Carew, T. J., Pinsker, H. M., & Kandel, E. R. (1972). Long-term habituation of a defensive withdrawal reflex in *Aplysia. Science,* **175**, 451–454.

Carlin, M. T., Soraci, S. A., Dennis, N. A., Chechile, N. A., & Loiselle, R. C. (2001). Enhancing free recall rates of individuals with mental retardation. *American Journal on Mental Retardation,* **106**, 314–326.

Cermak, L. S. (1975). *Improving your memory.* Düsseldorf, Germany: McGraw-Hill.

Cermak, L. S., Butters, N., & Gerrein, J. (1973). The extent of the verbal encoding ability of Korsakoff patients. *Neuropsycholgia,* **11**, 85–94.

Chamberland, J. R., & Chechile, R. A. (2005). *The role of monetary incentives on recognition memory.* Poster presented in November at the 46th annual meeting of the Psychonomic Society held in Toronto, Canada.

Chatley, N. (2013). *Development of the enactment effect: Examining individual differences in executive function to predict increased memory for action.* Unpublished master's thesis, University of North Carolina at Greensboro.

Chechile, R. A. (1973). *The relative storage and retrieval losses in short-term memory as a function of the similarity and amount of information processing in the interpolated task.* Unpublished doctoral dissertation, University of Pittsburgh.

Chechile, R. A. (1977a). Likelihood and posterior identification: Implications for mathematical psychology. *British Journal of Mathematical and Statistical Psychology,* **30**, 177–184.

Chechile, R. A. (1977b). Storage-retrieval analysis of acoustic similarity. *Memory & Cognition,* **5**, 535–540.

Chechile, R. A. (1978). Is d'_s a suitable measure of recognition memory "strength"? *Bulletin of the Psychonomic Society,* **12**, 152–154.

Chechile, R. A. (1987). Trace susceptibility theory. *Journal of Experimental Psychology: General,* **116**, 203–222.

Chechile, R. A. (1998). A new method for estimating model parameters for multinomial data. *Journal of Mathematical Psychology,* **42**, 432–471.

Chechile, R. A. (2003). Mathematical tools for hazard function analysis. *Journal of Mathematical Psychology,* **47**, 478–494.

Chechile, R. A. (2004). New multinomial models for the Chechile-Meyer task. *Journal of Mathematical Psychology,* **48**, 364–384.

Chechile, R. A. (2006). Memory hazard functions: A vehicle for theory development and test. *Psychological Review*, **113**, 31–56.

Chechile, R. A. (2007). A model-based storage-retrieval analysis of developmental dyslexia. In R. W. J. Neufeld (Ed.), *Advances in clinical cognitive science: Formal modeling of processes and symptoms* (pp. 51–79). Washington, DC: American Psychological Association.

Chechile, R. A. (2009). Pooling data versus averaging model fits for some prototypical multinomial processing tree models. *Journal of Mathematical Psychology*, **53**, 562–576.

Chechile, R. A. (2010a). A novel Bayesian parameter mapping method for estimating the parameters of an underlying scientific model. *Communications in Statistics: Theory and Methods*, **39**, 1190–1201.

Chechile, R. A. (2010b). Modeling storage and retrieval processes with clinical populations with applications examining alcohol-induced amnesia and Korsakoff amnesia. *Journal of Mathematical Psychology*, **54**, 150–166.

Chechile, R. A. (2011). Properties of reverse hazard functions. *Journal of Mathematical Psychology*, **55**, 203–222.

Chechile, R. A. (2013). A novel method for assessing rival models of recognition memory. *Journal of Mathematical Psychology*, **57**, 196–214.

Chechile, R. A. (2014). Using a multinomial tree model for detecting mixtures in perceptual detection. *Frontiers in Psychology*, **5**, Article 641, 1–11.

Chechile, R. A. (2016, August). *A new model for obtaining separate estimates for item and source information*. Paper presented at the meeting of the Society for Mathematical Psychology, New Brunswick, NJ.

Chechile, R. A., Anderson, J. E., Krafczek, S. A., & Coley, S. L. (1996). A syntactic complexity effect with visual patterns: Evidence for the syntactic nature of the memory representation. *Journal of Experimental Psychology: Learning, Memory, and Cognition*, **22**, 654–669.

Chechile, R., & Butler, K. (1975). Storage and retrieval changes that occur in the development and release of PI. *Journal of Verbal Learning and Verbal Behavior*, **14**, 430–437.

Chechile, R. A., & Butler, S. F. (2003). Reassessing the testing of generic utility models for mixed gambles. *Journal of Risk and Uncertainty*, **26**, 55–76.

Chechile, R. A., & Ehrensbeck, K. (1983). Long-term storage losses: A dilemma for multistore models. *Journal of General Psychology*, **109**, 15–30.

Chechile, R., & Fowler, H. (1973). Primary and secondary negative incentive contrast in differential conditioning. *Journal of Experimental Psychology*, **97**, 189–197.

Chechile, R. A., & Meyer, D. L. (1976). A Bayesian procedure for separately estimating storage and retrieval components of forgetting. *Journal of Mathematical Psychology*, **13**, 269–295.

Chechile, R. A., & Richman, C. L. (1982). The interaction of semantic memory with storage and retrieval processes. *Developmental Review*, 2, 237–250.

Chechile, R. A., Richman, C. L., Topinka, C., & Ehrensbeck, K. (1981). A developmental study of the storage and retrieval of information. *Child Development*, 52, 251–259.

Chechile, R. A., & Roder, B. (1998). Model-based measurement of group differences: An application directed toward understanding the information-processing mechanism of developmental dyslexia. In S. Soraci & W. J. McIlvane (Eds.), *Perspectives on fundamental processes in intellectual functioning: Volume 1. A survey of research approaches* (pp. 91–112). Stamford, CT: Ablex.

Chechile, R. A., & Sloboda, L. N. (2014). Reformulating Markov processes for learning and memory from a hazard function framework. *Journal of Mathematical Psychology*, 59, 65–81.

Chechile, R. A., Sloboda, L. N., & Chamberland, J. R. (2012). Obtaining separate measures for implicit and explicit memory. *Journal of Mathematical Psychology*, 56, 35–53.

Chechile, R. A., & Soraci, S. A. (1999). Evidence for a multiple-process account of the generation effect. *Memory*, 7, 483–508.

Cheng, R. C. H. (1978). Generating beta variates with non-integral parameters. *Communication for the Association for Computing Machinery*, 21, 317–322.

Chernev, I., & Reinfeld, F. (1949). *The fireside book of chess*. New York: Simon & Schuster.

Christianson, S-A. (1992). Emotional stress and eyewitness memory: A critical review. *Psychological Bulletin*, 112, 284–309.

Cieutat, V. J., Stockwell, F. E., & Noble, C. E. (1958). The interaction of ability and amount of practice with stimulus and response meaningfulness (m, m') in paired-associate learning. *Journal of Experimental Psychology*, 56, 193–202.

Clarke, K. A., & Parker, A. J. (1986). A quantitative study of normal locomotion in the rat. *Physiology & Behavior*, 38, 341–351.

Clapeyron, E. (1834). Mémoire sur la puissance motrice de la chaleur. *Journal de l'École Polytechnique*, 14, 153–190.

Clayton, E. C., & Williams, C. L. (2000). Noradrengic receptor blockage of the NTS attenuates the mnemonic effects of epinephrine in an appetitive light-dark discrimination learning task. *Neurobiology of Learning and Memory*, 74, 135–145.

Cliff, N. (1992). Abstract measurement theory and the revolution that never happened. *Psychological Science*, 3, 186–190.

Cohen, L. B. (1969). Observing response, visual preference, and habituation to visual stimuli in infants. *Journal of Experimental Child Psychology*, 7, 419–433.

Cohen, L. B., Gelber, E. R, & Lazar, M. A. (1971). Infant habituation and generalization to differing degrees of stimulus novelty. *Journal of Experimental Child Psychology*, 11, 379–389.

Cohen, N., & Squire, L. R. (1980). Preserved learning and retention of pattern analyzing skills in amnesia: Dissociation of knowing how and knowing that. *Science, 210,* 207–210.

Cohen, R. L. (1981). On the generality of some memory laws. *Scandinavian Journal of Psychology, 22,* 267–281.

Cohen, S. (1998). *A detailed field-based approach to assessing learning gains from instructional technology.* Unpublished doctoral dissertation, Tufts University.

Cohen, S., & Chechile, R. A. (1997). Probability distributions, assessment, and instructional software: Lessons learned from an evaluation of curricula software. In I. Gal & J. Garfield (Eds.), *The assessment challenge in statistics* (pp. 253–262). Amsterdam: IOS Press.

Cohen, S., Chechile, R., & Smith, G. (1996). A detailed multisite evaluation of curricular software. In T. W. Banta, J. P. Lund, K. E. Black, & F. W. Oblander (Eds.), *Assessment strategies that work* (pp. 220–222). San Francisco, CA: Jossey-Bass.

Cohen, S., Chechile, R., Smith, G., Tsai, F., & Burns, G. (1994). A method for evaluating the effectiveness of educational software. *Behavior Research Methods, Instruments, & Computers, 26,* 236–241.

Cohen, S., Smith, G., Chechile, R. A., Burns, G., & Tsai, F. (1996). Identifying impediments to learning probability and statistics from an assessment of instructional software. *Journal of Educational and Behavioral Statistics, 21,* 35–54.

Cohen, S., Tsai, F., & Chechile, R. (1995). A model for assessing student interaction with educational software. *Behavior Research Methods, Instruments, & Computers, 27,* 251–256.

Cole, L. W. (1911). The relation of strength of stimulus to rate of learning in the chick. *Journal of Animal Behavior, 1,* 111–124.

Coltheart, M. (1980). Iconic memory and visible persistence. *Perception & Psychophysics, 27,* 183–228.

Coltheart, M. (1985). Cognitive neuropsychology and the study of reading. In M. I. Posner & O. S. M. Marin (Eds.), *Attention and performance XI* (pp. 3–37). Hillsdale, NJ: Erlbaum.

Conrad, R. (1967). Interference or decay over short retention intervals? *Journal of Verbal Learning and Verbal Behavior, 6,* 49–54.

Cook, R., Chechile, R., & Roberts, S. (2003, March). *Mechanisms of recognition and recall in pigeons.* Paper presented at the Conference on Comparative Cognition, Melborne, FL.

Cooper, E. H., & Pantle, A. J. (1967). The total-time hypothesis in verbal learning. *Psychological Bulletin, 68,* 221–234.

Coppola, D., & Purves, D. (1996). The extraordinary rapid disappearance of entoptic images. *Proceedings of the National Academy of Science USA, 93,* 8001–8004.

Corbett, A. T., & Wickelgren, W. A. (1978). Semantic memory retrieval analysis by speed accuracy tradeoff functions. *Quarterly Journal of Experimental Psychology, 30,* 1–15.

Corkin, S. (1968). Acquisition of motor skill after bilateral medial temporal-lobe excision. *Neuropsychologia, 6*, 255–265.

Corkin, S. (2002). What's new with the amnesic patient H. M.? *Nature Reviews Neuroscience, 3*, 152–160.

Corkin, S. (2013). *Permanent present tense: The unforgettable life of the amnesic patient, H. M.*. New York: Basic Books.

Cornwell, B. R., Echiverri, A. M., & Grillion, C. (2007). Sensitivity to masked conditioned stimuli predicts conditioned response magnitude under masked conditions. *Psychophysiology, 44*, 403–406.

Corteen, R. S., & Wood, B. (1972). Autonomic responses to shock-associated words in an unattended channel. *Journal of Experimental Psychology, 94*, 308–313.

Cosmides, L., & Tooby, J. (2003). Evolutionary psychology: Theoretical Foundations. In L. Nadel (Ed.) *Encyclopedia of cognitive science* (pp. 54–64). London: Macmillan.

Cowan, N. (1984). On short and long auditory stores. *Psychological Bulletin, 96*, 341–370.

Cowan, N. (2015). George Miller's magical number of immediate memory in retrospect: Observations on the faltering progression of science. *Psychological Review, 122*, 536–541.

Cowles, J. T., & Nissen, H. W. (1937). Reward expectancy in delayed responses of chimpanzees. *Journal of Comparative Psychology, 24*, 345–358.

Craik, F. I. M., & Lockhart, R. S. (1972). Levels of processing: A framework for memory research. *Journal of Verbal Learning and Verbal Behavior, 11*, 671–684.

Craik, F. I. M., & Tulving, E. (1975). Depth of processing and the retention of words in episodic memory. *Journal of Experimental Psychology: General, 104*, 268–294.

Crannell, C. W., & Parrish, J. M. (1957). A comparison of immediate memory span for digits, letters, and words. *The Journal of Psychology: Interdisciplinary and Applied, 44*, 319–327.

Crowder, R. G. (1976). *Principles of learning and memory*. Hillsdale, NJ: Erlbaum.

Cutler, B. L., Penrod, S. D., & Martens, T. K. (1987). Improving the reliability of eyewitness identifications: Putting context into context. *Journal of Applied Cognition, 72*, 629–637.

Darley, C. F., Tinklenberg, J. R., Roth, W. T., & Atkinson, R. C. (1974). The nature of storage deficits and state-dependent retrieval under marihuana. *Psychopharmacologia, 37*, 139–149.

Darwin, C. (1859). *On the origin of species by means of natural selection*. London: John Murray.

Darwin, C. J., Turvey, M. T., & Crowder, R. G. (1972). An auditory analogue of the Sperling partial report procedure: Evidence for brief auditory storage. *Cognitive Psychology, 3*, 255–267.

Dawson, M. E., & Schell, A. M. (1982). Electrodermal responses to attended and nonattended significant stimuli during dichotic listening. *Journal of Experimental Psychology: Human Perception and Performance, 8*, 315–324.

Dawson, R. G., & McGaugh, J. L. (1969). Electroconvulsive shock effects on a reactivated memory trace: Further examination. *Science, 166,* 525–527.

Debner, J. A., & Jacoby, L. L. (1994). Unconscious perception: Attention, awareness, and control. *Journal of Experimental Psychology: Learning, Memory, and Cognition, 20,* 304–317.

De Boeck, P. (2008). Random item IRT models. *Psychometrika, 73,* 533–559.

Deese, J. (1959). On the prediction of occurrence of particular verbal intrusions in immediate recall. *Journal of Experimental Psychology, 58,* 17–22.

Derryberry, D., & Tucker, D. M. (1994). Motivating the focus of attention. In P. M. Neidenthal & S. Kitayama (Eds.), *The heart's eye: Emotional influences in perception and attention* (pp. 167–196). San Diego, CA: Academic Press.

Detterman, D. K. (1979). Memory in the mentally retarded. In N. R. Ellis (Ed.), *Handbook of mental deficiency* (2nd ed., pp. 729–760). Hillsdale, NJ: Erlbaum.

Deutsch, J. A., & Clarkson, J. K. (1959). Reasoning in the hooded rat. *Quarterly Journal of Experimental Psychology, 11,* 150–154.

deWinstanley, P. A., & Bjork, E. L. (2004). Processing strategies and the generation effect: Implications for making a better reader. *Memory & Cognition, 32,* 945–955.

Diamond, R., & Carey, S. (1986). Why faces are and are not special: An effect of expertise. *Journal of Experimental Psychology: General, 115,* 107–117.

Dickerson, S. S., & Kemeny, M. E. (2004). Acute stressors and cortisol responses: A theoretical integration and synthesis of laboratory research. *Psychological Bulletin, 130,* 355–391.

Diller, D. E., Nobel, P. A., & Shiffrin, R. M. (2001). An ARC-REM model for accuracy and response time in recognition and cued recall. *Journal of Experimental Psychology: Learning, Memory, and Cognition, 27,* 414–435.

Di Lollo, V. (1977). Temporal characteristics of iconic memory. *Nature, 267,* 241–243.

Dodson, C. S., & Johnson, M. K. (1996). Some problems with the process-dissociation approach to memory. *Journal of Experimental Psychology: General, 125,* 181–194.

Dodson, C. S., & Schacter, D. L. (2001). "If I had said it I would have remembered it": Reducing false memories with a distinctiveness heuristic. *Psychonomic Bulletin & Review, 8,* 155–161.

Dodson, J. D. (1915). The relation of strength of stimulus to rapidity of habit-formation in the kitten. *Journal of Animal Behavior, 5,* 330–336.

Doering, D. G., & Rabinovitch, M. S. (1969). Auditory abilities of the children with learning problems. *Journal of Learning Disabilities, 2,* 467–474.

Domber, E. A. (1971). *Facilitation and retardation of instrumental appetitive learning by prior Pavlovian aversive conditioning.* Unpublished doctoral dissertation, University of Pittsburgh.

Donaldson, W. (1996). The role of decision processes in remembering and knowing. *Memory & Cognition, 24,* 523–533.

Donaldson, W., & Bass, M. (1980). Relational information and memory for problem solutions. *Journal of Verbal Learning and Verbal Behavior, 19,* 26–35.

Doyle, A. C. (1892). *The adventures of Sherlock Holmes.* London: George Newnes Ltd.

Duane, A. (1922). Studies in monocular and binocular accommodation with their clinical applications. *American Journal of Ophthalmology, 5,* 865–877.

Dudai, Y. (2006). Reconsolidation: The advantage of being refocused. *Current Opinion in Neurobiology, 16,* 174–178.

Dulsky, S. G. (1935). The effect of a change of background on recall and relearning. *Journal of Experimental Psychology, 18,* 725–740.

Dunn, J. C. (2004). Remember-know: A matter of confidence. *Psychological Review, 111,* 524–542.

Dunwiddie, T., & Lynch, G. (1978). Long-term potentiation and depression of synaptic responses in the rat hippocampus: Localization and frequency dependency. *Journal of Physiology, 276,* 353–367.

Dywan, J. (1988). The imagery factor in hypnotic hypermnesia. *International Journal of Clinical and Experimental Hypnosis, 36,* 312–326.

Easterbrook, J. A. (1959). The effect of emotion on cue utilization and the organization of behavior. *Psychological Review, 66,* 183–201.

Ebbinghaus, H. (1885). *Memory: A contribution to experimental psychology.* New York: Teachers College Press.

Edeline, J. M., & Neuenschwander-El Massioui, N. (1988). Retention of CS-US association learned under ketamine anesthesia. *Brain Research, 457,* 274–280.

Egan, J. P. (1958). *Recognition memory and the operating characteristic* (Technical Note AFCRC-TN-58-51). Hearing and Communication Laboratory, Indiana University.

Ehrensbeck, K. A. (1979). *Analysis of storage and retrieval changes in long-term memory for overall retention, negative transfer, and retroactive interference.* Unpublished master's thesis, Tufts University.

Eich, E. (1985). Context, memory, and integrated item/context imagery. *Journal of Experimental Psychology: Learning, Memory, & Cognition, 11,* 764–770.

Eich, E., Macaulay, D., & Ryan, L. (1994). Mood dependent memory for events of the personal past. *Journal of Experimental Psychology: General, 123,* 201–215.

Eichenbaum, H. (2008). *Learning & memory.* New York: Norton.

Eichenbaum, H. (2012). *The cognitive neuroscience of memory: An introduction* (2nd ed.). New York: Oxford University Press.

Eichenbaum, H. (2014). Time cells in the hippocampus: A new dimension for mapping memories. *Nature Neuroscience*, **15**, 732–744.

Elandt-Johnson, R. C. (1971). *Probability models and statistical methods in genetics*. New York: Wiley.

Ellis, W. D. (1938). *Source book of gestalt psychology*. London: Routledge & Kegan Paul, Ltd.

Ellis, N. R., & Sloan, W. (1959). Oddity learning as a function of mental age. *Journal of Comparative and Physiological Psychology*, **52**, 228–230.

Ellis, A. W., & Young, A. (Eds.). (1988). *Human cognitive neuropsycholog*. Hove, UK: Erlbaum.

Engelkamp, J., & Krumnacker, H. (1980). Image- and motor-processes in the retention of verbal materials. *Zeitschrift für Experimentelle und Angewandte Psychologie*, **27**, 511–533.

Engelkamp, J., & Zimmer, H. D. (1985). Motor programs and their relation to semantic memory. *German Journal of Psychology*, **9**, 239–254.

Engelkamp, J., & Zimmer, H. D. (1989). Memory for action events: A new field of research. *Psychological Research*, **51**, 153–157.

Entwisle, D. R. (1966). *Word associations of young children*. Baltimore, MD: Johns Hopkins University Press.

Erdfelder, E., Auer, T-S., Hilbig, B. E., Aßfalg, A., Moshagen, M., & Nadarevic, L. (2009). Multinomial processing tree models: A review of the literature. *Zeithschrift für Psychologie/Journal of Psychology*, **217**, 108–124.

Eriksen, C. W., & Hoffman, J. E. (1972). Temporal and spatial characteristics of selective encoding from visual displays. *Perception & Psychophysics*, **12(2B)**, 201–204.

Eriksen, C. W., & St. James, J. D. (1986). Visual attention within and around the field of focal attention: A zoom lens model. *Perception & Psychophysics*, **40**, 225–240.

Estes, W. K. (1956). The problem of inference from curves based on grouped data. *Psychological Bulletin*, **53**, 134–140.

Estes, W. K. (1960). Learning theory and the new "mental chemistry." *Psychological Review*, **67**, 207–223.

Estes, W. K. (1961). New developments in statistical behavior theory: Differential tests of axioms for association learning. *Psychometrika*, **26**, 73–84.

Estes, W. K. (1980). Is human memory obsolete? *Scientific American*, **68**, 62–69.

Estes, W. K. (1994). *Classification and cognition*. New York: Oxford University Press.

Eysenck, M. W. (1978). Levels of processing: A critique. *British Journal of Psychology*, **69**, 157–169.

Fanselow, M. S. (1980). Factors governing one-trial contextual conditioning. *Animal Learning & Behavior*, **18**, 264–270.

Fantz, R. L. (1964). Visual experience in infants: Decreased attention to familiar patterns relative to novel ones. *Science*, **146**, (3644), 668–670.

Farah, M. J. (1984). The neurological basis of mental imagery: A component analysis. *Cognition*, **18**, 245–272.

Farrell, S. (2006). Mixed-list phonological similarity effects in delayed serial recall. *Journal of Memory and Language*, **55**, 587–600.

Fawcett, J. M. (2013). The production effect benefits performance in between-subjects designs: A meta-analysis. *Acta Psychologica*, **142**, 1–5.

Fechner, G. T. (1860). *Elemente der Psychophysik*. Leipzig, Germany: Breitkopf & Härtel.

Fellner, M-C., Bäuml, K-H. T., & Hanslmayr, S. (2013). Brain oscillatory subsequent memory effects differ in power and long-range synchronization between semantic and survival processing. *NeuroImage*, **79**, 361–170.

Fernandez, A., & Glenberg, A. M. (1985). Changing environmental context does not reliably affect memory. *Memory & Cognition*, **13**, 333–345.

Ferreira, C. T., Ceccaldi, M., Giusiano, B., & Poncet, M. (1998). Separate visual pathways for perception of actions and objects: Evidence from a case of apperceptive agnosia. *Journal of Neurology, Neurosurgery and Psychiatry*, **65**, 382–385.

Ferster, C. S., & Skinner, B. F. (1957). *Schedules of reinforcement*. New York: Appleton-Century-Crofts.

Fiacco, T. A., Agulhon, C., & McCarthy, K. D. (2009). Sorting out astrocyte physiology from pharmacology, *Annual Reviews: Pharmacology and Toxicology*, **49**, 151–174.

Finnegan, R., & Becker, S. (2015). Neurogenesis paradoxically decreases both pattern separation and memory interference. *Frontiers in System Neuroscience*, **9**, Article 136. doi:10.3389/fnsys.2015.00136.

Fischer, A., Sananbenesi, F., Schrick, C., Speiss, J., & Radulovic, J. (2004). Distinct roles of hippocampal de novo protein synthesis and actin rearrangement in extinction of contextual fear. *Journal of Neuroscience*, **24**, 1962–1966.

Fleischman, R. N. (1987). *Lexical access without search: Evidence from a speed-accuracy tradeoff paradigm*. Unpublished doctoral thesis, Tufts University.

Flexser, A. J., & Tulving, E. (1978). Retrieval independence in recognition and recall. *Psychological Review*, **85**, 153–171.

Foley, M. A., Foley, H. J., Wilder, A., & Rusch, L. (1989). Anagram solving: Does effort have an effect? *Memory & Cognition*, **17**, 755–758.

Forbach, G. B., Stanners, R. F., & Hochhaus, L. (1974). Repetition and practice effects in a lexical decision task. *Memory & Cognition*, **2**, 337–339.

Forgas, J. P., & Eich, E. (2012). Affective influences on cognition: Mood congruence, mood dependence, and mood effects on processing strategies. In I. B. Weiner, A. F. Healy, & R. W. Proctor (Eds.), *Handbook of psychology: Vol. 4. Experimental psychology* (2nd ed., pp. 61–82). Hoboken, NJ: Wiley.

Forrin, N. D., MacLeod, C. M., & Ozubko, J. D. (2012). Widening the boundaries of the production effect. *Memory & Cognition, 40*, 1046–1055.

Forster, K. I., & Bednall, E. S. (1976). Terminating and exhaustive search in lexical access. *Memory & Cognition, 4*, 53–61.

Fowler, H., Fago, G. C., Domber, E. A., & Hochhauser, M. (1973). Signaling and affective functions in Pavlovian conditioning. *Animal Learning & Behavior, 2*, 81–89.

Frankel, F., & Cole, M. (1971). Measures of category clustering in free recall. *Psychological Bulletin, 76*, 39–44.

Fredrickson, B. L. (1998). What good are positive emotions? *Review of General Psychology, 2*, 300–319.

Freed, D. M., & Corkin, S. (1988). Rate of forgetting in H. M.: 6-month recognition. *Behavioral Neuroscience, 102*, 823–825.

Freed, D. M., Corkin, S., & Cohen, N. J. (1987). Forgetting in H. M.: A second look. *Neuropsychologia, 25*, 561–571.

Freud, S. (1910). The origin and development of psychoanalysis. *American Journal of Psychology, 21*, 181–218.

Frey, U., Huang, Y.-Y., & Kandel, E. R. (1993). Effects of cAMP simulate a late stage of LTP in hippocampal CA1 neurons. *Science, 260*, 1661–1664.

Fu, K. S. (1985). Syntactic pattern recognition. In T. V. Young & K. S. Fu (Eds.), *Handbook of pattern recognition and image processing* (pp. 85–117). San Diego, CA: Academic Press.

Gabrieli, J. D. E., Corkin, S., Mickel, S. F., & Growdon, J. H. (1993). Intact acquisition and long-term retention of mirror-tracing skill in Alzheimer's disease and in global amnesia. *Behavioral Neuroscience, 107*, 899–910.

Gagliano, M., Renton, M., Depczynski, & Mancuso, S. (2014). Experience teaches plants to learn faster and forget slower in environments where it matters. *Oecologia, 175*, 63–72.

Gallistel, C. R., & Gibbon, J. (2002). *The symbolic foundations of conditioned behavior.* Mahwah, NJ: Erlbaum.

Gallistel, C. R., & King, A. P. (2009). *Memory and the computational brain: Why cognitive sciences will transform neuroscience.* West Sussex, UK: Wiley-Blackwell.

Ganellen, R. J., & Carver, C. S. (1985). Why does self-reference promote incidental encoding? *Journal of Experimental Social Psychology, 21*, 284–300.

Garcia, J., Ervin, F. R., & Koelling, R. A. (1972). Learning with prolonged delay of reinforcement. In M. E. P. Seligman & J. L. Hager (Eds.), *Biological boundaries of learning* (pp. 16–20). New York: Appleton-Century-Crofts.

Garcia, J., Kimeldorf, D. J., & Hunt, E. L. (1961). The use of ionizing radiation as a motivating stimulus. *Psychological Review, 68*, 383–395.

Garcia, J., & Koelling, R. A. (1972). Relation of cue to consequence in avoidance learning. In M. E. P. Seligman & J. L. Hager (Eds.), *Biological boundaries of learning* (pp. 10–15). New York: Appleton-Century-Crofts.

Gardiner, J. M., Gregg, V. H., & Hampton, J. A. (1988). Word frequency and generation effects. *Journal of Experimental Psychology: Learning, Memory, and Cognition, 14*, 687–693.

Gasper, K., & Clore, G. L. (2002). Attending to the BIG PICTURE: Mood and global versus local processing of visual information. *Psychological Science, 13*, 34–40.

Gathercole, S. E. (1998). The development of memory. *Journal of Child Psychology and Psychiatry, 39*, 3–27.

Gathercole, S. E., & Conway, M. A. (1988). Exploring long-term modality effects: Vocalization leads to best retention. *Memory & Cognition, 16*, 110–119.

Gaunt, R., Leynes, J., & Demoulin, S. (2002). Intergroup relations and the attribution of emotions: Control over memory for secondary emotions associated with the ingroup and outgroup. *Journal of Experimental Social Psychology, 38*, 508–514.

Gauthier, I., Skudlarski, P., Gore, J. C., & Anderson, A. W. (2000). Expertise for cars and birds recruits brain areas involved in face recognition. *Nature Neuroscience, 3*, 191–197.

Gazzaniga, M. S., & LeDoux, J. E. (1978). *The integrated mind.* New York: Plenum.

Geghman, K. D., & Multhaup, K. S. (2004). How generation affects source memory. *Memory & Cognition, 32*, 819–823.

George, M. S., Parekh, P. I., Rosinsky, N., Ketter, T. A., Heilman, K. M., Herscovitch, P., & Post, R. M. (1996). Understanding emotional prosody activates right hemisphere regions. *Archives of Neurology, 53*, 665–670.

Geraci, L., McCabe, D. P., & Guillory, J. J. (2009). On the interpreting the relationship between remember-know judgments and confidence: The role of instructions. *Consciousness and Cognition, 18*, 701–709.

Gerrein, J. R. (1976). *Organization, storage, and retrieval factors in alcohol-induced amnesia.* Unpublished doctoral dissertation, Tufts University.

Gerrein, J. R., & Chechile, R. A. (1977). Storage and retrieval processes of alcohol-induced amnesia. *Journal of Abnormal Psychology, 86*, 285–294.

Ghiselli, W. B., & Fowler, H. (1976). Signaling and affective functions of conditioned aversive stimuli in an appetitive choice discrimination: US intensity effects. *Learning and Motivation*, **1**, 81–89.

Ghoneim, M. M., & Block, R. I. (1992). Learning and consciousness during general anesthesia. *Anesthesiology*, **76**, 279–305.

Ghoneim, M. M., Chen, P., El-Zahaby, H. M., & Block, R. I. (1994). Ketamine: Acquisition and retention of classically conditioned responses during treatment with large doses. *Pharmacology Biochemistry and Behavior*, **49**, 1061–1066.

Gillund, G., & Shiffrin, R. M. (1984). A retrieval model for both recognition and recall. *Psychological Review*, **91**, 1–67.

Giray, E. F., Altkin, W. M., Roodin, P. A., Yoon, G., & Flagg, P. (1978). *The incidence of eidetic imagery in adulthood and old age.* Paper presented at the September American Psychological Association meeting, Toronto.

Girden, E., & Culler, E. A. (1937). Conditioned responses in curarized striate muscle in dogs. *Journal of Comparative Psychology*, **23**, 261–274.

Glanzer, M., & Adams, J. K. (1985). The mirror effect in recognition memory. *Memory & Cognition*, **13**, 8–20.

Glanzer, M., Adams, J. K., & Iverson, G. (1991). Forgetting and the mirror effect in recognition memory: Concentering of underlying distributions. *Journal of Experimental Psychology: Learning, Memory, and Cognition*, **17**, 81–93.

Glanzer, M., Adams, J. K., Iverson, G. J., & Kim, K. (1993). The regularities of recognition memory. *Psychological Review*, **100**, 546–567.

Glanzer, M., & Cunitz, A. R. (1966). Two storage mechanisms in free recall. *Journal of Verbal Learning and Verbal Behavior*, **5**, 351–360.

Glanzer, M., Kim, K., & Adams, J. K. (1998). Response distribution as an explanation of the mirror effect. *Journal of Experimental Psychology: Learning, Memory, and Cognition*, **24**, 633–644.

Glaze, J. A. (1928). The association value of non-sense syllables. *Journal of Genetic Psychology*, **35**, 255–269.

Glenberg, A. M. (1976). Monotonic and non monotonic lag effects in paired-associate and recognition memory paradigms. *Journal of Verbal Learning and Verbal Behavior*, **15**, 1–16.

Glisky, E. L., & Marquine, M. J. (2009). Semantic and self-referential processing of positive and negative trait adjectives in older adults. *Memory*, **17**, 144–157.

Gloor, P., Olivier, A., Quesney, L. F., Andermann, F., & Horowitz, S. (1982). The role of the limbic system in experiential phenomena of temporal lobe epilepsy. *Annals of Neurology*, **12**, 129–144.

Gluck, M. A., Mercado, E., & Myers, C. E. (2008). *Learning and memory: From brain to behavior.* New York: Worth.

Godden, D., & Baddeley, A. (1980). When does context influence recognition memory? *British Journal of Psychology, 71*, 99–104.

Godden, D. R., & Baddeley, A. D. (1975). Context-dependent memory in two natural environments: On land and underwater. *British Journal of Psychology, 66*, 325–331.

Goggin, J., & Wickens, D. D. (1971). Proactive interference and language change in short-term memory. *Journal of Verbal Learning and Verbal Behavior, 10*, 453–458.

Gold, P. E. (1986). Glucose modulation of memory storage processing. *Behavioral and Neural Biology, 45*, 342–349.

Gold, P. E., & Korol, D. L. (2012). Making memories matter. *Frontiers in Integrative Neuroscience, 6*, Article 116, 1–11.

Gold, P. E., & Van Buskirk, R. B. (1975). Facilitation of time-dependent memory processes with posttrial epinephrine injections. *Behavioral Biology, 13*, 145–153.

Gollin, E. S., Saravo, A., & Salten, C. (1967). Perceptual distinctiveness and oddity problem solving in children. *Journal of Experimental Child Psychology, 5*, 586–596.

Gonzalez, R. C., & Thomason, M. G. (1978). *Syntactic pattern recognition: An introduction.* Reading, MA: Addison-Wesley.

Good, M. (2002). Spatial memory and hippocampal function: Where are we now? *Psicológica, 23*, 109–138.

Goodman, J. H., & Fowler, H. (1976). Transfer of the signaling properties of aversive CSs to an instrumental appetitive discrimination. *Learning and Motivation, 7*, 446–457.

Goodwin, D. W., Powell, B., Bremer, D., Hoine, H., & Stern, J. (1969). Alcohol and recall: State dependent effects in man. *Science, 163*, 1358–1360.

Gorfein, D. S., & Blair, C. (1971). Factors affecting multiple trial free recall. *Journal of Educational Psychology, 62*, 17–24.

Graf, P. (1980). Two consequences of generating: Increasing inter- and intraword organization of sentences. *Journal of Verbal Learning and Verbal Behavior, 19*, 316–327.

Graf, P. (1982). The memorial consequences of generation and transformation. *Journal of Verbal Learning and Verbal Behavior, 21*, 539–548.

Graf, P., & Schacter, D. L. (1985). Implicit and explicit memory for new associations in normal and amnesic subjects. *Journal of Experimental Psychology: Learning, Memory, and Cognition, 11*, 501–518.

Graf, P., Squire, L. R., & Mandler, G. (1984). The information that amnesic patients do not forget. *Journal of Experimental Psychology: Learning, Memory, and Cognition, 10*, 164–178.

Gray, C. R., & Gummerman, K. (1975). The enigmatic eidetic image: A critical examination of methods, data, and theories. *Psychological Bulletin, 82*, 383–407.

Green, D. M., & Swets, J. A. (1966). *Signal detection theory and psychophysics*. New York: Wiley.

Greene, R. J. (1990). Memory for pair frequency. *Journal of Experimental Psychology: Learning, Memory, & Cognition*, **16**, 110–116.

Greenfield, D. B. (1985). Facilitating mentally retarded children's relational learning through novelty-familiarity training. *American Journal of Mental Deficiency*, **90**, 342–348.

Greeno, J. G., James, C. T., DaPolito, F., & Polson, P. G. (1978). *Associate learning: A cognitive analysis*. Englewood Cliffs, NJ: Prentice-Hall.

Greenough, W. T., & Bailey, C. H. (1988). The anatomy of a memory: Convergence of results across a diversity of tests. *Trends in Neurosciences*, **11**, 142–147.

Griffiths, D. (1987). *Introduction to elementary particles*. New York: Wiley.

Grissom, N., & Bhatnagar, S. (2009). Habituation to repeated stress: Get used to it. *Neurobiology of Learning and Memory*, **92**, 215–224.

Grossenbacher, P. G., & Lovelace, C. T. (2001). Mechanisms of synesthesia: Cognitive and physiological constraints. *Trends in Cognitive Sciences*, **5**, 36–41.

Grover, L. M., Kim, E., Cooke, J. D., & Holmes, W. R. (2009). LTP in hippocampal area CA1 is induced by burst stimulation over a broad frequency range centered around delta. *Learning & Memory*, **16**, 69–81.

Groves, P. M., & Thompson, R. F. (1970). Habituation: A dual-process theory. *Psychological Review*, **77**, 419–450.

Grünwald, P. D., Myung, I. J., & Pitt, M. A. (2005). *Advances in minimum description length*. Cambridge, MA: MIT Press.

Guthrie, E. R. (1935). *The psychology of learning*. New York: Harper & Row.

Guthrie, E. R. (1952). *The psychology of learning* (Rev. ed.). New York: Harper & Row.

Gutowski, W. E., & Chechile, R. A. (1987). Encoding, storage, and retrieval components of associative memory deficits of mildly mentally retarded adults. *American Journal of Mental Deficiency*, **92**, 85–93.

Haber, R. N. (1979). Twenty years of haunting eidetic imagery: Where's the ghost? *Behavioral and Brain Sciences*, **2**, 583–629.

Haber, R. N. & Haber, R. B. (1964). Eidetic imagery: I. Frequency. *Perceptual and Motor Skills*, **19**, 131–138.

Haines, D. E. (1991). *Neuroanatomy: An atlas of structures sections and systems*. Baltimore, MD: Urban & Schwarzenberg.

Halgren, E., Walter, R. D., Cherlow, A. G., & Crandall, P. H. (1978). Mental phenomena evoked by electrical stimulation of the human hippocampal formation and amygdala. *Brain*, **101**, 83–117.

Hall, J. W., Grossman, L. R., & Elwood, K. D. (1976). Differences in encoding for free recall vs. recognition. *Memory & Cognition, 4,* 507–513.

Halpin, J. A., Puff, R., Mason, H. F., & Marston, S. P. (1984). Self-reference encoding and incidental recall by children. *Bulletin of the Psychonomic Society, 22,* 87–89.

Han, X., Chen, M., Wang, F., Windrem, M., Wang, S., Shanz, S., Xu, Q., Nedergaard, M. (2013). Forebrain engraftment by human glial progenitor cells enhances synaptic plasticity and learning in adult mice. *Cell Stem Cell, 12,* 342–353.

Hancock, P. A., & Ganey, N. C. N. (2003). From the inverted-U to the extended-U: The evolution of a law of psychology. *Journal of Human Performance in Extreme Environments, 7,* 5–13.

Hanley, J. A., & McNeil, B. J. (1982). The meaning and use of the area under the Receiver Operating Characteristic (ROC) curve. *Radiology, 143,* 29–36.

Harmon-Jones, E, Gable, P. A., & Price, T. F. (2013). Does negative affect always narrow and positive affect always broaden the mind? Considering the influence of motivation intensity on cognitive scope. *Current Directions in Psychological Science, 22,* 301–307.

Hartigan, J. A. (1983). *Bayes theory.* New York: Springer-Verlag.

Hastings, W. K. (1970). Monte Carlo sampling methods using Markov chains and their applications. *Biometrika, 57,* 97–109.

Hebb, D. O. (1949). *The organization of behavior: A neuropsychological theory.* New York: John Wiley.

Hebb, D. O. (1955). Drives and the C.N.S. (Conceptual Nervous System). *Psychological Review, 62,* 243–254.

Heilman, K. M. (2002). *Matters of mind.* New York: Oxford University Press.

Held, R., Ostrovsky, Y., de Gelder, B., Gandhi, T., Ganesh, S., Mathur, U., & Sinha, P. (2011). The newly sighted fail to match seen with felt. *Nature Neuroscience, 14,* 551–553.

Hellyer, S. (1962). Supplementary report: Frequency of stimulus presentation and short-term decrement in recall. *Journal of Experimental Psychology, 64,* 650.

Henson, R. N. A. (1998). Short-term memory for serial order: The start-end model. *Cognitive Psychology, 36,* 73–137.

Higbee, K. L. (1977). *Your memory: How it works and how to improve it.* Englewood Cliffs, NJ: Prentice-Hall.

Hilgard, E. R. (1975). Hypnosis. *Annual Reviews in Psychology, 26,* 19–44.

Hilts, P. F. (1995). *Memory's ghost: The strange tale of Mr. M and the nature of memory.* New York: Simon & Schuster.

Hinkle, D. J., & Wood, D. C. (1994). Is tube-escape learning by protozoa associative learning? *Behavioral Neuroscience, 108,* 94–99.

Hintzman, D. L. (1984). Minerva 2: A simulation model of human memory. *Behavior Research Methods, Instruments, and Computers, 16*, 96–101.

Hintzman, D. L. (1988). Judgments of frequency and recognition memory in a multiple-trace memory model. *Psychological Review, 95*, 528–551.

Hintzman, D. L., & Rogers, M. K. (1973). Spacing effects in picture memory. *Memory & Cognition, 1*, 430–430.

Hirschman, E., & Bjork, R. A. (1988). The generation effect: Support for a two-factor theory. *Journal of Experimental Psychology: Learning, Memory and Cognition, 13*, 484–494.

Hochner, B., Brown, E. R., Langella, M., Shomrat, T., & Fiorito, G. (2003). A learning and memory area in the octopus brain manifests a vertebrate-like long-term potentiation. *Journal of Neurophysiology, 90*, 3547–3554.

Hofstadter, D., & Sander, E. (2013). *Surfaces and essences: Analogy as the fuel and fire of thinking.* New York: Basic Books.

Hopfield, J. J. (1982). Neural networks and physical systems with emerging collective computational abilities. *Proceedings of the National Academy of Science, 79*, 2554–2558.

Hopkins, R. H., & Edwards, R. E. (1972). Pronunciation effects in recognition memory. *Journal of Verbal Learning and Verbal Behavior, 11*, 534–537.

Howard, M. W., & Eichenbaum, H. (2015). Time and space in the hippocampus. *Brain Research, 1621*, 345–355.

Howard, M. W., MacDonald, C. J., Tiganj, Z., Shankar, K. H., Du, Q., Hasselmo, M. E., & Eichenbaum, H. (2014). A unified mathematical framework for coding time, space, and sequences in the hippocampal region. *Journal of Neuroscience, 34*, 4692–4707.

Howard, M. W., Shankar, K. H., Aue, W. R., & Criss, A. H. (2015). A distributed representation of internal time. *Psychological Review, 122*, 24–53.

Howe, M. L. (2000). *The fate of early memories: Developmental science and the retention of childhood experiences.* Washington, DC: American Psychological Association.

Hu, X. (1999). Multinomial processing tree models: An implementation. *Behavior Research Methods, Instruments, & Computers, 31*, 689–695.

Hu, X. (2001). Extending general processing tree models to analyze reaction time experiments. *Journal of Mathematical Psychology, 45*, 603–634.

Hu, X., & Phillips, G. A. (1999). GPT.EXE: A powerful tool for the visualization and analysis of general processing tree models. *Behavior Research Methods, Instruments, & Computers, 31*, 220–234.

Hubel, D. H., & Wiesel, T. N. (1962). Receptive fields, binocular interaction, and functional architecture in the cat's visual cortex. *Journal of Physiology, 160*, 106–154.

Hubel, D. H., & Wiesel, T. N. (2005). *Brain and visual perception: The story of a 25-year collaboration.* New York: Oxford University Press.

Hull, C. L. (1931). Goal attraction and directing ideas conceived as habit phenomena. *Psychological Review*, **38**, 487–506.

Hull, C. L. (1933). The meaningfulness of 320 selected nonsense syllables. *American Journal of Psychology*, **45**, 730–734.

Hummel, J. E., & Biederman, I. (1992). Dynamic binding in a neural network for shape recognition. *Psychological Review*, **99**, 480–517.

Humphrey, G. (1930). Extinction and negative adaptation. *Psychological Review*, **37**, 361–363.

Hunt, R. R. (1995). The subtlety of distinctiveness: What von Restorff really did. *Psychonomic Bulletin & Review*, **2**, 105–112.

Icht, M., Mama, Y., & Algom, D. (2014). The production effect in memory: Multiple species of distinctiveness. *Frontiers in Psychology*, **5**, Article 886, 1–7.

Isen, A. M., Clark, M., Shalker, T. E., & Karp, L. (1978). Affect, accessibility of material in memory, and behavior: A cognitive loop? *Journal of Personality and Social Psychology*, **36**, 1–12.

Jacoby, L. L. (1978). On interpreting the effects of repetition: Solving a problem versus remembering a solution. *Journal of Verbal Learning and Verbal Behavior*, **17**, 649–667.

Jacoby, L. L. (1991). A process dissociation framework: Separating automatic from intentional uses of memory. *Journal of Memory and Language*, **30**, 513–541.

James, W. (1890a). *The principles of psychology: Vol. 1*. New York: Holt and Company.

James, W. (1890b). *The principles of psychology: Vol. 2*. New York: Holt and Company.

James, W. (1892). *Psychology the briefer course*. New York: Holt and Company.

Jamieson, R. K., Crump, J. C., & Hannah, S. D. (2012). An instance theory of associative learning. *Learning & Behavior*, **40**, 61–82.

Jamieson, R. K., & Spear, J. (2014). The offline production effect. *Canadian Journal of Experimental Psychology*, **68**, 20–28.

Jeffreys, H. (1961). *Theory of probability* (3rd ed.). Oxford, UK: Oxford University Press.

Jenkins, J. G., & Dallenbach, E. B. (1924). Obliviscence during sleep and waking. *American Journal of Psychology*, **35**, 605–612.

Jenkins, L. J., & Ranganath, C. (2010). Prefrontal and medial temporal lobe activity at encoding predicts temporal context memory. *Journal of Neuroscience*, **30**, 15558–15565.

Jennings, H. S. (1906). *Behavior of the lower organisms*. New York: Columbia University Press.

Jersild, A. (1929). Primacy, recency, frequency, and vividness. *Journal of Experimental Psychology*, **12**, 58–70.

Johnson, M. K. (1983). A multiple-entry, modular memory system. In G. H. Bower (Ed.), *The psychology of learning and motivation* (Vol. 17, pp. 81–123). New York: Academic Press.

Jones, A. C., & Pyc, M. A. (2014). The production effect: Cost and benefit in free recall. *Journal of Experimental Psychology: Learning, Memory, and Cognition, 40*, 300–305.

Jones, M., & Love, B. C. (2007). Beyond common features: The role of roles in determining similarity. *Cognitive Psychology, 55*, 196–231.

Jones, M. N., & Mewhort, D. J. K. (2007). Representing word meaning and order information in a composite holographic lexicon. *Psychological Review, 114*, 1–37.

Jost, A. (1897). Die assoziationsfestigkeit in ihrer abhängigkeit von der verteilung der wiederholungen. *Zeitschrift für Psychologie, 14*, 436–472.

Julesz, B. (1971). *Foundations of cyclopean perception.* Chicago: University of Chicago Press.

Jurica, P. J., & Shimamura, A. P. (1999). Monitoring item an source information: Evidence for a negative generation effect in source memory. *Memory & Cognition, 27*, 648–656.

Kaernbach, C. (2001). Adoptive threshold estimation with unforced-choice tasks. *Perception & Psychophysics, 63*, 1377–1388.

Kahana, M. J. (1996). Associative retrieval processes in free recall. *Memory & Cognition, 24*, 103–109.

Kahana, M. J. (2012). *Foundations of human memory.* New York: Oxford University Press.

Kahana, M. J., Howard, M. W., & Polyn, S. M. (2008). Associative retrieval processes in episodic memory. In J. Byrne & H. L. Roediger III (Eds.), *Learning and memory: A comprehensive reference* (Vol. 2, pp. 467–490). Amsterdam: Elsevier.

Kahana, M. J., Rizzuto, D. S., & Schneider, A. R. (2005). Theoretical correlations and measured correlations: Relating recognition and recall in four distributed memory models. *Journal of Experimental Psychology: Learning, Memory, and Cognition, 31*, 933–953.

Kamin, L. T. (1969). Predictability surprise, attention and conditioning. In B. A. Campbell & R. M. Church (Eds.), *Punishment and aversive behavior* (pp. 279–296). New York: Appleton-Century-Crofts.

Kandel, E. R. (2000). Cellular mechanisms of learning and the biological basis of individuality. In E. R. Kandel, J. H. Schwartz, & T. M. Jessell (Eds.), *Principles of neural science* (4th ed., pp. 1248–1279). New York: McGraw-Hill.

Kandel, E. R. (2006). *In search of memory: The emergence of a new science of mind.* New York: Norton.

Kandel, E. R. (2009). The biology of memory: A forty-year perspective. *The Journal of Neuroscience, 29*, 12748–12756.

Kanwisher, N., McDermott, J., & Chun, M. M. (1997). The fusiform face area: A module in human extrastriate cortex specialized for face perception. *Journal of Neuroscience, 17*, 4302–4311.

Karpicke, J. D., Lehman, M., & Aue, W. R. (2014). Retrieval-based learning: An episodic context account. In B. H. Ross (Ed.), *The psychology of learning and motivation* (Vol. 61, pp. 237–284). Waltham, MA: Academic Press.

Karpicke, J. D., & Roediger, H. L., III (2006). *Repeated retrieval during learning is the key to enhancing later retention.* Unpublished manuscript, Washington University in St. Louis.

Kass, R. E., & Raftery, A. E. (1995). Bayes factors. *Journal of the American Statistical Association, 90,* 773–795.

Katz, B. (1966). *Nerve, muscle, and synapse.* New York: McGraw-Hill.

Kausik, S., Choi, Y-B., White-Grindley, E., Majumdar, A., & Kandel, E. R. (2010). *Aplysia* CPEB can form prion-like multimers in sensory neurons that contribute to long-term facilitation. *Cell, 140,* 421–435.

Keenan, J. M., & Baillet, S. D. (1980). Memory for personality and socially significant effects. In R. S. Nickerson (Ed.), *Attention and performance* (Vol. 8, pp. 651–669). Hillsdale, NJ: Erlbaum.

Keil, F. C. (1989). *Concepts, kinds, and cognitive development.* Cambridge, MA: MIT Press.

Kellen, D., & Klauer, K. C. (2015). Signal detection and threshold modeling of confidence-rating ROCs: A critical test with minimal assumptions. *Psychological Review, 122,* 542–557.

Kelvin, W. T. (1871). *Report of the forty-first meeting of the British Association for the Advancement of Science.* London: John Murray.

Kelvin, W. T. (1889). Electrical units of measurement. *Popular Lectures, 1,* 73–136.

Kemeny, M. E. (2003). The psychobiology of stress. *Current Directions in Psychological Science, 12,* 124–129.

Kempermann, G., Kuhn, H. G., & Gage, F. H. (1997). More hippocampal neurons in adult mice living in an enriched environment. *Nature, 386,* 493–495.

Kendzierski, D. (1980). Self-schemata and scripts: The recall of self-referent and scriptal information. *Personality and Social Psychology Bulletin, 6,* 23–29.

Kensinger, E. A., & Schacter, D. L. (2007). Remembering the specific visual details of presented objects: Neuroimaging evidence for effects of emotion. *Neuropsychologia, 45,* 2951–2962.

Kesner, R. P., Hopkins, R. O., & Fineman, B. (1994). Item and order dissociation in humans with prefrontal cortex damage. *Neuropsychologia, 32,* 881–891.

Kihlstrom, J. F. (1985). Hypnosis. *Annual Review in Psychology, 36,* 385–418.

Kihlstrom, J. F. (1997). Hypnosis, memory and amnesia. *Philosophical Transactions of the Royal Society London B, 352,* 1727–1732.

Kimura, D. (1963). Right temporal-lobe damage. *Archives of Neurology, 8,* 264–271.

Kinder, A., & Shanks, D. R. (2003). Neuropsychological dissociations between priming and recognition: A single-system connectionist account. *Psychological Review, 110,* 728–744.

Kinjo, H., & Snodgrass, J. G. (2000). Does the generation effect occur for pictures? *American Journal of Psychology, 113,* 95–121.

Kirsch, I., & Lynn, S. J. (1995). The altered state of hypnosis: Changes in the theoretical landscape. *American Psychologist, 50,* 846–858.

Kirsner, K., & Smith, M. C. (1974). Modality effects in word identification. *Memory & Cognition, 2,* 637–640.

Kis, A., Huber, L., & Wilkinson, A. (2015). Social learning by imitation in a reptile (*Pogona vitticeps*). *Animal Cognition, 18,* 325–331.

Klein, S. B., & Kihlstrom, J. F. (1986). Elaboration, organization, and the self-reference effect in memory. *Journal of Experimental Psychology: General, 115,* 26–38.

Klein, S. B., Loftus, J., & Burton, H. A. (1989). Two self-reference effects: The importance of distinguishing between self-descriptiveness judgments and autobiographical retrieval in self-referent encoding. *Journal of Personality and Social Psychology, 56,* 853–865.

Klug, W. S., & Cummings, M. R. (1991). *Concepts of genetics.* New York: Macmillan.

Knight, D. C., Nguyen, H. T., & Bandettini, P. A. (2003). Expression of conditional fear with and without awareness. *Proceedings of the National Academy of Science, 100,* 15280–15283.

Knight, D. C., Nguyen, H. T., & Bandettini, P. A. (2006). The role of awareness in delay and trace conditioning in humans. *Cognition, Affective, & Behavioral Neuroscience, 6,* 157–162.

Kobasigawa, A. (1974). Utilization of retrieval by children in recall. *Child Development, 45,* 127–134.

Kohonen, T. (1977). *Associative memory.* Berlin: Springer-Verlag.

Kolmogorov, A. (1933). *Grundbegriffe der Wahrscheinlichkeitsrechnug.* Berlin: Springer.

Konheim, A. G. (2010). *Hashing in computer science: Fifty years of slicing and dicing.* Hoboken, NJ: Wiley.

Koppenaal, R. J. (1963). Time changes in the strength of A-B, A-C lists; Spontaneous recovery? *Journal of Verbal Learning and Verbal Behavior, 2,* 310–319.

Kornell, N., Bjork, R. A., & Garcia, M. A. (2011). Why tests appear to prevent forgetting: A distribution-based bifurcation model. *Journal of Memory and Language, 65,* 85–97.

Kosslyn, S. M. (1976). *Image and mind.* Cambridge, MA: Harvard University Press.

Kosslyn, S. M. (1994). *Image and brain.* Cambridge, MA: MIT Press.

Kounious, J., & Beeman, M. (2009). The aha! moment: The cognitive neuroscience of insight. *Current Directions in Psychological Science, 18,* 210–216.

Kramer, D. A., & Kahlbaugh, P. E. (1994). Memory for a dialectical and a non dialectical prose passage in young and older adults. *Journal of Adult Development, 1,* 13–26.

Kramer, J., Buckhout, A., & Eugenio, P. (1989). Weapon focus, arousal, and eyewitness memory: Attention must be paid. *Applied Cognitive Psychology, 14,* 167–184.

Krantz, D. H., Luce, R. D., Suppes, P., & Tversky, A. (1971). *Foundations of measurement: Vol. 1. Additive and polynomial representations*. New York: Academic Press.

Krantz, D. H., Luce, R. D., Suppes, P., & Tversky, A. (1989). *Foundations of measurement: Vol. 2. Geometrical, threshold, and probabilistic representations*. San Diego, CA: Academic Press.

Kringelbach, M. L. (2005). The human orbitofrontal cortex: Linking reward to hedonic experience. *Nature Reviews Neuroscience*, **6**, 691–702.

Kroneisen, M., & Erdfelder, E. (2011). On the plasticity of the survival processing effect. *Journal of Experimental Psychology: Learning, Memory, and Cognition*, **37**, 1553–1562.

Krueger, W. C. F. (1934). The relative difficulty of nonsense syllables. *Journal of Experimental Psychology*, **17**, 145–153.

Kruschke, J. K. (1992). ALCOVE: An exemplar-based connectionist model of category learning. *Psychological Review*, **99**, 22–44.

Kruskal, J. B., & Wish, M. (1977). *Multidimensional scaling*. Beverly Hills, CA: Sage.

Kucera, H., & Francis, W. N. (1967). *Computational analysis of present-day American English*. Providence, RI: Brown University Press.

Kutas, M., & Federmeier, K. D. (2011). Thirty years and counting: Finding meaning in the N400 component of the event related brain potential (ERP). *Annual Review of Psychology*, **62**, 621–647.

Kutas, M., & Hillyard, S. A. (1980). Reading senseless sentences: Brain potentials reflect semantic incongruity. *Science*, **207**, 203–208.

Labar, K. S., & Cabeza, R. (2006). Cognitive neuroscience of emotional memory. *Nature Reviews Neuroscience*, **7**, 54–64.

Lacey, J. I. (1967). Somatic response patterning and stress: Some revisions of activation theory. In M. H. Appley & R. Trumbull (Eds.), *Psychological stress: Issues in research* (pp. 14–37). New York: Appleton-Century Crofts.

Lacroix, J. P. W., Murre, J. M. J., Postma, E. O., & van den Herik, H. J. (2006). Modeling recognition memory using the similarity structure of natural input. *Cognitive Science*, **30**, 121–145.

Lad, F. (1996). *Operational statistical methods*. New York: Wiley.

Laming, D. (1973). *Mathematical psychology*. London: Academic Press.

Landauer, T. K. (1986). How much do people remember? Some estimates of quantity of learned information in long-term memory. *Cognitive Science*, **10**, 477–493.

Landauer, T. K., & Dumais, S. T. (1997). A solution to Plato's problem: The latent semantic analysis theory of acquisition, induction, and representation of knowledge. *Psychological Review*, **104**, 211–240.

Larner, G. (1979). *A storage-retrieval analysis of the memory deficit in the alcoholic Korsakoff syndrome*. Unpublished doctoral dissertation, Sydney University.

Larson, J., & Lynch, G. (1986). Induction of synaptic potentiation in hippocampus by patterned stimulation involves two events. *Science, 232*, 985–988.

Larson, J., Wong, D., & Lynch, G. (1986). Patterned stimulation at the theta frequency is optimal for the induction of hippocampal long-term potentiation. *Brain Research, 368*, 347–350.

Lashley, K. S. (1929). Learning: I. Nervous mechanisms in learning. In C. Murchison (Ed.), *The foundations of experimental psychology* (pp. 524–563). Worcester, MA: Clark University Press.

Lattal, K. M., & Abel, T. (2004). Behavioral impairments caused by injections of the protein synthesis inhibitor anisomycin after context retrieval reverse with time. *Proceedings of the National Academy of Science USA, 101*, 4667–4672.

Laurence, J.-R., & Perry, C. (1983). Hypnotically created memory among highly hypnotizable subjects. *Science, 222*, 523–524.

Lazareva, O. F. (2012). Relational learning in a context of transposition: A review. *Journal of the Experimental Analysis of Behavior, 97*, 231–248.

Leask, J., Haber, R. N., & Haber, R. B. (1969). Eidetic imagery in children: II. Longitudinal and experimental results. *Psychonomic Monograph Supplements, 3*, 25–48.

Le Cam, L. (1986). *Asymptotic methods in statistical decision theory*. New York: Springer.

Lee, P. M. (1989). *Bayesian statistics: An introduction*. New York: Oxford University Press.

Leeuwenberg, E., & van der Helm, P. (1991). Unity and variety in visual form. *Perception, 20*, 595–622.

Lewandowsky, S., & Farrell, S. (2008). Short-term memory: New data and a model. In B. H. Ross (Ed.), *The psychology of learning and motivation: Advances in research and theory* (pp. 1–48). Amsterdam: Academic Press.

Lewandowsky, S., & Farrell, S. (2011). *Computational modeling in cognition: Principles and practice*. Los Angeles: Sage.

Light, L. L., & Carter-Sobell, L. (1970). Effects of changed semantic context on recognition memory. *Journal of Verbal Learning and Verbal Behavior, 9*, 1–11.

Lindell, A. K. (2006). In your right mind: Right hemisphere contributions to language processing and production. *Neuropsychology Review, 16*, 131–148.

Liss, P. (1968). Does backward masking by visual noise stop stimulus processing? *Perception & Psychophysics, 4*, 328–330.

Liu, X., Crump, M. J. C., & Logan, G. D. (2010). Do you know where your fingers have been? Explicit knowledge of the spatial layout of the keyboard in skilled typists. *Memory & Cognition, 38*, 474–484.

Locke, J. (1694). *An essay concerning human understanding*. London: Awnsham, Churchill, & Manship.

Loftus, E. F. (1979). *Eyewitness testimony*. Cambridge, MA: Harvard University Press.

Loftus, E. F., Loftus, G. R., & Messo, J. (1987). Some facts about "weapon focus." *Law and Human Behavior*, **11**, 55–62.

Loftus, G. R., & Mackworth, N. H. (1978). Cognitive determinants of fixation location during picture viewing. *Journal of Experimental Psychology: Human Perception and Performance*, **4**, 565–572.

Loftus, G. R., Oberg, M. A., & Dillon, A. M. (2004). Linear theory, dimensional theory, and the face-inversion effect. *Psychological Review*, **111**, 835–863.

Logan, G. D. (1990). Repetition priming and automaticity: Common underlying mechanisms? *Cognitive Psychology*, **22**, 1–35.

Logan, G. D., & Crump, M. J. C. (2011). Hierarchical control of cognitive processes: The case for skilled typewriting. In B. H. Ross (Ed.), *The psychology of learning and motivation: Advances in research an theory* (Vol. 54, pp. 1–27). Burlington, VT: Academic Press.

Løhre, E. (2011). *Context-dependent memory and mood*. Unpublished master's thesis, University of Oslo.

Lømo, T. (1966). Frequency potentiation of excitatory synaptic activity in the dentate area of the hippocampal formation [abstract]. *Acta Physiologica Scandinavica*, **68**, 128.

Long, J., Nimmo-Smith, I., & Whitefield, A. (1983). Skilled typing: A characterization based on the distribution of times between responses. In W. E. Cooper (Ed.), *Cognitive aspects of skilled typewriting* (pp. 145–195). New York: Springer-Verlag.

Lorayne, H. & Lucas, J. (1974). *The memory book*. New York: Ballantine Books.

Lord, F. M., & Novick, M. R. (1986). *Statistical theories of mental test scores*. Reading, MA: Addison-Wesley.

Lorenz, K. (1965). *Evolution and modification of behavior*. Chicago: University of Chicago Press.

Lorenz, K. (1966). *On aggression*. New York: Grosset & Dunlap.

Lorrain, D. S., Baccel, C. S., Bristow, L. J., Anderson, J. J., & Varney, M. A. (2003). Effects of ketamine and N-methyl-D-aspartate on glutamate and dopamine release in the rat prefrontal cortex: Modulation by a group II selective metabotropic glutamate receptor agonist LY379268. *Neuroscience*, **117**, 697–706.

Lovibond, P. F., & Shanks, D. R. (2002). The role of awareness in Pavlovian conditioning: Empirical evidence and theoretical implications. *Journal of Experimental Psychology: Animal Behavior Processes*, **28**, 3–26.

Luce, R. D. (1963). A threshold theory for simple detection experiments. *Psychological Review*, **70**, 61–79.

Luce, R. D., Krantz, D. H., Suppes, P., & Tversky, A. (1990). *Foundations of measurement: Vol. 3. Representations, axiomatization, and invariance*. New York: Academic Press.

Luce, R. D., & Suppes, P. (2002). Representational measurement theory. In H. Pashler & J. Wixted (Eds.), *Stevens' handbook of experimental psychology* (3rd ed., Vol. 4, pp. 1–41). New York: Wiley.

Luria, A. R. (1968). *The mind of a mnemonist.* New York: Basic Books.

Maass, A., & Köhnken, G. (1989). Eyewitness identification: Simulating the "weapon effect." *Law and Human Behavior, 13,* 397–408.

Maatsch, J. L. (1959). Learning and fixation after a single shock trial. *Journal of Comparative and Physiological Psychology, 52,* 408–410.

MacCorquodale, K., & Meehl, P. E. (1953). Preliminary suggestions as to a formalization of expectancy theory. *Psychological Review, 60,* 55–63.

MacCorquodale, K., & Meehl, P. E. (1954). Edward C. Tolman. In W. K. Estes, S. Koch, K. Mac-Corquodale, P. E. Meehl, C. G. Mueller, W. N. Schoenfeld, & W. S. Verplanck (Eds.), *Modern learning theory* (pp. 177–266). New York: Appleton-Century Crofts.

MacDonald, C. J., Lepage, K. Q., Eden, U. T., & Eichenbaum, H. (2011). Hippocampal "time cells" bridge the gap in memory for discontiguous events. *Neuron, 71,* 737–749.

MacDonald, P. A., & MacLeod, C. M. (1998). The influence of attention at encoding on direct and indirect remembering. *Acta Psychologia, 98,* 291–310.

MacKay, D. G., Burke, D. M., & Stewart, R. (1998). H. M.'s language production deficits: Implications for relations between memory, semantic binding, and the hippocampal system. *Journal of Memory and Language, 38,* 28–69.

MacKay, D. G., James, L. E., Taylor, J. K., & Marian, D. E. (2007). Amnesic H. M. exhibits parallel deficits and sparing in language and memory: Systems versus binding theory accounts. *Language and Cognitive Process, 22,* 377–452.

Mackintosh, N. J. (1975). A theory of attention: Variations in the associability of stimuli with reinforcement. *Psychological Review, 82,* 276–298.

Macknik, S. L., King, M., Randi, J., Robbins, A., Teller, Thompson, J., & Martinez-Conde, S. (2008). Attention and awareness in stage magic: Turning tricks into research. *Nature Reviews in Neuroscience, 9,* 871–879.

MacLeod, C. M., Gopie, N., Hourihan, K. L., Neary, K. R., & Ozubko, J. D. (2010). The production effect: Delineation of a phenomenon. *Journal of Experimental Psychology: Learning, Memory, and Cognition, 36,* 671–685.

Macmillan, N. A., & Creelman, C. D. (2005). *Detection theory: A user's guide* (2nd ed.). Mahwah, NJ: Erlbaum.

Malenka, R. C., & Bear, M. F. (2004). LTP and LTD: An embarrassment of riches. *Neuron, 44,* 5–21.

Malmberg, K. J. (2008). Recognition memory: A review of the critical findings and an integrated theory for relating them. *Cognitive Psychology, 57,* 335–384.

Mandler, G. (1955). Associative frequency and associative prepotency as measures of response to nonsense syllables. *American Journal of Psychology, 68,* 662–665.

Mandler, G. (1967). Organization and memory. In K. W. Spence & J. T. Spence (Eds.), *The psychology of learning and motivation: Advances in research and theory* (Vol. 1, pp. 327–372). New York: Academic Press.

Mangels, J. A. (1997). Strategic processing and memory for temporal order in patients with frontal lobe lesions. *Neuropsychology, 11,* 207–221.

Manns, J. R., Clark, R. E., & Squires, L. R. (2002). Standard delay eyeblink classical conditioning is independent of awareness. *Journal of Experimental Psychology: Animal Behavior Processes, 28,* 32–37.

Maren, S. (1999). Long-term potentiation in the amygdala: A mechanism for emotional learning and memory. *Trends in Neuroscience, 22,* 561–567.

Markowitsch, H. J. (2000). The anatomical bases of memory. In M. S. Gazzaniga (Ed.), *The new cognitive neurosciences* (2nd ed., pp. 781–795). Cambridge, MA: MIT Press.

Marr, D. (1971). Simple memory: A theory for archicortex. *Philosophical Transactions of the Royal Society of London, B, Biological Sciences, 262,* 23–81.

Marschark, M., & Surian, L. (1989). Why does imagery improve memory? *European Journal of Cognitive Psychology, 1,* 251–263.

Marsh, E. J. (2006). When does generation enhance memory for location? *Journal of Experimental Psychology: Learning, Memory, and Cognition, 32,* 1216–1220.

Marsh, E. J., Edelman, G., & Bower, G. H. (2001). Demonstrations of a generation effect in context memory. *Memory & Cognition, 29,* 795–805.

Martinez-Conde, S., Macknik, S. L., & Hubel, D. H. (2004). The role of fixational eye movements in visual perception. *Nature Reviews: Neuroscience, 5,* 229–240.

Mather, M. (2007). Emotional arousal and memory binding. *Perspectives on Psychological Science, 2,* 33–52.

Mather, M., & Knight, M. R. (2006). Angry faces get noticed quickly: Threat detection is not impaired among older adults. *Journal of Gerontology, Series B: Psychological Sciences, 61B,* P54–P57.

Matzke, D., Dolan, C. V., Batchelder, W. H., & Wagenmakers, E-J. (2015). Bayesian estimation of multinomial processing tree models with heterogeneity in participants and items. *Psychometrika, 80,* 205–235.

McCabe, D. P., & Geraci, L. D. (2009). The influence of instructions and terminology on the accuracy of remember-know judgments. *Consciousness and Cognition, 18,* 401–413.

McClelland, J. L., McNaughton, B. L., & O'Reilly, R. C. (1995). Why there are complementary learning systems in the hippocampus and neocortex: Insights from the successes and failures of connectionist models of learning and memory. *Psychological Review, 102,* 419–457.

McClelland, J. L., & Rumelhart, D. E. (1981). An interactive activation model of context effects in letter perception, Part 1: An account of basic findings. *Psychological Review*, **88**, 375–405.

McDaniel, M. A., & Fisher, R. P. (1991). Tests and test feedback as learning sources. *Contemporary Educational Psychology*, **16**, 192–201.

McDaniel, M. A., Kowitz, M. D., & Dunay, P. K. (1989). Altering memory through recall: The effects of cue-guided retrieval processing. *Memory & Cognition*, **17**, 423–434.

McDaniel, M. A., Waddill, P. J., & Einstein, G. O. (1988). A contextual account of the generation effect: A three-factor theory. *Journal of Memory and Language*, **27**, 521–536.

McDermott, T. (2010). *101 Theory Drive: A neuroscientist's quest for memory*. New York: Pantheon.

McElroy, L. A., & Slamecka, N. J. (1982). Memorial consequences of generating nonwords: Implications for semantic-memory interpretations of the generation effect. *Journal of Verbal Learning and Verbal Behavior*, **21**, 249–259.

McGaugh, J. L. (2004). The amygdala modulates the consolidation of memories of emotionally arousing experiences. *Annual Review of Neuroscience*, **27**, 1–28.

McGaugh, J. L., Cahill, L., & Roozendaal, B. (1996). Involvement of the amygdala in memory storage: Interaction with other brain systems. *Proceedings of the National Academy of Science USA*, **93**, 13508–13514.

McGaugh, J. L., & Herz, M. J. (1972). *Memory consolidation*. San Francisco: Albion.

McGaugh, J. L., & Roozendaal, B. (2002). Role of adrenal stress hormones in forming lasting memories. *Current Opinion in Neurobiology*, **12**, 205–210.

McGee, T. D. (1988). *Principles and methods of temperature measurement*. New York: Wiley.

McGeoch, J. A. (1932). Forgetting and the law of disuse. *Psychological Review*, **39**, 352–370.

McGeoch, J. A. (1942). *The psychology of human learning*. New York: Longmans, Green.

McGovern, J. B. (1964). Extinction of associations in four transfer paradigms. *Psychological Monographs*, **78**, Whole No. 593.

McGrady, H. J., & Olson, D. A. (1970). Visual and auditory learning processes in normal children and children with specific learning disabilities. *Exceptional Children*, **36**, 581–589.

McKoon, G., & Ratcliff, R. (1992). Spreading activation versus compound cue accounts of priming: Mediated priming revisited. *Journal of Experimental Psychology: Learning, Memory, and Cognition*, **18**, 1155–1172.

McKoon, G., Ratcliff, R., & Dell, G. S. (1986). A critical evaluation of the semantic-episodic distinction. *Journal of Experimental Psychology: Learning, Memory, and Cognition*, **12**, 295–306.

McNally, R. J. (2003). *Remembering trauma*. Cambridge, MA: Belknap.

McNally, R. J. (2005). Debunking myths about trauma and memory. *Canadian Journal of Psychiatry*, **50**, 817–822.

McNamara, D. S., & Healy, A. F. (1995). A procedural explanation of the generation effect: The use of an operand retrieval strategy for multiplication and addition problems. *Journal of Memory & Language, 34*, 399–416.

Medin, D. L., & Schaffer, M. M. (1978). Context theory of classification learning. *Psychological Review, 85*, 207–238.

Merikle, P. M. (1980). Selection from visual persistence by perceptual groups and category membership. *Journal of Experimental Psychology: General, 109*, 279–295.

Merikle, P. M., & Daneman, M. (1996). Memory for unconsciously perceived events: Evidence from anesthetized patients. *Consciousness and Cognition, 5*, 525–541.

Metcalfe-Eich, J. (1982). A composite holographic associative recall model. *Psychological Review, 89*, 627–661.

Metropolis, N., Rosenbluth, A. W., Rosenbluth, M. N., Teller, A. H., & Teller, E. (1953). Equations of state calculations by fast computing machines. *Journal of Chemical Physics, 21*, 301–311.

Meyer, B. J. F., & Rice, E. (1981). Information recalled from prose by young, middle, and old adult readers. *Experimental Aging Research, 7*, 253–268.

Meyer, D. E., & Schvaneveldt, R. W. (1971). Facilitation in recognizing pairs of words: Evidence of a dependence between retrieval operations. *Journal of Experimental Psychology, 90*, 227–234.

Michell, J. (2000). Normal science, pathological science and psychometrics. *Theory & Psychology, 10*, 639–667.

Michell, J. (2004). Item response models, pathological science and the shape of error: Reply to Borsboom and Mellenbergh. *Theory & Psychology, 14*, 121–129.

Miller, G. A. (1956). The magic number seven, plus or minus two: Some limits on our capacity of for processing information. *Psychological Review, 63*, 81–97.

Miller, J. (1991). Threshold variability in subliminal perception experiments: Fixed threshold estimates reduce power to detect subliminal effects. *Journal of Experimental Psychology: Human Perception and Performance, 17*, 841–851.

Miller, R. R., Barnet, R. C., & Grahame, N. J. (1995). Assessment of the Rescorla-Wagner model. *Psychological Bulletin, 117*, 363–386.

Miller, R. R., & Springer, A. D. (1973). Amnesia, consolidation, and retrieval. *Psychological Review, 80*, 69–79.

Miller, S., & Konorski, J. (1928). Sur une forme particuliere des reflexes conditionnels. *Comptes Rendus des Séances de la Société de Biologie, 99*, 1155–1157.

Milner, B., Corkin, S., & Teuber, H.-L. (1968). Further analysis of the hippocampal amnesic syndrome: 14-year follow-up study of H. M. *Neuropsychologia, 6*, 215–234.

Milner, B., Corsi, P., & Leonard, G. (1991). Frontal-lobe contribution to recency judgements. *Neuropsychologia, 29*, 601–618.

Misanin, J. R., Miller, R. R., & Lewis, D. J. (1968). Retrograde amnesia produced by electroconvulsive shock after reactivation of a consolidated memory trace. *Science, 160,* 554–555.

Mitchell, C. J., De Houwer, J., & Lovibond, P. F. (2009). The propositional nature of human associative learning. *Behavioral and Brain Science, 32,* 183–246.

Mitchell, K. J., Livosky, M., & Mather, M. (1998). The weapon focus effect revisited: The role of novelty. *Legal and Criminological Psychology, 3,* 287–303.

Mohr, G., Engelkamp, J., & Zimmer, H. D. (1989). Recall and recognition of self-performed acts. *Psychological Research, 51,* 181–187.

Moon, L. E., & Harlow, H. F. (1955). Analysis of oddity learning by rhesus monkeys. *Journal of Comparative Psychology, 48,* 188–194.

Morgan, C. L. (1894). *An introduction to comparative psychology.* London: Walter Scott, Ltd.

Morris, C. D., Bransford, J., & Franks, J. (1977). Levels of processing versus transfer appropriate processing. *Journal of Verbal Learning and Verbal Behavior, 16,* 519–533.

Morris, R. G. M. (1981). Spatial localization does not depend on the presence of local cues. *Learning & Motivation, 12,* 239–260.

Morrison, F. J., Giordani, B., & Nagy, J. (1977). Reading disability: An information-processing analysis. *Science, 196,* 77–79.

Moser, E. I., Kropff, E., & Moser, M-B. (2008). Place cells, grid cells, and the brain's spatial representation system. *Annual Review of Neuroscience, 31,* 69–89.

Moscovitch, A., & LoLordo, V. M. (1968). Role of safety in the Pavlovian backward fear conditioning procedure. *Journal of Comparative and Physiological Psychology, 66,* 673–678.

Moshagen, M. (2010). MultiTree: A computer program for the analysis of multinomial processing tree models. *Behavior Research Methods, 42,* 42–54.

Moutoussis, K., & Zeki, S. (2002). The relationship between cortical activation and perception investigated with invisible stimuli. *Proceedings of the National Academy of Sciences USA, 99,* 9527–9532.

Müller, G. J., & Pilzecker, A. (1900). Experimentelle beiträge zur lehre von gedächintz. *Zeitschrift für Psychologie, 1,* 1–300.

Mulligan, N. W. (2002). The generation effect: Dissociating enhanced item memory and disrupted order memory. *Memory & Cognition, 30,* 850–861.

Mulligan, N. W. (2004). Generation and memory for contextual detail. *Journal of Experimental Psychology: Learning, Memory, and Cognition, 30,* 838–855.

Mulligan, N. W., & Duke, M. D. (2002). Positive and negative generation effects, hypermnesia, and total recall time. *Memory & Cognition, 30,* 1044–1053.

Mulligan, N. W., Lozito, J. P., & Rosner, Z. A. (2006). Generation and context memory. *Journal of Experimental Psychology: Learning, Memory, and Cognition, 32*, 836–846.

Münsterberg, H., & Campbell, W. W. (1894). Studies from the Harvard Psychological Laboratory II. *Psychological Review, 1*, 441–495.

Murdock, B. B., Jr. (1960). The immediate retention of unrelated words. *Journal of Experimental Psychology, 60*, 222–234.

Murdock, B. B., Jr. (1974). *Human memory: Theory and data*. New York: Wiley.

Murdock, B. B. (1995). Developing TODAM: Three models for serial-order information. *Memory & Cognition, 23*, 631–645.

Murphy, G. L. (2004). *The big book of concepts*. Cambridge, MA: MIT Press.

Murphy, G. L., & Medin, D. L. (1985). The role of theories in conceptual coherence. *Psychological Review, 92*, 289–316.

Murray, D. J. (1968). Articulation and acoustic confusability in short-term memory. *Journal of Experimental Psychology, 4*, 679–684.

Myung, I. J., Forster, M. R., & Browne, M. W. (Eds.). (2000). Introduction: Special issue on model selection. *Journal of Mathematical Psychology, 44*, 1–2.

Nader, K. (2003). Memory traces unbound. *Trends in Neuroscience, 26*, 65–72.

Nader, K., Schafe, G. E., & Le Doux, J. E. (2000). Fear memories require protein synthesis in the amygdala for reconsolidating after retrieval. *Nature, 406*, 722–726.

Nagel, E., & Newman, J. R. (1958). *Gödel's proof*. New York: New York University Press.

Nairne, J. S., & Pandeirada, J. N. S. (2008). Adaptive memory: Is survival processing special? *Journal of Memory and Language, 59*, 377–385.

Nairne, J. S., Pandeirada, J. N. S., & Thompson, S. R. (2008). Adaptive memory: The comparative value of survival processing. *Psychological Science, 19*, 176–180.

Nairne, J. S., Pusen, C., & Widner, R. L. (1985). Representation in the mental lexicon: Implications for theories of the generation effect. *Memory & Cognition, 13*, 183–191.

Nairne, J. S., Riegler, G. J., & Serra, M. (1991). Dissociative effects of generation on item and order information. *Journal of Experimental Psychology: Learning, Memory, & Cognition, 17*, 702–709.

Nairne, J. S., Thompson, S. R., & Pandeirada, J. N. S. (2007). Adaptive memory: Survival processing enhances retention. *Journal of Experimental Psychology: Learning, Memory, and Cognition, 33*, 263–273.

Narens, L. (1985). *Abstract measurement theory*. Cambridge, MA: MIT Press.

Narens, L., & Luce, R. D. (1986). Measurement: The theory of numerical assignments. *Psychological Bulletin, 99*, 166–180.

Naya, Y., & Suzuki, W. A. (2011). Integrating what and when across the primate medial temporal lobe. *Science*, **333**, 773–776.

Neath, I., & Brown, G. D. A. (2006). SIMPLE: Further applications of a local distinctiveness model of memory. *The Psychology of Learning and Motivation*, **46**, 201–243.

Neely, J. H., & Balota, D. A. (1981). Test-expectancy and semantic-organization effects in recall and recognition. *Memory & Cognition*, **9**, 283–300.

Neiss, R. (1988). Reconceptualizing arousal: Psychological states in motor performance. *Psychological Bulletin*, **103**, 345–366.

Nelson, D. L., Walling, J. R., & McEvoy, C. L. (1979). Doubts about depth. *Journal of Experimental Psychology: Human Learning and Memory*, **5**, 24–44.

Nelson, T. O. (1977). Repetition and depth of processing. *Journal of Verbal Learning and Verbal Behavior*, **16**, 151–171.

Nickerson, R. S. (1968). A note on long-term recognition memory for pictorial material. *Psychonomic Science*, **11**, 58.

Nishijo, H., Ono, T., & Hishijo, H. (1988). Single neuron responses in amygdala of alert monkey during complex sensory stimulation with affective significance. *The Journal of Neuroscience*, **8**, 3570–3583.

Noble, C. E., Stockwell, F. E., & Pryer, M. W. (1957). Meaningfulness (m') and association (a) in paired-associate syllable learning. *Psychological Report*, **3**, 441–452.

Noble, D. W. A., Byrne, R. W., & Whiting, M. J. (2014). Age-dependent social learning in a lizard. *Biology Letters*, **10**, doi: 10.1098/rsbl.2014.0430.

Nosofsky, R. M. (1986). Attention, similarity, and the identification categorization relationship. *Journal of Experimental Psychology: General*, **115**, 39–57.

Nunnally, J. C., & Bernstein, I. H. (1994). *Psychometric theory*. New York: McGraw-Hill.

Nyberg, L. (1993). *The enactment effect: Studies of a memory phenomenon*. Unpublished doctoral dissertation, University of Umeå.

Oakes, L. M. (2010). Using habituation of looking time to assess mental processes in infancy. *Journal of Cognition and Development*, **11**, 255–268.

Öhman, A., & Mineka, S. (2001). Fears, phobias, and preparedness: Toward an evolved module of fear and fear learning. *Psychological Review*, **108**, 483–522.

Öhman, A., & Soares, J. J. F. (1993). On the automatic nature of phobic fear: Conditioned electrodermal responses to masked fear-related stimuli. *Journal of Abnormal Psychology*, **102**, 121–132.

Öhman, A., & Soares, J. J. F. (1994). "Unconscious anxiety": Phobic responses to masked stimuli. *Journal of Abnormal Psychology*, **103**, 231–240.

O'Keefe, J., & Dostrovsky, J. (1971). The hippocampus as a spatial map. Preliminary evidence from unit activity in the freely-moving rat. *Brain Research, 34,* 171–175.

O'Keefe, J., & Nadel, L. (1978). *The hippocampus as a spatial map.* Oxford, England: Oxford University Press.

Okuda, S., Roozendaal, B., & McGaugh, J. L. (2004). Glucocorticoid effects on object recognition memory require training-associated emotional arousal. *Proceedings of the National Academy of Science USA, 101,* 853–858.

Olton, D. S., Becker, J. T., & Handelmann, G. E. (1980). Hippocampal function: Working memory or cognitive mapping? *Physiological Psychology, 8,* 239–246.

Onifer, W., & Swinney, D. A. (1981). Accessing lexical ambiguities during sentence comprehension: Effects of frequency of meaning and contextual bias. *Memory & Cognition, 9,* 225–236.

O'Reilly, R. C., & Rudy, J. W. (2001). Conjunctive representations in learning and memory: Principles of cortical and hippocampal function. *Psychological Review, 108,* 311–345.

Ott, R., Curio, I., & Scholz, O. B. (2000). Implicit memory for auditorily presented threatening stimuli: A process-dissociation approach. *Perceptual & Motor Skills, 90,* 131–146.

Overton, D. A. (1964). State-dependent or "dissociated" learning produced with pentobarbital. *Journal of Comparative and Physiological Psychology, 57,* 3–12.

Ozubko, J. D., & MacLeod, C. M. (2010). The production effect in memory: Evidence that distinctiveness underlies the benefit. *Journal of Experimental Psychology: Learning, Memory, and Cognition, 36,* 1543–1547.

Page, M. P. A., & Norris, D. (1998). The primacy model: A new model of immediate serial recall. *Psychological Review, 105,* 761–781.

Palmer, S. E. (1977). Hierarchical structure in perceptual representation. *Cognitive Psychology, 9,* 441–474.

Pang, R., Turndorf, H., & Quartermain, D. (1996). Pavlovian fear conditioning in mice anesthetized with halothane. *Physiology & Behavior, 59,* 873–875.

Parker, E. S., Cahill, L., & McGaugh, J. L. (2006). A case of unusual autobiographical remembering. *Neurocase, 12,* 35–49.

Parkin, A. J. (1993). Implicit memory across the lifespan. In P. Graf & M. E. J. Masson (Eds.), *Implicit memory: New directions in cognition, development, and neuropsychology* (pp. 191–206). Hillsdale, NJ: Erlbaum.

Pastalkova, E., Itskov, V., Amarasingham, A., & Buzsáki, G. (2008). Internally generated cell assembly sequences in the rat hippocampus. *Science, 321,* 1322–1327.

Pavlov, I. P. (1960). *Conditioned reflex: An investigation of the physiological activity of the cerebral cortex* (an unabridged republication of a translated 1927 Oxford University Press book). New York: Dover.

Payne, D. G., Neely, J. H., & Burns, D. J. (1986). The generation effect: Further tests of the lexical activation hypothesis. *Memory & Cognition*, **14**, 246–252.

Peace, K. A., Porter, S., & ten Brinke, L. (2008). Are memories for sexually traumatic events "special"? A within-subjects investigation of trauma and memory in a clinical sample. *Memory*, **16**, 10–21.

Pearce, J. M., & Hall, G. (1980). A model for Pavlovian learning: Variations in the effectiveness of conditioned but not of unconditioned stimuli. *Psychological Review*, **87**, 532–552.

Penfield, W. (1958). *The excitable cortex in conscious man*. Springfield, IL: Thomas.

Penfield, W. W., & Jasper, H. (1954). *Epilepsy and the functional anatomy of the human brain*. Boston, MA: Little, Brown.

Penfield, W., & Perot, P. (1963). The brain's record of auditory and visual experience. *Brain*, **86**, 595–696.

Peterson, D. J. (2009). *Enactment and retrieval*. Unpublished master's thesis, University of North Carolina at Chapel Hill.

Peterson, L. R., & Johnson, S. T. (1971). Some effects of minimizing articulation on short-term retention. *Journal of Verbal Learning and Verbal Behavior*, **10**, 346–354.

Peterson, L. R., & Peterson, M. J. (1959). Short-term retention of individual verbal items. *Journal of Experimental Psychology*, **58**, 193–198.

Peynircioğlu, Z. F. (1989). The generation effect with pictures and nonsense figures. *Acta Psychologica*, **70**, 153–160.

Pickel, K. L. (2009). The weapon focus effect on memory for female versus male perpetrators. *Memory*, **17**, 664–678.

Piéron, H. (1913). Reserches expérimentales sur la phénonènes de mémoire. *Année Psychologie*, **19**, 91–193.

Pinsker, H., Hening, W. A., Carew, T. J., & Kandel, E. R. (1973). Long-term sensitization of a defensive withdrawal reflex in *Aplysia*. *Science*, **182**, 4116, 1039–1042.

Pond, R. (1987, April). Fun in metals. *Johns Hopkins Magazine*, pp. 60–68.

Poon, C-S., & Young, D. L. (2006). Non associative learning as gated neural integrator and differentiator in stimulus-response pathways. *Behavioral and Brain Functions*, **8**, doi: 10.1186/1744-9081-2-29.

Posner, M. I., & Snyder, C. R. R. (1975). Attention and cognitive control. In R. L. Solso (Ed.), *Information processing and cognition: The Loyola symposium* (pp. 55–85). Hillsdale, NJ: Erlbaum.

Postle, B. R., & Corkin, S. (1998). Impaired word-stem completion priming but intact perceptual identification priming with novel words: Evidence from the amnesic patient H. M. *Neuropschologia*, **36**, 421–440.

Power, A. E., Berlau, D. J., McGaugh, J. L., & Steward, O. (2006). Anisomycin infused into the hippocampus fails to block "reconsolidating" but impairs extinction: The role of re-exposure duration. *Learning & Memory*, **13**, 27–34.

Prescott, S. A. (1998). Interactions between depression and facilitation with neural networks: Updating the dual-process theory of plasticity. *Learning and Memory*, **5**, 446–466.

Prescott, S. A., & Chase, R. (1999). Sites of plasticity in the neural circuit mediating tentacle withdrawal in the snail *Helix apersa*: Implications for behavioral change and learning kinetics. *Learning & Memory*, **6**, 363–380.

Press, S. J. (1989). *Bayesian statistics: Principles, models, and applications*. New York: Wiley.

Proctor, R. W., & Vu, K-P. L. (1999). Index of norms and ratings published in the Psychonomic Society journals. *Behavior Research Methods, Instruments, & Computers*, **31**, 659–667.

Prosser, C. L., & Hunter, W. S. (1936). The extinction of startle responses and spinal reflexes in the white rat. *American Journal of Physiology*, **117**, 609–618.

Purves, D., Augustine, G. J., Fitzpatrick, D., Hall, W. C., LaMantia, A-S., & White, L. E. (2012). *Neuroscience* (5th ed.). Sunderland, MA: Sinauer Associates.

Purves, D., & Beau Lotto, R. (2003). *Why we see what we do: An empirical theory of vision*. Sunderland, MA: Sinauer Associates.

Putnam, W. H. (1979). Hypnosis and distortions in eyewitness memory. *International Journal of Clinical and Experimental Hypnosis*, **27**, 437–448.

Pylyshyn, Z. W. (1973). What the mind's eye tells the mind's brain: A critique of mental imagery. *Psychological Bulletin*, **80**, 1–24.

Pylyshyn, Z. W. (1979). The rate of "mental rotation" of images. *Memory & Cognition*, **7**, 19–28.

Pylyshyn, Z. W. (1981). The imagery debate: Analog media versus tacit knowledge. *Psychological Review*, **88**, 16–45.

Pylyshyn, Z. (2003). *Seeing and visualizing: It's not what you think*. Cambridge, MA: MIT Press.

Qiu, J., Li, H., Jou, J., Liu, J., Luo, Y., Feng, T., Wu, Z., & Zhang, Q. (2010). Neural correlates of the "Aha" experiences: Evidence from an FMRI study of insight problem solving. *Cortex*, **46**, 397–403.

Quirarte, G. L., Roozendaal, B., & McGaugh, J. L. (1997). Glucocorticoid enhancement of memory storage involves noradrenergic activation in the basolateral amygdala. *Proceedings of the National Academy of Science USA*, **94**, 14048–14053.

Raaijmakers, J. G. W. (1979). *Retrieval from long-term store: A general theory and mathematical models*. Unpublished doctoral dissertation, University of Nijmegen.

Raaijmakers, J. G. W., & Shiffrin, R. M. (1980). SAM: A theory of probabilistic search of associative memory. In G. H. Bower (Ed.), *The psychology of learning and motivation: Advances in research and theory* (Vol. 14, pp. 207–262). New York: Academic Press.

Raaijmakers, J. G. W., & Shiffrin, R. M. (1981). Search of associative memory. *Psychological Review*, **88**, 93–134.

Radvansky, G. (2006). *Human memory*. Boston: Pearson.

Ramón y Cajal, S. (1913). *Degeneration and regeneration of the nervous system*. London: Oxford University Press.

Rapp, B. (Ed.). (2001). *The handbook of cognitive neuropsychology*. Philadelphia, PA: Psychology Press/Taylor and Francis.

Ratcliff, R., & McKoon, G. (1988). A retrieval theory of priming in memory. *Psychological Review*, **95**, 385–408.

Ratcliff, R., McKoon, G., & Tindall, M. (1994). Empirical generality of data from recognition memory receiver-operating characteristic functions and implications for the global memory models. *Journal of Experimental Psychology: Learning, Memory, and Cognition*, **20**, 763–785.

Ratcliff, R., Sheu, C.-F., & Gronlund, S. D. (1992). Testing global memory models using ROC curves. *Psychological Review*, **99**, 518–535.

Read, T. R. C., & Cressie, N. A. C. (1988). *Goodness-of-fit statistics for discrete multivariate data*. New York: Springer-Verlag.

Reader, S. M., & Laland, K. N. (2002). Social intelligence, innovation, and enhanced brain size in primates. *Proceedings of the National Academy of Science USA*, **99**, 4436–4441.

Reitz, J. R., & Milford, F. J. (1960). *Foundations of electromagnetic theory*. Reading, MA: Addison-Wesley.

Rescorla, R. A. (1967). Pavlovian conditioning and its proper control procedures. *Psychological Review*, **74**, 71–80.

Rescorla, R. A., & Heth, C. D. (1975). Reinstatement of fear to an extinguished conditioned stimulus. *Journal of Experimental Psychology: Animal Behavior Processes*, **1**, 88–96.

Rescorla, R. A., & Wagner, A. R. (1972). A theory of Pavlovian conditioning: Variations in the effectiveness of reinforcement and nonreinforcement. In A. H. Black & W. F. Prokasy (Eds.), Classical conditioning II: Current research and theory (pp. 64–99). New York: Appleton-Century-Crofts.

Resnick, R., Halliday, D., & Krane, K. S. (2002). *Physics* (5th ed.). New York: Wiley.

Restle, F. (1962). The selection of strategies in cue learning. *Psychological Review*, **69**, 329–343.

Restle, F., & Greeno, J. G. (1970). *Introduction to mathematical psychology*. Reading, MA: Addison-Wesley.

Reutener, D. B. (1972). Class shift, symbolic shift, and background shift in short-term memory. *Journal of Experimental Psychology*, **93**, 90–94.

Rhodes, G. (1996). *Superportraits: Caricatures and recognition*. New York: Psychology Press.

Ribot, T. (1882). *Diseases of the memory: An essay in the positive psychology*. New York: Appleton.

Richman, C. L., Nida, S., & Pittman, L. (1976). Effects of meaningfulness on child free-recall learning. *Developmental Psychology, 12*, 460–465.

Riefer, D. M., & Batchelder, W. H. (1988). Multinomial modeling and the measurement of cognitive processes. *Psychological Review, 95*, 318–339.

Riefer, D. M., & Rouder, J. N. (1992). A multinomial modeling analysis of the mnemonic benefits of bizarre imagery. *Memory & Cognition, 20*, 601–611.

Rips, L. J., Shoben, E. J., & Smith, E. E. (1973). Semantic distance and the verification of semantic relations. *Journal of Verbal Learning and Verbal Behavior, 12*, 1–20.

Rizzuto, D. S., & Kahana, M. J. (2001). An auto associative neural network model of paired-associate learning. *Neural Computation, 13*, 2075–2092.

Robbins, S. J. (1990). Mechanisms underlying spontaneous recovery in autoshaping. *Journal of Experimental Psychology: Animal Behavior Process, 16*, 235–249.

Roberts, W. A. (1972). Free recall of word lists varying in length and rate of presentation: A test of total-time hypotheses. *Journal of Experimental Psychology, 92*, 365–372.

Rock, I. (1957). The role of repetition in associative learning. *American Journal of Psychology, 70*, 186–193.

Robleto, K., Poulos, A. M., & Thompson, R. F. (2004). Brain mechanisms of extinction of the classically conditioned eyeballing response. *Learning & Memory, 11*, 517–524.

Roediger, H. L., III (1990). Implicit memory: Retention without remembering. *American Psychologist, 45*, 1043–1056.

Roediger, H. L., III, & Arnold, K. M. (2012). The one-trial learning controversy and its aftermath: Remembering Rock (1957). *American Journal of Psychology, 125*, 127–143.

Roediger, H. L., III, & Karpicke, J. D. (2006a). Test-enhanced learning: Taking memory tests improves long-term retention. *Psychological Science, 17*, 249–255.

Roediger, H. L., III, & Karpicke, J. D. (2006b). The power of testing memory: Basic research and implications for educational practice. *Perspectives on Psychological Science, 1*, 181–210.

Roediger, H. L., III, & McDermott, K. B. (1993). Implicit memory in normal human subjects. In H. Spinnler & F. Boller (Eds.), *Handbook of neuropsychology* (Vol. 8, pp. 63–131). Amsterdam: Elsevier.

Roediger, H. L., III, & McDermott, K. B. (1995). Creating false memories: Remembering words not presented in lists. *Journal of Experimental Psychology: Learning, Memory, & Cognition, 21*, 803–814.

Roediger, H. L., III, Putnam, A. L., & Smith, M. A. (2011). Ten benefits of testing and their applications to educational practice. In J. P. Mestre & B. H. Ross (Eds.), *The psychology of learning and motivation: Cognition in education* (Vol. 55, pp. 1–36). San Diego, CA: Academic Press.

Roenker, D. L., Thompson, C. P., & Brown, S. C. (1971). Comparison of measures for the estimation of clustering in free recall. *Psychological Bulletin, 76*, 45–48.

Roenker, D. L., Wenger, S. K., Thompson, C. P., & Watkins, B. (1978). Depth of processing: When the principle of congruity fails. *Memory & Cognition, 6*, 288–295.

Rogers, S. L., & Friedman, R. B. (2008). The underlying mechanism of semantic memory loss in Alzheimer's disease and semantic dementia. *Neuropsychologia, 46*, 12–21.

Rogers, T. B. (1977). Self-reference in memory: Recognition of personality items. *Journal of Research in Personality, 11*, 295–305.

Rogers, T. B., Kuiper, N. A., & Kirker, W. S. (1977). Self-reference and the encoding of personal information. *Journal of Personality and Social Psychology, 35*, 677–688.

Rosch, E. H. (1973). On the internal structure of perceptual and semantic categories. In T. E. Moore (Ed.), *Cognitive development and the acquisition of language* (pp. 111–144). New York: Academic Press.

Rosch, E., & Mervis, C. B. (1975). Family resemblance: Studies in the internal structure of categories. *Cognitive Psychology, 7*, 573–605.

Rosenthal, R. (1991). *Meta-analytic procedures for social research.* Newbury Park, CA: Sage.

Rothkegel, R. (1999). AppleTree: A multinomial processing tree modeling program for Macintosh computers. *Behavior Research Methods, Instruments, & Computers, 31*, 696–700.

Rouder, J. N., & Batchelder, W. H. (1998). Multinomial models for measuring storage and retrieval processes in paired associate learning. In C. E. Dowling, F. S. Roberts, & P. Theuns (Eds.), *Recent progress in mathematical psychology* (pp. 195–225). Mahwah, NJ: Erlbaumn.

Rouder, J. N., Lu, J., Morey, R. D., Sun, D., & Speckman, P. L. (2008). A hierarchical process-dissociation model. *Journal of Experimental Psychology: General, 137*, 370–389.

Rovee-Collier, C. (1997). Dissociations in infant memory: Rethinking the development of implicit and explicit memory. *Psychological Review, 104*, 467–498.

Rowe, G., Hirsh, J. B., & Anderson, A. K. (2007). Positive affect increases the breadth of attention selection. *Proceedings of the National Academy of Sciences USA, 104*, 383–388.

Rudy, J. W. (2008). *The neurobiology of learning and memory.* Sunderland, MA: Sinauer Associates.

Rudy, J. W., Biedenkapp, J. C., Moineau, J., & Bolding, K. (2006). Anisomycin and the reconsolidation hypothesis. *Learning & Memory, 13*, 1–3.

Rugg, M. D. (1985). The effect of semantic priming and word repetition of event-related potentials. *Psychophysiology, 22*, 642–647.

Rugg, M. D. (1987). Dissociation of semantic priming, word and non-word repetition effects by event-related potentials. *Quarterly Journal of Experimental Psychology, 39A*, 123–148.

Rumelhart, D. E. (1990). Brain style computation: Learning and generalization. In S. F. Zornetzer, J. L. Davis, & C. Lau (Eds.), *An introduction to neural and electrical networks* (pp. 405–420). San Diego, CA: Academic Press.

Rumelhart, D. E., & McClelland, J. L. (1982). An interactive activation model of context effects in letter perception: Part 2. The contextual enhancement effect and some test and extensions of the model. *Psychological Review*, **89**, 60–94.

Runquist, W. N., & Freeman, M. (1960). Role of association value and syllable familiarization in verbal discrimination learning. *Journal of Experimental Psychology*, **59**, 396–401.

Ryback, R. S. (1971). The continuum and specificity of the effects of alcohol on memory: A review. *Quarterly Journal of Studies on Alcohol*, **32**, 995–1016.

Ryle, G. (1960). *Dilemmas: The Tarner lectures 1953*. Cambridge, UK: Cambridge University Press.

Sacks, O. (1985). *The man who mistook his wife for a hat and other clinical tales*. New York: HarperCollins.

Sacks, O. (1995). *An anthropologist on Mars*. New York: Knopf.

Saltz, E., & Donnenwerth-Nolan, S. (1981). Does motoric imagery facilitate memory for sentences? A selective interference test. *Journal of Verbal Learning & Verbal Behavior*, **20**, 322–332.

Samet, J. (1999). History of nativism. In R. A. Wilson & F. C. Keil (Eds.), *The MIT encyclopedia of the cognitive sciences* (pp. 586–587). Cambridge, MA: MIT Press.

Sandkühler, S., & Bhattacharya, J. (2008). Deconstructing insight: EGG correlates of insightful problem solving. *PLoS ONE*, **3**, e1459.

Scarborough, D. L., Cortese, C., & Scarborough, H. (1977). Frequency and repetition effects in lexical memory. *Journal of Experimental Psychology: Human Perception and Performance*, **3**, 1–17.

Schacter, D. L. (1982). *Stranger behind the engram: Theories of memory and the psychology of science*. Hillsdale, NJ: Erlbaum.

Schacter, D. L. (1987). Implicit memory: history and current status. *Journal of Experimental Psychology: Learning, Memory, and Cognition*, **13**, 501–518.

Schacter, D. L. (1989). On the relation between memory and consciousness: Dissociable interactions and conscious experience. In H. L. Roediger & F. I. M. Craik (Eds.), *Varieties of memory and consciousness: Essays in honour of Endel Tulving* (pp. 355–389). Hillsdale, NJ: Erlbaum.

Schacter, D. L., Chin, C.-Y. P., & Ochsner, K. N. (1993). Implicit memory: A selective review. *Annual Review of Neuroscience*, **16**, 159–182.

Schacter, D. L., & Graf, P. (1986). Effects of elaborative processing on implicit and explicit memory for new associations. *Journal of Experimental Psychology: Learning, Memory, and Cognition*, **12**, 432–444.

Schare, M. L., Lisman, S. A., & Spear, N. E. (1984). The effect of mood variation on state-dependent retention. *Cognitive Therapy and Research*, **8**, 387–408.

Schelach, L., & Nachson, I. (2001). Memory of Auschwitz survivors. *Applied Cognitive Psychology*, **15**, 119–132.

Schmidt, S. R., & Cherry, K. (1989). The negative generation effect: Delineation of a phenomenon. *Memory & Cognition*, **17**, 359–369.

Schmidt-Hieber, C., Jonas, P., & Bischofberger, J. (2004). Enhanced synaptic plasticity in newly generated granule cells of the adult hippocampus. *Nature*, **429**, 184–187.

Schneider, W., & Shiffrin, R. M. (1977). Controlled and automatic human information processing: I. Detection, search, and attention. *Psychological Review*, **84**, 1–66.

Schoenbaum, G., Chiba, A. A., & Gallagher, M. (1998). Orbitofrontal cortex and basolateral amygdala encode expected outcome during learning. *Nature Neuroscience*, **1**, 155–159.

Schoenbaum, G., Setlow, B., & Ramus, S. J. (2003). A systems approach to orbitofrontal cortex function: Recording in rat orbitofrontal cortex reveal interactions with different learning systems. *Behavioural Brain Research*, **146**, 19–29.

Schulman, A. I. (1974). Memory for words recently classified. *Memory & Cognition*, **2**, 47–52.

Schwabe, L., Joëls, M., Roozendaal, B., & Oitzl, M. S. (2012). Stress effects on memory: An update and integration. *Neuroscience and Biobehavioral Reviews*, **36**, 1740–1749.

Schwartz, R. (1999). Rationalism vs. empiricism. In R. A. Wilson & F. C. Keil (Eds.), *The MIT encyclopedia of the cognitive sciences* (pp. 703–705). Cambridge, MA: MIT Press.

Schweickert, R. (1993). A multinomial processing tree model for degradation and redintegration in immediate recall. *Memory & Cognition*, **21**, 168–175.

Schweickert, R., & Chen, S. (2008). Tree inference with factors selectively influencing processes in a processing tree. *Journal of Mathematical Psychology*, **52**, 158–183.

Scoville, W. B., & Milner, B. (1957). Loss of recent memory after bilateral hippocampal lesions. *Journal of Neurology, Neurosurgery, and Psychiatry*, **20**, 11–21.

Sederberg, P. B., Miller, J. F., Howard, M. W., & Kahana, M. J. (2010). The temporal contiguity effect predicts episodic memory performance. *Memory & Cognition*, **38**, 689–699.

Seidenberg, M. S., Tanenhaus, M. K., Leiman, J. M., & Bienkowski, M. (1982). Automatic access of the meanings of ambiguous words in context: Some limitations of knowledge-based processing. *Cognitive Psychology*, **14**, 489–527.

Semon, R. (1921). *The mneme*. London: George Allen, & Unwin.

Sergent, J., Ohta, S., & MacDonald, B. (1992). Functional neuroanatomy of face and object processing: A positron emission tomography study. *Brain*, **115**, 15–36.

Shallice, T. (1988). *From neuropsychology to mental structure*. Cambridge, UK: Cambridge University Press.

Shiffrin, R. M. (1970). Memory search. In D. A. Norman (Ed.), *Models of human memory* (pp. 375–447). New York: Academic Press.

Shiffrin, R. M. (1975). Short-term store: The basis for a memory system. In F. Restle, R. M. Shiffrin, N. J. Castellan, H. R. Lindman, & D. B. Pisoni (Eds.), *Cognitive theory* (Vol. 1, pp. 193–218). Hillsdale, NJ: Wiley.

Shiffrin, R. M., & Atkinson, R. C. (1969). Storage and retrieval processes in long-term memory. *Psychological Review, 76*, 179–193.

Shiffrin, R. M., & Steyvers, M. (1997). A model for recognition memory: REM: Retrieving effectively from memory. *Psychonomic Bulletin and Review, 4*, 145–166.

Shimamura, A. P., Janowsky, J. S., & Squire, L. R. (1990). Memory for the temporal order of events in patients with frontal lobe lesions and amnesic patients. *Neuropsychologia, 28*, 803–813.

Shmuelof, L., & Zohar, E. (2006). Dissociation between ventral and dorsal fMRI activation during object and action recognition. *Neuron, 47*, 467–470.

Siegel, S. (1956). *Nonparametric statistics for the behavioral sciences*. New York: McGraw-Hill.

Siegel, S., & Allan, L. G. (1996). The widespread influence of the Rescorla-Wagner model. *Psychonomic Bulletin & Review, 3*, 314–321.

Simon, D. J., & Chabris, C. F. (2011). What people believe about how memory works: A representative survey of the U.S. population. *PLoS ONE, 6*, e22757, 1–7.

Simpson, G. B. (1984). Lexical ambiguity and its role in models of word recognition. *Psychological Bulletin, 96*, 316–340.

Singmann, H., & Kellen, D. (2013). MPTinR: Analysis of multinomial processing tree models in R. *Behavior Research Methods, 45*, 560–575.

Skinner, B. F. (1935). Two types of conditioned reflex and a pseudo type. *Journal of General Psychology, 12*, 66–77.

Skinner, B. F. (1960). Pigeons in a pelican. *American Psychologist, 15*, 28–37.

Skoff, B., & Chechile, R. A. (1977). Storage and retrieval processes in the serial position effect. *Bulletin of the Psychonomic Society, 9*, 265–268.

Skotko, B. G., Kensinger, E. A., Locascio, J. J., Einstein, G., Rubin, D. C., Tubler, L. A., Krendl, A., & Corkin, S. (2004). Puzzling thoughts for H. M.: Can new semantic information be anchored to old semantic memories? *Neuropsychology, 18*, 756–769.

Slamecka, N. J., & Graf, P. (1978). The generation effect: Delineation of a phenomenon. *Journal of Experimental Psychology: Human Learning and Memory, 4*, 592–604.

Slamecka, N. J., & Katsaiti, L. T. (1987). The generation effect as an artifact of selective displaced rehearsal. *Journal of Memory and Language, 26*, 589–607.

Sloboda, L. N. (2008). *The repetition effect, the spacing effect, and the two-trace hazard model.* Unpublished master's thesis, Tufts University.

Sloboda, L. N. (2012). *The quantitative measurement of explicit and implicit memory and its application to an aging population.* Unpublished doctoral dissertation, Tufts University.

Sloboda, L. N., & Chechile, R. A. (2008). *Applying hazard function theory to learning.* Poster presented at the November meeting of the Psychonomic Society, Chicago, IL.

Smith, J. O. (2007). *Mathematics of the discrete Fourier transform (DFT): With audio applications* (2nd ed.). Stanford, CA: W3K Publishing.

Smith, S. M. (1985). Background music an context-dependent memory. *The American Journal of Psychology, 98*, 591–603.

Smith, G. E., Cohen, S., Brown, L. C., Chechile, R. A., Garman, D., Cook, R. G., Ennis, J. G., & Lewis, S. (1997). *ConStats: Software for conceptualizing statistics, a user's manual.* Upper Saddle River, NJ: Prentice Hall.

Smith, A. M., Floerke, V. A., & Thomas, A. K. (2016). Retrieval practice protects memory against acute stress. *Science, 354*, 1046–1048.

Smith, S. M., Glenberg, A., & Bjork, R. A. (1978). Environmental context and human memory. *Memory & Cognition, 6*, 342–353.

Snyder, K. M., Ashitaka, Y., Shimada, H., Ulrich, J. E., & Logan, G. D. (2014). What skilled typists don't know about the QWERTY keyboard. *Attention, Perception, & Psychophysics, 76*, 162–171.

Soderstrom, N. C., & McCabe, D. P. (2011). Are survival processing memory advantages based on ancestral priorities? *Psychonomic Bulletin & Review, 18*, 564–569.

Solstad, T., Boccara, C. N., Kropff, E., Moser, M-B., & Moser, E. I. (2008). Representation of geometric borders in the entorhinal cortex. *Science, 322*, 1865–1868.

Soraci, S. A., Jr., Alpher, V. S., Deckner, C. W., & Blanton, R. L. (1983). Oddity performance and the perception of relational information. *Psychologia, 26*, 175–184.

Soraci, S. A., Carlin, M. T., Chechile, R. A., Franks, J. J., Wills, T., & Watanabe, T. (1999). Encoding variability and cuing in generative processing. *Journal of Memory and Language, 41*, 541–559.

Soraci, S. A., Carlin, M. T., Toglia, M. P., Chechile, R. A., & Neuschatz, J. S. (2003). Generative processing and false memories: When there is no cost. *Journal of Experimental Psychology: Learning, Memory, & Cognition, 29*, 511–523.

Soraci, S. A., Jr., Deckner, C. W., Haenlein, M., Baumeister, A. A., Murata-Soraci, K., & Blanton, R. L. (1987). Oddity performance in preschool children at risk for mental retardation: Transfer and maintenance. *Research in Developmental Disabilities, 8*, 137–151.

Soraci, S. A., Jr., Franks, J. J., Bransford, J. D., Chechile, R. A., Belli, R. F., Carr, M., & Carlin, M. T. (1994). Incongruous item generation effects: A multiple-cue perspective. *Journal of Experimental Psychology: Learning, Memory, and Cognition, 20,* 1–12.

Spear, N. E. (1978). *The processing of memories: Forgetting and retention.* Hillsdale, NJ: Erlbaum.

Spence, K. W. (1937). The differential response in animals to stimuli varying within single dimension. *Psychological Review,* **44,** 430–444.

Sperling, G. (1960). The information available in brief visual presentations. *Psychological Monographs: General and Applied,* **74,** Whole No. 498, 1–29.

Squire, L. R. (1986). Mechanisms of memory. *Science,* **232,** 1612–1619.

Squire, L. R. (1987). *Memory and brain.* New York: Oxford University Press.

Squire, L. R. (1992). Memory and the hippocampus: A synthesis from findings with rats, monkeys, and humans. *Psychological Review,* **99,** 195–231.

Squire, L. R., Berg, B., Bloom, F. L., du Lac, S., Ghosh, A., & Spitzer, N. C. (2013). *Fundamental neuroscience* (4th ed.). Amsterdam: Elsevier.

Squire, L. R., & Cohen, N. (1979). Memory and amnesia: Resistance to disruption develops for years after learning. *Behavioral and Neural Biology,* **25,** 115–125.

Squire, L. R., & Kandel, E. R. (1999). *Memory: From mind to molecules.* New York: Scientific American Library.

Squire, L. R., Slater, P. C., & Chace, P. M. (1975). Retrograde amnesia: Temporal gradient in very long-term memory following electroconvulsive therapy. *Science,* **187,** 77–79.

Squire, L. R., & Spanis, C. W. (1984). Long gradient of retrograde amnesia in mice: Continuity with the findings in humans. *Behavioral Neuroscience,* **98,** 345–348.

Stahl, C., & Klauer, K. C. (2007). HMMTree: A computer program for latent-class hierarchical multinomial processing tree models. *Behavior Research Methods,* **39,** 267–273.

Standing, L. (1973). Learning 10,000 pictures. *Quarterly Journal of Experimental Psychology,* **25,** 207–222.

Stanislaw, H., & Todorov, N. (1999). Calculation of signal detection theory measures. *Behavior Research Methods, Instruments, & Computers,* **31,** 137–149.

Steffensen, J. F. (1930). *Some recent researches in the theory of statistics and actuarial science.* New York: Cambridge University Press.

Steffens, M. C., & Erdfelder, E. (1998). Determinants of positive and negative generation effects in free recall. *Quarterly Journal of Experimental Psychology,* **51A,** 705–733.

Steiner, G. Z., & Barry, R. J. (2014). The mechanism of dishabituation. *Frontiers in Integrative Neuroscience,* **8,**14. doi: 10.3389/fnint.2014.00014.

Sternberg, R. J., & Tulving, E. (1977). The measurement of subjective organization in free recall. *Psychological Bulletin*, **84**, 539–556.

Sternberg, S. (1966). High speed scanning in human memory. *Science*, **153**, 652–654.

Stevens, S. S. (1936). A scale for the measurement of a psychological magnitude, loudness. *Psychological Review*, **43**, 405–416.

Stevens, S. S. (1951). Mathematics, measurement, and psychophysics. In S. S. Stevens (Ed.), *Handbook of experimental psychology* (pp. 1–49). New York: Wiley.

Stevens, S. S. (1957). On the psychophysical law. *Psychological Review*, **64**, 163–181.

Stevens, S. S. (1961). Toward a resolution of the Fechner-Thurstone legacy. *Psychometrika*, **26**, 35–47.

Stickgold, R., & Walker, M. P. (2013). Sleep-dependent memory triage: Evolving generalization through selective processing. *Nature Neuroscience*, **16**, 139–145.

Storm, T., & Caird, W. K. (1967). The effects of alcohol on serial verbal learning in chronic alcoholics. *Psychonomic Science*, **9**, 43–44.

Stromeyer, C. F., & Psotka, J. (1970). The detailed texture of eidetic images. *Nature* **225**, 346–349.

Stuart, A., & Ord, J. K. (1991). *Kendall's advance theory of statistics* (Vol. 2). New York: Oxford University Press.

Swets, J. A., Tanner, W. P., & Birdsall, T. G. (1961). Decision processes in perception. *Psychological Review*, **68**, 301–340.

Symons, C. S., & Johnson, B. T. (1997). The self-reference effect in memory: A meta-analysis. *Psychological Bulletin*, **121**, 371–394.

Szubko-Sitarek, W. (2015). *Multilingual lexical recognition in the mental lexicon of third language users*. Berlin: Springer-Verlag.

Takahashi, M. (1991). The role of choice in memory as a function of age: Support for a metamemory interpretation of the self-choice effect. *Psychologia*, **34**, 254–258.

Talland, G. A. (1965). *Deranged memory: A psychonomic study of the amnesia syndrome*. New York: Academic Press.

Tanner, W. P., & Swets, J. A. (1954). A decision-making theory of visual detection. *Psychological Review*, **61**, 401–409.

Taube, J. S. (2007). The head direction signal: origins and sensory-motor integration. *Annual Review of Neuroscience*, **30**, 181–207.

Teigen, K. H. (1994). Yerkes-Dodson: A law for all seasons. *Theory & Psychology*, **4**, 525–547.

Tell, P. M. (1972). The role of certain acoustic and semantic features at short and long retention intervals. *Journal of Verbal Learning and Verbal Behavior*, **11**, 455–464.

Teuber, H.-L., Milner, B., & Vaughan, H. G. (1968). Persistent anterograde amnesia after stab wound of the basal brain. *Neuropsychologia, 6,* 267–282.

Teyler, T. J., & DiScenna, P. (1986). The hippocampal memory indexing theory. *Behavioral Neuroscience, 100,* 147–154.

Teyler, T. J., & Rudy, J. W. (2007). The hippocampal indexing theory and episodic memory: Updating the index. *Hippocampus, 17,* 1158–1169. doi 10.1002/hippo.20350.

Theios, J., & Dunaway, J. E. (1964). One-way versus shuttle avoidance conditioning. *Psychonomic Science, 1,* 251–252.

Theios, J., Lynch, A. D., & Lowe, W. F. (1966). Differential effects of shock intensity on oneway and shuttle avoidance conditioning. *Journal of Experimental Psychology, 72,* 294–299.

Thompson, P. (1980). Margaret Thatcher: A new illusion. *Perception, 9,* 483–484.

Thompson, R. F., & Spencer, W. A. (1966). Habituation: A model phenomenon for the study of neuronal substrates of behavior. *Psychological Review, 73,* 16–43.

Thomson, A. D. (2000). Mechanisms of vitamin deficiency in chronic alcohol misusers and the development of the Wernicke-Korsakoff syndrome. *Alcohol & Alcoholism, Supplement, 35,* 2–7.

Thorndike, E. L. (1911). *Animal intelligence: Experimental studies.* New York: Macmillan.

Tinklepaugh, O. L. (1928). An experimental study of representative factors in monkeys. *Journal of Comparative Psychology, 8,* 197–236.

Toglia, M. P., Neuschatz, J. S., & Goodwin, K. A. (1999). Recall accuracy and illusory memories: When more is less. *Memory, 7,* 233–256.

Tolman, E. C. (1932). *Purposive behavior in animals and men.* New York: Century.

Tolman, E. C. (1945). A stimulus-expectancy need-cathexis psychology. *Science, 101,* 160–166.

Tolman, E. C., & Honzik, C. H. (1930a). "Insight" in rats. *University of California Publications in Psychology, 4,* 215–232.

Tolman, E. C., & Honzik, C. H. (1930b). Introduction and removal of reward, and maze performance in rats. *University of California Publications in Psychology, 4,* 257–275.

Tolman, E. C., Ritchie, B. F., & Kalish, D. (1946). Studies in spatial learning: II. Place learning versus response learning. *Journal of Experimental Psychology, 36,* 221–229.

Tolman, E. C., Ritchie, B. F., & Kalish, D. (1947). Studies in spatial learning: V. Response learning vs. place learning by the non-correction method. *Journal of Experimental Psychology, 37,* 285–292.

Torgerson, W. S. (1952). Multidimensional scaling: I. Theory and method. *Psychometrika, 17,* 401–419.

Treisman, A. M., & Gelade, G. (1980). A feature-integration theory of attention. *Cognitive Psychology, 12,* 97–136.

Trendler, G. (2009). Measurement theory, psychology and the revolution that cannot happen. *Theory & Psychology*, **19**, 579–599.

Tulving, E. (1962). Subjective organization in free recall of "unrelated" words. *Psychological Review*, **69**, 344–354.

Tulving, E. (1964). Intratrial and intertrial retention: Notes towards a theory of free recall verbal learning. *Psychological Review*, **71**, 219–237.

Tulving, E. (1972). Episodic and semantic memory. In E. Tulving & W. Donaldson (Eds.), *Organization of memory* (pp. 381–403). New York: Academic Press.

Tulving, E. (1974). Recall and recognition of semantically encoded words. *Journal of Experimental Psychology*, **102**, 778–787.

Tulving, E. (1983). *Elements of episodic memory*. Oxford, UK: Clarendon.

Tulving, E. (1985a). How many memory systems are there? *American Psychologist*, **40**, 385–398.

Tulving, E. (1985b). Memory and consciousness. *Canadian Psychology*, **26**, 1–12.

Tulving, E., & Thomson, D. M. (1973). Encoding specificity and retrieval processes in episodic memory. *Psychological Review*, **80**, 352–373.

Tulving, E., & Watkins, O. C. (1977). Recognition failure of words with a single meaning. *Memory & Cognition*, **5**, 513–522.

Tussing, A. A., & Greene, R. L. (2001). Effects of familiarity level and repetition on recognition accuracy. *American Journal of Psychology*, **114**, 31–41.

Tversky, A. (1977). Features of similarity. *Psychological Review*, **84**, 327–352.

Tye, M. (1988). The picture theory of mental images. *Philosophical Review*, **97**, 497–520.

Underwood, B. J. (1961). Ten years of massed practice on distributed practice. *Psychological Review*, **68**, 229–247.

Underwood, B. J. (1977). *Temporal codes for memory: Issues and problems*. Hillsdale NJ: Erlbaum.

Underwood, B. J. (1983). *Attributes of memory*. Glenview, IL: Scott, Foreman, & Co.

Underwood, B. J., Runquist, W. N., & Schultz, R. W. (1959). Response learning in paired-associate lists as a function intralist similarity. *Journal of Experimental Psychology*, **58**, 70–78.

Underwood, B. J., & Schulz, R. W. (1960). *Meaningfulness and verbal learning*. Philadelphia, PA: Lippincott.

Underwood, B. J., & Schulz, R. W. (1961). Studies of distributed practice: XXI. Effect of interference from language habits. *Journal of Experimental Psychology*, **62**, 571–575.

Ungerleider, L. G., & Mishkin, M. (1982). Two cortical visual systems. In D. J. Ingle, M. A. Goodale, & R. J. W. Mansfield (Eds.), *Analysis of visual behavior* (pp. 549–586). Boston: MIT Press.

Vallar, G. (1999). The methodological foundations of neuropsychology. In G. Denes & L. Pizzamiglio (Eds.), *Handbook of clinical and experimental neuropsychology* (pp. 95–131). Hove, UK: Psychology Press.

Van Buskirk, W. L. (1932). An experimental study of vividness in learning and retention. *Journal of Experimental Psychology, 15,* 563–573.

van der Helm, P. A., & Leeuwenberg, E. L. J. (1991). Accessibility: A criterion for regularity and hierarchy in visual pattern codes. *Journal of Mathematical Psychology, 35,* 151–213.

van der Linden, W. J., & Hambleton, R. K. (1996). *Handbook of modern item response theory.* New York: Springer.

van Praag, H., Kempermann, G., & Gage, F. H. (1999). Running increases cell proliferation and neurogenesis in the adult mouse dentate gyrus. *Nature Neuroscience, 2,* 266–270.

van Praag, H., Schinder, A. F., Christie, B. R., Toni, N., Palmer, T. D., & Gage, F. H. (2002). Functional neurogenesis in the adult hippocampus. *Nature, 415,* 1030–1034.

Vellutino, F. R. (1977). Alternative conceptualizations of dyslexia: Evidence in support of a verbal-deficit hypothesis. *Harvard Educational Review, 47,* 334–354.

Vitti, J. J. (2013). Cephalopod cognition in an evolutionary context: Implications for ethology. *Biosemiotics, 6,* 393–401. doi 10.1007/s12304-013-9175-7.

Vogel, S., & Schwabe, L. (2016). Learning and memory under stress: Implications for the classroom. *Science of Learning,* doi 10.1038/npjscilearn.2016.11.

von Restorff, H. (1933). Über die Wirkung von Bereichsbildungen im Spurenfeld. *Psychologische Forschung, 18,* 299–342.

von Wright, J. M., Anderson, K., & Stenman, U. (1975). Generalization of conditioned GSRs in dichotic listening. In P. M. A. Rabbitt & S. Dornic (Eds.), *Attention and performance V* (pp. 194–204). New York: Academic Press.

Wagenaar, W. A., & Groeneweg, J. (1990). The memory of concentration camp survivors. *Applied Cognitive Psychology, 4,* 77–87.

Wang, D. (1994). A neural model of synaptic plasticity underlying short-term and long-term habituation. *Apaptive Behavior, 2,* 111–129.

Ward, R. D., Gallistel, C. R., & Balsam, P. D. (2013). It's the information! *Behavioural Processes, 95,* 3–7.

Warren, E. L. (2015). *Memory capacity and storage for images and shapes.* Unpublished doctoral dissertation, Tufts University.

Warrington, E. K., & Weiskrantz, L. (1970). Amnesic syndrome: Consolidation or retrieval? *Nature, 228,* 628–630.

Warrington, E. K., & Weiskrantz, L. (1974). The effect of prior learning on subsequent retention in amnesic patients. *Neuropsychologia*, **12**, 419–428.

Watanabe, T. (2001). *The self-choice effect: Multiple-cue mechanism at encoding and retrieval.* Unpublished doctoral dissertation, Tufts University.

Watanabe, T., & Soraci, S. A. (2004). The self-choice effect from a multiple-cue perspective. *Psychonomic Bulletin & Review*, **11**, 168–172.

Weibull, W. (1951). A statistical distribution of wide applicability. *Journal of Applied Mechanics*, **18**, 293–297.

Weinberger, N. M., Gold, P. E., & Sternberg, D. B. (1984). Epinephrine enables Pavlovian fear conditioning under anesthesia. *Science*, **223**, 605–607.

Weingartner, H. (1978). Human state dependent learning. In B. T. Ho, D. W. Richards, & D. L. Chute (Eds.), *Drug discrimination and state dependent learning* (pp. 361–382). New York: Academic Press.

Weinstein, Y., Bugg, J. M., & Roediger, H. L., III. (2008). Can the survival recall advantage be explained by basic memory processes? *Memory & Cognition*, **36**, 913–919.

Weiskrantz, L. (1987). Neuroanatomy of memory and amnesia: A case for multiple memory systems. *Human Neurobiology*, **6**, 93–105.

Weiskrantz, L. (1989). Remembering dissociations. In H. L. Roediger & F. I. M. Craik (Eds.), *Varieties of memory and consciousness: Essays in honour of Endel Tulving* (pp. 101–120). Hillsdale, NJ: Erlbaum.

Wetzler, S. (1985). Mood state-dependent retrieval: A failure to replicate. *Psychological Reports*, **56**, 759–765.

Whitehouse, W. G., Dinges, D. F., Orne, E. C. & Orne, M. T. (1988). Hypnotic hypermnesia: Enhanced memory accessibility or report bias? *Journal of Abnormal Psychology*, **97**, 289–295.

Whitty, C. W. M., & Lewin, W. A. (1960). A Korsakoff syndrome in the postcingulectony confusion state. *Brain*, **83**, 648–653.

Wickelgren, W. A. (1968). Sparing of short-term memory in an amnesic patient: Implications for strength theory of memory. *Neuropschologia*, **6**, 235–244.

Wickelgren, W. A. (1975). Alcohol intoxication and memory storage dynamics. *Memory & Cognition*, **3**, 385–389.

Wickens, D. D. (1972). Characteristics of word encoding. In A. W. Melton & E. Martin (Eds.), *Coding processes in human memory* (pp. 191–215). Washington, DC: Winston.

Wickens, D. D. (1973). Some characteristics of word encoding. *Memory & Cognition*, **1**, 485–490.

Wickens, T. D. (2002). *Elementary signal detection theory.* New York: Oxford University Press.

Wickens, D. D., & Clark, S. E. (1968). Osgood dimensions as an encoding category in short-term memory. *Journal of Experimental Psychology, 78,* 580–584.

Wiens, S., & Öhman, A. (2002). Unawareness is more than a chance event: Comment on Lovibond and Shanks (2002). *Journal of Experimental Psychology: Animal Behavior Processes, 28,* 27–31.

Williams, J. P. (1961). A selection artifact in Rock's study of the role of repetition. *Journal of Experimental Psychology, 62,* 627–628.

Wills, T. W., Soraci, S. A., Chechile, R. A., & Taylor, H. A. (2000). "Aha" effects in the generation of pictures. *Memory & Cognition, 28,* 939–948.

Willshaw, D. J., Buneman, O. P., & Longuet-Higgins, H. C. (1969). Non-holographic associative memory. *Nature, 222,* 960–962.

Wilson, M. A., & McNaughton, B. L. (1993). Dynamics of the hippocampal ensemble code for space. *Science, 261,* 1055–1058.

Winston, P. H. (1975). Learning structural descriptions from examples. In P. H. Winston (Ed.), *The psychology of computer vision* (pp. 157–209). New York: McGraw-Hill.

Winton, W. M. (1987). Do introductory textbooks present the Yerkes-Dodson law correctly? *American Psychologist, 42,* 202–203.

Wiseman, S., & Tulving, E. (1976). Encoding specificity: Relation between recall superiority and recognition failure. *Journal of Experimental Psychology: Learning, Memory, and Cognition, 2,* 349–361.

Witmer, L. R. (1935). The association value of three-place consonant syllables. *Journal of Genetic Psychology, 47,* 337–360.

Wittlinger, R. P. (1967). *Phasic arousal in short-term memory.* Unpublished doctoral dissertation, The Ohio State University.

Wixted, J. T., & Stretch, V. (2004). In defense of the signal-detection interpretation of remember/know judgments *Psychonomic Bulletin & Review, 11,* 616–641.

Wolfe, J. M., Cave, K. R., & Franzel, S. L. (1989). Guided search: An alternative to the feature integration model for visual search. *Journal of Experiment Psychology: Human Perception and Performance, 15,* 419–433.

Wood, D. C. (1970). Parametric studies of the response decrement produced by mechanical stimuli in the protozoan, *Stentor coeruleus. Journal of Neurobiology, 1,* 345–360.

Wood, D. C. (1988). Habituation in *Stentor*: A response-dependent process. *The Journal of Neuroscience, 8,* 2248–2253.

Woodruff-Pak, D. S. (1993). Eyeblink classical conditioning in H. M.: Delay and trace paradigms. *Behavioral Neuroscience, 107,* 911–925.

Woodworth, R. S., & Schlosberg, H. (1954). *Experimental psychology* (Rev. ed.). New York: Holt.

Wright, A. A., Cook, R. G., Rivera, J. J., Sands, S. F., & Deliux, J. D. (1988). Concept learning by pigeons: Matching to sample with trial-unique video picture stimuli. *Animal Learning and Behavior*, **16**, 436–444.

Yang, T., & Poovaiah, B. W. (2003). Calcium/calmodulin-mediated signal networks in plants. *Trends in Plant Science*, **62**, 3727–3735.

Yerkes, R. M., & Dodson, J. D. (1908). The relation of strength of stimulus to rapidity of habit formation. *Journal of Comparative and Neurological Psychology*, **18**, 459–482.

Yin, R. K. (1969). Looking at upside-down faces. *Journal of Experimental Psychology*, **81**, 141–145.

Yin, R. K. (1970). *Face recognition: A special process?* Unpublished doctoral dissertation, Massachusetts Institute of Technology.

Yonelinas, A. P. (1994). Receiver-operating characteristics in recognition memory: Evidence for a dual-process model. *Journal of Experimental Psychology: Learning, Memory, and Cognition*, **20**, 1341–1354.

Yonelinas, A. P. (1997). Recognition memory ROCs for item and associative information: The contribution of recollection and familiarity. *Memory & Cognition*, **25**, 747–763.

Yonelinas, A. P. (2002). The nature of recollection and familarity: A review of 30 years of research. *Journal of Memory and Language*, **46**, 441–517.

Yonelinas, A. P., & Jacoby, L. L. (1996). Response bias and the process-dissociation procedure. *Journal of Experimental Psychology: General*, **125**, 422–434.

Young, J. Z. (1991). Computation in the learning system of cephalopods. *Biological Bulletin*, **180**, 200–208.

Youtz, A. C. (1941). An experimental evaluation of Jost's laws. *Psychological Monographs*, **53**, Whole No. 238, 1–54.

Yu, J., & Bellezza, F. S. (2000). Process dissociation as source monitoring. *Journal of Experimental Psychology: Learning, Memory, & Cognition*, **26**, 1518–1533.

Zečevi, D., Wu, J-Y., Cohen, L. B., London, J. H., Höpp, H-P., & Falk, C. X. (1989). Hundreds of neurons in the *Aplysia* abdominal ganglion are active during the gill-withdrawal reflex. *The Journal of Neuroscience*, **9**, 3681–3689.

Zeelenberg, R. (2005). Encoding specificity manipulations do affect *retrieval* from memory. *Acta Psychologia*, **119**, 107–121.

Zeman, A. Z. J., Della Sala, S., Torrens, L. A., Gountouna, V.-E., McGonigle, D. J., & Logie, R. H. (2010). Loss of imagery phenomenology with intact visuo-spatial task performance: A case of 'blind imagination'. *Neuropsychologia*, **48**, 145–155.

Zeman, A. Z. J., Dewar, M., & Della Sala, S. (2015). Lives without imagery-congenital aphantasia. *Cortex*, **73**, 378–380.

Zentall, T. R., Hogan, D. E., Edwards, C. A., & Hearst, E. (1980). Oddity learning in the pigeon as a function of the number of incorrect alternatives. *Journal of Experimental Psychology: Animal Behavior Processes*, **6**, 278–299.

Zimmer, H. D. (1991). Memory after motoric encoding in a generation-recognition model. *Psychological Research*, **53**, 226–231.

Zimmer, H. D., Cohen, R. L., Guynn, M. J., Engelkamp, J., Kormi-Nouri, R., & Foley, M. A. (2001). *Memory for action: A distinct form of episodic memory?* New York: Oxford University Press.

Index